Cattle Health: Preventive Care and Disease Management

Cattle Health: Preventive Care and Disease Management

Editor: Eleanor Clark

R CALLISTO
REFERENCE

www.callistoreference.com

Callisto Reference,
118-35 Queens Blvd., Suite 400,
Forest Hills, NY 11375, USA

Visit us on the World Wide Web at:
www.callistoreference.com

ISBN: 978-1-64116-288-3 (Hardback)

Cataloging-in-Publication Data

Cattle health : preventive care and disease management / edited by Eleanor Clark.
    p. cm.
Includes bibliographical references and index.
ISBN 978-1-64116-288-3
1. Cattle--Health aspects. 2. Cattle--Diseases. 3. Cattle--Diseases--Treatment.
4. Veterinary medicine. I. Clark, Eleanor.
SF197 .C38 2020
636.2--dc23

Table of Contents

Preface

This book aims to highlight the current researches and provides a platform to further the scope of innovations in this area. This book is a product of the combined efforts of many researchers and scientists, after going through thorough studies and analysis from different parts of the world. The objective of this book is to provide the readers with the latest information of the field.

Cattle are the large domesticated animals that are primarily raised as livestock for meat, milk and leather. Cattle health is looked after under the field of animal welfare. It includes the observation of both mental and physical state of the cattle. It is ensured that the animals are safe, healthy, and are expressing their innate behavior. Cattle can develop and catch various serious diseases such as bluetongue, foot rot and blackleg. These diseases are studied under the veterinary discipline known as buiatrics. Disease prevention, nutrition, appropriate shelter and veterinary treatment are a few of the major concerns that are associated with cattle health. The topics included in this book on cattle health are of utmost significance and bound to provide incredible insights to readers. It provides comprehensive insights into the field of preventive care and disease management for cattle. It will serve as a valuable source of reference for graduate and post graduate students.

I would like to express my sincere thanks to the authors for their dedicated efforts in the completion of this book. I acknowledge the efforts of the publisher for providing constant support. Lastly, I would like to thank my family for their support in all academic endeavors.

Editor

Molecular detection of *Mycobacterium bovis* in cattle herds of the state of Pernambuco, Brazil

Renata Duarte da Silva Cezar[1], Norma Lucena-Silva[2], Antônio Fernando Barbosa Batista Filho[3], Jonas de Melo Borges[3], Pollyane Raysa Fernandes de Oliveira[1], Érica Chaves Lúcio[3], Maíra Arruda-Lima[2], Vania Lucia de Assis Santana[4] and José Wilton Pinheiro Junior[1*]

Abstract

Background: The present study aimed to direct detect *Mycobacterium bovis* in milk ($n = 401$) and blood ($n = 401$) samples collected from 401 dairy cows of 20 properties located in the state of Pernambuco, Brazil, by real-time quantitative PCR (qPCR) targeting the region of difference 4 (RD4). Risk factors possibly associated with bovine tuberculosis (BTB) were also evaluated.

Results: Of the 802 samples analyzed, one milk (0.25 %) and eight blood (2 %) samples were positive for *M. bovis* in the qPCR and their identities were confirmed by sequencing. Animals positive for *M. bovis* were found in six (30 %) of the 20 properties visited. None of the risk factors evaluated were statistically associated with BTB.

Conclusions: *M. bovis* DNA was detected in one milk sample what may pose a risk to public health because raw milk is commonly consumed in Brazil.

Keywords: Tuberculosis, Cows, PCR, Milk

Background

Bovine tuberculosis (BTB) is caused by *Mycobacterium bovis*, a member of the *Mycobacterium tuberculosis* complex that affects mammals, including humans [1].

M. bovis has been isolated from milk and colostrum samples what can be important to perpetuate BTB infection in a herd through the digestive route [2]. Raw milk is commonly consumed in Brazil [3] and clandestine milk is an important public health issue in the country [4, 5].

Despite the fact that Brazil has a National Program for Control and Eradication of Tuberculosis and Brucellosis (Programa Nacional de Controle e Erradicação da Tuberculose e da Brucelose - PNCETB) supervised by a public agency, its implementation is not mandatory [6]. The tuberculin skin test is the diagnostic method recommended by the PNCETB and it must be followed by

bacterial isolation for result confirmation [7]. Efforts to reduce the risk of *M. bovis* infection must include sanitary measures to ensure a healthy cattle herd.

Molecular techniques such as Polymerase Chain Reaction (PCR) have been used for BTB diagnosis in several clinical samples such as blood, milk and nasal exudates [2]. Standardization of direct methods for detection of *M. bovis* in clinical samples will enable a more accurate BTB diagnosis and facilitate epidemiological studies on *M. bovis* prevalence [1, 8].

In the present study, we used qPCR for direct detection of *Mycobacterium bovis* in milk and blood samples of cattle from the state of Pernambuco, Brazil.

Methods
Sampling

The sample size was calculated as recommended by Thrusfield [9] using the following parameters: bovine population of 336,221 animals in the micro region of Garanhuns, state of Pernambuco, Brazil [10], 95 % confidence interval and 5 % sampling error margin using a

* Correspondence: wiltonjuniorufrpe@gmail.com
[1]Federal Rural University of Pernambuco (Universidade Federal Rural de Pernambuco - UFRPE), Rua Dom Manuel de Medeiros, s/n, Dois Irmãos, Recife, Pernambuco CEP 52171-900, Brazil
Full list of author information is available at the end of the article

prevalence of 50 %, since there is no official data on BTB prevalence in the studied region. According to the calculation, the minimum sample size should be 385 dairy cattle.

From January to February 2014, a total of 802 milk and blood samples were collected from 401 dairy cows of 20 properties distributed in the municipalities of Angelim, Bom Conselho, Brejão, Caetés, Calçado, Canhotinho, Correntes, Garanhuns, Iati, Jucati, Jupi, Jurema, Lagoa do Ouro, Lajedo, Palmeirina, Paranatama, Saloá, São João and Terezinha, state of Pernambuco, Brazil.

Written informed consent was obtained from the farmers to take samples from the cattle. The blood samples ($n = 401$) were collected by caudal venipuncture, stored in tubes containing citrate, properly identified and sent to the Garanhuns Laboratories Center (Central de Laboratórios de Garanhuns - CENLAG), located in the Garanhuns Academic Unit (Unidade Acadêmica de Garanhuns - UAG) of the Federal Rural University of Pernambuco (Universidade Federal Rural de Pernambuco - UFRPE), Brazil.

The milk samples ($n = 401$), which consisted of 50 ml of a pool of milk from the four quarters of each cow, were collected during milking after the udder disinfection with 70 % alcohol and the first jets of milk were discarded. Then the samples were stored in sterile bottles, cooled and sent to the CENLAG (UAG- UFRPE).

An epidemiological questionnaire containing multiple-choice questions concerning animal production characteristics, hygiene and sanitary aspects of the herd, and reproductive management was applied in each property. The questionnaire comprised 11 possible risk factors for *M. bovis* infection, as follows: herd size (less than 50 animals, 51–100 animals, 101–200 animals, more than 201 animals), rearing system (intensive, extensive, semi-intensive), origin of replacement animals (farm's own herd, another farm, both), conducting quarantine after animal's purchase, performance of BTB diagnostic tests upon animals' acquisition, water source (stagnant or running), milking procedure (manual or mechanic), frequency of cleaning the farm facilities, udder disinfection, feeding colostrum to calves and history of BTB in the herd.

DNA Extraction

M. bovis DNA was extracted from milk samples using the QIamp® kit (Qiagen Inc.) following the manufacturer's instructions. Leukocyte DNA was isolated by a modified phenol-chloroform extraction method [11], 100 μl of white blood cells were used, ammonium acetate and phenol-chloroform for DNA extraction from the blood samples.

Positive control

The *Mycobacterium bovis* ATCC 19274 strain was gently provided by the Oswaldo Cruz Foundation (Fundação Osvaldo Cruz - FIOCRUZ, Rio de Janeiro, Brazil) and was used for construction of a plasmid harboring the target sequence, which was the positive control in the molecular assays. *M. bovis* genomic DNA was extracted and the fragment corresponding to the region of difference 4 (RD4) was amplified with the specific primers reported by Sales et al. [12]. The target fragment was cloned using *Escherichia coli* XL1 blue strain and TA cloning kit® (Invitrogen) according to the manufacturer's instructions. The recombinant plasmid pRD4-TA was sequenced by the Sanger method using an ABI3100 Genetic Analyzer (Applied Biosystems).

Real-time PCR

The molecular detection of *M. bovis* DNA in the milk and blood samples was performed in the Laboratory of Immunogenetics (Laboratório de Imunogenética) of FIOCRUZ, state of Pernambuco, Brazil. Quantitative real time PCR (qPCR) was performed using the same primer set used for amplification of the RD4 fragment of the positive control and a fluorescent probe that discriminates *M. bovis* from other *M. tuberculosis* complex members since it hybridizes with both the 5′ and 3′ RD4 deletion flanking sequences, which only occur directly adjacent to each other in *M. bovis*. The probe was designed using the software Primer Express® Software and targeted a region in the amplicon in between the primer pair. The probe showed 100 % homology to *M. bovis* in BLAST/ncbi. Probe sequence: 5'- /56-FAM/AGCCG-TAGTCGTGCAGAAGCGCA/3BHQ_1/- 3'. The total reaction volume was 25 μL comprising 2.0 μL of DNA, 12.5 μL of TaqMan®Universal PCR Master Mix, 1.0 μL each primer (5 pmol), 0.5 μL of probe (5 pmol) and water (8 μL). The amplification conditions were 95 °C for 15 min (denaturation) followed by 40 cycles of 94 °C for 15 s and 60 °C for 60 s. In all PCR runs, standard curves were obtained using the positive control, plasmid DNA encompassing the mycobacteria RD4 sequence, which was prepared in triplicate by serial dilution of 10x plasmid DNA from 200 ng (Quantification cycle - Cq = 11.8) to 0.0002 ng (Cq = 32.2). The qPCR was performed in an ABI 7500 Real-Time PCR system set for absolute quantification The slope of the standard curve was −3.40 and R = 0.999, with 97 % efficiency.

Spiked samples were not used. The standard curve and the detection limit were determined using a serial 10X dilution from 100 to 10^{-10} ng/ul in triplicate and the positive control was detected in samples with up to 10^{-6} ng/ul. The reaction was repeated four times in different days and the same results were obtained in each day.

To verify the presence of inhibitors in the samples, a few blood samples of 1000 ng/μl DNA were randomly selected and diluted them to 800, 600, 400, 200 and

50 ng/µl of DNA. Then 20 ng were added of positive control to the diluted samples and performed a qPCR. All the dilutions had the same Cq in the qPCR; therefore, there were no inhibitors in the samples.

Sequencing

DNA sequencing was performed in the Center of Technological Platforms (Núcleo de Plataformas Tecnológicas - NPT), of the Research Center Aggeu Magalhães (Centro de Pesquisas Aggeu Magalhães-CPqAM), FIOCRUZ, state of Pernambuco, Brazil.

The commercial kit ABI PRISM BigDye Terminator Cycle Sequencing Ready Reaction v3.1 (Applied Biosystems®) was used for DNA sequencing following the manufacturer's recommendations. The RD4 fragments were sequenced by the Sanger method and the reaction products were analyzed in the ABI 3500xL Genetic Analyzer (Applied Biosystems).

All the sequences obtained in the present study were compared with the RD4 fragment of the reference genome (88 bp) (GenBank Access number BX248339.1) using the software Blast-N (http://www.ncbi.nlm.nih.gov) and MEGA6 [13].

Ethical considerations

The Ethics Committee on Animal Use (Comissão de Ética no Uso de Animais – CEUA) of UFRPE provided scientific and ethical clearance for the present study (reference number 23082.004671/2013, license number 028/2013).

Statistical analysis

The absolute and relative prevalence of *M. bovis* in the milk and blood samples were determined by descriptive analysis. Univariate analysis using chi-square test, Pearson's test or Fisher's exact test were used to evaluate the possible risk factors associated with BTB. All statistical analyzes were performed in the Epi Info 3.5.1 software.

Results

Of the 802 samples analyzed, one milk (0.25 %) and eight blood (2 %) samples were positive for *M. bovis* in the qPCR and their identities were confirmed by sequencing. All positive samples were from different animals (Table 1).

Six (30 %) of the 20 properties visited had animals positive for *M. bovis* (Table 1). None of the risk factors evaluated in the present study were statistically associated with BTB as shown in Table 2.

Discussion

This is the first report of direct detection of *M. bovis* DNA in milk and blood samples from cattle of the region of Garanhuns, state of Pernambuco, Brazil.

Table 1 Results of qPCR of milk and blood samples collected from cattle of the micro region of Garanhuns, state of Pernambuco, Brazil, 2014

	Number of animals	Positive		Negative	
Municipality[a]		Milk	Blood	Milk	Blood
Bom Conselho	25	-	01	25	24
Lagoa do Ouro	40	-	01	40	39
Paranatama	18	-	01	18	17
Iati	13	-	01	13	12
Caetés	18	01	-	17	18
Palmerina	10	-	04	10	06
Total[b]	401	01	08	400	393

[a]Municipalities which herds had only negative results in the qPCR are not shown
[b]Refers to the total number of animals evaluated in the study; it is not a sum of each column

The prevalence of *M. bovis* DNA in milk samples ranges from 2 to 87 % according to different studies [14–20], which evaluated the mycobacteria presence by PCR. The different prevalence rates observed by them may be related to management characteristics [21], sampling methods [22] and disease-control measures adopted in each location [23].

The presence of *M. bovis* in milk may pose a risk to public health, because humans can become infected by *M. bovis* through exposure to infected animals, consumption of infected raw milk and dairy products [24, 25]. The presence of *M. bovis* in milk samples is a concern because it is estimated that 41 % of all milk consumed in Brazil is not pasteurized [12] being a source of infection of human TB. Since the clinical symptoms of human TB caused by *M. bovis* are indistinguishable from those caused by *M. tuberculosis* [23, 26], a detailed epidemiological investigation considering the patients eating habits and professional occupation must be performed by the health surveillance service to determine the TB causal agent. In Brazil, a study performed by Silva et al. [27] with 189 TB patients identified co-infection with *M. bovis* in three patients. In two of these patients, consumption of cheese made with raw milk was the probable cause of infection. The other patient used to work in a slaughterhouse, so the infection was related to labor risk. Other study conducted in Brazil also identified *M. bovis* in humans, although with lower prevalence. It is believed that *M. bovis* prevalence in Brazil is underestimated [28].

In the present study, the milk sample positive for *M. bovis* did not belong to any of the eight animals that had positive blood samples what can be explained by various reasons: collection of only one milk sample per dairy cow, interaction between the bacillus and the bovine immune system cells, which may have decreased the amount of bacillus in milk [2, 20, 29], and presence of

Table 2 Analysis of risk factors associated with prevalence of M. bovis in cattle herds of the micro region of Garanhuns, state of Pernambuco, Brazil, 2014

Risk factors	n	Positive		Negative		OR (95 % CI)	p value
		AF	RF%	AF	RF%		
Herd size[a]							
< 50 animals	9	2	22.2	7	77.8	-	0.467
51–100 animals	6	3	50.0	3	50.0	3.50 (0.37–32.97)	
101–200 animals	2	1	50.0	1	50.0	1.00 (0.04–24.55)	
> 200 animals	2	-	-	2	100	-	
Rearing system							
Intensive	3	1	33.3	2	66.7	-	0.788
Extensive	2	1	50.0	1	50.0	2.00 (0.05–78.25)	
Semi-Intensive	15	4	26.7	11	73.3	0.36 (0.02–7.30)	
Origin of replacement animals							
Farm's own herd	12	4	33.3	8	66.7	0.67 (0.09–4.92)	0.544
Other farms	8	2	25.0	6	75.0		
Quarantine							
Yes	4	1	25.0	3	75.0	1.36 (0.11–16.57)	0.657
No	16	5	31.3	11	68.8		
BTB diagnostic tests upon animals' acquisition[a]							
Yes	10	3	30.0	7	70.0	1.16 (0.16–8.0)	0.630
No	9	3	33.3	6	66.7		
Water source[a]							
Stagnant	14	4	28.6	10	71.4	1.66 (0.19–14.0)	0.520
Running	5	2	40.0	3	60.0		
Milking procedure							
Manual	12	4	33.3	8	66.7	0.66 (0.09–4.92)	0.544
Mechanic	8	2	25.0	6	75.0		
Frequency of cleaning the farm facilities[a]							
Daily	12	3	25.0	9	75.0	-	0.360
Weekly	3	2	66.7	1	33.3	6.00 (0.39–92.28)	
Monthly	2	1	50.0	1	50.0	0.50 (0.01–19.56)	
Udder disinfection							
Yes	7	1	14.3	6	85.7	3.75 (0.34–41.0	0.276
No	13	5	38.5	8	61.5		
Feeding colostrum to calves							
Yes	17	5	29.4	12	70.6	1.20 (0.08–16.44)	0.370
No	3	1	33.3	2	66.7		
History of bovine tuberculosis in the herd[a]							
Yes	2	-	-	2	100.0	-	0.456
No	17	6	35.3	11	64.6		

AF absolute frequency, RF relative frequency, OR odds ratio, 95 % CI 95 % confidence interval
[a]Not all the respondents answered the question

milk proteins and fat that may have impaired the extraction of M. bovis DNA from the milk samples [14, 30]. Despite these limitations, studies using experimentally contaminated milk have demonstrated that PCR can detect the mycobacteria in milk samples with much lower concentrations of M. bovis DNA than the concentration usually present in natural infections [2, 20, 29, 31]. The mycobacteria has been more frequently identified in blood than in milk samples [20, 32, 33].

Another factor that can hinder the detection of *M. bovis* in milk is its intermittent release during a short period post-infection [20, 29]. Pardo et al. [34] evaluated the mycobacteria secretion pattern in 780 milk samples collected from 52 animals for 15 consecutive days and *M. bovis* showed an intermittent and irregular release pattern in 26.5 % of the samples [34].

In the present study, of the six farms that had animals positive for *M. bovis*, only one have a history of performing tuberculin tests upon acquisition of new animals. According to the farmer, all the animals were negative for *M. bovis* in the tuberculin tests. This data shows the importance of using tests that are more sensitive in enzootic areas, as the state of Pernambuco, Brazil, where cases of BTB have already been identified using tuberculin skin test [35, 36].

Although there was no statistical association between herd size and *M. bovis* positivity, the mycobacteria was more prevalent in larger herds (101–200 animals). Herd size can influence BTB epidemiology because a high population density favors a more frequent contact between animals, facilitating the mycobacteria dissemination [37, 38].

According to Skuce et al. [39], *M. bovis* can survive in water what favors its dissemination. *M. bovis* DNA was identified in water samples experimentally contaminated even 11 months after contamination [40]. In Uganda, Africa, where cattle commonly drinks running water from rivers or streams, a study evaluated the risk factors associated with BTB and concluded that the water source was statistically associated with the disease [41]. However, in the present study, no correlation was found between water source and presence of *M. bovis* in milk and blood samples (Table 1). The fact that water sources could be implicated with BTB transmission may be a concern to health authorities, because control measures would also have to consider this contamination source besides slaughter of positive animals.

Despite the higher number of animals positive for *M. bovis* in herds with low frequency of cleaning the farm facilities and lack of udder disinfection before milking, no positive correlation was found between pathogen presence and these risk factors. Roxo [42] reported that cleaning, disinfection and hygiene are risk factors for TB. Waste management and treatment of organic matter can influence *M. bovis* prevalence in areas with previous cases of TB [43].

Conclusion

M. bovis DNA was detected in one milk sample what may pose a risk to public health. We suggest that environmental control measures should be implemented in farms at high risk of TB transmission because environmental factors contribute to bacteria perpetuation and dissemination.

Competing interests

The authors declare that they have no competing interests.

Authors' contributions

RDSC performed the study, analyzed the data and drafted the manuscript. NLS, VLAS and JWPJ JG conceived and designed the study, critically revised the paper and acted as the first author's study supervisors. AFBBF, JMB, PRF, RCL. MAL performed some of the data analysis and critically revised the paper. All authors read and approved the final manuscript.

Author details

Federal Rural University of Pernambuco (Universidade Federal Rural de Pernambuco - UFRPE), Rua Dom Manuel de Medeiros, s/n, Dois Irmãos, Recife, Pernambuco CEP 52171-900, Brazil. ²Department of Immunology (Departamento de Imunologia), Research Center Aggeu Magalhães (Centro de Pesquisas Aggeu Magalhães-CPqAM) Oswaldo Cruz Foundation (Fundação Oswaldo Cruz - Fiocruz), Av. Professor Moraes Rego, s/n, Cidade Universitária, Recife, Pernambuco CEP 50.740-465, Brazil. ³Academic Unit of Garanhuns (Unidade Acadêmica de Garanhuns), Federal Rural University of Pernambuco (Universidade Federal Rural de Pernambuco – UFRPE). Avenida Bom Pastor, s/n, Boa Vista, Garanhuns, Pernambuco CEP 55292-270, Brazil. National Agricultural Laboratory of Pernambuco (Laboratório Nacional Agropecuário de Pernambuco- Lanagro/PE), Ministry of Agriculture, Livestock and Food Supply of Brazil (Ministério da Agricultura, Pecuária e Abastecimento – MAPA). Rua Manoel de Medeiros, s/n, Dois Irmãos, Recife, Pernambuco CEP 52171-030, Brazil.

References

1. Taylor GM, Worth DR, Palmer S, Jahans K, Hewinson RG. Rapid detection of Mycobacterium bovis DNA in cattle lymph nodes with visible lesions using PCR. BMC Vet Res. 2007;3:12. doi:10.1186/1746-6148-3-12.
2. Serrano-Moreno BA, Romero TA, Arriaga C, Torres RA, Pereira-Suárez AL, García-Salazar JA, et al. High frequency of Mycobacterium bovis DNA in colostra from tuberculous cattle detected by nested PCR. Zoonoses Public Health. 2008;55:258–66. doi:10.1111/j.1863-2378.2008.01125.x.
3. Milka CDA, Grace BDS, Fernando Z, Maurcio CH, Francesca SD, Mateus MDC. Microbiological evaluation of raw milk and coalho cheese commercialised in the semi-arid region of Pernambuco, Brazil. African J Microbiol Res. 2014; 8:222–9. doi:10.5897/AJMR2013.6476.
4. Bermudez HR, Henteira ET, Medina BG, Hori-Oshima S, De La Mora VA, Lopez VG, et al. Correlation between histopathological, Bacteriological and PCR diagnosis of bovine tuberculosis. J Anim Vet Adv. 2010;9:2082–4.
5. Sgarioni SA, Dominguez R, Hirata C, Hirata MH, Queico C, Leite F, et al. Occurrence of Mycobacterium bovis and non-tuberculous mycobacteria (NTM) in raw and pasteurized milk in the northwestern region of Paraná, Brazil. Braz J Microbiol. 2014;711:707–11.
6. BRASIL. Ministério da Agricultura, Pecuária e Abastecimento. *Programa Nacional de Controle e Erradicação da Brucelose e da Tuberculose Animal (PNCEBT)*. Brasília, 2006. http://www.agricultura.gov.br/arq_editor/file/Aniamal/programa%20nacional%20sanidade%20brucelose/Manual%20do%20PNCEBT%20-%20Original.pdf. Accessed 15 Jan 2015.
7. Araujo CP, Osório ALAR, Jorge KSG, Ramos CAN, Filho AFS, Vidal CES, Vargas APC, et al. Direct detection of Mycobacterium tuberculosis complex in bovine and bubaline tissues through nested-PCR. Braz J Microbiol. 2014;45:633–40.
8. Majeed MA-M, Ahmed WA, Manki A. Amplification of a 500 base-pair fragment from routinely identified isolates of m . Bovis from cow's milk in Baghdad. IJABR. 2013;3:163–7.
9. Thrusfield MV. Epidemiologia Veterinária. 2ª ed. São Paulo: ROCA; 2004.
10. IBGE – Instituto Brasileiro de Geografia e Estatística. 2005. Available from: http://www.sidra.ibge.gov.br/bda/territorio/unit.

asp?e=c&t=71&p=CA&v=2057&codunit=6186&z=t&o=4&i=P. A. Accessed 15 Jan 2015.

11. Maniatis T, Fritsch EF, Sambrook J. Molecular cloning: a laboratory manual. New York: Cold Spring Harbor Laboratory Press; 1989. p. 3.

12. Sales ML, Jr AAF, Orzil L, Alencar AP, Hodon MA. Validation of two real-time PCRs targeting the PE-PGRS 20 gene and the region of difference 4 for the characterization of Mycobacterium bovis isolates. Genet Mol Res. 2014;13:4607–16.

13. Tamura K, Stecher G, Peterson D, Filipski A, Kumar S. MEGA6: molecular evolutionary genetics analysis version 6.0. Mol Biol Evol. 2013;30:2725–9.

14. Zumárraga MJ, Soutullo A, García MI, Marini R, Abdala A, Tarabla H, et al. Infected dairy herds using PCR in bulk tank milk samples. Foodborne Pathog Dis. 2012;9:132–7. doi:10.1089/fpd.2011.0963.

15. Franco MMJ, Paes AC, Ribeiro MG, de Figueiredo Pantoja JC, Santos ACB, Miyata M, et al. Occurrence of mycobacteria in bovine milk samples from both individual and collective bulk tanks at farms and informal markets in the southeast region of Sao Paulo, Brazil. BMC Vet Res. 2013;9:2–8. doi:10.1186/1746-6148-9-85.

16. Zarden CFO, Marassi CD, Figueiredo EEES, Lilenbaum W. Mycobacterium bovis detection from milk of negative skin test cows. Vet Rec. 2013;172:130. doi:10.1136/vr.101054.

17. El-Gedaway AA, Ahmed HA, Awadallah MAI. Occurrence and molecular characterization of some zoonotic bacteria in bovine milk, milking equipment and humans in dairy farms, Sharkia, Egypt. Int Food Res J. 2014;21:1813–23.

18. Mohamed SA, Aggour MG, Ahmed HA, Selim SA. Detection and differentiation between mycobacterium bovis and mycobacterium tuberculosis in cattle milk and lymph nodes using multiplex real-time PCR. Global Veterinaria. 2014;13:794–800. doi:10.5829/idosi.gv.2014.13.05.86179.

19. Senthil NR, Ranjani MR, Vasumathi K. Comparative diagnosis of Mycobacterium bovis by Polymerase chain reaction and Ziel- Neilson staining technique using Milk and nasal washing. J Res Agri Animal Sci. 2014;2:1–3.

20. Carvalho RCT, Castro VS, Fernandes DVGS, Moura GF, Santos ECC, Paschoalin VMF, et al. Use of the Pcr for detection mycobacterium bovis in milk. Proc XII Lat Am Congr Food Microbiol Hyg. 2014;1:359–60. doi:10.5151/foodsci-microal-180.

21. Evangelista TBR, De Anda JH. Tuberculosis in dairy calves: risk of Mycobacterium spp. exposure associated with management of colostrum and milk. Prev Vet Med. 1996;27:23–7. doi:10.1016/0167-5877(95)00573-0.

22. Katale BZ, Mbugi EV, Karimuribo ED, Keyyu JD, Kendall S, Kibiki GS, et al. Prevalence and risk factors for infection of bovine tuberculosis in indigenous cattle in the Serengeti ecosystem, Tanzania. BMC Vet Res. 2013;9:267. doi:10.1186/1746-6148-9-267.

23. Silaigwana B, Green E, Ndip RN. Molecular detection and drug resistance of Mycobacterium tuberculosis complex from cattle at a dairy farm in the Nkonkobe region of South Africa: A pilot study. Int J Environ Res Public Health. 2012;9:2045–56. doi:10.3390/ijerph9062045.

24. Rowe MT, Donaghy J. Mycobacterium bovis: The importance of milk and dairy products as a cause of human tuberculosis in the UK. A review of taxonomy and culture methods, with particular reference to artisanal cheeses. Int J Dairy Technol. 2008;61:317–26. doi:10.1111/j.1471-0307.2008.00433.x.

25. Ereqat S, Nasereddin A, Levine H, Azmi K, Al-Jawabreh A, Greenblatt CL, et al. First-time detection of Mycobacterium bovis in livestock tissues and milk in the West Bank. Palestinian Territories PLoS Negl Trop Dis. 2013;7:e2417. doi:10.1371/journal.pntd.0002417.

26. Malama S, Muma JB, Olea-Popelka F, Mbulo G. Isolation of mycobacterium bovis from human sputum in Zambia: public health and diagnostic significance. J Infect Dis Ther. 2013;1:2. doi:10.4172/jidt.1000114.

27. Silva MR, Rocha AS, da Costa RR, de Alencar AP, de Oliveira VM, Fonseca Junior AA, et al. Tuberculosis patients co-infected with Mycobacterium bovis and Mycobacterium tuberculosis in an urban area of Brazil. Mem Inst Oswaldo Cruz. 2013;108:321–7.

28. Kantor IN, Ambroggib M, Poggib S, Morcillo N, Telles MAS, Marta Ribeiro MO, Torres MDG, et al. Human Mycobacterium bovis infection in ten Latin American countries. Tuberculosis. 2008;88:358–65.

29. Figueiredo EES, Junior CQC, Adam C, Furlaneto LV, Silvestre FG, Duarte RS. et al. Molecular techniques for identification of species of the Mycobacterium tuberculosis complex: the use of multiplex PCR and an adapted HPLC method for identification of Mycobacterium bovis and diagnosis of bovine tuberculosis. Underst Tuberc - Glob Exp Innov Approaches to Diagnosis 2012:411–432. Available in available in http://cdn.intechopen.com/pdfs-wm/28550.pdf. Accessed 15 Jan 2015.

30. Kich DM, Kreling CS, Pozzobon A. Análise da presença de mycobacterium

31. bovis através da técnica de reação em cadeia da polimerase (pcr) em amostras de leite bovino in natura na região do Vale do Raquari, RS. Revista Destaques Acadêmicos. 2012;4:19–26.

31. Jordão-Junior CM, Lopes FCM, Pinto MRA, Roxo E, Leite CQF. Delevopment of a PCR assay for the direct detection of Mycobacterium bovis in milk. Brazilian J Food Nutr. 2005;16:51–5.

32. Romero RE, Garzón DL, Mejía GA, Monroy W, Patarroyo ME, Murillo LA. Identification of Mycobacterium bovis in bovine clinical samples by PCR species-specific primers. Can J Vet Res. 1999;63:101–6.

33. Srivastava K, Chauhan DS, Gupta P, Singh HB, Sharma VD, Yadav VS, et al. Isolation of Mycobacterium bovis & M. tuberculosis from cattle of some farms in north India–possible relevance in human health. Indian J Med Res. 2008;128:26–31.

34. Pardo RB, Langoni H, Mendonça LJP, Chi KD. Isolation of Mycobacterium spp. in milk from cows suspected or positive to tuberculosis. Braz J Vet Res Anim Sci. 2001;38:284–7.

35. Izael MA, Silva STG, Costa NA. Estudo retrospectivo da ocorrência dos casos de tuberculose bovina diagnosticados na clínica de bovinos de Garanhuns - PE, de 2000 A 2009. Brazilian Animal Sci. 2009;1:452–7.

36. Mendes EI, Melo LEH, Tenório TGS, Sá LM, Souto RJC, Fernandes ACC, et al. Intercorrência entre leucose enzoótica e tuberculose em bovinos leiteiros do estado de Pernambuco. Arq Inst Biol. 2011;78:1–8.

37. Humblet MF, Gilbert M, Govaerts M, Fauville-Dufaux M, Walravens K, Saegerman C. New assessment of bovine tuberculosis risk factors in Belgium based on nationwide molecular epidemiology. J Clin Microbiol. 2010;48:2802–8. doi:10.1128/JCM.00293-10.

38. Humblet MF, Boschiroli ML, Saegerman C. Classification of worldwide bovine tuberculosis risk factors in cattle: A stratified approach. Vet Res 2009; 40. doi:10.1051/vetres/2009033.

39. Skuce RA, Allen AR, McDowell SWJ. 2011. Bovine tuberculosis (TB): a review of cattle-to-cattle transmission, risk factors and susceptibility. https://www.dardni.gov.uk/sites/default/files/publications/dard/afbi-literature-review-tb-review-cattle-to-cattle-transmission.pdf. Accessed 15 Jan 2015.

40. Adams AP, Bolin SR, Fine AE, Bolin CA, Kaneene JB. Comparison of PCR versus culture for detection of Mycobacterium bovis after experimental inoculation of various matrices held under environmental conditions for extended periods. Appl Environ Microbiol. 2013;79:6501–6. doi:10.1128/AEM.02032-13.

41. Kazoora HB, Majalija S, Kiwanuka N, Kaneene JB. Prevalence of Mycobacterium bovis skin positivity and associated risk factors in cattle from Western Uganda. Trop Anim Health Prod. 2014;46:1383–90. doi:10.1007/s11250-014-0650-1.

42. Roxo E. Bovine tuberculosis: review. Arquivos do Instituto Biológico. 1996; 63(2):91–7.

43. Lilenbaum W, Souza GN, Fonseca LDS. Fatores de manejo associados à ocorrência de tuberculose bovina em rebanhos leiteiros do Rio de Janeiro, Brasil. R Bras Ci Vet. 2007;14:98–100.

2

Immunization with one *Theileria parva* strain results in similar level of CTL strain-specificity and protection compared to immunization with the three-component Muguga cocktail in MHC-matched animals

Lucilla Steinaa[1]* (iD), Nicholas Svitek[1], Elias Awino[1], Thomas Njoroge[1], Rosemary Saya[1], Ivan Morrison[2] and Philip Toye[1]

Abstract

Background: The tick-borne protozoan parasite *Theileria parva* causes a usually fatal cattle disease known as East Coast fever in sub-Saharan Africa, with devastating consequences for poor small-holder farmers. Immunity to *T. parva*, believed to be mediated by a cytotoxic T lymphocyte (CTL) response, is induced following natural infection and after vaccination with a live vaccine, known as the Infection and Treatment Method (ITM). The most commonly used version of ITM is a combination of parasites derived from three isolates (Muguga, Kiambu 5 and Serengeti-transformed), known as the "Muguga cocktail". The use of a vaccine comprising several strains is believed to be required to induce a broad immune response effective against field challenge. In this study we investigated whether immunization with the Muguga cocktail induces a broader CTL response than immunization with a single strain (Muguga).

Results: Four MHC haplotype-matched pairs of cattle were immunized with either the trivalent Muguga cocktail or the single Muguga strain. CTL specificity was assessed on a panel of five different strains, and clonal responses to these strains were also assessed in one of the MHC-matched pairs. We did not find evidence for a broader CTL response in animals immunized with the Muguga cocktail compared to those immunized with the Muguga strain alone, in either the bulk or clonal CTL analyses. This was supported by an in vivo trial in which all vaccinated animals survived challenge with a lethal dose of the Muguga cocktail vaccine stabilate.

Conclusion: We did not observe any substantial differences in the immunity generated from animals immunized with either Muguga alone or the Muguga cocktail in the animals tested here, corroborating earlier results showing limited antigenic diversity in the Muguga cocktail. These results may warrant further field studies using single *T. parva* strains as future vaccine candidates.

Keywords: *Theileria parva*, Live vaccine, Cytotoxic T cells, Immunity, Strain specificity

* Correspondence: l.steinaa@cgiar.org
[1]International Livestock Research Institute, P.O. Box 30709, Nairobi 00100, Kenya
Full list of author information is available at the end of the article

Background

Theileria parva is a tick-borne protozoan parasite which causes an acute and usually fatal cattle disease, known as East Coast fever, in eastern, central and southern Africa. The parasite infects bovine lymphocytes, which subsequently undergo blast transformation and rapid multiplication [1]. In susceptible animals, this usually results in overwhelming parasitosis and death within 2 to 4 weeks of infection.

Cattle which recover from natural infection can develop a strong immunity to subsequent challenge. This has been exploited to develop a vaccination procedure known as the "Infection and Treatment Method" (ITM) in which live sporozoites are administered simultaneously with oxytetracycline. The main protective mechanism in both vaccinated and naturally recovered animals is believed to be CD8+ cytotoxic T lymphocyte (CTL) killing of infected lymphocytes. Thus, adoptive transfer of CD8$^+$ cells from immunized animals has been demonstrated to protect naïve animals from challenge [2]. Furthermore, it has been shown that the time point of recovery correlates with a peak of CD8$^+$ cells in the blood of the infected animals [3].

Early experiments in the development of the ITM vaccine revealed the presence of strain specificity through infection and challenge trials where animals were immunized with one strain and challenged with a heterologous strain. While good protection was often obtained following the inoculation of single strains, it did not always extend to heterologous challenge [4]. It was subsequently shown that a combination of three strains (Muguga, Serengeti-transformed and Kiambu 5) provided better protection than single strains. The mixture, known as the "Muguga cocktail", is the basis of a commercial ITM vaccine which appears to provide broad protection against *T. parva* in the field, so far best explored in Tanzania [5, 6]. It should be noted that in experiments involving single isolates, some cross-protection was observed with a notable exception being if animals vaccinated with the Muguga strain or Kilifi strain were challenged with Marikebuni [4, 7, 8], suggesting that Marikebuni is antigenically quite distinct.

Strain specificity has also been observed in in vitro CTL assays and reflects the in vivo immune status [3, 9–11]. A possible explanation for the strain specificity of the CTL response to *T. parva* is based on two phenomena – immunodominance and antigenic diversity. Immunodominance, where the immune response is directed to a very limited number of antigens in individual animals, is commonly observed in viral infections [12–15] and also in other infectious diseases and cancers [16–18]. In *T. parva* infections, indications of immunodominance came with initial analyses of the immune response where, in some cases, it was possible to show that CTL restriction was mediated by a single class I MHC molecule [19, 20] despite the presence of many potential epitopes expressed by the parasite genome, which is predicted to encode 4034 genes [21, 22], and a T cell receptor (TCR) repertoire capable of reacting to a wide variety of epitopes. More recently, with the discovery of CTL antigens [23], a study was performed to shed light on the issue of immunodominance [24]. Findings from this work suggest that one or a few MHC alleles in individual animals govern the specificity of the immune response elicited by infection or vaccination and are crucial for the outcome of a later challenge with genotypically different parasites.

The most plausible explanation for the broad protection offered by the Muguga cocktail, is that the mixture of the parasites in the cocktail, provides a more diverse set of antigens which potentially can induce a CTL response of broader antigenic specificity than inoculation with any of the individual components or single strains, and thus provide better protection against heterologous parasites encountered in the field. To test this hypothesis, we compared the specificities for different parasite strains of CTL induced in animals immunized by ITM with the Muguga cocktail with those generated by the Muguga stabilate alone. To minimize any effect of the MHC background in individual animals, which is known to influence the selection of antigens recognized by the CTLs, the responses were compared in MHC haploidentical pairs of cattle. Analysis of the CTL response at the clonal level was also undertaken for one haplotype matched pair. All haploidentical pairs of cattle were challenged with the Muguga cocktail to uncover any antigenic differences in the in vivo CTL response.

Methods

Animals and MHC class I typing

All animal experiments were reviewed and approved by the Institutional Animal Care and Use Committee at International Livestock Research Institute (ILRI). Eight *Bos taurus* cattle (five Friesian and three Ayrshire) were bought from farms in the Nyeri area in Kenya. They were screened free for tickborne diseases including *T. parva*, and BoLA typed using a combination of serology (ELISA using antibodies defining particular MHC haplotypes), IFNγ ELISPOT assay using PBMC from the cattle pulsed with Tp1$_{214-224}$ and a peptide-specific CTL line, and by PCR using haplotype specific primers followed by sequencing, essentially as described before [25]. Four haploidentical pairs of the following MHC haplotypes were selected for the study: A10/A12, A12/A14, A15/A18, A11/A15 (Table 1). Animals were kept in standard pens and were fed normally. At the end of the experiment, animals were returned to the farm and eventually slaughtered for meat. Two control animals, which developed disease, were euthanized for humane reasons using

Table 1 Cattle used in the study

Calf	Breed	MHC Class 1 Haplotype	Alleles	Immunization
BG033	Friesian	A10/A12	N*00201, N*01901	Muguga (3308)
BG042	Friesian	A10/A12	N*00201, N*01901	Muguga cocktail (0801)
BG053	Ayrshire	A12/A14	N*01901, N*02301	Muguga (3308)
BG051	Ayrshire	A12/A14	N*01901, N*02301	Muguga cocktail (0801)
BG052	Friesian	A15/A18	N*00901, N*01302	Muguga (3308)
BG056	Friesian	A15/A18	N*00901, N*01302	Muguga cocktail (0801)
BH055	Ayrshire	A11/A15	N*01802, N*00902	Muguga (3308)
BH047	Friesian	A11/A15	N*01802, N*00902	Muguga cocktail (0801)

The breed, MHC Class I haplotype and associated MHC alleles of the cattle included in the study are shown. Within each of the haplotype matched pairs of cattle, one was immunized with the single *T. parva* strain Muguga 3308 and the other was immunized with the Muguga cocktail 0801 using ITM

an overdose of Euthatal (Pentobarbital sodium, 200 mg/ml), 1 ml Euthatal per 1.4 kg body weight, given intravenously, after restraining the animals.

Immunization

Each of the haplotype-matched pairs of cattle was inoculated subcutaneously in front and below the right ear with either the Muguga stabilate 3308 or the Muguga cocktail vaccine stabilate ILRI0801 [26] and treated immediately with long acting oxytetracyclin.

Parasitized cell lines

Cell lines infected with *T. parva* were established by infection of autologous PBMC in vitro with sporozoites as described previously [27]. The sporozoites were from the cloned stabilates Marikebuni 3292, Muguga 3308, Boleni 3230, Uganda 3645, derived from 3569 [28], and Mariakani 3212 (unpublished). In addition, cell lines were established using sporozoites of the ILRI0801 reference stabilates: Muguga (4230), Serengeti (4229), Kiambu 5 (4228). These are stabilates of the individual Muguga cocktail components made from the same production ticks as the ILRI0801 vaccine [26].

Generation of CTL

CTL bulk cultures were generated and maintained in RPMI 1640 (Sigma-Aldrich, St.Louis, MO, USA) supplemented with 10% FBS (Thermo Fisher Scientific, Waltham, MA, USA), 2 mM L-glutamin (Sigma-Aldrich, St.Louis, MO, USA), 50 μM 2-mercaptoethanol (Sigma-Aldrich, St.Louis, MO, USA), 100 IU of penicillin/ml (Sigma-Aldrich, St.Louis, MO, USA), 100 μg of streptomycin/ml (Sigma-Aldrich, St.Louis, MO, USA), 50 μg of gentamicin/ml, (Sigma-Aldrich, St.Louis, MO, USA) 10% TCGF (Conditioned media from ConA blasts).

CTL were generated essentially as described [27, 29]. Briefly, PBMC were re-stimulated three times with irradiated autologous *T. parva*-infected cell lines. CTL from animals immunized with the vaccine were generated by stimulating with equal fractions of three cell lines infected with one of the three vaccine reference stabilates. CTL from the Muguga- immunized animals were stimulated with the Muguga-infected cell line only. The remainder of the procedure was as described previously [27, 29].

Generation of CD8+ CTL clones

Clones were generated by purifying CD8+ cells after two restimulations as described above. Enrichment of CD8+ cells was achieved by incubating the CTL line for 30 min with mouse-anti-bovine CD8 mAb ILA105 (ILRI) diluted 1:500 in PBS + 2% FBS. Cells were washed twice in PBS + 2%FBS, labeled with magnetic beads attached to goat-anti-mIgG (Miltenyi Biotec, Bergish Gladbach, Germany) and purified according to the protocol provided by the manufacturer. Purified CD8+ cells were then seeded by limiting dilutions in 96-well plates using 2×10^4 irradiated PBMC as filler cells as previously described [27].

CTL lines and clones were generated and maintained in RPMI 1640 medium (Sigma-Aldrich, St. Louis, MO, USA) supplemented with 10% FBS (Thermo Fisher Scientific, Waltham, MA, USA), 2 mM L-glutamine, 50 μM 2-mercaptoethanol, 100 IU/ml of penicillin, 100 μg/ml of streptomycin, 50 μg/ml of gentamicin, (all from Sigma-Aldrich, St. Louis, MO, USA) and 10% inactivated ConA supernatant (conditioned media from ConA blasts).

Cytotoxicity assay

A standard 4 h release assay using ^{51}Cr-labeled target cells was used to measure cytotoxicity. ^{51}Cr was obtained from American Radiolabeled Chemicals, Inc., St. Louis, MO, USA. Supernatants were counted using Lumaplates (PerkinElmer, Waltham, MA, USA) in a TopCounter (PerkinElmer, Waltham, MA, USA). The cytotoxicity was calculated as: (experimental release-spontaneous release/total release-spontaneous release). Target cells were either *T. parva*-infected autologous PBMC using MHC-mismatched *T. parva*-infected cell lines as controls. In one case (BH055), autologous uninfected PBMC were used as control. Each CTL was tested in dilutions using a fixed number of target cells. Each dilution was tested in triplicate. Cytotoxicity at an effector:target ratio of 10:1 was used to compare between CTL derived from different animals and between different target cells.

Challenge experiment

Cattle were challenged with the Muguga cocktail vaccine stabilate ILRI0801, without oxytetracycline. The vaccine (2 ml neat) was injected subcutaneously in front of and below the right ear. Animals were monitored for the required clinical parameters to determine the severity of disease according to the Rowlands index [30].

Results

Similar strain specificities of CTL from haplotype-matched animals immunized by ITM with either *T. parva* Muguga or the Muguga cocktail

Eight cattle comprising four haploidentical pairs were immunized by ITM with either the Muguga strain (3308) or the trivalent Muguga cocktail (ILRI0801), as detailed in Table 1.

CTL were generated from all animals and tested against autologous cell lines infected with five cloned sporozoite stabilates and an MHC-mismatched infected cell line as control. Figure 1 shows an example of the results for a CTL assay of one of the animals. For simplicity, the cytotoxicity for all animals at the same effector/target ratio of 10:1 was deduced from dilution curves, as shown in Fig. 1. The results for all animals, as shown in Table 2, allows a comparison of the breadth of specificity to five different parasite strains exhibited by CTL from the Muguga-immunized and the Muguga cocktail-immunized calves. ANOVA analysis confirmed that there were no statistically difference ($P = 0.421$). Importantly, CTL from the Muguga-immunized animals killed all five targets to various degrees and no clear strain specificity was observed. CTL from BH047 showed a consistently lower level of killing compared to the other CTL, despite several attempts to establish a bulk culture showing higher cytotoxicity.

CTL from the Muguga cocktail-immunized animals were also assessed for recognition of all three

Table 2 Cytotoxic T cell responses in immunized cattle assayed on target cells infected with different cloned *T. parva* strains

Muguga	Uganda 3308	Mariakani 3645	Boleni 3212	Marikebuni 3230	Control 3292	
BG033 (M)	34	13	35	16	ND	0
BG042 (C)	26	18	17	9	18	0
BG053 (M)	59	40	14	20	34	0
BG051 (C)	51	25	38	22	28	3
BG052 (M)	38	33	25	23	17	4
BG056 (C)	3	22	20	11	13	3
BH055 (M)	40[a]	30	26	13	20	0[a]
BH047 (C)	10	8	8	7	4	0

Cytotoxic T cell responses using autologous PBMC infected with different cloned *T. parva* strains as target cells. The specific cytotoxicity at effector/ target ratios of 10:1 are shown. Each value is deduced from an effector (CTL) titration curve as shown in Fig. 1. (M) Muguga 3308 immunized; (C) Muguga cocktail immunized. (ND) Not determined, ([a]) Control was autologous PBMC. Relative differences of the CTL specificities, between M and C, were tested using ANOVA analysis ($P = 0.421$)

components present in the Muguga cocktail. As seen in Table 3, all Muguga cocktail-immunized animals showed killing of the three components. Interestingly, the Muguga component was killed more effectively in all cases ($P < 0.001$ for each of the cattle) except BH047, where the killing of the Serengeti component was slightly greater (not statistically significant). As there can be variability in CTL assays, these experiments were done at least twice, and for some animals three times, and the results were similar.

Cloned CTL from a haplotype-matched pair show similar strain specificities

To examine the CTL specificity in more detail, we generated CTL clones from one haploidentical pair, namely the A12/A14 pair of BG053 and BG051. We also successfully generated clones from BG052 (results not shown) but not from the rest of the animals. Each clone

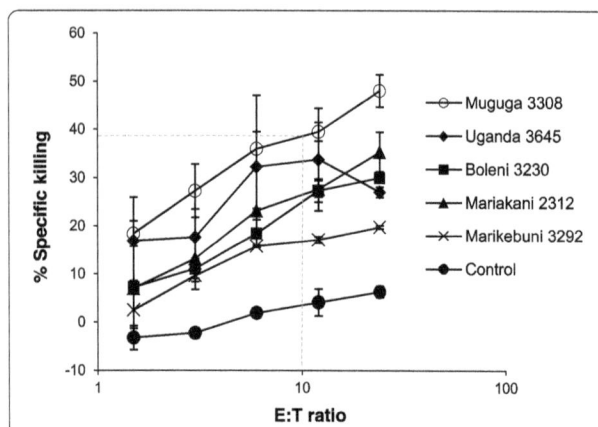

Fig. 1 Example of the result from a CTL assay. Serial dilutions of CTL from BG052 were tested for lysis of fixed numbers of the various target cells shown in the figure. The dashed line represents the deduced specific killing at an effector:target ratio (E:T ratio) of 10:1 for the target Muguga 3308. This method was used to compare killing of the various target cells as listed in Table 2. Each point represents the average of a double-determination with the SD shown

Table 3 Cytotoxic T cell responses in cattle immunized with the Muguga cocktail on the three *T. parva* component strains as targets

	Muguga 4230	Serengeti 4229	Kiambu 5 4228	Control
BG042	26[a]	18	13	0
BG051	40[a]	28	34	0
BG056	27[a]	22	15	0
BH047	9	11	5	0

Cytotoxic T cell responses by vaccine immunized animals using autologous PBMC infected with the three different *T. parva* component strains as targets. Each CTL was titrated on a fixed number of target cells. The effector/target ratio of 10:1 is shown and each value is deduced from an effector (CTL) titration curve as shown in Fig. 1. Statistical comparison of parameters from fitted curves were used to test differences in cytotoxicity between targets. ([a]) significant higher than for both other targets for each animal ($p < 0.001$)

was tested for cytotoxicity using the five different cloned sporozoite strains as target cells. Figure 2a shows a heat map of the clonal analysis of BG053 which was immunized with Muguga 3308. It is clear from the pattern of recognition that nine different clonotypes from BG053 were identified, with some clones recognizing only one or two strains, and others recognizing three or four of the five strains. The least recognized strain was Marikebuni, and only one clone was found that recognized Muguga alone. In general, the clonal analysis corresponded well with the bulk analysis, where strongest killing was also observed with the Muguga and Uganda stabilates. On the other hand, few of the clones recognized Marikebuni in contrast to the results observed with the bulk cell lines.

The clonal analysis of the Muguga cocktail-immunized animal BG051 is shown in Fig. 2b. Eight clonotypes were identified, with a surprisingly high number of clones specific for Kiambu5 only. There were also clones which recognized a broader set of targets. As observed in the analysis of bulk CTL lines, there was no evidence from the clonal analysis of a wider set of clonal reactivities in the Muguga cocktail-immunized animal compared to the Muguga only-immunized animal. Table 4 shows the number of clones with the percentages in brackets, that recognize indicated numbers of different cloned strains (upper part), and the number of clones with percentage in brackets recognizing a particular parasite strain (lower part). It is evident that a large fraction of clones from the Muguga cocktail-immunized animal recognized one strain only, whereas more clones from the Muguga only-immunized animal recognized many of the strains, which is opposite of what was expected. All clones from the Muguga only-immunized animal recognized Muguga as expected, and 89% recognized Uganda, implying that this strain is quite similar to Muguga. On the other hand, Marikebuni was the least recognized strain, being specifically lysed by only 17% of the clones. In the Muguga cocktail-immunized animal, 87% of the clones recognized Kiambu5, one of the components in the vaccine, compared to 5% and 16% that recognized the Muguga and Serengeti-transformed components,

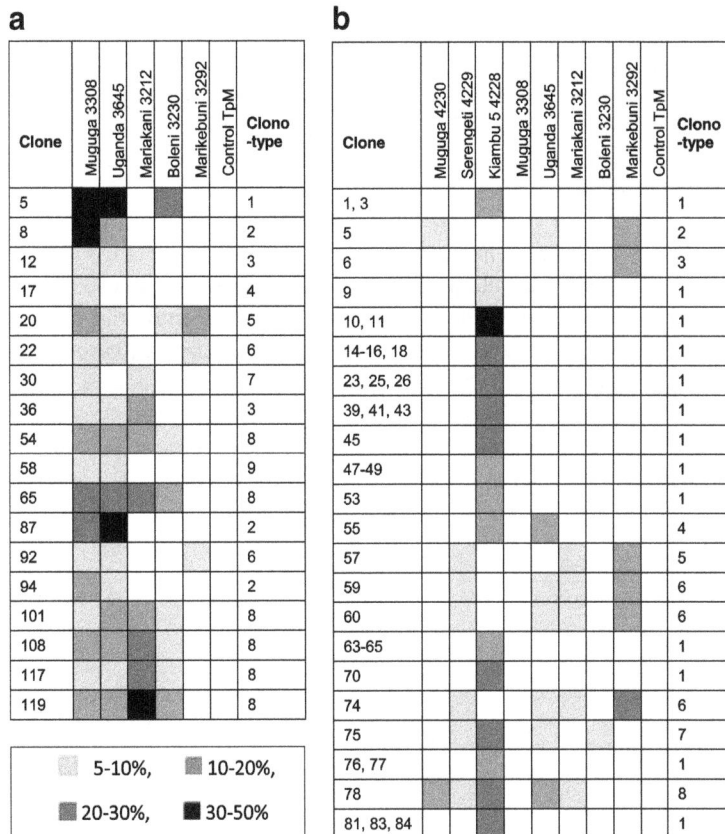

a

Clone	Muguga 3308	Uganda 3645	Mariakani 3212	Boleni 3230	Marikebuni 3292	Control TpM	Clono-type
5							1
8							2
12							3
17							4
20							5
22							6
30							7
36							3
54							8
58							9
65							8
87							2
92							6
94							2
101							8
108							8
117							8
119							8

5–10%,　10–20%,　20–30%,　30–50%

b

Clone	Muguga 4230	Serengeti 4229	Kiambu 5 4228	Muguga 3308	Uganda 3645	Mariakani 3212	Boleni 3230	Marikebuni 3292	Control TpM	Clono-type
1, 3										1
5										2
6										3
9										1
10, 11										1
14-16, 18										1
23, 25, 26										1
39, 41, 43										1
45										1
47-49										1
53										1
55										4
57										5
59										6
60										6
63-65										1
70										1
74										6
75										7
76, 77										1
78										8
81, 83, 84										1

Fig. 2 Cytotoxicity obtained by T cell clones on a panel of target cells infected with different cloned *T. parva* strains. The cutoff value was 5% cytotoxicity. Clones were categorized into clonotypes based on their pattern of reactivity. The level of cytotoxicity is visualized as a heat map – colour codes are shown. **a** Calf BG053 immunized with Muguga 3308. CD8 T cell clones were generated from CD8-purified bulk cultures and tested for cytotoxicity to 5 different strains (as shown) and a MHC-mismatched control TpM. **b** The haplotype-matched calf BG051 immunized with the Muguga cocktail 0801. Clones were tested for cytotoxicity on the same target cells as BG053 and the additional components of the Muguga Cocktail

Table 4 Number and percentages of CTL clones from BH053 and BH051 recognizing multiple *T. parva* strains

Calf	BG053 (M)	BG051 (C)
Total	18	38
1 strain	1 (6)	29 (76)
2 strains	5 (28)	2 (5)
3 strains	5 (28)	2 (5)
4 strains	7 (39)	4 (11)
5 strains	0 (0)	1 (3)
Muguga (4230)	ND	2 (5)
Serengeti (4229)	ND	6 (16)
Kiambu-5 (4228)	ND	33 (87)
Muguga (3308)	18 (100)	2 (5)
Uganda (3645)	16 (89)	7 (18)
Mariakani (3212)	9 (50)	5 (13)
Boleni (3230)	8 (44)	1 (3)
Marikebuni (3292)	3 (17)	6 (16)

Upper part of the table: The total number of clones with percentages in brackets, that recognize multiple cloned *T. parva* strains, with a cutoff value of 5% cytotoxicity, for the haplotype matched pair, BG053 (Muguga immunized) and BG051 (Muguga cocktail immunized), is shown. Each clone was analyzed for the number of strains that it recognized. The left numbers in each column represents the actual number of clones recognizing 1 strain, 2 strains, etc. The numbers in brackets represents the corresponding percentages of clones recognizing 1 strain, 2 strains etc. Percentages have been rounded. Lower part of the table: The total number of clones and percentages in brackets from BG053 and BG051 recognizing the different *T. parva* strains. Percentages have been rounded. BH051 was tested on the Muguga cocktail component reference stabilate strains in addition to the cloned strains. (ND) Not determined

respectively. Interestingly, there was a difference in the two Muguga-infected targets, which may reflect qualitative differences in presented epitopes in these two target cells.

In summary, the results of the CTL assays do not support the hypothesis that immunization with the Muguga cocktail induces CTL with a broader reactivity against *T. parva*-infected cell lines compared to immunization with the Muguga stabilate alone, at least with the parasite strains and MHC haplotypes of the animals assessed here.

Similar protection to the vaccine strains in animals vaccinated with Muguga or the Muguga cocktail

The animals were challenged with the Muguga cocktail in order to investigate if there were any differences between the Muguga-immunized animals and the Muguga cocktail-immunized animals in their immunity to parasite strains present in the Muguga cocktail vaccine but not in the Muguga stabilate. Two non-immunized control animals were used to confirm a sufficient challenge had been delivered and the clinical outcome was assessed with the Rowlands ECF index [30]. The experiment was stopped at day 14 as there was no development of disease in the animals except for the two control animals. As seen in Table 5, there was no substantive difference in the protection to the Muguga cocktail between animals immunized with Muguga only or the Muguga cocktail. Only one animal from each immunized group developed pyrexia, and schizonts were detected in three of the seven animals, two from the Muguga-immunized group and one from the Muguga cocktail-immunized group. This contrasts strongly with the control animals, both of which developed pyrexia and had detectable schizonts and piroplasms. Interestingly, the animal with the highest ECF score, BH047, was the animal with the weakest CTL response.

Discussion

It is of major importance to fully understand the mechanisms underlying the protective immune response to *T. parva*, both to underpin the deployment of the current live Muguga cocktail vaccine and to guide the development of a subunit vaccine.

The Muguga cocktail vaccine is composed of parasites from three different isolates (Muguga, Kiambu 5 and Serengeti-transformed), and it is known to protect

Table 5 Challenge of cattle with the Muguga cocktail 0801

Animal	Immunization	Days with Schizonts	Days with pyrexia	Days with piroplasms	Day treated	ECF index
BG033	Mug 3308 (M)	5				1.04
BG053	Mug 3308 (M)	2				2.21
BH055	Mug 3308 (M)		2			2.43
BG051	Vac 0801 (C)					1.04
BG042	Vac 0801 (C)					1.04
BH047	Vac 0801 (C)	6	4			2.98
BG056	Vac 0801 (C)					1.04
BG039	None	7	5	2	13	6
BH043	None	7	4	2	13	5.93

All 7 immunized cattle (BG052 succumbed for unknown reasons) and two naïve control cattle BG039, BH043 were challenged with the Muguga cocktail vaccine stabilate 0801. Number of days with schizonts, pyrexia, piroplasms and the days until treatment are indicated. The ECF index is also indicated

animals efficiently when deployed in the field [5], despite the presence of antigenic variation in field populations of the parasite [31]. The experiments reported in this article were undertaken to demonstrate that the trivalent Muguga cocktail vaccine provided a more complete protection against heterologous challenge than the Muguga stabilate alone by inducing CTLs capable of recognizing a broader range of parasite strains.

Surprisingly, there was no substantive difference in the CTL responses to the different *T. parva* strains between MHC-matched animals immunized with the Muguga cocktail and the Muguga only. The clonal analysis of one of the haplotype matched pairs supported these results (unfortunately we were not able to generate clones from the other haplotype matched pairs). We were surprised to see the many Kiambu 5-specific clones from the vaccine immunized BG051 animal. This result did not corroborate with the bulk result (Table 2), where there were good responses to Muguga and to Serengeti. It is possible that there is a bias in the cloning process in some cases. Some clones could be easier to expand than others, which would limit the interpretation of the clonal analyses in general. Nevertheless, the bulk results also did not favour a broader CTL specificity in the Muguga cocktail immunized animals compared to those immunized with the Muguga stabilate. An interesting observation was that certain clones recognized many different target cells, which shows that there are indeed broadly cross-reactive antigenic determinants which induce a CTL response following ITM immunization, either with Muguga alone or with the Muguga cocktail. Future work will aim to map these epitopes, which could be useful in a subunit vaccine.

Previous research on the strain specificity of the CTL response in animals immune to *T. parva* suggests that the response in each animal is dominated by a small number of antigens [10, 11, 19, 20]. Clonal analysis of the response in MHC-homozygous animals confirmed this immunodominance by showing that over 60% of the clones from the animals recognized single epitopes in the two respective antigens presented by the MHC haplotypes [24]. The relatively high number of clonotypes observed in the A12/A14 animals analysed here suggests that this is not the case in these animals, unless there is an immunodominant antigen which displays antigenic diversity. In other words, the differential reactivity of the clonotypes is a consequence of the different levels of cross-reactivity of each clone with variant forms of a dominant epitope. The *T. parva* antigens recognized by the various MHC alleles in the A12 and A14 animals have not been fully identified so it is currently not possible to assess the recognition of specific epitopes and the level of diversity displayed by such epitopes. The focus of the response in the Muguga cocktail-immunized animal on

Kiambu 5 is interesting and may reflect an immunodominant antigen found predominantly in Kiambu and not shared with many other strains. In this respect, it would be interesting to assess whether CTL from Muguga-only immunized animals are cytotoxic towards the three components in the Muguga cocktail, and to assess whether CTL derived from animals immunized with any one of the components comprising the Muguga cocktail recognize the cloned strains as this may reveal if there are immunodominant, cross-reactive epitopes in the stabilates, and whether this outcome is dependent on the MHC background of the animals.

In order to show if the lack of differences in the CTL responses was reflected in vivo, all animals were challenged with a lethal dose of the Muguga cocktail. There were no major differences in the protection observed in the animals, which indicates that there are no important differences in the antigenic composition of the Muguga stabilate and the Muguga cocktail, at least in animals of the MHC types studied here. This result is perhaps not surprising, as the orginal studies of Radley et al., (1975a) showed that the Muguga stabilate provided good protection against both the Kiambu 5 and Serengeti transformed stabilates, and *vice versa*. In addition, recent deep-sequencing results from our group have also shown that the Muguga cocktail does not contain a great amount of diversity in the known CTL antigens which were examined [32]. Five of the nine antigen genes sequenced were present as a single version, with three present in two forms and the final as three variants. These results suggest that the three components are antigenically very similar and they invoke the question of why the Muguga cocktail provided better protection in the original experiments and why the Muguga cocktail has been so successful in protecting animals against field challenge, where heterogeneous challenges will be far more predominant. It may be, as argued elsewhere [33], that antigenic diversity in the vaccine stabilate is not as essential as originally believed. A note of caution is that most of the antigens which have been sequenced are those presented by predominantly European cattle, and may not reflect important antigens recognized in breeds where the vaccine has been deployed.

These results are somewhat contradictory to the earlier study, where it was shown that a mixture of Muguga, Kiambu5 and Serengeti transformed stabilates provided better protection than individual stabilates [34]. It should be noted that a different challenge strain (Kiambu 1) was used in the earlier experiment. Thus, a possible explanation for the difference in results is that the CTL induced by the Muguga isolate do not cross-react with antigens present in Kiambu 1, at least in the animals of the MHC types used in the earlier experiment. This experiment was performed before the role of

CTL as the mediators of immunity and the influence of the MHC were established, and the MHC types of the animals are not available.

However, the difference in the results does indicate that, although we have shown that Muguga-immunized animals of the MHC types used here can generate CTL of similar strain specificities as those immunized with the Muguga cocktail, we cannot with certainty state that use of the Muguga isolate alone will provide the same broad protection in the field as the Muguga cocktail appears to provide. In this respect, it would also be interesting to test if cattle immunized with either Muguga only or the Muguga cocktail would be protected if challenged with the strains, that previously were shown to break through single-strain immunization, such as Marikebuni and Kiambu 1, this time with a larger number of cattle per group.

The field situation is far more complicated than the experimental conditions employed here, due to the presence of a much more heterogenous population of parasites and the diversity of MHC types in outbred cattle populations of several breeds. Both of these factors threaten the success of vaccines composed of a limited number of parasite strains. Particularly at risk are cattle populations in the buffalo-cattle interface, as parasites derived from buffalo show much greater antigenic diversity than those from cattle [31]. Indeed, it has been shown that immunization with the Muguga cocktail does not protect cattle in areas of close interaction with buffalo [35], although it has not been established that this is due to antigenic diversity. Close monitoring of break-through incidences in the field with use of the Muguga cocktail versus single strains would show whether or not single strains protect as well as the Mugaga cocktail.

Conclusion

There were no indications that a broader immune response was induced by immunization with the Muguga cocktail compared to the Muguga strain only, in the haplotypes examined in this study. In agreement with this, previous studies on antigenic diversity in the Muguga cocktail found limited diversity. As the original studies using single strain and Muguga cocktail for induction of protection were performed with limited number of animals and with high doses of needle challenges, this may warrant for testing single vaccine strains in field settings where the load of parasites during challenge will be much lower.

Abbreviations
ConA: Concanavalin A; CTL: Cytotoxic T Lymphocytes; FBS: Fetal Bovine Serum; IFNγ: Interferon Gamma; ITM: Infection and Treatment Method; mAb: Monoclonal Antibody; MHC: Major Histocompatibility Complex; PBMC: Peripheral Blood Mononuclear Cells; SD: Standard Deviation; TCGF: T-Cell Growth Factor; TCR: T Cell Receptor

Acknowledgements
We wish to thank Stephen Mwaura for assisting with the immunization and monitoring of the cattle and Jane Poole, Research and Methods Group, ILRI, for performing the statistical analyses.

Funding
This work was partially supported by a grant awarded jointly by the Department for International Development (UK Government) and the Biotechnology and Biological Sciences Research Council (BBSRC) UK grant number BB/H009515/1 of the Combating Infectious Diseases of Livestock for International Development (CIDLID) program. Additional support was received from the CGIAR Research Program on Livestock and Fish, led by ILRI. We also acknowledge the CGIAR Fund Donors (http://www.cgiar.org/funders).

Authors' contributions
LS designed the study, interpreted results and wrote the manuscript. NS performed the MHC Class I typing, EA performed the CTL assay, TN sourced animals and did the clinical laboratory assessments, RS was involved in generation of cell lines and CTL assays. IM was involved in the overall conceptualization, PT was involved in the design, result interpretation and helped with writing of the manuscript. All authors read and approved the final manuscript.

Competing interests
The authors declare that they have no competing interests.

Author details
[1]International Livestock Research Institute, P.O. Box 30709, Nairobi 00100, Kenya. [2]The Roslin Institute, The University of Edinburgh, Midlothian EH25 9RG, UK.

References
1. Hulliger L, Wilde KH, Brown CG, Turner L. Mode of multiplication of Theileria in cultures of bovine lymphocytic cells. Nature. 1964;203:728–30.
2. McKeever DJ, Morrison WI. Immunity to a parasite that transforms T lymphocytes. Curr Opin Immunol. 1994;6:564–7.
3. Morrison WI, Goddeeris BM, Teale AJ, Groocock CM, Kemp SJ, Stagg DA. Cytotoxic T-cells elicited in cattle challenged with Theileria parva (Muguga): evidence for restriction by class I MHC determinants and parasite strain specificity. Parasite Immunol. 1987;9:563–78.
4. Radley DE, Brown CGD, Cunningham MP. East Coast fever: 1. Chemprophylactic immunization of cattle against Theileria parva (Muguga) and five Thelerial strains. Vet Parasitol. 1975;1:35–41.
5. Di Giulio G, Lynen G, Morzaria S, Oura C, Bishop R. Live immunization against East Coast fever–current status. Trends Parasitol. 2009;25:85–92.
6. Martins SB, Di Giulio G, Lynen G, Peters A, Rushton J. Assessing the impact of East Coast fever immunisation by the infection and treatment method in Tanzanian pastoralist systems. Prev Vet Med. 2010;97:175–82.
7. Irvin AD, Mwamachi DM. Clinical and diagnostic features of East Coast fever (Theileria parva) infection of cattle. Vet Rec. 1983;113:192–8.
8. Irvin AD, Dobbelaere DA, Mwamachi DM, Minami T, Spooner PR, Ocama JG. Immunisation against East Coast fever: correlation between monoclonal antibody profiles of Theileria parva stocks and cross immunity in vivo. Res Vet Sci. 1983;35:341–6.
9. Emery DL, Eugui EM, Nelson RT, Tenywa T. Cell-mediated immune responses to Theileria parva (East Coast fever) during immunization and lethal infections in cattle. Immunology. 1981;43:323–36.
10. Goddeeris BM, Morrison WI, Teale AJ. Generation of bovine cytotoxic cell lines, specific for cells infected with the protozoan parasite Theileria parva and restricted by products of the major histocompatibility complex. Eur J Immunol. 1986;16:1243–9.

11. Goddeeris BM, Morrison WI, Teale AJ, Bensaid A, Baldwin CL. Bovine cytotoxic T-cell clones specific for cells infected with the protozoan parasite Theileria parva: parasite strain specificity and class I major histocompatibility complex restriction. Proc Natl Acad Sci U S A. 1986;83:5238–42.

12. Allen TM, Sidney J, del Guercio MF, Glickman RL, Lensmeyer GL, Wiebe DA, DeMars R, Pauza CD, Johnson RP, Sette A, Watkins DI. Characterization of the peptide binding motif of a rhesus MHC class I molecule (Mamu-a*01) that binds an immunodominant CTL epitope from simian immunodeficiency virus. J Immunol. 1998;160:6062–71.

13. Altfeld MA, Trocha A, Eldridge RL, Rosenberg ES, Phillips MN, Addo MM, Sekaly RP, Kalams SA, Burchett SA, McIntosh K, et al. Identification of dominant optimal HLA-B60- and HLA-B61-restricted cytotoxic T-lymphocyte (CTL) epitopes: rapid characterization of CTL responses by enzyme-linked immunospot assay. J Virol. 2000;74:8541–9.

14. McMurtrey CP, Lelic A, Piazza P, Chakrabarti AK, Yablonsky EJ, Wahl A, Bardet W, Eckerd A, Cook RL, Hess R, et al. Epitope discovery in West Nile virus infection: identification and immune recognition of viral epitopes. Proc Natl Acad Sci U S A. 2008;105:2981–6.

15. Provenzano M, Mocellin S, Bettinotti M, Preuss J, Monsurro V, Marincola FM, Stroncek D. Identification of immune dominant cytomegalovirus epitopes using quantitative real-time polymerase chain reactions to measure interferon-gamma production by peptide-stimulated peripheral blood mononuclear cells. J Immunother. 2002;25:342–51.

16. Bakker AB, Schreurs MW, Tafazzul G, de Boer AJ, Kawakami Y, Adema GJ, Figdor CG. Identification of a novel peptide derived from the melanocyte-specific gp100 antigen as the dominant epitope recognized by an HLA-A2. 1-restricted anti-melanoma CTL line. Int J Cancer. 1995;62:97–102.

17. Ghosh A, Wolenski M, Klein C, Welte K, Blazar BR, Sauer MG. Cytotoxic T cells reactive to an immunodominant leukemia-associated antigen can be specifically primed and expanded by combining a specific priming step with nonspecific large-scale expansion. J Immunother. 2008;31:121–31.

18. Tzelepis F, de Alencar BC, Penido ML, Claser C, Machado AV, Bruna-Romero O, Gazzinelli RT, Rodrigues MM. Infection with Trypanosoma cruzi restricts the repertoire of parasite-specific CD8+ T cells leading to immunodominance. J Immunol. 2008;180:1737–48.

19. Taracha EL, Goddeeris BM, Morzaria SP, Morrison WI. Parasite strain specificity of precursor cytotoxic T cells in individual animals correlates with cross-protection in cattle challenged with Theileria parva. Infect Immun. 1995;63:1258–62.

20. Taracha EL, Goddeeris BM, Teale AJ, Kemp SJ, Morrison WI. Parasite strain specificity of bovine cytotoxic T cell responses to Theileria parva is determined primarily by immunodominance. J Immunol. 1995;155:4854–60.

21. Bishop R, Nene V, Staeyert J, Rowlands J, Nyanjui J, Osaso J, Morzaria S, Musoke A. Immunity to East Coast fever in cattle induced by a polypeptide fragment of the major surface coat protein of Theileria parva sporozoites. Vaccine. 2003;21:1205–12.

22. Gardner MJ, Bishop R, Shah T, de Villiers EP, Carlton JM, Hall N, Ren Q, Paulsen IT, Pain A, Berriman M, et al. Genome sequence of Theileria parva, a bovine pathogen that transforms lymphocytes. Science. 2005;309:134–7.

23. Graham SP, Pelle R, Honda Y, Mwangi DM, Tonukari NJ, Yamage M, Glew EJ, de Villiers EP, Shah T, Bishop R, et al. Theileria parva candidate vaccine antigens recognized by immune bovine cytotoxic T lymphocytes. Proc Natl Acad Sci U S A. 2006;103:3286–91.

24. MacHugh ND, Connelley T, Graham SP, Pelle R, Formisano P, Taracha EL, Ellis SA, McKeever DJ, Burrells A, Morrison WI. CD8+ T-cell responses to Theileria parva are preferentially directed to a single dominant antigen: implications for parasite strain-specific immunity. Eur J Immunol. 2009;39:2459–69.

25. Ellis SA, Staines KA, Stear MJ, Hensen EJ, Morrison WI. DNA typing for BoLA class I using sequence-specific primers (PCR-SSP). Eur J Immunogenet. 1998; 25:365–70.

26. Patel EH, Lubembe DM, Gachanja J, Mwaura S, Spooner P, Toye P. Molecular characterization of live Theileria parva sporozoite vaccine stabilates reveals extensive genotypic diversity. Vet Parasitol. 2011;179:62–8.

27. Goddeeris BM, Morrison WI. Techniques for generation, cloning, and charachterization of bovine cytotoxic T cells specific for the protozoan Theileria parva. J Tiss Culture Methods. 1988;11:101.

28. Morzaria SP, Dolan TT, Norval RA, Bishop RP, Spooner PR. Generation and characterization of cloned Theileria parva parasites. Parasitology. 1995;111(Pt 1):39–49.

29. Svitek N, Taracha EL, Saya R, Awino E, Nene V, Steinaa L. Analysis of the cellular immune responses to vaccines. Methods Mol Biol. 2016;1349:247–62.

30. Rowlands GJ, Musoke AJ, Morzaria SP, Nagda SM, Ballingall KT, McKeever DJ. A statistically derived index for classifying East Coast fever reactions in cattle challenged with Theileria parva under experimental conditions. Parasitology. 2000;120(Pt 4):371–81.

31. Pelle R, Graham SP, Njahira MN, Osaso J, Saya RM, Odongo DO, Toye PG, Spooner PR, Musoke AJ, Mwangi DM, et al. Two Theileria parva CD8 T cell antigen genes are more variable in buffalo than cattle parasites, but differ in pattern of sequence diversity. PLoS One. 2011;6:e19015.

32. Hemmink JD, Weir W, MacHugh ND, Graham SP, Patel E, Paxton E, Shiels B, Toye PG, Morrison WI, Pelle R. Limited genetic and antigenic diversity within parasite isolates used in a live vaccine against Theileria parva. Int J Parasitol. 2016;46:495–506.

33. Morrison WI, Connelley T, Hemmink JD, MacHugh ND. Understanding the basis of parasite strain-restricted immunity to Theileria parva. Annu Rev Anim Biosci. 2015;3:397–418.

34. Radley DE, Brown CG, Cunningham MP, Kimber CD, Musisi FL, Payne RC, Purnell RE, Stagg DA, Young AS. East cost fever:3. Chemoprophylactic immunization of cattle using oxytetracycline and a combination of Theileria strains. Vet Parasitol. 1975;1:51–60.

35. Sitt T, Poole EJ, Ndambuki G, Mwaura S, Njoroge T, Omondi GP, Mutinda M, Mathenge J, Prettejohn G, Morrison WI, Toye P. Exposure of vaccinated and naive cattle to natural challenge from buffalo-derived Theileria parva. Int J Parasitol Parasites Wildl. 2015;4:244–51.

3

Interaction between *Pasteurella multocida* B:2 and its derivatives with bovine aortic endothelial cell (BAEC)

Nuriqmaliza M. Kamal[1], M. Zamri-Saad[2], Mas Jaffri Masarudin[1] and Sarah Othman[1,3]*

Abstract

Background: *Pasteurella multocida* B:2 causes bovine haemorrhagic septicaemia (HS), leading to rapid fatalities in cattle and buffaloes. An attenuated derivative of *P. multocida* B:2 GDH7, was previously constructed through mutation of the *gdhA* gene and proved to be an effective live attenuated vaccine for HS. Currently, only two potential live attenuated vaccine candidates for HS are being reported; *P. multocida* B:2 GDH7 and *P. multocida* B:2 JRMT12. This study primarily aims to investigate the potential of *P. multocida* B:2 GDH7 strain as a delivery vehicle for DNA vaccine for future multivalent applications.

Results: An investigation on the adherence, invasion and intracellular survival of bacterial strains within the bovine aortic endothelial cell line (BAEC) were carried out. The potential vaccine strain, *P. multocida* B:2 GDH7, was significantly better ($p \leq 0.05$) at adhering to and invading BAEC compared to its parent strain and to *P. multocida* B:2 JRMT12 and survived intracellularly 7 h post treatment, with a steady decline over time. A dual reporter plasmid, pSRGM, which enabled tracking of bacterial movement from the extracellular environment into the intracellular compartment of the mammalian cells, was subsequently transformed into *P. multocida* B:2 GDH7. Intracellular trafficking of the vaccine strain, *P. multocida* B:2 GDH7 was subsequently visualized by tracking the reporter proteins via confocal laser scanning microscopy (CLSM).

Conclusions: The ability of *P. multocida* B:2 GDH7 to model bactofection represents a possibility for this vaccine strain to be used as a delivery vehicle for DNA vaccine for future multivalent protection in cattle and buffaloes.

Keywords: *Pasteurella multocida*, Haemorrhagic septicaemia, Bovine aortic endothelial cell, Bactofection, DNA vaccine

Background

Haemorrhagic septicemia (HS) is a major disease in cattle and buffaloes caused by the infection of *Pasteurella multocida* [1]. In Asia, *P. multocida* B:2 is the serotype responsible for the disease [2]. Transmission occurs from diseased animals or carriers by means of intranasal and oral routes [1]. Invasion of the bacteria through endothelial cells result in rapid infiltration of the animal bloodstream [3]. Vaccination against HS is usually conducted prior to rainy seasons using oil-adjuvant vaccines or alum-precipitated vaccines [4]; however both bacterin vaccines only confer short-term protection. Commonly used live attenuated vaccines against HS consist of live organisms, such as the attenuated bacteria with reduced virulence compared to the wild-type [5]. *P. multocida* B:2 GDH7 is an attenuated derivative of the wild-type *P. multocida* B:2 isolated from a previous outbreak in Malaysia, that upon intranasal administration is an efficient vaccine for HS [6]. This strain was genetically modified by the disruption of the wild-type *gdhA* gene with the insertion of a kanamycin cassette [7]. This resulted in an interference of bacterial metabolism hence arresting its pathogenicity. Subsequently, it was also reported that a mutant strain known as *P. multocida* B:2 JRMT12, derived from a parent strain of Sri Lanka

* Correspondence: sarahothman@upm.edu.my
[1]Department of Cell and Molecular Biology, Faculty of Biotechnology and Biomolecular Sciences, Universiti Putra Malaysia, 43400 UPM Serdang, Selangor, Malaysia
[3]Present address: Department of Cell and Molecular Biology, Faculty of Biotechnology and Biomolecular Sciences, Universiti Putra Malaysia, 43400 UPM Serdang, Selangor, Malaysia
Full list of author information is available at the end of the article

origin, *P. multocida* B:2 85,020 was developed and can be administered intramuscularly to confer a high degree of protection as a live vaccine in a mouse model of HS [2]. Since alum precipitated vaccine and oil-adjuvant vaccines are less effective against HS, an alternative is therefore crucially needed. The aforementioned mutants (*P. multocida* B:2 GDH7 and *P. multocida* B:2 JRMT12) have been found to be good candidates for attenuated *P. multocida* B:2 vaccine development in vivo [2, 8]. In this study, the interaction rate of both attenuated vaccine strains, *P. multocida* B:2 GDH7 and *P. multocida* B:2 JRMT12 towards bovine aortic endothelial cells (BAEC) was assessed. Moreover, the ability and efficiency of *P. multocida* B:2 GDH7 to persist in the intracellular environment of the host cells and to transfer plasmid DNA intracellularly was investigated. To assess this interaction, a dual-reporter plasmid that expresses in both prokaryotic and eukaryotic cells was used. It is crucial to understand the bacterial pathogenesis during progression of this disease particularly towards the fate of the plasmid carried by the bacterium after it enters into mammalian cells to further strengthen the ability of *P. multocida* B:2 GDH7 as a vaccine.

Methods
Bacterial strains and growth condition
Bacterial strains used in this study were: *P.multocida* B:2 wild-type, a local isolate from a previous outbreak of haemorrhagic septicaemia in Malaysian cattle, *P.multocida* B:2 GDH7, Δ*gdhA* derivative *P.multocida* B:2 wild-type and *P. multocida* B:2 JRMT12, an Δ*aroA* mutant of strain *P. multocida* B:2 85,020 from an outbreak in Sri Lanka. In previous studies, stability test for both mutant strains *P.multocida* B:2 GDH7 and *P. multocida* B:2 JRMT12 have been reported [2, 7]. To determine the stability of all bacterial strains, each strain was passaged several times and growth studies were conducted prior to the interaction assays.

The bacterial strains are classified as biosafety level 2. All strains were cultured using Brain Heart Infusion (BHI) agar and broth (Oxoid) at 37 °C and shaken at 180 rpm. Whenever required, a total concentration of 50 µg/ml kanamycin and 60 µg/ml of streptomycin were added.

Preparation of bovine aortic endothelial cell (BAEC)
BAEC (Cells applications. Inc., catalogue no. B304–05) was cultured in Dulbecco's Modified Eagle's Medium (DMEM 08459, Nacalai Tesque) supplemented with 10% fetal bovine serum (FBS, I-DNA), 1% glutamine and antibiotics (100 µg/ml of streptomycin and 100 U/ml of penicillin). BAEC was passaged accordingly and was maintained in complete DMEM medium with incubation at humidified environment of 5% (*v*/v) CO_2 and

95% (*v*/v) air at 37 °C. All experimental assays were performed at the third cell passage.

BAEC cell viability assessment
Cell viability was assessed using the trypan blue exclusion method [9]. A mixture (1:1) of the cell suspension with trypan blue (0.4% *w*/v) was placed in an improved Neubauer slide (1/400 mm^2 × 0.1 mm depth). The slide was viewed under a light microscope, where viable cells will confer a clear cytoplasm whereas nonviable cells will be stained blue.

Adherence assay
All *Pasteurellaceae* strains were harvested from 18 h cultures to achieve the optimum multiplicity of infection MOI (100 bacteria/mammalian cell). The washing step was performed twice using phosphate-buffered saline (PBS) before was resuspension of bacterial pellet in DMEM without antibiotics. Trypan blue exclusion assay was performed to monitor the concentration of BAEC in each well of a 24-well tissue culture plate before addition of the appropriate amount of bacterial suspension. The plate was then centrifuged at 300 x g for 5 min using an Eppendorf 5430 R centrifuge to disperse the bacteria onto each cell. The plate was then incubated for 2 h at 37 °C supplemented with 5% (*v*/v) CO_2. Loosely bound bacteria were washed twice using PBS prior to trypsinization with 1 ml of 0.5% (*w*/v) trypsin-EDTA (pH 7.0) per well and incubated for 5 min at 37 °C with 5% (*v*/v) CO_2. BAEC were harvested into the centrifuge tubes using PBS and the remaining cells in the plate were recovered using PBS. The suspensions were then centrifuged at room temperature for 5 min. In order to remove any remaining bacteria and any traces of trypsin-EDTA, the pellets were washed once using PBS before suspended in DMEM without antibiotics. An aliquot of cells were removed from each tubes to assess cell viability. By adding digitonin, the remaining cells was lysed during incubation at 37 °C in 5% (*v*/v) CO_2 for 30 min. The cell suspensions were serially diluted and plated onto BHI agar followed by overnight incubation at 37 °C. Adhesion is expressed as the average value of bacteria/BAEC after incubation of bacteria with BAEC at MOI of 100:1 for 2 h. Data represent the means (±SEM) of three independent assays with duplicate samples.

Invasion assay
The invasion assay was conducted according to the adherence assay with an additional incubation step. After 2 h incubation of the bacteria/BAEC cell mixture in the 24-well plate, the wells were washed twice with PBS to remove loosely bound bacteria. Approximately 1 mL of DMEM with polymyxin B (50 µg/mL) and gentamicin

(50 μg/mL) was added in each well to eliminate the remaining extracellular bacteria. The plate was then incubated for 1 h at 37 °C in 5% (v/v) CO_2. BAEC cell viability assessment were performed according to Section 2.3. Invasion rate was expressed as the average value of bacteria/BAEC that survive after 1 h exposure of polymyxin B and gentamicin with each final concentration of 50 μg/mL after 2 h infection at MOI 100:1. The data represented as the means (± SEM) of three independent assays with duplicated samples.

Intracellular survival assay

Invasion assays were performed towards all three strains with an additional incubation period after the antibiotic treatment step, and before the viable cells were counted. In order to determine the potential intracellular growth of all strains, the number of viable BAEC was counted at various stages. The mixture of bacteria and BAEC were further incubated up to 4 h after the standard invasion assay without the presence of antibiotics, polymyxin B and gentamicin (P&G) at different concentrations of either 10 mg/ml or 50 mg/ml. These are to avoid the risk of any remaining extracellular bacteria to replicate after the initial antibiotic treatment without the presence of P&G. The method for assessment of BAEC cell viability was conducted according to the invasion assay above.

Construction of dual-reporter plasmid, pSRGM

A plasmid to track the location and viability of bacterial cells when moving from extracellular into intracellular compartment of mammalian cells, known as pSRG was previously developed by Othman et al. [5]. In this study, pSRG was slightly modified to enable selection in *P. multocida* B:2 GDH7. Previously, *P. multocida* B:2 GDH7 was constructed by the insertion of a kanamycin cassette into one of the housekeeping genes to confer resistance to kanamycin [7]. Similar to *P. multocida* B:2 GDH7, pSRG is also resistance towards kanamycin. Therefore, the ampicillin cassette (amplified by PCR from pDSRed-Monomer, Clonetech, USA) was inserted into pSRG at *Afl*II sites and used for antibiotic selection of the plasmid (Fig. 1). The modified plasmid, pSRGM was then electroporated into *P. multocida* B:2 GDH7.

Intracellular trafficking of *P. multocida* B:2 GDH7 pSRGM within BAEC

BAEC were prepared similar to the invasion assay. The only difference on preparation of slides for confocal laser scanning microscopy (CLSM) was that the BAEC were seeded in a removable chamber-slide (SPL Life Sciences, Korea) rather than a 24-well plate. The reason for that because when the well was removed from the chamber slide, the cells can be fixed directly onto the slide. The monolayer was washed twice with PBS after the antibiotic

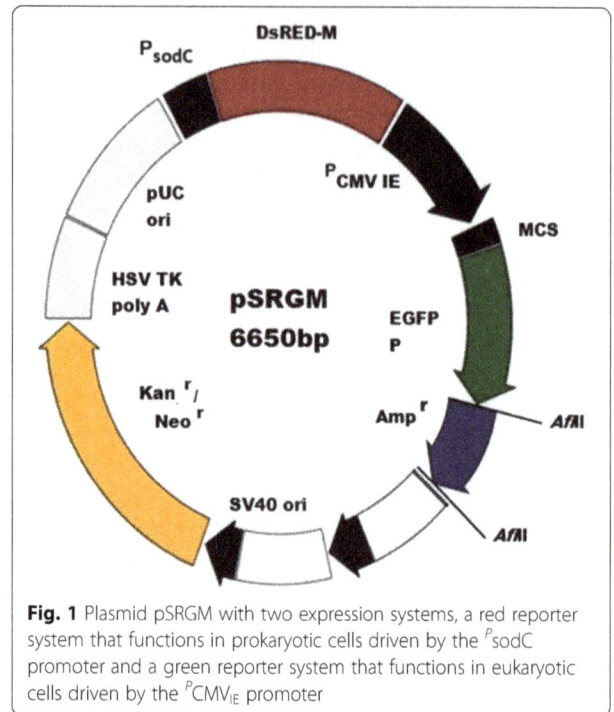

Fig. 1 Plasmid pSRGM with two expression systems, a red reporter system that functions in prokaryotic cells driven by the ^{P}sodC promoter and a green reporter system that functions in eukaryotic cells driven by the ^{P}CMV$_{IE}$ promoter

treatment. Approximately 200 μl of pre-warmed 4% (v/v) paraformaldehyde (PFA) was then added to each well and incubated at 37 °C with 5% (v/v) CO_2 for 30 min. The cells were washed three times in order to remove PFA. An aliquot of HCS CellMask™ blue stain (Life technologies, USA) (5 mg/ml) was added to each well to counterstain the cells. The chamber-slide was then incubated for 7 min at same conditions as PFA. The staining solution was removed from the chamber-slide and washed three times with PBS. The wells were carefully removed from the chamber-slide and the slide was left to dry. Prolong gold antifade reagent (Life technologies, USA) was used as a mounting medium, followed by the coverslip. Slides were then labelled and stored in the dark until viewing.

Confocal laser scanning microscopy (CLSM)

Fluorescence imaging was done with a Leica TCS SP5 II microscope (Leica microsystem, Germany) that was connected to Las AF software to capture images. The system allows visualization of fluorescence at resolution (1024 × 1024 pixels) and bit depth of 8-bit gray scale. The images were then processed with Leica application suite X (Las X) software and the 3D images were processed using LAS X 3D visualization software.

Statistical data

In this study, Microsoft excel was used for calculating mean values, standard deviations and standard errors. For adherence, invasion and intracellular survival assay, student's t-test and one-way ANOVA were applied.

Results

Cytotoxicity effects of *P. multocida* B:2 on viability of BAEC

In this experiment, the cytotoxicity effect of *P. multocida* B:2 wild-type as as well as two attenuated strains, *P. multocida* B:2 GDH7 and *P. multocida* B:2 JRMT12 towards the viability of BAEC were determined. Figure 2 showed that all bacterial strains possessed no cytotoxicity effects towards the BAEC at 3 h post-infection. All control and experimental cells were found to be at least 80% viable at the end of the assay.

Adherence of *Pasteurellaceae* strains to BAEC

The ability of *Pasteurellaceae* strains to adhere to BAEC were investigated and compared in the adherence assay. Using one-way ANOVA test, *P. multocida* B:2 strain GDH7 showed significantly ($p < 0.05$) higher adherence to BAEC (27.46 ± 1.14 bacteria/BAEC) compared to *P. multocida* B:2 wild-type and *P. multocida* B:2 strain JRMT12 (12.30 ± 1.34 and 17.57 ± 1.32 bacteria/BAEC, respectively) (Table 1).

Invasion rate of *Pasteurellaceae* strains towards BAEC

The adherence assay showed that all three strains were able to adhere to BAEC. Similarly shown in Table 2, all strains were detected intracellularly during the invasion assay. *P. multocida* B:2 GDH7 showed the highest invasion rate of 3.12 ± 0.06, followed by the wild-type and Sri Lankan strains (1.43 ± 0.17 and 1.38 ± 0.05 bacteria/BAEC, respectively). These findings suggested that the bovine *Pasteurellaceae* strains were able to adhere to and persist intracellularly in BAEC.

Intracellular survival of the *Pasteurellaceae* strains in BAEC

The number of viable intracellular bacteria in BAEC was determined in order to demonstrate intracellular

bacteria survivability, or the potential of intracellular replication of the bacterial strains within the cytoplasmic compartment of BAEC. Table 3 indicates that all three *Pasteurellaceae* strains at three different conditions showed a consistent decline in the number of viable intracellular bacteria per cell. Moreover, the pattern across all strains were considered not significant ($p > 0.05$). This may be attributed by an inability of the intracellular bacteria to replicate within the cell and is gradually eradicated by the mammalian cells with time.

Intracellular trafficking of *P. multocida* B:2 GDH7 pSRGM within BAEC

The functionality of each expression cassettes in pSRGM was individually assessed. Both prokaryotic and eukaryotic expression cassettes were shown to independently express the reporter proteins in their respective hosts (data not shown). In order to visualize the intracellular trafficking of *P. multocida* B:2 GDH7 pSRGM within BAEC, both the mammalian and bacteria cells were prepared as previously described in Section 2.4. Slides shown in Figs. 3, 4 and 5 were viewed under the 60× objective lens of the CLSM. Images from different light paths were captured at the same field of the slides. Images from Fig. 3 were obtained from slides prepared during 3 h invasion time. *P. multocida* B:2 GDH7 pSRGM (red) shown some internalization into BAEC, while no expression of GFP was detected. Therefore, this indicated that the plasmid was not released by the bacteria into the intracellular compartment of BAEC 3 h post-invasion.

Table 1 Comparison of adherence rate of BAEC by *Pasteurellaceae* strains

Bacterial strains	Bacterial adhesion (No. of bacteria/BAEC)
Pasteurella multocida B:2 wild-type	12.30 ± 1.34*
Pasteurella multocida B:2 GDH7	27.46 ± 1.14*
Pasteurella multocida B:2 JRMT12	17.57 ± 1.32*

Adhesion is expressed as (no. of bacteria/BAEC) that has been deducted from the internalized bacteria by bacterial count after overnight incubation. Data represent the means (±SEM) of three independent assays with duplicate samples
* Signify significant differences ($p < 0.05$) with other strain

Table 2 Invasion rate of the *Pasteurellaceae* strains towards BAEC at MOI 100:1

Bacterial strains	Bacterial invasion (No. of bacteria/BAEC)
Pasteurella multocida B:2 wild-type	1.43 ± 0.17*
Pasteurella multocida B:2 GDH7	3.12 ± 0.06*
Pasteurella multocida B:2 JRMT12	1.38 ± 0.05*

Invasion is expressed as no. of bacteria/BAEC that resisted exposure to polymyxin B and gentamicin at a final concentration of 50 µg/ml for 1 h after an infection period of 2 h. Data represent the means (±SEM) of three independent assays with duplicate samples
* Signify significant differences ($p < 0.05$) with other strain

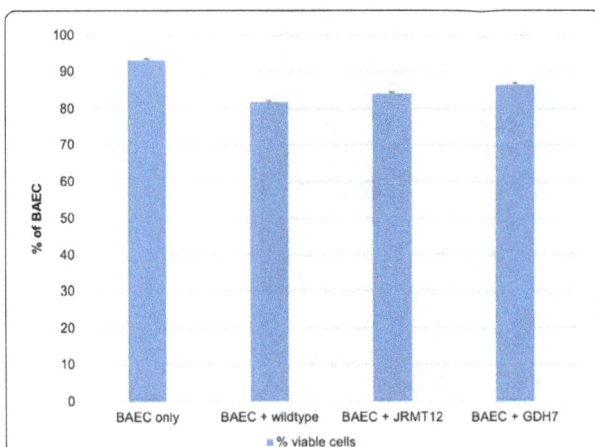

Fig. 2 Percentage of viable and non-viable BAEC at 3 h post-invasion. Data represent the means (±SEM) of three independent assays with duplicate samples

Table 3 Intracellular survival of *P. multocida* B:2 GDH7, *P. multocida* B:2 JRMT12 and *P. multocida* B:2 wild-type in BAEC at MOI 100:1

| Bacterial strain | Invasion time (H) | No. of viable intracellular bacteria/BAEC | | | |
		Standard invasion assay	No antibiotic added after washing	P&G 10 µg/ml added after washing	P&G 50 µg/ml added after washing
P. multocida B:2 GDH7	3	3.00 ± 0.13	-	-	-
	5	-	2.62 ± 0.12	1.87 ± 0.17	1.90 ± 0.06
	7	-	1.73 ± 0.04	1.61 ± 0.04	1.59 ± 0.05
P. multocida B:2 JRMT12	3	1.60 ± 0.09	-	-	-
	5	-	0.93 ± 0.03	0.82 ± 0.06	0.67 ± 0.05
	7	-	0.78 ± 0.03	0.75 ± 0.01	0.63 ± 0.01
P. multocida B:2 Wild type	3	1.22 ± 0.21	-	-	-
	5	-	1.11 ± 0.04	0.94 ± 0.01	0.71 ± 0.04
	7	-	0.93 ± 0.01	0.83 ± 0.01	0.78 ± 0.03

Internalized viable bacteria were evaluated as no. of bacteria/BAEC after incubation of BAEC with bacteria for 2 h followed by exposure to polymyxin B and gentamicin (P&G) each at a final concentration of 50 µg/ml for 1 h prior to further incubation for 2 and 4 h in the presence of either 0, 10 or 50 µg/ml each of polymyxin B and gentamicin in the medium. Data represent the means (±SEM) of three independent assays with duplicate samples

Figure 4 showed a single BAEC expressing GFP within the cell membrane. This data was subsequently corroborated with an optical sectioning of the sample at 1.73 µm intervals from top to bottom with a total section stacking of 88.23 µm, as shown in Fig. 5.

From these images, intracellular bacteria was found to be lysed and undetectable 5 h post-invasion. However, plasmid pSRGM was found to be released into the BAEC cytoplasm from the expression of GFP by the cells.

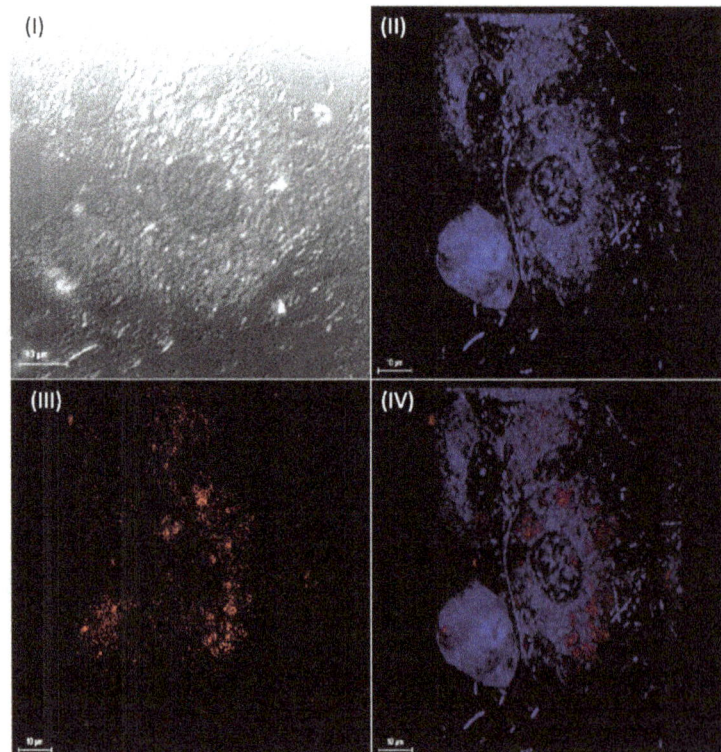

Fig. 3 Localization of prokaryotic protein expression. Images showed BAEC with *P. multocida* B:2 GDH7 pSRGM at 3 h post-invasion. Cells were viewed under 60× objective lens with a Leica confocal laser scanning microscope (CLSM). Images were taken of the same field under three different light paths; (*I*): Image taken under DIC (*white*) light. (*II*): Image taken under filtered fluorescent light (Exc. = 415 nm, Ems. = 493 nm) (*blue*) indicating the plasma membrane of BAEC. (*III*): Image taken under filtered fluorescent light (Exc. = 570 nm, Ems. = 700 nm) (*red*) which are referring to *P. multocida* B:2 GDH7. (*IV*): Overlay of image (*II*) and (*III*)

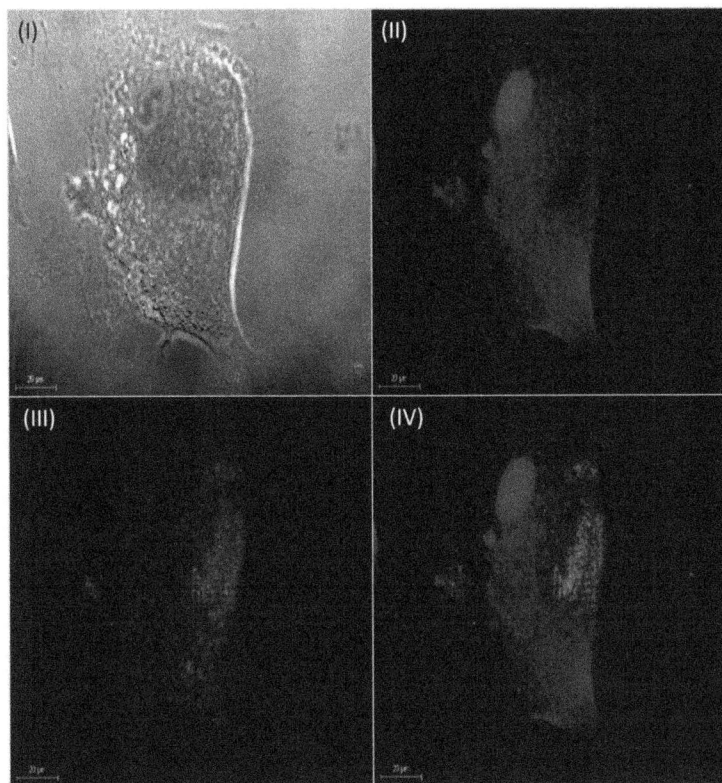

Fig. 4 Localization of eukaryotic protein expression. Images showed BAEC with *P. multocida* B:2 GDH7 pSRGM at 5 h post-invasion. Cells were viewed under 60× objective lens with a Leica confocal laser scanning microscope. Images were taken of the same field under three different light paths; (*I*): Image taken under DIC (*white*) light. (*II*): Image taken under filtered fluorescent light (Exc. = 415 nm, Ems. = 493 nm) (*blue*) indicating the plasma membrane of BAEC. (*III*): Image taken under filtered fluorescent light (Exc. = 498 nm, Ems. = 540 nm) (*green*) indicating that the plasmid has been released into BAEC cytoplasm. (*IV*): Overlay of image (*II*) and 4.13 (*III*)

Discussion

Live-attenuated vaccines confer the advantage of mimicking early stages of the natural route for infection. Intranasal administrations not only stimulates the mucosal immunity of the exposed host, but also transmit the organism to the in-contact host to stimulate mucosal immunity. Rafidah et al. [8] proved that intranasal (i.n.) inoculation with *P. multocida* B:2 GDH7 induced strong immunity in exposed and in-contact buffaloes that protected the buffaloes against wild-type *P. multocida* B:2 challenge. This may be pivotal in overcoming difficulties faced with current vaccines, such as the difficulty in gathering free-ranging animals for vaccination [10].

In this study, the attenuated derivative *P. multocida* B:2 GDH7 strain was shown to adhere to BAEC significantly better ($p < 0.05$) than its parent strain and *P. multocida* B:2 JRMT12 [2]. This is important especially in the development of a potential vaccine strain as a delivery vehicle for DNA vaccine. A higher number of viable intracellular bacteria will increase the chances of higher accumulation of DNA in vivo. The significant differences between the two potential vaccine strains was most probably due to the different mutations introduced in

the *aroA* gene of the JRMT12 and the *gdhA* gene of the GDH7 derivatives. Mutation of *gdhA* gene of the GDH7 derivative renders an inability of this strain to convert α-ketoglutarate to glutamate [11] involved in the amino acid transfer in the bacteria. Hence, the bacteria has limited access to glutamate in the host which restricts bacterial replication before clearance from the host.

Assessment of the pathogen movement to the vascular system through the epithelial tissues is one of the important stages in pathogenesis of infectious *P. multocida* B:2. To be able to invade the bloodstream efficiently, cells that translocate through the blood vessel by adhering and invading may have a higher potential to model this mechanism. Previously, Othman et al. [12] demonstrated the capacity of the Sri Lankan *P. multocida* B:2 strains to be internalized and to persist up to 7 h in embryonic bovine lung cells. Similarly in this study, the Malaysian *P. multocida* B:2 strain and its derivative was internalized into BAEC 3 h post-infection and persisted up to 7 h post-infection, although a stable degeneration in viability with time was observed. During cytotoxicity assessment by trypan blue exclusion method, BAEC did not show any cytotoxicity effects towards any bacterial strains that were tested. Similarly, Galdiero et al. [3]

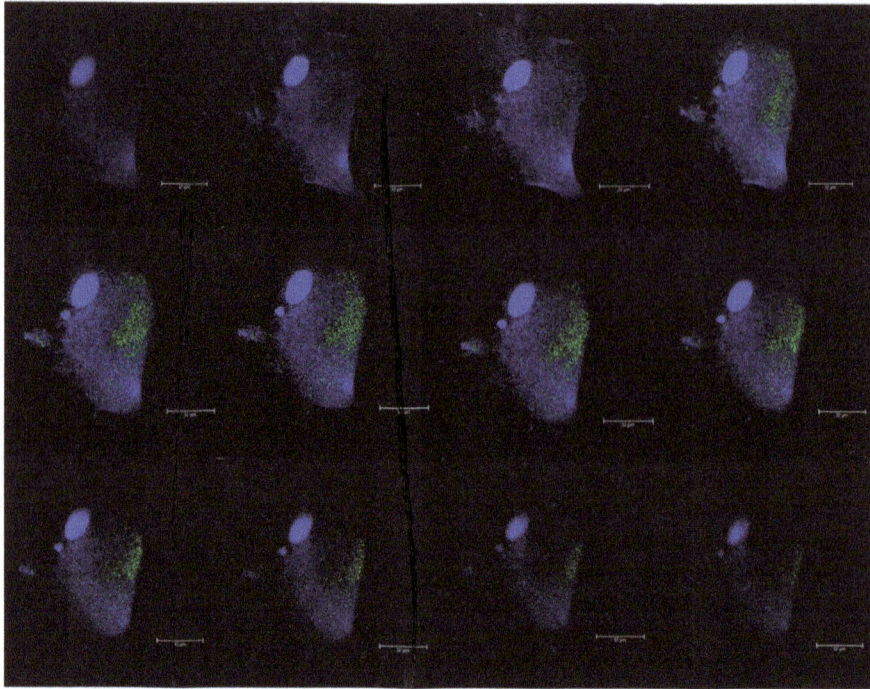

Fig. 5 Z-stack images on GFP expression during bactofection. Optical sectioning of BAEC at 1.73 μm intervals confirmed that the cell was expressing GFP at 5 h post-invasion. The images were arranged accordingly from *top* to *bottom* view (*left* to *right*)

evaluated cytotoxicity effects by quantifying the amount of lactic dehydrogenase released from BAEC after infection with *P. multocida* serotype B. The strain was reported to have almost no cytotoxicity towards BAEC.

The capacity of this vaccine strain to invade shows a promise to be developed as a candidate of a bactofection-derived system. This is an alternate approach to express plasmid-encoded heterologous proteins in a large subset of cells. However, not all attenuated bacterial strains have the same pattern to deliver foreign antigens. For example, in *Shigella spp.* and *L. monocytogenes* exogenous DNA was lysed after delivery to the cytosol of infected hosts [13, 14]. Other findings that described localisation of released DNA in the nuclei and phagosomal compartments of infected hosts of *Salmonella*, *Yersinia* and *E. coli* occurred to mechanisms that are unknown [13, 15, 16]. According to Othman et al. [5], bactofection using an attenuated *P. multocida* B:2 JRMT12 pSRG towards embryonic bovine lung cell (EBL) showed plasmid localisation in the nuclear compartment of the cells from the vacuoles through the cytoplasm. This corroborated with data obtained from this study, where bacterial RFP expression were observed intracellularly 3 h post-invasion (Fig. 3). This also suggested that the bacterial cell was not lysed, and the plasmid remained within the bacterial cytoplasm. However Othman et al. [5] reported that as early as 3 h post-infection, some of the EBL cells managed to express GFP and may indicate that the bacteria was lysed, leading to a transfer of the plasmid

into the host cytoplasm. It was also observed that bacterial lysis and the released of plasmid into the host cytoplasm occurred only 5 h post invasion, as shown in Figs. 4 and 5. In this study, the internalized bacteria remained viable and detectable from 3 to 7 h after infection in adherence assays, but however was difficult to locate 5 h post-infection in the bactofection assay.

Conclusions

In conclusion, the capacity of *P. multocida* B:2 strains to be internalized and persist up to 7 h in mammalian cell was demonstrated. However, this study shows that the local attenuated strain, *P. multocida* B:2 GDH7 were significantly more efficient in internalizing within cells as compared to its parent strain and the Sri Lankan attenuated strain, *P. multocida* B:2 JRMT12. Additionally, not only does the Malaysian attenuated strain, *P. multocida* B:2 GDH7 efficiently vaccinated via the intranasal routes [17] but this study showed that the vaccine strain is a promising candidate as a delivery vehicle for DNA vaccine (bactofection). In the future, both reporter proteins in the pSRGM can be replaced with other antigenic genes from different bovine diseases in order to promote multivalent protection against cattle diseases. It will also be interesting to further investigate the transfer mechanism of plasmid in vivo from the bacteria to the host cells. This could provide evidence towards enhancement of the efficacy of DNA vaccine delivery via bactofection.

Abbreviations
AP: Ampicillin; *aro*A: Aromatic amino acid metabolism gene; B:2: *Pasteurella multocida* biotype B capsular serotype 2; BAEC: Bovine Aortic Endothelial Cell; BHI: Brain Heart Infusion; CD: Cytochalasin D; CFU: Colony forming unit; DNA: Deoxyribonucleic acid; EBL: Embryonic Bovine Lung; EMS: Emission; Exc: Excitation; FBS: Fetal Bovine Serum; FI: Fluorescence intensity; *gdh*A: Glutamate dehydrogenase gene; GFP: Green fluorescent protein; GM: Gentamicin; HS: Haemorrhagic septicaemia; i.m: Intramuscular; i.n.: Intranasal; MOI: Multiplicity of infection; Ng: Nanogram; PBS: Phosphate buffer saline; PCR: Polymerase chain reaction; PFA: Paraformaldehyde; RFP: Red fluorescent protein; rpm: Revolutions per minute; Sm: Streptomycin; TAE: Tris acetate EDTA; UV: Ultraviolet

Acknowledgements
Special thanks to Ms. Yuen Aoi Chee from Hi-Tech Instrument SDN BHD, Malaysia for assistance with the CLSM.

Funding
Nuriqmaliza M.K. was supported by a University of Putra Malaysia Research Grant (IPS-9438744) from the Ministry of Higher Education, Government of Malaysia. The funding sponsors had no role in the design of the study; in the collection, analyses, or interpretation of data; in the writing of the manuscript, and in the decision to publish the results.

Authors' contributions
SO and MZS conceived and designed the experiments; NMK performed the experiments; SO and NMK analyzed the data; MJM contributed reagents/materials/analysis tools; SO, MZS, MJM and NMK wrote the paper. All authors have read and approved the final version of the manuscript.

Competing interests
The authors declare that they have no competing interests.

Consent for publication
Not applicable.

Author details
[1]Department of Cell and Molecular Biology, Faculty of Biotechnology and Biomolecular Sciences, Universiti Putra Malaysia, 43400 UPM Serdang, Selangor, Malaysia. [2]Research Centre for Ruminant Diseases, Faculty of Veterinary Medicine, Universiti Putra Malaysia, 43400 UPM Serdang, Selangor, Malaysia. [3]Present address: Department of Cell and Molecular Biology, Faculty of Biotechnology and Biomolecular Sciences, Universiti Putra Malaysia, 43400 UPM Serdang, Selangor, Malaysia.

References
1. Abubakar MS, Zamri-Saad M. Clinico-pathological changes in buffalo calves following oral exposure to *Pasteurella multocida* B:2. Basic Appl Pathol. 2011; 4:130–5.
2. Tabatabaei M, Liu Z, Finucane A, Parton R, Coote J. Protective immunity conferred by attenuated *aro*A derivatives of *Pasteurella multocida* B:2 strains in a mouse model of hemorrhagic septicemia. Infect Immun. 2002;70:3355–62.
3. Galdiero M, De Martino L, Pagnini U, Pisciotta MG, Galdiero E. Interactions between bovine endothelial cells and *Pasteurella multocida*: association and invasion. Res Microbiol. 2001;152:57–65.
4. Zamri-Saad M, Annas S. Vaccination against hemorrhagic septicemia of bovines: a review. Pak Vet J. 2016;36:1–5.
5. Othman S, Roe AJ, Parton R, Coote JG. Use of a dual reporter plasmid to demonstrate bactofection with an attenuated AroA(−) derivative of *Pasteurella multocida* B:2. Plos One. 2013;8:e71524.
6. Hazwani SO, Saad MZ, Rosfarizan M, Siti-Khairani B. In vitro and in vivo survivality of *gdh*A derivative *Pasteurella multocida* B:2. Malays J Microbiol. 2013;9:166–75.
7. Sarah SO, Zamri-Saad M, Zunita Z, Raha AR. Molecular cloning and sequence analysis of *gdh*A gene of *Pasteurella multocida* B:2. J Anim Vet Adv. 2006;5:1146–9.
8. Rafidah O, Zamri-Saad M. Effect of dexamethasone on protective efficacy of live *gdh*A derivative *Pasteurella multocida* B:2 vaccine. Asian J Anim Vet Adv. 2013;8:548–54.
9. Strober W. Trypan blue exclusion test of cell viability. Curr Protoc Immunol. 2015;111:1–3.
10. Saharee AA, Salim NB. The epidemiology of haemorrahgic septicemia in cattle and buffaloes in Malaysia. Proc 4th Intl Workshop Haemorrhagic Septicemia, Kandy, Sri Lanka. 1991; 109–12.
11. Hashim S, Kwon DH, Abdelal A, Lu CD. The arginine regulatory protein mediates repression by arginine of the operons encoding glutamate synthase and anabolic glutamate dehydrogenase in *Pseudomonas aeruginosa*. J Bacteriol. 2004;186:3848–54.
12. Othman S, Parton R, Coote J. Interaction between mammalian cells and *Pasteurella multocida* B:2. Adherence, invasion and intracellular survival. Microb Pathog Elsevier Ltd; 2012;52:353–358.
13. Grillot-Courvalin C, Goussard S, Courvalin P. Wild-type intracellular bacteria deliver DNA into mammalian cells. Cell Microbiol. 2002;4:177–86.
14. Hense M, Domann E, Krusch S, Wachholz P, Dittmar KE, Rohde M, et al. Eukaryotic expression plasmid transfer from the intracellular bacterium *Listeria monocytogenes* to host cells. Cell Microbiol. 2001;3:599–609.
15. Pilgrim S, Stritzker J, Schoen C, Kolb-Mäurer A, Geginat G, Loessner MJ, et al. Bactofection of mammalian cells by *Listeria monocytogenes*: improvement and mechanism of DNA delivery. Gene Ther. 2003;10:2036–45.
16. Al-mariri A, Tibor A, Lestrate P, De Bolle X, Letesson J, Mertens P. *Yersinia enterocolitica* as a vehicle for a naked DNA vaccine encoding *Brucella abortus* bacterioferritin or P39 antigen. Infect Immun. 2002;70:1915–23.
17. Rafidah O, Zamri-Saad M, Shahirudin S, Nasip E. Efficacy of intranasal vaccination of field buffaloes against haemorrhagic septicaemia with a live *gdh*A derivative *Pasteurella multocida* B:2. Vet Rec. 2012;171:175.

4

1,25-Dihydroxyvitamin D3 modulates the phenotype and function of Monocyte derived dendritic cells in cattle

Yolanda Corripio-Miyar[1] (iD), Richard J. Mellanby[2], Katy Morrison[1] and Tom N. McNeilly[1*]

Abstract

Background: The active form of the vitamin D_3, 1,25-Dihydroxyvitamin D_3 (1,25-$(OH)_2D_3$) has been shown to have major effects not only on physiological processes but also on the regulation of the immune system of vertebrates. Dendritic cells are specialised antigen presenting cells which are in charge of the initiation of T-cell dependant immune responses and as such are key regulators of responses towards pathogens. In this study we set out to evaluate the effects of 1,25-$(OH)_2D_3$ on the phenotype of cattle monocyte-derived dendritic cells (MoDCs) and how the conditioning with this vitamin affects the function of these myeloid cells.

Results: MoDCs were generated from $CD14^+$ monocytes with bovine IL-4 and GM-CSF with or without 1,25-$(OH)_2D_3$ supplementation for 10 days. Vitamin D conditioned MoDCs showed a reduced expression of co-stimulatory and antigen presenting molecules, as well as a reduced capability of endocytose ovalbumin. Furthermore, the capacity of MoDCs to induce proliferation in an allogeneic mixed leukocyte reaction was abolished when MoDCs were generated in presence of 1,25-$(OH)_2D_3$. LPS induced maturation of 1,25-$(OH)_2D_3$conditioned MoDCs resulted in lower secretion of IL-12 and higher IL-10 than that observed in MoDCs.

Conclusions: The typical immunotolerant phenotype observed in cattle DCs after exposure to 1,25-$(OH)_2D_3$ has a significant effect on the functionality of these immune cells, inhibiting the T-cell stimulatory capacity of MoDCs. This could have profound implications on how the bovine immune system deals with pathogens, particularly in diseases such as tuberculosis or paratuberculosis.

Keywords: Monocyte-derived Dendritic cells, 1,25$(OH)_2D_3$ conditioned MoDCs, Cattle

Background

Dendritic cells (DCs) are professional antigen-presenting cells (APC) which act as a bridge between innate and adaptive immune responses. DCs patrol the periphery of the immune system in an immature state where they use their unique specialised function to capture and process antigens. Once they receive this maturation signal, they migrate to lymphoid organs and, by secreting cytokines and chemokines, expressing co-stimulatory molecules in their surface and presenting antigens to T cells, are able to initiate a cascade of immune responses [1]. However, DCs not only activate primary immune responses against foreign antigens, they also differentiate non-self from self-antigens and by inducing immunological tolerance in T cells, are able to control autoimmune responses which have the potential to damage the host. The depletion of DCs in animal models breaks self-tolerance of $CD4^+$ T cells and is associated with the onset of fatal autoimmune diseases [2].

Given that the induction of immunotolerance could be used to prevent undesirable immune responses for individuals with autoimmune diseases or undergoing transplantations, these cells are being investigated for their potential use as therapeutic targets [3–5]. However, studying the function of DCs in different species can be hampered by the difficulties which arise during the isolation and identification of these cells. Not only are there various DC phenotypes depending on the tissue where they are found, but they also express different markers in different species [6–9]. However, monocytes can be differentiated into dendritic cells (MoDCs) by the addition of

* Correspondence: Tom.McNeilly@moredun.ac.uk
[1]Moredun Research Institute, Pentlands Science Park, Bush Loan, Midlothian, UK
Full list of author information is available at the end of the article

IL-4 and GM-CSF [10, 11]. As their function and phenotype are typical of immature DCs, MoDCs are a good substitute to freshly isolated DCs and can be used as in vitro models of innate immune responses.

Recent studies have revealed that the active form of the vitamin D_3, 1,25-Dihydroxyvitamin D_3 (1,25-$(OH)_2D_3$), can play an important role in calcium and bone homeostasis and the regulation of the immune system [12–18]. The immune modulatory functions of vitamin D include the regulation of genes involved in host immune defense such as antimicrobial peptides [15], the prevention of the onset of autoimmune diseases models [19], or inhibition of the production of inflammatory cytokines in response to inflammatory or infectious stimuli [20]. However, as an immunomodulator, this hormone can also affect the function and development of monocytes and dendritic cells. As professional antigen presenting cells, DCs require the production of IL-12 to drive naïve T cell to a Th1 phenotype and recent studies have shown a reduction in secretion of IL-12 by DCs which have been treated with high doses of 1,25-$(OH)_2D_3$, leading to an overall reduced functionality [16, 21]. Furthermore, 1,25-$(OH)_2D_3$ is able to down-regulate the expression of MHC-II and co-stimulatory molecules in DCs as well as modulate their cytokine production.

The majority of studies investigating the effects of 1,25-$(OH)_2D_3$ on DCs and the immune system have been carried out in humans [5, 16] or mice [22, 23], with only a few studies carried out in veterinary species. In poultry, Vitamin D has been shown to inhibit proliferation and IFN-γ production by lymphocytes [14], while the addition of 1,25-$(OH)_2D_3$ to chicken macrophages up-regulates the production of nitric oxide by 5-fold, consequently improving the phagocytic nature of these cells [24]. In ruminants, the majority of the work investigating the effects of vitamin D has been focused on nutritional aspects [25], although the role of vitamin D in reproductive fitness [26] and infectious diseases such as bovine tuberculosis [27] and mastitis [28] has also been investigated. In *Mycobacterium bovis* infected cattle it has been shown that 1,25-$(OH)_2D_3$ can inhibit *M. bovis*-specific proliferation of CD4+ and $\gamma\delta$ T cells [29, 30]. Furthermore, activation of the vitamin D pathway using a monoclonal antibody to the vitamin D receptor has been shown to suppress *M. bovis*-specific proliferation and interferon gamma (IFN-γ) production of peripheral blood mononuclear cells [27]. In this latter study the suppressive effects were associated with a down-regulation of CD80 expression, suggesting that activation of the vitamin D pathway was associated with a deficiency in antigen presenting cell (APC) function. However, in contrast to work in mice and humans there have been no studies specifically addressing the effects of vitamin D on ruminant APC phenotype and functionality.

In this study we investigate the immunomodulatory effects of 1,25-$(OH)_2D_3$ on cattle DCs. We generated Vitamin D conditioned MoDCs by differentiating bovine CD14 monocytes with IL-4 and GM-CSF in the presence of 1,25-$(OH)_2D_3$ (VitD-MoDCs). We then examined the phenotype and functionality of these VitD-MoDCs and compared to that of standard MoDCs.

Methods
Animals
Healthy 6-months old Holstein-Friesian and Ayrshire calves were purchased from two Scottish commercial dairy farms and maintained at Moredun Research Institute (MRI), UK. All animals were kept off pasture for the duration of the experiments. All experiments were approved by the Ethics Committee at MRI and were performed to Home Office Guidelines under Project Licence (PPL 60/3854).

Isolation of bovine PBMC and in vitro generation of Monocyte derived Dendritc cells (MoDCs)
Blood was collected aseptically into 350 ml blood bags containing 45 ml of Citrate phosphate dextrose-adenine 1 (CPDA-1) stabiliser (Sarstedt, Germany). PBMC were isolated as previously described [31] using density gradient centrifugation by layering whole blood diluted with phosphate buffered saline (PBS) onto Ficoll-Paque™ PLUS (GE Healthcare Life Sciences). Buffy coat was collected and washed three times with PBS and re-suspended in MACS Buffer [PBS + 0.5% foetal bovine serum (FBS, from USA supplied by Sigma-Aldrich, UK)]. CD14 monocytes were positively selected by incubation of PBMC with CD14 MicroBeads (clone TÜK4, Miltenyi Biotech, Germany) in cold MACS buffer for 15 min at 4 °C. Cell-microbead complex were washed twice, resuspended in 3 ml of MACS buffer and purified over an LS column as per manufacturer's instructions. Purified CD14 monocytes were washed to eliminate any residual microbeads and resuspended in tissue culture medium (RPMI-1640 medium) supplemented with 10% FBS and 50 µM 2-mercaptoethanol, 2 mM L-glutamine, 100 U/ml Penicillin and 100 µg/ml Streptomycin (all from Sigma-Aldrich, UK). Enrichment purity was consistently above 95% as assessed by flow cytometry. Monocyte derived dendritic cells (MoDC) were generated in the presence of bovine GM-CSF and IL-4 (Bovine Dendritic Cell Growth kit, Bio-Rad) for 10 days as previously described [13, 31] with some modifications. Briefly, purified cattle monocytes were seeded in 6-well plates at a concentration of 10^6 cells/ml. Cells were cultured in the presence of 50 µl/ml of Bovine DC Growth Kit and 10 nM 1,25-$(OH)_2D_3$ for Vitamin D_3 conditioned MoDCs (VitD-MoDCs), or 50 µl/ml of Bovine DC Growth Kit only for MoDCs. Fresh medium, cytokines and 1,25-$(OH)_2D_3$ where applicable were replenished on day 3 and 6 of culture. Cells were harvested on day 10.

Phenotype of MoDCs and VitD-MoDCs

Single colour flow cytometric analysis was carried out to phenotype bovine MoDCs and VitD conditioned MoDCs. After 10 days of culture as detailed above, cells were harvested and following a 10 min blocking step with 20% normal goat serum (NGS, Bio-Rad) in PBS, incubated with the following un-conjugated monoclonal antibodies: CD1b (CC14, Bio-Rad), CD80 (IL-A159, Bio-Rad), CD86 (IL-A190, Bio-Rad) and MHCII-DR (CC108, Bio-Rad) at pre-optimised concentrations for 20 min. Cells were then washed twice with FACS buffer (PBS + 5%FBS + 0.05%NaN$_3$) and further incubated for 20 min with a secondary anti-mouse IgG mAb conjugated to Alexa-Fluor® 647 (Invitrogen, Life Technologies, US). After two washes, cells were resuspended in the dead cell stain Sytox Blue (Invitrogen, Life Technologies, US) and immediately a minimum of 10,000 events were acquired using a MACSQuant® Analyzer 10 (Miltenyi Biotech, Germany). Post acquisition gating, including dead cell and doublet cell discrimination, and analysis were carried out using FlowJo vX for Windows 7. Phase contrast images of MoDC and VitD-MoDCs were captured using an Axiovert 200 M inverted microscope (Carl Zeiss Ltd., UK).

Endocytosis by MoDCs and VitD-MoDCs

One of the main roles of DCs is the sampling of their environment for antigens through endocytosis. Consequently, to assess the effect of 1,25-(OH)$_2$D$_3$ enrichment on endocytosis, we used DQ-Ovalbumin (DQ-OVA, Molecular Probes, Life Technologies, US), a self-quenched conjugate which exhibits bright green fluorescence upon proteolytic degradation yielding a low background signal. Hence, 10^5 MoDCs or VitD conditioned MoDCs were incubated with 10 μg/ml of DQ-OVA or medium only for 1 h at 37 °C. After incubation, cells were washed 3 times with FACS buffer and resuspended in Sytox Blue (ThermoFisher) prior to flow cytometry analysis as detailed above.

Mixed leukocyte reaction

In order to define the capability of VitD conditioned MoDCs to stimulate an alloreactive mixed leukocyte reaction (allo-MLR) an experiment was set up using two animals from different breeds which showed high proliferation in preliminary MLRs. PBMC obtained from an Ayrshire calf were used as responder cells, whilst MoDCs generated from a Holstein-Friesian calf were used as the allogeneic stimulators. MoDCs and VitD-MoDCs were generated for 10 days as described earlier and stimulated with or without LPS for 24 h. Cells were harvested and irradiated (60Gy) to ensure that detected proliferation was derived only from PBMCs. Irradiated MoDCs were then resuspended in complete medium with 10% FBS at a concentration of 10^{5}cells/ml. Responder PBMC

(10^5 per well) were incubated with irradiated MoDCs/ VitDMoDCs at the following ratios, 1:10, 1:20, 1:100, 1:200, 1:400 and 1:1000. Reactions were set up in quadruplicate in U-well microtitre plates in a total volume of 200 μl. Controls consisted of responder PBMC or irradiated stimulator cells in medium alone or with 5 μg/ml of Concanavalin A (ConA, Sigma-Aldrich). Cells were incubated at 37 °C for 4 days after which 50 μl of medium was collected from each replicate/treatment and replenished with fresh complete medium containing methyl-3H thymidine (0.5 μCi per well; Amersham Biosciences UK Ltd., Chalfont St. Giles, Buckinghamshire). Proliferation was measured by the incorporation of methyl-3H thymidine during the final 18 h of culture as previously described [32]. Data are presented as the corrected counts per minute (ccpm) averaged over 3 min.

Cytokine secretion after LPS stimulation

MoDCs and 1,25(OH)$_2$D$_3$ conditioned MoDCs were harvested on day 10, incubated with 1 μg/ml of LPS (0111:B4, Sigma UK) or medium only for 24 h. After stimulation supernatants were collected to analyse cytokine secretion induced by LPS. Capture ELISAs were performed to examine the secretion of IL-1β (Anti-Bovine IL-1β polyclonal capture and detection antibodies, BioRad), IL-10 (clones CC318 and CC320b, BioRad) and IL-12 (clones CC301 and CC326b, BioRad). Standard curves were constructed using recombinant bovine IL-1β (BioRad) and transfected COS-7 cells supernatants for IL-10 [33] or IL-12 [34]. All incubations were carried out at room temperature unless stated and washing steps were performed 6 times with 350 μl washing buffer (PBS + 0.05% Tween 20) using a Thermo Scientific Wellwash™ Versa (ThermoFisher). Briefly, high-binding capacity ELISA plates (Immunolon™ 2HB 96-well microtiter plates, Thermo-Fisher) plates were coated with capture antibodies at pre-optimised concentrations and incubated over night at 4 °C. Plates were then washed and blocked for 1 h with PBS+ 3% Bovine Serum Albumin (BSA, Sigma). Following a further washing step, 50 μl of supernatants or standards were added in duplicate for 1 h. Plates were then washed and detection antibodies added for 1 h. This was followed by washing and addition of Streptavidin-HRP (Sigma) for 45 min. After the final washing step, 50 μl of SureBlue TMB substrate (Insight Biotechnology, London, UK) was added and the reaction was stopped by the addition of H$_2$SO$_4$. Absorbance values were read at O.D. 450 nm. All values were blanked corrected and concentrations determined from standard curve.

Statistical analysis

Statistical analyses were performed using the non-parametric test Kruskal-Wallis and a Dunns multiple comparison post hoc test in the case of not normally distributed data or One-Way ANOVA and a Tukey post hoc test when data were normally distributed. All analyses were carried out within the Minitab version 17 statistical package, with $p < 0.05$ considered significant.

Results

1,25(OH)$_2$D$_3$ conditioning influences the phenotype and endocytic capabilities of bovine MoDCs

Monocytes were incubated with boIL-4 and boGM-CSF alone or with 1,25-(OH)$_2$D$_3$ for a period of 10 days. All cultures were set up with the same starting monocyte number, but interestingly, the supplementation of MoDCs with 1,25-(OH)$_2$D$_3$ during differentiation clearly improved the survival of the MoDCs. Typically, after the 10 day culture period, around 4.6 times more cells were recovered from the 1,25-(OH)$_2$D$_3$ conditioned cultures than those differentiated with IL-4 + GM-CSF only (cell recovery expressed as percentage of starting population: 6–30% in MoDCs vs 25–100% in VitD-MoDCs; Fig. 1), showing the need for supplementation with 1,25-(OH)$_2$D$_3$ throughout the differentiation process. In order to determine if 1,25(OH)$_2$D$_3$ conditioning affected the phenotype and function of MoDCs, we initially investigated the expression of antigen presenting and co-stimulatory molecules which are typically up-regulated in CD14 monocytes cultured with GM-CSF and IL-4 alone. As expected, MoDCs expressed high levels of CD1b, CD80, CD86 and MHC-II (Fig. 2). However, when MoDCs were supplemented with 1,25(OH)$_2$D$_3$ during differentiation, the expression of all four markers remained consistently lower than in MoDCs. Although this reduced MFI was observed for all four markers, only CD1b was significantly down-regulated in VitD-MoDCs when compared to MoDCs ($p = 0.016$).

Immature DCs have the ability to efficiently uptake antigens by endocytosis, consequently in order to assess the endocytic capacity of the VitD-MoDCs, we investigated the uptake of the model antigen ovalbumin (OVA), a protein taken up by clathrin-coated pits in dendritic cells [35].

1,25(OH)$_2$D$_3$ conditioned MoDCs and MoDCs and were incubated with DQ-OVA for 1 h, after which cells were harvested and analysed by flow cytometry. A significantly lower level of internalisation and processing of OVA was observed in VitD-MoDCs when compared MoDCs (Fig. 3, $p = 0.025$).

Impaired ability of 1,25(OH)$_2$D$_3$ conditioned MoDC to induce lymphocyte proliferation

We then tested the influence of 1,25(OH)$_2$D$_3$ conditioning in the ability of MoDCs to induce lymphocyte proliferation in an allo-MLR following LPS maturation. Unstimulated MoDCs were able to significantly increase spontaneous proliferation of PBMC at ratios of 1:10, 1:20 and 1:100 of MoDCs to PBMC ($p = 0.009$, $p = 0.001$, $p = 0.021$ respectively, Fig. 4). In contrast, the incorporation of thymidine by PBMC incubated with unstimulated MoDCs conditioned with 1,25(OH)$_2$D$_3$ remained below the level of spontaneous proliferation of PBMC only controls (dashed line in Fig. 4), even at high stimulator:responder ratios. This difference was enhanced up to 7-fold when both cell types were matured with LPS prior to setting up the MLR. While LPS stimulated MoDCs were able to generate a high level of proliferation in lymphocytes with as little as 10^3 MoDCs per 10^5 PBMC, LPS-stimulated VitD-MoDCs were unable to induce a convincing PBMC proliferative response at any of the MoDC:PBMC ratios tested (LPS VitD conditioned MoDCs vs LPS MoDCS; at 1:10, $p < 0.001$, at 1:20, $p < 0.001$, and at 1:100, $p = 0.021$).

1,25(OH)$_2$D$_3$ enhances LPS driven IL-10 production by MoDCs

As professional APCs, the main function of DCs is the activation of naive T cells. In order to do so, DCs process antigens and present them on their surface to the T cells. Furthermore, the secretion of cytokines by MoDCs is able to influence the phenotype of the T cells they activate [36]. Consequently, we analysed the LPS induced cytokine secretion by MoDCs differentiated in the

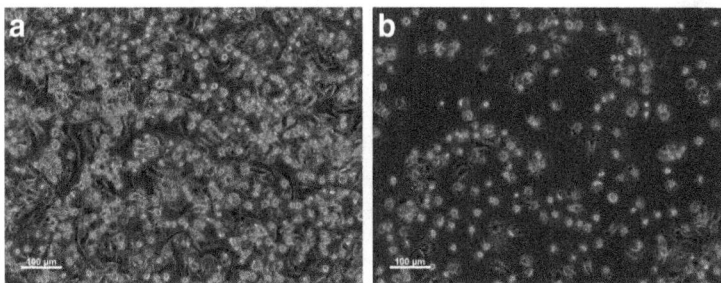

Fig. 1 Cattle CD14 + ve monocytes differentiated in the presence of 1,25 (OH)$_2$D$_3$. MoDCs were generated as detailed in Material and Methods with or without 1,25(OH)$_2$D$_3$. All treatments were seeded at a 10^6cells/ml and supplemented with boIL-4, boGM-CSF with or without 10 nM of 1,25(OH)$_2$D$_3$ on day 1, day 3 and day 6 of culture. Cells were harvested on day 10. Phase contrast images of MoDCs with (**a**) or without (**b**) 1,25 (OH)$_2$D$_3$ depicts an increased proportion of cells in wells containing cells differentiated in the presence of 1,25(OH)$_2$D$_3$

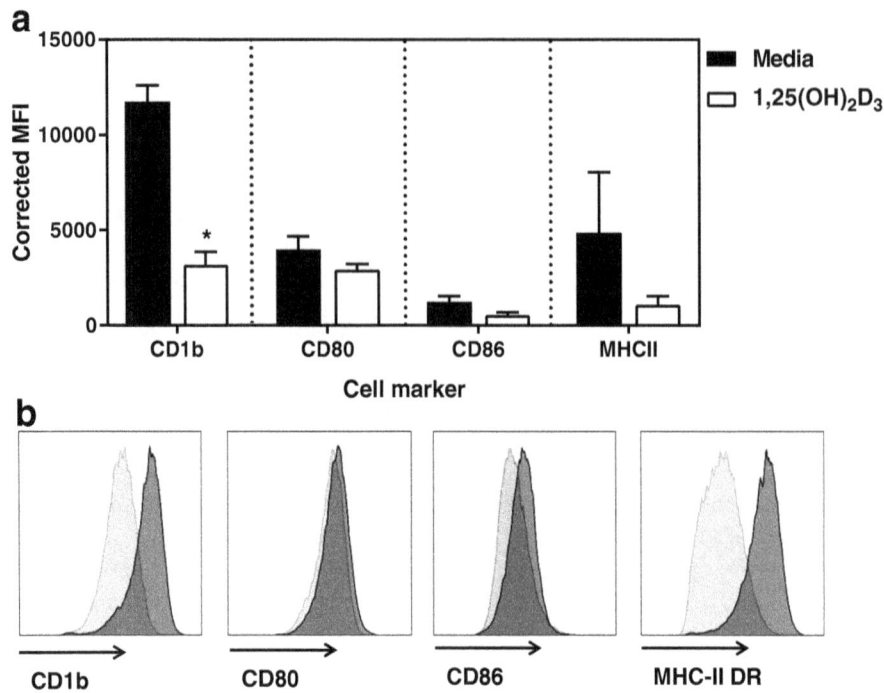

Fig. 2 Phenotype of cattle MoDCs differentiated in the presence or absence of 1,25 $(OH)_2D_3$. Phenotype of MoDCs or 1,25 $(OH)_2D_3$ conditioned MoDCS derived from cattle CD14 + ve monocytes was analysed by flow cytometry. MoDCs were single stained with primary mAbs to CD1b, CD80, CD86 or MHC-II DR and then stained with Alexa Fluor 647 IgG secondary antibody. Live, single gated cells were assessed for expression of these markers. **a** Data shown are the average ± the SE of the corrected median fluorescence intensity (MFI) for MoDCs (black bars) and 1,25 $(OH)_2D_3$ conditioned MoDCS (white bars). **b** Histogram for a representative animal showing the level of uptake of expression of chosen markers for MoDCs (black histograms) or 1,25 $(OH)_2D_3$ conditioned MoDCS (grey histograms). *denotes statistical significance for $p < 0.05$

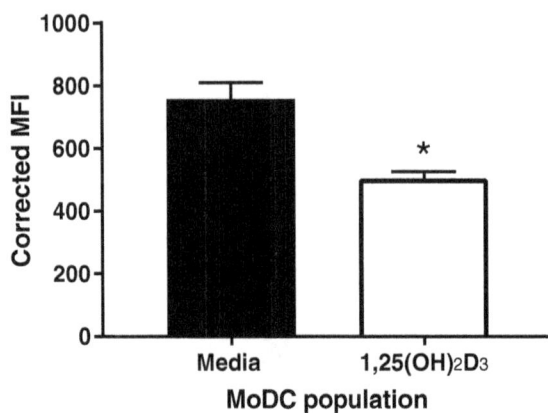

Fig. 3 The endocytic capabilities of cattle MoDCs are diminished by conditioning with 1,25$(OH)_2D_3$. MoDCs were differentiated from cattle CD14 monocytes in the absence (black bars) or presence (white bars) of 1,25$(OH)_2D_3$ for 10 days. After harvesting, cells were incubated for 1 h with DQ-OVA and uptake analysed by flow cytometry. Live, single cells were gated and the MFI of each treatment calculated. The results shown are average ± SE of two animals representing the DQ MFI for uptake by MoDCs and 1,25$(OH)_2D_3$ conditioned MoDCs. * denotes statistical significance for $p < 0.05$

presence or absence of 1,25$(OH)_2D_3$ after a stimulation period of 24 h. There was no significant difference in the secretion of any of the cytokines between resting MoDCs and VitD-MoDCs (Fig. 5). However, after a 24 h stimulation with LPS, the secretion of IL-1β, IL-10 and IL-12 was significantly upregulated both by MoDCs or VitD-MoDCs (Fig. 5a, b, c; all $p = 0.009$). This LPS driven cytokine secretion was consistently higher in VitD-MoDCs for all three cytokines (all $p = 0.0122$) when compared to MoDCs. When expressed as a fold increase in cytokine release relative to the un-stimulated controls, the fold increase in IL-1β secretion by 1,25-$(OH)_2D_3$ conditioned MoDCs was 21.4, while the fold increase for MoDCs was only 6.5 (Fig. 5e, $p<0.0001$). The opposite could be seen for IL-12 secretion, where the MoDCs produced significantly more cytokine than its vitamin D conditioned counterpart (Fig. 5e, $p=0.01$). No significant differences were observed in the fold increase in IL-10 secretion between MoDCs and VitD-MoDCs (Fig. 5e). However, when expressed as a ratio of IL-10/IL-12, MoDCs cells secreted higher levels of IL-12 in comparison to IL-10 after LPS stimulation compared to the 1,25-$(OH)_2D_3$ conditioned MoDCs (Fig. 5d; $p < 0.0001$).

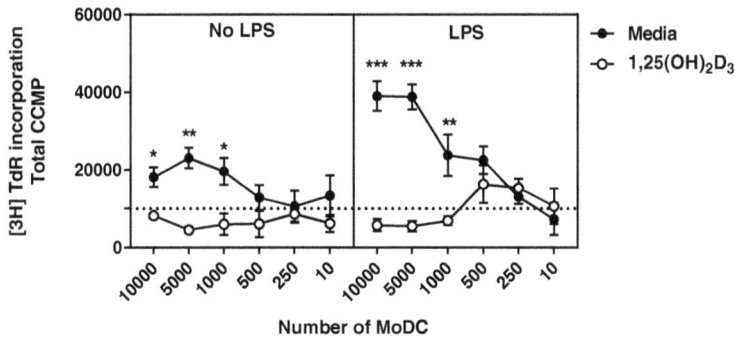

Fig. 4 MoDCs differentiated in the presence of 1,25(OH)₂D₃ are not able to induce an allogeneic mixed leukocyte reaction. MoDCs differentiated from cattle CD14 monocytes in the presence or absence of 1,25(OH)₂D₃ for 10 days. After harvesting, cells incubated with or without LPS for 24 h. Following stimulation, cells were irradiated and cultured in quadruplicate at different ratios with 10⁵ responder PBMC for 5 days. Responder PBMC and stimulator MoDCs were incubated in medium only or with ConA (5μg/ml) as controls of proliferation. Proliferation was measured by the incorporation of methyl-3H thymidine ([3H]TdR; 0.5 μCi per well) for the final 18 h of culture. Data are presented as the corrected counts per minute (ccpm) averaged over 3 min. Data shown are the representative of two independent experiments with error bars denoting ± SE. Dashed line denotes the spontaneous proliferation of PBMC with no ConA/stimulator cells. * denotes statistical significance for $p < 0.05$, ** p value between 0.001 and 0.01 and *** for $p < 0.001$

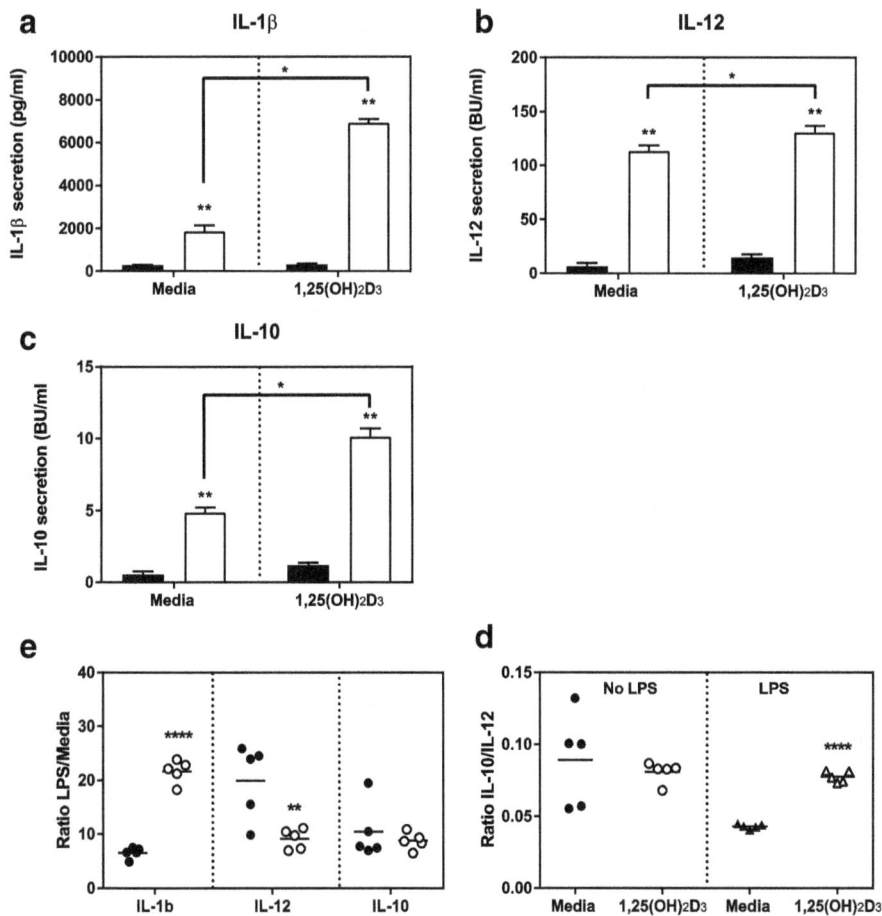

Fig. 5 Cytokine secretion by cattle MoDCs differentiated in the presence or absence of 1,25(OH)₂D₃. MoDCs differentiated from cattle CD14 + ve monocytes in the presence or absence of 1,25(OH)₂D₃ for 10 days. After harvesting, cells incubated with (white bars) or without LPS (black bars) for 24 h. Following stimulation, secretion of IL-1β (a), IL-12 (b) and IL-10 (c) into culture supernatants were measured by ELISA. Data are expressed as the concentration of cytokine in picograms (pg) or biological units (BU) per ml of supernatant. (d) Fold increase of cytokine secretion. (e) Ratio of IL10/IL12 secretion. Results are shown as the mean values with error bars indicating ± SE from four animals. * denotes statistical significance for $p < 0.05$, ** p value between 0.001 and 0.01 and *** for $p < 0.001$

Discussion

The influence of $1,25\text{-}(OH)_2D_3$ on the function of immune cells has been widely discussed, from their effects on the central nervous system [13], to the modulation of innate immune responses by macrophages [28], or the induction of tolerogenic DCs [16, 18, 22]. However, little information is available on the immunomodulatory effects of $1,25\text{-}(OH)_2D_3$ in ruminants or how they obtain this hormone from the environment.

When exploring the effect of vitamin D_3 at a cellular level, some studies have focused on its ability to inhibit mitogen or antigen induced secretion of IFNγ in bovine lymphocytes [30, 37], while others have shown that when $1,25\text{-}(OH)_2D_3$ is added to bovine monocyte cultures infected with *Mycobacterium bovis*, NO production is enhanced and apoptosis of antigen-stimulated cells reduced [30]. The production of $1,25\text{-}(OH)_2D_3$ by bovine monocytes has also been reported to modulate iNOS and RANTES expression in LPS stimulated monocytes [38]. However, there is currently little evidence regarding how $1,25\text{-}(OH)_2D_3$ affects other key immune cells such as DCs, which are required to activate naive T cells in order to trigger an effective immune response.

Here we show the profound effects that are caused by $1,25\text{-}(OH)_2D_3$ conditioning during the differentiation of bovine MoDCs. Bovine MoDCs have a distinct phenotype when compared to afferent lymph DCs [35, 39], they express co-stimulatory molecules such as CD1b and MHCII at a higher level than CD14 monocytes all of which are required for antigen presentation. When we investigated the phenotype of MoDCs differentiated in the presence or absence of $1,25\text{-}(OH)_2D_3$, the expression of both markers was lower in the $1,25\text{-}(OH)_2D_3$ conditioned MoDCs, particularly CD1b. This is in agreement with work carried out in other mammalian species, such as human [16, 40] or mice [21], indicating that the presence of $1,25\text{-}(OH)_2D_3$ in the culture medium is able to hinder the complete differentiation of monocytes into MoDCs. Reports on the expression of CD86 and CD80 have been more inconsistent. In some cases CD86 was lowly expressed by VitD-MoDCs and CD80 was unaffected [16]; in other cases, the expression of both cell surface markers is lower when MoDCs are differentiated with $1,25\text{-}(OH)_2D_3$ [23]. During the present study the addition of $1,25\text{-}(OH)_2D_3$ from day 0 impeded the same level of upregulation of CD80 and CD86 seen in MoDCs, a trend seen for all four markers investigated.

The ability to take up antigens is a crucial biological function of dendritic cells. When encountering an antigen, APCs are able to process antigens via the endocytic pathway and present them to quiescent naive T cells, initiating a cascade of immune responses [41]. Consequently, in order to examine if $1,25\text{-}(OH)_2D_3$ conditioned MoDCs are able to endocytose antigens, we investigated the uptake of OVA by clathrin-coated pits [31, 35]. As in human [16, 42], bovine $1,25\text{-}(OH)_2D_3$ conditioned MoDCs are functionally impaired for endocytosis, as a significantly lower level of internalisation of OVA could be observed when compared to cells incubated without vitamin D during the differentiation process.

The key function of DCs, antigen presentation, was not only affected phenotypically by the supplementation of MoDCs with $1,25\text{-}(OH)_2D_3$, but also functionally as seen by the suppression of the T-cell stimulatory capacity in $1,25\text{-}(OH)_2D_3$ conditioned MoDCs. Five days after incubation with allogeneic PBMC, MoDCs were able to induce proliferation with numbers as low as 10^3 MoDCs per 10^5 PBMC. However, $1,25\text{-}(OH)_2D_3$ conditioned MoDCs were never able to induce a proliferation higher than background proliferation measured by PBMC incubated in medium only. When maturation was driven by stimulation with LPS for a period of 24 h, this T-cell stimulatory capacity was enhanced in MoDCs while conditioned MoDCs remained lower than the background proliferation. This correlation between phenotype and T-cell stimulatory capacity has been seen in other species [16, 23, 40, 43] and confirms that vitamin D_3 also fails to activate cattle dendritic cells. Upon TLR4 activation, we also observed a clear up-regulation of IL-12 secretion both in MoDCs and in $1,25\text{-}(OH)_2D_3$ conditioned MoDCs. However, when investigated further, the fold increase in IL-12 with MoDCs was double that seen in $1,25\text{-}(OH)_2D_3$ conditioned MoDCs, a difference also reflected by the low IL-10/IL-12 ratio in MoDCs. Vitamin D3 has been shown to have a negative effect on the production of IL-12 MoDCs after exposure to LPS in human studies [16, 43]. As IL-12 is the main cytokine which drives Th1 differentiation in naive T cells [36], the reduction we observed in IL-12 secretion by $1,25\text{-}(OH)_2D_3$ conditioned MoDCs after LPS stimulation, suggests that these Vitamin D conditioned cells may induce a reduced Th1 phenotype in the T cells they activate.

Secretion of IL-10 by DCs has an important role in immunosuppressive responses and is key to the differentiation of $CD4^+$ type 1 T-regulatory (Tr1) cells [44, 45]. Consistent with other murine and human studies [16, 22, 23, 43], we demonstrated that $1,25\text{-}(OH)_2D_3$ conditioned MoDCs secrete relatively higher levels of IL-10 and lower of IL-12 than MoDCs. This indicates a clear suppressing action of vitamin D_3 on DC development which is able to drive a typical immunotolerant phenotype on cattle DCs. The implications of this on adaptive immune responses in vivo is unclear, although it is known that co-immunization of antigens with supplementary vitamin D results in class-switching of B cells to IgA, suggesting this vitamin can play an important role in modulating bovine adaptive immune responses in vivo [46].

Cattle obtain vitamin D_3 from either the diet or from photoconversion of 7-dehydrocholesterol in the skin following exposure to UV light from sunlight [47]. As common grassland plants do not contain vitamin D_3, skin is the principle source of this vitamin in grazing cattle [48]. However, in current agricultural systems a significant proportion of cattle are house under conditions with little or no sunlight and therefore dietary supplementation of vitamin D, usually in the form of vitamin D_3, is required [25, 49]. Supplementation guidelines are available for cattle which provide daily vitamin D requirements for different classes and ages of cattle [50]. However, these recommendations are largely based on levels of vitamin D required to maintain calcium balance rather than immune function. As vitamin D acts in an endocrine manner for calcium homeostasis, but an intracrine and paracrine manner for many of the non-calcaemic functions of vitamin D [51], it is possible that the requirements of vitamin D for calcium homeostasis and immune function may differ. Consequently current recommendations for vitamin D supplementation in cattle may not be sufficient for optimal immune function. Given the growing body of evidence that vitamin D can modulate immunity in cattle, future research should focus on determining the optimum concentrations of vitamin D_3 required for immune function, and how variables associated with vitamin D_3 synthesis in the skin, such as quantity and UV light exposure to the skin, the levels of skin 7-dehydrocholesterol, and skin pigmentation [52], as well as different dietary levels of dietary supplementation of vitamin D, affects the immunity and health status of cattle.

Conclusion

In summary, the present work demonstrates that conditioning of monocytes with the hormone $1,25\text{-}(OH)_2D_3$ during the monocyte to DC maturation process induces a semi-mature or immunotolerant DC phenotype. As a consequence, the antigen presenting capabilities of these cells is hampered as shown by the reduced ability to endocytose ovalbumin and the inability to induce lymphocyte proliferation in the context of a mixed leukocyte reaction. The effects of Vitamin D_3-mediated modulation of DC function (both MoDCs and the recently described bovine blood DCs [53]) on pathogen-specific T cell responses should now be investigated, particularly in the context of diseases such as bovine tuberculosis for which a key role for Vitamin D_3 has been proposed [27].

Abbreviations

1,25-(OH)2D3: 1,25-Dihydroxyvitamin D3; allo-MLR: Allogeneic Mixed Leukocyte Reaction; ANOVA: Analysis of variance; APC: Antigen presenting cell; BSA: Bovine serum albumin; CPDA-1: Citrate phosphate dextrose-adenine 1; DC: Dendritic cells; FBS: Foetal bovine serum; GM-CSF: Granulocyte macrophages colony-stimulating factor; HRP: Horseradish peroxidase; IFN-γ: Interferon gamma; IgA: Immunoglobulin A; IL-10: Interleukin-10; IL-12: Interleukin-12; IL-1β: Interleukin-1 beta; IL-4: Interleukin 4; iNOS: Inducible nitric oxide synthase; LPS: Lipopolysaccharide; MACS: Magnetic-activated cell sorting; MFI: Median fluorescence intensity; MHC-II: Major histocompatibility complex type II; MoDC: Monocyte derived dendritic cells; NGS: Normal goat serum; OVA: Ovalbumin; PBMC: Peripheral mononuclear cells; PBS: Phosphate buffed saline; RANTES: Regulated on activation, normal t cell expressed and secreted; Th1: T helper 1; Tr1: T-regulatory; UV: Ultraviolet; VitD-MoDCs: Vitamin D conditioned monocyte derived

Acknowledgements
Not applicable

Funding
YCM was supported by European Union's Horizon 2020 research and innovation programme under grant agreement No. 635408. TNM received funding from the Rural & Environment Science & Analytical Services Division of the Scottish Government. RJM was supported by a 'Wellcome Trust Intermediate Clinical Fellowship'.

Authors' contributions
YC-M, RM and TMN conceived the study and participated in its design. YC-M performed the experimental work and wrote the manuscript. KM assisted in the phenotyping and cytokine analysis. RM and TMN participated in the writing of the manuscript and its critical review. All co-authors revised the manuscript and approved the final submitted version.

Consent for publication
Not applicable

Competing interests
The authors declare that they have no competing interests.

Author details
[1]Moredun Research Institute, Pentlands Science Park, Bush Loan, Midlothian, UK. [2]The Roslin Institute, Royal (Dick) School of Veterinary Studies, The University of Edinburgh, Midlothian, UK.

References
1. Banchereau J, Steinman RM. Dendritic cells and the control of immunity. Nature. 1998;392
2. Ohnmacht C, Pullner A, King SBS, Drexler I, Meier S, Brocker T, Voehringer D. Constitutive ablation of dendritic cells breaks self-tolerance of CD4 T cells and results in spontaneous fatal autoimmunity. J Exp Med. 2009;206(3):549–59.
3. Lüssi F, Zipp F, Witsch E. Dendritic cells as therapeutic targets in neuroinflammation. Cell Mol Life Sci. 2016;73(13):2425–50.
4. Khan S, Greenberg JD, Bhardwaj N. Dendritic cells as targets for therapy in rheumatoid arthritis. Nat Rev Rheumatol. 2009;5(10):566–71.
5. Naranjo-Gómez M, Raïch-Regué D, Oñate C, Grau-López L, Ramo-Tello C, Pujol-Borrell R, Martínez-Cáceres E, Borràs FE. Comparative study of clinical grade human tolerogenic dendritic cells. J Transl Med. 2011;9(1):89.
6. Contreras V, Urien C, Guiton R, Alexandre Y, Vu Manh T-P, Andrieu T, Crozat K, Jouneau L, Bertho N, Epardaud M, et al. Existence of CD8α-like Dendritic

cells with a conserved functional specialization and a common molecular signature in distant mammalian species. J Immunol. 2010;185(6):3313–25.

7. Marquet F, Bonneau M, Pascale F, Urien C, Kang C, Schwartz-Cornil I, Bertho N. Characterization of Dendritic cells subpopulations in skin and afferent lymph in the swine model. PLoS One. 2011;6(1):e16320.

8. Vu Manh T-P, Bertho N, Hosmalin A, Schwartz-Cornil I, Dalod M. Investigating evolutionary conservation of Dendritic cell subset identity and functions. Front Immunol. 2015;6:260.

9. Guilliams M, Henri S, Tamoutounour S, Ardouin L, Schwartz-Cornil I, Dalod M, Malissen B. From skin dendritic cells to a simplified classification of human and mouse dendritic cell subsets. Eur J Immunol. 2010;40(8):2089–94.

10. Sallusto F, Lanzavecchia A. Efficient presentation of soluble antigen by cultured human dendritic cells is maintained by granulocyte/macrophage colony-stimulating factor plus interleukin 4 and downregulated by tumor necrosis factor alpha. J Exp Med. 1994;179(4):1109–18.

11. Chan SSM, McConnell I, Blacklaws BA. Generation and characterization of ovine dendritic cells derived from peripheral blood monocytes. Immunology. 2002;107(3):366–72.

12. Adorini L. Tolerogenic dendritic cells induced by vitamin D receptor ligands enhance regulatory T cells inhibiting autoimmune diabetes. Ann N Y Acad Sci. 2003;987:258–61.

13. Besusso D, Saul L, Leech MD, O'Connor RA, MacDonald AS, Anderton SM, Mellanby RJ. 1,25-Dihydroxyvitamin D(3)-conditioned CD11c+ Dendritic cells are effective initiators of CNS autoimmune disease. Front Immunol. 2015;6:575.

14. Boodhoo N, Sharif S, Behboudi S. 1α,25(OH)2 vitamin D3 modulates avian T lymphocyte functions without inducing CTL unresponsiveness. PLoS One. 2016;11(2):e0150134.

15. Heulens N, Korf H, Mathyssen C, Everaerts S, De Smidt E, Dooms C, Yserbyt J, Gysemans C, Gayan-Ramirez G, Mathieu C, et al. 1,25-Dihydroxyvitamin D modulates antibacterial and inflammatory response in human cigarette smoke-exposed macrophages. PLoS One. 2016;11(8):e0160482.

16. Piemonti L, Monti P, Sironi M, Fraticelli P, Leone BE, Dal Cin E, Allavena P, Di Carlo V. Vitamin D3 affects differentiation, maturation, and function of human Monocyte-derived Dendritic cells. J Immunol. 2000;164(9):4443.

17. Xing N, L Maldonado ML, Bachman LA, DJ MK, Kumar R, Griffin MD. Distinctive dendritic cell modulation by vitamin D(3) and glucocorticoid pathways. Biochem Biophys Res Commun. 2002;297(3):645–52.

18. Adorini L, Penna G. Induction of Tolerogenic dendritic cells by vitamin D receptor agonists. In: Lombardi G, Riffo-Vasquez Y, editors. Dendritic cells. Berlin, Heidelberg: Springer Berlin Heidelberg; 2009. p. 251–73.

19. Lemire JM. Immunomodulatory actions of 1,25-Dihydroxyvitamin D3. J Steroid Biochem Mol Biol. 1995;53(1–6):599–602.

20. Zhang Y, Leung DYM, Richers BN, Liu Y, Remigio LK, Riches DW, Goleva E. Vitamin D inhibits Monocyte/macrophage Proinflammatory cytokine production by targeting MAPK Phosphatase-1. J Immunol. 2012;188(5):2127–35.

21. Griffin MD, Lutz W, Phan VA, Bachman LA, McKean DJ, Kumar R. Dendritic cell modulation by 1α,25 dihydroxyvitamin D3 and its analogs: a vitamin D receptor-dependent pathway that promotes a persistent state of immaturity in vitro and in vivo. Proc Natl Acad Sci. 2001;98(12):6800–5.

22. Ferreira GB, Gysemans CA, Demengeot J, da Cunha JPMCM, Vanherwegen A-S, Overbergh L, Van Belle TL, Pauwels F, Verstuyf A, Korf H, et al. 1,25-Dihydroxyvitamin D3 promotes Tolerogenic Dendritic cells with functional migratory properties in NOD mice. J Immunol. 2014;192(9):4210–20.

23. Ferreira GB, van Etten E, Verstuyf A, Waer M, Overbergh L, Gysemans C, Mathieu C. 1,25-Dihydroxyvitamin D3 alters murine dendritic cell behaviour in vitro and in vivo. Diabetes Metab Res Rev. 2011;27(8):933–41.

24. Morris A, Selvaraj RK. In vitro 25-hydroxycholecalciferol treatment of lipopolysaccharide-stimulated chicken macrophages increases nitric oxide production and mRNA of interleukin- 1beta and 10. Vet Immunol Immunopathol. 2014;161(3–4):265–70.

25. Nelson CD, Lippolis JD, Reinhardt TA, Sacco RE, Powell JL, Drewnoski ME, O'Neil M, Beitz DC, Weiss WP. Vitamin D status of dairy cattle: outcomes of current practices in the dairy industry. J Dairy Sci. 2016;99(12):10150–60.

26. Handel I, Watt KA, Pilkington JG, Pemberton JM, Macrae A, Scott P, McNeilly TN, Berry JL, Clements DN, Nussey DH, Mellanby RJ. Vitamin D status predicts reproductive fitness in a wild sheep population. Scientific Reports 2016; 6:18986.

27. Rhodes SG, Terry LA, Hope J, Hewinson RG, Vordermeier HM. 1,25-Dihydroxyvitamin D(3) and development of tuberculosis in cattle. Clin Diagn Lab Immunol. 2003;10(6):1129–35.

28. Nelson CD, Reinhardt TA, Beitz DC, Lippolis JD. In vivo activation of the Intracrine vitamin D pathway in innate immune cells and mammary tissue during a bacterial infection. PLoS One. 2010;5(11):e15469.

29. Waters WR, Nonnecke BJ, Foote MR, Maue AC, Rahner TE, Palmer MV, Whipple DL, Horst RL, Estes DM. Mycobacterium Bovis bacille Calmette–Guerin vaccination of cattle: activation of bovine CD4+ and γδ TCR+ cells and modulation by 1,25-dihydroxyvitamin D3. Tuberculosis. 2003; 83(5):287–97.

30. Waters WR, Nonnecke BJ, Rahner TE, Palmer MV, Whipple DL, Horst RL. Modulation of Mycobacterium Bovis-specific responses of bovine peripheral blood mononuclear cells by 1,25-Dihydroxyvitamin D3. Clin Diagn Lab Immunol. 2001;8(6):1204–12.

31. Corripio-Miyar Y, Hope J, McInnes CJ, Wattegedera SR, Jensen K, Pang Y, Entrican G, Glass EJ. Phenotypic and functional analysis of monocyte populations in cattle peripheral blood identifies a subset with high endocytic and allogeneic T-cell stimulatory capacity. Vet Res. 2015;46(1):1–19.

32. McNeilly TN, Rocchi M, Bartley Y, Brown JK, Frew D, Longhi C, McLean L, McIntyre J, Nisbet AJ, Wattegedera S, et al. Suppression of ovine lymphocyte activation by Teladorsagia circumcincta larval excretory-secretory products. Vet Res. 2013;44(1):1–18.

33. Kwong LS, Hope JC, Thom ML, Sopp P, Duggan S, Bembridge GP, Howard CJ. Development of an ELISA for bovine IL-10. Vet Immunol Immunopathol. 2002;85

34. Hope JC, Kwong LS, Entrican G, Wattegedera S, Vordermeier HM, Sopp P, Howard CJ. Development of detection methods for ruminant interleukin (IL)-12. J Immunol Methods. 2002;266(1–2):117–26.

35. Werling D, Hope JC, Chaplin P, Collins RA, Taylor G, Howard CJ. Involvement of caveolae in the uptake of respiratory syncytial virus antigen by dendritic cells. J Leukoc Biol. 1999;66(1):50–8.

36. Macatonia SE, Hosken NA, Litton M, Vieira P, Hsieh CS, Culpepper JA, Wysocka M, Trinchieri G, Murphy KM, O'Garra A. Dendritic cells produce IL-12 and direct the development of Th1 cells from naive CD4+ T cells. J Immunol. 1995;154(10):5071–9.

37. Ametaj BN, Beitz DC, Reinhardt TA, Nonnecke BJ. 1,25-Dihydroxyvitamin D3 inhibits secretion of interferon-γ by mitogen- and antigen-stimulated bovine mononuclear leukocytes. Vet Immunol Immunopathol. 1996;52(1):77–90.

38. Nelson CD, Reinhardt TA, Thacker TC, Beitz DC, Lippolis JD. Modulation of the bovine innate immune response by production of 1α,25-dihydroxyvitamin D3 in bovine monocytes. J Dairy Sci. 2010;93(3):1041–9.

39. Howard CJ, Sopp P, Brownlie J, Kwong LS, Parsons KR, Taylor G. Identification of two distinct populations of dendritic cells in afferent lymph that vary in their ability to stimulate T cells. J Immunol. 1997; 159(11):5372–82.

40. Canning M, Grotenhuis K, de Wit H, Ruwhof C, Drexhage H. 1-alpha,25-Dihydroxyvitamin D3 (1,25(OH)(2)D(3)) hampers the maturation of fully active immature dendritic cells from monocytes. Eur J Endocrinol. 2001; 145(3):351–7.

41. Mellman I, Turley SJ, Steinman RM. Antigen processing for amateurs and professionals. Trends Cell Biol. 1998;8(6):231–7.

42. Berer A, Stöckl J, Majdic O, Wagner T, Kollars M, Lechner K, Geissler K, Oehler L. 1,25-Dihydroxyvitamin D3 inhibits dendritic cell differentiation and maturation in vitro. Exp Hematol. 2000;28(5):575–83.

43. Chamorro S, García-Vallejo JJ, Unger WWJ, Fernandes RJ, Bruijns SCM, Laban S, Roep BO, 't Hart BA, van Kooyk Y: TLR triggering on Tolerogenic Dendritic cells results in TLR2 up-regulation and a reduced Proinflammatory immune program. J Immunol 2009, 183(5):2984-2994.

44. Corthay A. How do regulatory T cells work? Scand J Immunol. 2009;70(4):326–36.

45. Levings MK, Gregori S, Tresoldi E, Cazzaniga S, Bonini C, Roncarolo MG. Differentiation of Tr1 cells by immature dendritic cells requires IL-10 but not CD25+CD4+ Tr cells. Blood. 2005;105(3):1162–9.

46. Vilte DA, Larzábal M, Garbaccio S, Gammella M, Rabinovitz BC, Elizondo AM, Cantet RJC, Delgado F, Meikle V, Cataldi A, et al. Reduced faecal shedding of Escherichia Coli O157:H7 in cattle following systemic vaccination with γ-intimin C280 and EspB proteins. Vaccine. 2011;29(23):3962–8.

47. Hymøller L, Jensen SK. 25-Hydroxycholecalciferol status in plasma is linearly correlated to daily summer pasture time in cattle at 56°N. Br J Nutr. 2012; 108(4):666–71.

48. Hymøller L, Jensen SK. Vitamin D3 synthesis in the entire skin surface of dairy cows despite hair coverage. J Dairy Sci. 2010;93(5):2025–9.

49. Hidiroglou M, Karpinski K. Providing vitamin D to confined sheep by oral supplementation vs ultraviolet irradiation. J Anim Sci. 1989;67(3):794–802.
50. NRC: National Research Council (U.S.) subcommittee on dairy cattle nutrition: nutrient requirements of dairy cattle. In. Edited by Press NA, 7th revised edition edn. Washington, D.C.; 2001.
51. Hewison M. Vitamin D and the intracrinology of innate immunity. Mol Cell Endocrinol. 2010;321(2):103–11.
52. Norman AW, Henry HC. Vitamin D. In: Zempleni J, Rucker RB, DB MC, Suttle JW, editors. Handbook of vitamins. 4th ed. Boca Raton: CRC Press; 2007.
53. Park KT, ElNaggar MM, Abdellrazeq GS, Bannantine JP, Mack V, Fry LM, Davis WC. Phenotype and function of CD209+ bovine blood Dendritic cells, Monocyte-derived-Dendritic cells and Monocyte-derived macrophages. PLoS One. 2016;11(10):e0165247.

5

Plants of the Cerrado with antimicrobial effects against *Staphylococcus* spp. and *Escherichia coli* from cattle

Izabella Carolina de O. Ribeiro[1], Emanuelly Gomes A. Mariano[1], Roberta T. Careli[1], Franciellen Morais-Costa[1], Felipe M. de Sant'Anna[2], Maximiliano S. Pinto[1], Marcelo R. de Souza[2] and Eduardo R. Duarte[1,3*]

Abstract

Background: Both diarrhea in calves and mastitis in cows limit cattle production. The bacteria involved in these diseases have shown multi-resistance to antimicrobials, however plant metabolites therefore can provide an alternative method of control. This study selected and characterized Cerrado plant extracts showing inhibitory effects against *Escherichia coli* and *Staphylococcus* spp. from cattle. Thirteen leaf extracts were initially screened and diameters of inhibition zones produced against the pathogens were recorded using an agar disk diffusion method. Total condensed tannin contents were determined and antibacterial activities were analyzed after tannin removal from the five selected extracts. The minimum inhibitory concentrations (MIC) and minimum bactericidal concentrations (MBC) were evaluated by macro-dilution antimicrobial susceptibility tests, and the extracts were characterized by high performance liquid chromatography.

Results: Inter- and intra-specific bacterial variations in the susceptibility to the extracts were detected. The aqueous extract (AE) from *Caryocar brasiliense* Cambess. leaves produced larger inhibition zones against *E. coli* strains than did other selected extracts. However, the AE from *Schinopsis brasiliensis* was the most effective against *Staphylococcus* spp. strains ($P < 0.001$). The MIC of ethanolic extracts (EE) from *C. brasiliense* (0.27 mg/mL) and *S. brasiliensis* (0.17 mg/mL) were lower than those of other extracts. The MIC and MBC of the *Annona crassiflora* EE were 6.24 mg/mL for all bacteria. Flavonoids were the main metabolites detected in the *A. crassiflora* EE as well as in the AE and EE from *C. brasiliense*, while tannins were the main metabolites in the *S. brasiliensis* leaf extracts.

Conclusion: The AE from *C. brasiliense* was more effective against Gram-negative bacteria, while the AE from *S. brasiliensis* was more effective against Gram-positive bacteria. *A. crassiflora* EE and *S. brasiliensis* extracts are potent bactericide. After removal of the tannins, no antimicrobial effects were observed, indicating that these metabolites are the main active antibacterial components.

Keywords: Antibacterial, Brazilian savannah, Colibacillosis, Mastitis, Medicinal plants, *Staphylococcus Aureus*, *Staphylococcus haemolyticus*

Background

Diseases must be prevented or controlled in order to achieve a sustainable and viable production of ruminants. Diarrhea is the most common pathology in young calves and *Escherichia coli* represents one of its main etiological agents [1, 2]. Mastitis, caused by the *Staphylococcus* spp., is the most important disorder in cows, and leads to reduced milk production and increased production costs [3, 4]. These bacteria have shown multi-resistance to antimicrobials in different continents and present a public health risk [2, 5–7].

Plant metabolites are considered alternative control agents for the reduction of resistant microorganisms and antimicrobial residues in foods of animal origin [5–7].

* Correspondence: duartevet@hotmail.com
[1]Instituto de Ciências Agrárias, Universidade Federal de Minas Gerais, Avenida Universitária, 1000, Bairro Universitário, Montes Claros, Minas Gerais CEP 39401-790, Brazil
[3]Instituto de Ciências Agrárias, Universidade Federal de Minas Gerais, Av Universitária 1000, Bairro Universitario, Montes Claros, MG 39400-006, Brazil
Full list of author information is available at the end of the article

Scientific literature has frequently reported the inhibitory action of certain plant extracts against bacteria in humans [7, 8]. However, few studies have showed effective plant extracts inhibiting microorganisms from ruminants or other animals. Extracts from *Solanum paniculatum* L. (Jurubeba) and *Punica granatum* L. (Romã) display antibacterial effects against microorganisms that cause bovine mastitis [9], and *Rhodomyrtus tomentosa* L. (rose myrtle) leaf extract shows potent antibacterial activity against *Staphylococcus aureus* in milk [10]. Tannins are the main antimicrobial metabolites in vegetal extracts, and also inhibit enzymes and alter metabolism via membrane or cell wall interactions [11].

The Cerrado, a type of savannah present in South America, is native to more than 10,000 plant species that contain natural products for phytotherapy [12]. With regards to alternative antibacterial agents, four medicinal plants from the Brazilian Cerrado have been shown to inhibit the growth of *S. aureus* [13], and the leaf extract from *Schinopsis brasiliensis* Engl. is effective against multidrug-resistant *S. aureus* [14].

However, the antimicrobial activity of plant species from the Brazilian Cerrado against animal pathogens has not been fully explored. Extracts from these plants that show antibacterial effects could favor the alternative controls, thereby reducing pathogen multi-resistance. In organic animal production systems, the use of these extracts would enable the production of foods free from antimicrobial residues, thus increasing the value of these animal products.

In this study, plant species commonly found in the Cerrado were selected and their antimicrobial activities were evaluated against isolates of *E. coli* and *Staphylococcus* spp. from cattle. To identify the main antibacterial components, these extracts were characterized by high performance liquid chromatography (HPLC), and inhibitory activity was analyzed after tannin removal.

Methods

Microorganisms

The antibacterial effects of the plant extracts were evaluated against three *Staphylococcus* isolates (S178, S135, and S182) from cows with mastitis. These bacteria were isolated and cultured on mannitol salt agar, and evaluated based on colony characteristics, Gram staining, catalase reaction, and coagulase test. We also assessed the inhibitory effects of the extracts on two *E. coli* isolates (E2 and E3) from the feces of dairy calves with diarrhea. These were isolated and cultured on MacConkey agar and colony characteristics, Gram staining, and catalase reactions were evaluated. These animals were raised on an experimental farm in northern Minas Gerais, Brazil. In addition, the human clinical isolates *S. aureus* ATCC 25923 and *E. coli* ATCC 25922 were included as

reference strains. All bacteria were cultured in Brain Heart Infusion (BHI) broth, and subsamples were stored at – 80 °C after glycerol inclusion (1:1).

DNA from these bovine isolates was extracted and amplified as described by Chapaval et al. [15]. DNA samples were amplified via polymerase chain reaction (PCR) using primers 27F (5′-AGAGTTTGATCCTGGC TCAG-3′) and 1492R (5′-GGTTACCTTGTTACGA CTT-3′), as described by Lane [16]. 16S ribosomal RNA (rRNA) was sequenced following the method described by Sanger [17], using the automatic sequencer Mega-BACE® 1000 (GE Life Sciences, USA), according to Rey-senbach et al. [18]. 16S rRNA gene sequencing was verified using the SeqScanner Software® v1.0 (Applied Biosystems, USA), and the results were compared online by BLAST (database from NCBI - https://blast.ncbi.nlm. nih.gov/Blast.cgi). The bacteria species were identified with a similarity level of at least 99%.

Antibacterial susceptibility

The procedure for the agar disk diffusion method was performed in triplicate, according to recommendations by the National Committee for Clinical Laboratory Standards (NCCLS) [19]. For *Staphylococcus* strains, the following antimicrobial discs were added onto the medium surface: chloramphenicol, 30 µg; erythromycin, 15 µg; vancomycin, 30 µg; oxacillin, 1 µg; gentamicin, 10 µg; tetracycline, 30 µg; clindamycin, 2 µg; and penicillin, 10 µg. For *E. coli* strains, the following were used: chloramphenicol, 30 µg; ampicillin, 10 µg; gentamicin, 10 µg; ciprofloxacin, 5 µg; tetracycline, 30 µg; and norfloxacin, 10 µg, as described by the NCCLS [19]. For quality control purposes, the strains ATCC 25923 and ATCC 25922 were used. All plates were incubated at 35 °C for 24 h, and inhibition zones (mm) were measured and the bacteria were classified as resistant or sensitive according to the NCCLS guidelines [19].

Plant extracts

Plant leaves were collected from April to June at the Institute of Agricultural Sciences, UFMG, in Montes Claros, Minas Gerais, Brazil. This region is located at latitude 16°51′ and longitude 44°55′, and the climate is tropical and humid with dry summers (A) according to Köppen classification [20].

Vegetal materials were collected from *Caryocar brasiliense* Camb. (Caryocaraceae), *Annona crassiflora* Mart. (Annonaceae), *S. brasiliensis* Engl. (Anacardiaceae), *Piptadenia viridiflora* (Kunth) Benth. (Fabaceae), *Serjania lethalis* A.St.-Hil. (Sapindaceae), *Casearia sylvestris* (Flacourtiaceae), and *Ximenia americana*L. (Olacaceae). Plant samples were deposited in the Montes Claros Herbarium of Universidade Estadual de Montes Claros, as voucher specimens 338, 1492, 377, 2283, 2249, 3008, and 211, respectively.

The leaves were carefully inspected, and those with gross lesions or damage were discarded. Selected leaves were dehydrated under forced air circulation (TE 394/4, Tecnal Equipamentos Científicos Tecnal, Piracicaba, SP, Brazil) at 38 °C for 72 h, crushed in a blender, and stored inside paper bags in the dark at − 4 °C [11, 21].

Aqueous extracts (AEs) were produced by placing the ground dried leaves in a distilled water bath at 40 °C for 60 min. Ethanolic extracts (EEs) were obtained from macerated dried leaves held in absolute ethanol in amber-colored glass containers in the dark for seven days. Extracts were filtered through a gauze funnel and subsequently evaporated at 40 °C for 48 h under forced air circulation until completely dry and stored at 4 °C until use [11]. In this study, both EEs and AEs were completely soluble in distilled water and did not require any other solvents for antimicrobial analysis.

Subsamples of extracts were subjected to tannin extraction according to the method described by Nyman et al. [22]. Extracts were dissolved in water (1 g per 20 mL) at 90 °C and cooled to room temperature. After reaching a temperature of 30 °C, 0.2 μL 10% NaCl was added, and 1 mL of this solution was combined with 4 mL 1% gelatin solution before centrifuging at 1800×g for 6 min. Supernatants were used to assess the effects of tannin-free extracts.

Characterization of extracts

A Waters Alliance 2695 HPLC system comprising a quaternary pump, auto-sampler, photodiode array detector (DAD) 2996, and Waters Empower Pro data handling system (Waters Corporation, Milford, Connecticut, USA) were used for extract characterization. Analyses were performed on a LiChrospher 100 RP-18 column (250 × 4 mm, 5 mm; Merck, Darmstadt, Germany) combined with a LiChrospher 100 RP-18 guard column (4 × 4 mm, 5 mm; Merck) at 40 °C. Water (A) and acetonitrile (B) were used as eluents, both containing 0.1% (v/v) H_3PO_4 at a flow rate of 1.0 mL/min as follows: 0 min, 95% A and 5% B; 60 min, 5% A, 95% B, followed by 10 min isocratic elution. Solvents used were of HPLC grade (Merck, Germany) and were degassed by sonication before use. Chromatograms were obtained at 210 nm, and the UV spectra were recorded online from 190 to 400 nm.

The dried crude extracts were dissolved in methanol (HPLC-grade), ultrapure water, or hydroethanolic solutions according to their solubility, to concentrations of 10 mg/mL. After centrifugation at 8400×g for 10 min, 10 mL sample were automatically injected into the apparatus.

The total condensed tannin (proanthocyanidins) content of the extracts was determined by measuring the absorbance of cyanidin chloride resulting from acid-catalyzed solvolysis with n-BuOH/HCl 12 M (95:5) at 540 nm, according to the method described by Hiermann et al. [23]. Each sample was analyzed in triplicate and the total condensed tannin content, expressed as cyanidin chloride, was calculated using the following formula:

Condensed tannins % = Absorbance (sample) − Absorbance (blank) × 4.155/sample weight (g).

Selection of plant extracts with inhibitory effects

Seven ethanolic and six aqueous extracts were diluted with distilled water at 0.1 g extract/mL and vortexed for 3 min. Extracts were used immediately after this preparation. Antibacterial activity was determined using the agar disk diffusion method [19, 24].

A loopful of bacteria was inoculated onto BHI agar under sterile conditions, and incubated at 37 °C for 24 h. Turbidity equivalent to a 0.5 McFarland standard was used as a reference to adjust for approximately 10^8 colony-forming units (CFU)/mL. One hundred microliters of freshly prepared inoculum suspension were spread on Mueller-Hinton agar using sterile swabs [19]. Eight microliters of extract solution was added to 6-mm paper filter disks, allocated onto the surface of the seeded plates, and incubated at 35 °C for 24 h. Inhibition zones were then measured using a digital caliper [8]. In this screening assay, all procedures were performed in duplicate.

Based on the largest zones of inhibition and the broadest spectrum of action, five extracts were selected, both with and without tannins. These extracts were filtered through a 0.2-μm Millipore membrane and aliquots were then submitted to dry matter (dm) determination in an oven at 105 °C, in order to standardize them at 1.58 mg dm/mL. Paper filter disks with sterile saline solution (without extract) and discs containing the extracts incubated without bacteria were used as controls [8, 24]. The experiment was designed in a factorial arrangement (5 extracts × 7 bacterial strains) and all procedures were performed in triplicate. The inhibition zone averages were compared by the analysis of variance using Scott-Knott's test at the 5% significance level, using the System for Statistical Analysis software (SAEG 9.1).

Minimum inhibitory concentration (MIC) and minimum bactericidal concentration (MBC)

After filtration of the extracts, we determined the MIC necessary to inhibit the growth of the microorganism by macro-dilution in Mueller-Hinton broth, as described by the NCCLS [25].

Using a 1:2 dilution with an equal volume of medium, the final concentrations (6.24–0.01 mg/mL) were evaluated. However, for MBC determinations of the *C. brasiliensis* extracts were also evaluated concentrations up to 40 mg/mL. Extract solutions were therefore prepared at double the final concentration [25]. Completing at final volume of 5 mL, 120 μL of bacteria inoculum prepared as reported above were added together with 2.48 mL Mueller Hinto broth and 2.5 mL of extract solution.

For the controls, we used growth control tubes containing broth without extract for each bacterium tested, tubes without bacteria containing broth alone or added of extracts. All tubes were incubated at 35 °C for 24 h in a thermo-shaker incubator (Novatécnica, São Paulo, SP, Brazil) to ensure homogenization. After this period, bacterial growth was assessed using 125 μL 0.5% triphenyl tetrazolium chloride (TTC) solution, which indicates cellular multiplication through the development of a reddish color in the presence of viable cells, thus enabling MIC determination [26].

Subsequently, the MBC was determined, which represents the lowest concentration of the extract necessary to achieve complete suppression of bacterial growth. One hundred-microliter aliquots from the tubes used for the MIC assay and the control tubes without extracts were inoculated on Mueller-Hinton agar and incubated at 37 °C for 24 h. The absence of bacterial growth on the agar plate was evaluated to determine MBC, and the experiment was carried out in triplicate.

Results

Characterization of bacterial isolates

Using molecular analysis, we successfully identified isolate S178 as *S. aureus* and isolates S135 and S182 as *S. haemolyticus* (99.9% similarity). Strains E2 and E3 were identified as *E. coli*, as shown in Table 1.

Isolate S178 (*S. aureus*) was resistant to erythromycin, clindamycin, oxacillin, penicillin, tetracycline, and vancomycin. However, *S. haemolyticus* isolates were sensitive to all antimicrobials tested in grampositive strains. All *E. coli* strains were resistant to tetracycline, while isolate E3 was also resistant to ampicillin and gentamicin (Table 2).

Table 2 Antimicrobial sensitivity profiles for *Staphylococcus aureus, Staphylococcus haemolyticus,* and *Escherichia coli* isolates from cattle and standard strains

	S. haemolyticus		S. aureus		E. coli		
Antibacterial	S135	S182	S178	ATCC	E2	E3	ATCC
Chloramphenicol	S	S	S	S	S	S	S
Erythromycin	S	S	R	S	–	–	–
Ampicillin	–	–	–	–	S	I	S
Vancomycin	S	S	R	R	–	–	–
Oxacillin	S	S	R	R	–	–	–
Gentamicin	S	S	S	S	S	I	I
Ciprofloxacin	–	–	–	–	S	S	S
Tetracycline	S	S	R	S	R	I	R
Clindamycin	S	S	R	R	–	–	–
Penicillin	S	S	R	R	–	–	–
Norfloxacin	–	–	–	–	S	S	S

chloramphenicol 30 μg, erythromycin 15 μg, ampicillin 10 μg, vancomycin 30 μg, oxacillin 1 μg, gentamicin 10 μg, ciprofloxacin 5 μg, tetracycline 30 μg, clindamycin 2 μg, penicillin 10 μg e norfloxacin 10 μg. *S* Sensitive, *I* Intermediate, *R* Resistant, according to NCCLS (2005)

Selection of antimicrobial extracts from Cerrado plants

The initial screen revealed that all extracts presented inhibitory effects on at least one of the evaluated bacterial strains. However, the EE from *X. americana* did not show any inhibitory effects on *E. coli* or *S. haemolyticus* strains (Table 3). The EE and AE from *C. brasiliense* and EEs from *A. crassiflora, S. brasiliensis,* and *S. lethalis* presented antagonism against all *Staphylococcus* spp. strains. The different phenological stages of *C. brasiliense* produced leaf extracts showing inhibitory effects against *S. aureus* and *S. haemolyticus* (Table 3). Inhibition zone measurements were not associated with tannin content in the extracts tested (*P* > 0.05, Pearson correlation).

The EE from *A. crassiflora*, and EEs and AEs from *S. brasiliensis* and *C. brasiliense* were selected owing to their inhibitory effects, which produced large inhibition zones. We also considered whether plants acted on one bacterial species or on both, the latter showing a broader spectrum of action. These extracts were sterilized by filtration and concentrations were standardized to 1.58 mg/mL to compare their effects via diffusion tests in agar.

Table 1 Bacterial identification by 16S rDNA sequencing and classification according to BLAST (NCBI database)

Isolates	Origen	Number of analyzed nucleotides	Identification with similarity > 99%
S178	Cow whit matitis	551	*Staphylococcus aureus subsp. aureus* strain *NCTC 8325*
S135	Cow whit matitis	553	*Staphylococcus haemolyticus* strain *JCSC1435*
S182	Cow whit matitis	550	*Staphylococcus haemolyticus* strain v *JCSC1435*
E2	Calf with diarrhea	590	*Escherichia coli str. K-12 substr.* strain *MG1655*
E3	Calf with diarrhea	574	*Escherichia coli str. K-12 substr.* strain *MG1655*

Table 3 Selection of vegetal extracts according to inhibition zones (mm) produced in *Staphylococcus aureus*, *Escherichia coli*, and *Staphylococcus haemolyticus* after addition of extracts from Cerrado plant leaves in an agar diffusion test

Vegetal species	Extracts	Tannin content (%)	S. haemolyticus		S. aureus		Escherichia coli		
			135 AE	182	178	ATCC	E2	E3	ATCC
Caryocar brasiliense in flowering	Ethanolic	1.99 ± 0.12	26.7 ± 2.5	14.1 ± 1.9	30.2 ± 3.4	20.2 ± 3.3	0.0	0.0	19.0 ± 1.9
C.brasiliense in flowering	Aqueous	1.37 ± 0.08	23.5 ± 3.5	14.3 ± 1.8	21.8 ± 2.4	19.3 ± 2.3	0.0	0.0	14.2 ± 2.7
C. brasiliense in fruiting	Aqueous	1.25 ± 0.02	24.3 ± 2.6	14.0 ± 2,7	21.0 ± 2.0	23.8 ± 2.0	0.0	0.0	16.4 ± 3.7
C. brasiliense flowerless and fruitless	Aqueous	2.66 ± 0.22	20.2 ± 2.0	11.1 ± 2.6	16.2 ± 4.1	15.8 ± 3.1	0.0	0.0	13.8 ± 5.0
Annona crassiflora	Aqueous	2.59 ± 1.47	12.8 ± 2.3	0.0	15.9 ± 2.7	13.1 ± 4.0	0.0	8.0 ± 3.0	0.0
Annona crassiflora	Ethanolic	4.20 ± 2.38	9.3 ± 2.1	14.5 ± 1.2	15.9 ± 3.1	14.6 ± 3.4	0.0	19.8 ± 4.1	11.1 ± 3.4
Piptadenia viridiflora	Aqueous	0.23 ± 0.01	16.4 ± 2.3	0.0	19.1 ± 3.1	0.0	0.0	11.5 ± 2.7	29.9 ± 5.5
Piptadenia viridiflora	Ethanolic	1.75 ± 0.21	9.0 ± 0.95	0.0	13.3 ± 3.3	17.1 ± 3.1	0.0	0.0	27.0 ± 4.9
Schinopsis brasiliensis	Aqueous	0.16 ± 0.37	16.4 ± 2.3	18.0 ± 0.3	24.1 ± 2.9	20.8 ± 3.1	9.4 ± 1.0	0.0	22.3 ± 3.3
Schinopsis brasiliensis	Ethanolic	0.72 ± 0.34	21.7 ± 3.6	21.2 ± 4.3	22.5 ± 2.7	22.8 ± 2.3	0.0	16.2 ± 3.4	20.6 ± 5.6
Serjania lethalis	Ethanolic	6.37 ± 0.29	13.2 ± 2.5	11.0 ± 2.2	14.5 ± 3.0	18.5 ± 4.7	0.0	10.0 ± 3.0	11.7 ± 4.6
Casearia sylvestris	Ethanolic	7.36 ± 0.54	8.2 ± 1.3	0.0	10.9 ± 2.0	7.5 ± 4.3	0.0	0.0	15.0 ± 3.8
Ximenia americana	Ethanolic	0.29 ± 0.02	0.0	0.0	0.0	24.2 ± 4.5	0.0	0.0	0.0

When evaluating the effects of the five selected extracts on the seven bacterial strains, differences between the type of extract, the plant species, and the strain and species of bacteria were detected ($P < 0.001$). Consequently, interactions between the type of extract evaluated and the bacterial strain were also significant ($P < 0.01$, Table 4). After removal of the tannins, no inhibitory effects were observed for the selected extracts (Table 4).

The AE from *S. brasiliensis* produced larger inhibition zones against *Staphylococcus* spp. strains than other extracts. However, considering the *E. coli* strains, the *C. brasiliense* AE produced the largest areas of inhibition among all the extracts (Table 4, $P < 0.001$).

The MICs observed for the EEs from *S. brasiliensis* and *C. brasiliense* were lower than *A. crassiflora* EE. Both the MIC and MBC of the *A. crassiflora* EE were 6.24 mg/mL for all bacterial strains (Table 5).

Reversed-phase HPLC characterization of selected plant extracts

According to the UV spectra observed, the presence of flavonoids was detected in the region between 261 and 279.3 nm for the EE from *A. crassiflora* (Fig. 1), and EE and AE from *C. brasiliense* (Fig. 2). Tannins were detected in the EE and AE of *S. brasiliensis*, with absorbance at 257–263 nm for the respective retention times (Fig. 3).

Table 4 Average inhibition zones (mm) produced in an agar diffusion test in *Staphylococcus aureus*, *Staphylococcus haemolyticus*, and *Escherichia coli* treated with leaf extracts from *Annona Crassiflora*, *Caryocar brasiliense*, and *Schinopsis brasiliensis* with (TA) or without (WT) tannins (1.58 mg/mL)

Bacteria Strains[a]	A. crassiflora Ethanolic-		C. brasiliense Ethanolic		C. brasiliense Aqueous		S. brasiliensis Ethanolic		S. brasiliensis Aqueous	
	TA	WT	TA	WT	TA	WT	TA	WT	TA	WT
S135	7.3 ± 0,67 Dd	0	8.1 ± 0,11 Ce	0	6.1 ± 0,08 Ed	0	9.4 ± 0,27 Bb	0	10.0 ± 0,18 Ac	0
S182	6.9 ± 0,38 Ce	0	6.5 ± 0,20 Dg	0	6.1 ± 0,11 Ed	0	9.3 ± 0,22 Bb	0	10.8 ± 0,33 Ac	0
S178	8.6 ± 0,47 Dc	0	8.4 ± 0,04 Dd	0	6.1 ± 0,11 Ed	0	9.1 ± 0,29 Bb	0	9.8 ± 0,11 Ac	0
ATCC 25923	7.0 ± 0,06 Ee	0	7.3 ± 0,18 Df	0	9.4 ± 0,4 Cc	0	10.2 ± 0,22 Ba	0	15.1 ± 0,18 Aa	0
E2	9.8 ± 0,31 Ba	0	8.9 ± 0,18 Dc	0	14.1 ± 0,18 Ab	0	6.2 ± 0,07 Ec	0	9.3 ± 0,22 Cc	0
E3	9.5 ± 0,76 Cb	0	9.4 ± 0,27 Db	0	14.8 ± 0,27 Aa	0	6.1 ± 0,15 c	0	12.4 ± 1,44 Bb	0
ATCC 25922	10.1 ± 0,53 Da	0	11.3 ± 0,2 Ca	0	14.5 ± 0,44 Aa	0	8.5 ± 0,29 Ec	0	11.9 ± 0,22 Bb	0

Lowercase letters in lines indicate significant difference between bacteria strains and uppercase letters in columns indicate significant difference between plant extracts as determined by Scoott-Knott'test with a 5% significance
[a] *S. haemolyticus* (S135 and S182); *S. aureus* (S178 and ATCC 25923) and *E. coli* (E2,E3 and ATCC25922)

Table 5 Minimum inhibitory concentration (MIC) and minimum bacterial concentration (MBC) of leaf extracts from *Annona Crassiflora, Caryocar brasiliense,* and *Schinopsis brasiliensis* tested on *Escherichia coli* and *Staphylococcus* spp. from cattle

| Bacteria strains[a] | Annona crassiflora | | Caryocar brasiliense | | | | Schinopsis brasiliensis | | | |
| | Ethanolic mg/mL | | Ethanolic mg/mL | | Aqueous mg/mL | | Ethanolic mg/mL | | Aqueous mg/mL | |
	MIC	MBC	MIC	MBC	MIC	MBC	MIC	MBC	MIC	MBC
Staphylococcus spp.										
S135	6.24	6.24	0.27	> 40.0	0.71	0.71	0.17	0.34	0.42	0.42
S182	6.24	6.24	0.27	> 40.0	0.71	0.71	0.17	0.68	0.42	0.84
S178	6.24	6.24	0.27	> 40.0	0.71	> 40.0	0.17	0.34	0.42	0.42
ATCC 25923	6.24	6.24	0.27	> 40.0	0.71	0.71	0.17	0.68	0.42	0.84
Escherichia coli										
E2	6.24	6.24	0.27	> 40.0	0.71	> 40.0	0.17	0.34	0.10	0.42
E3	6.24	6.24	0.27	> 40.0	0.71	0.71	0.17	0.34	0.42	0.42
ATCC 25922	6.24	6.24	0.27	30.0	0.71	> 40.0	0.17	0.34	0.42	0.42

[a]*S. haemolyticus* (S135 and S182); *S. aureus* (S178 and ATCC 25923) and *E. coli* (E2, E3 and ATCC25922)

Discussion

In this study, six plants showed inhibitory effect against the three bacterial species evaluated, indicating the importance of bioprospecting studies on Cerrado vegetation. We observed intra- and inter-species differences in the bacterial susceptibility to plant extracts, which should be clarified in future investigations.

The differences in inhibition zones produced by the five selected extracts could be associated with their biochemical compositions, polarity, and solubility in Muller-Hinton agar. Additionally the interactions and the differences of extract constituents of the extracts could explain their spectra of action against Gram-positive or Gram-negative bacteria.

In agar diffusion test, the best inhibitory action against the Gram-negative *E. coli* strains was promoted by *C. brasiliense* AE which contained flavonoids. However, the AE from *S. brasiliensis,* containing tannins, produced larger inhibition zones in the Gram-positive *Staphylococcus* spp. strains than did other extracts. We suggested intra-specific variations in bacterial response to the extracts, considering the different effects of these extracts against human clinical strains (*E. coli* ATCC 25922 and *S. aureus* ATCC 25923) and respective bovine isolates. The different hosts could lead to the selection of external or internal variations in these bacteria, which could influence their susceptibly to these extracts. However, this would need to be elucidated in future research.

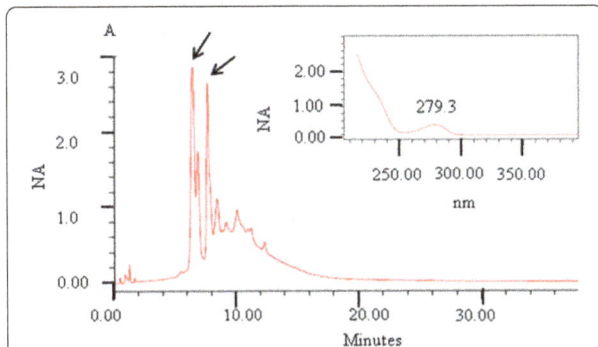

Fig. 1 HPLC chromatographic profile, retention times (RT), and UV (279.3 nn) spectrum characteristics of flavonoids, in panels inside the image, in the ethanolic extract from *Annona crassiflora* (first RT = 6.484 min).

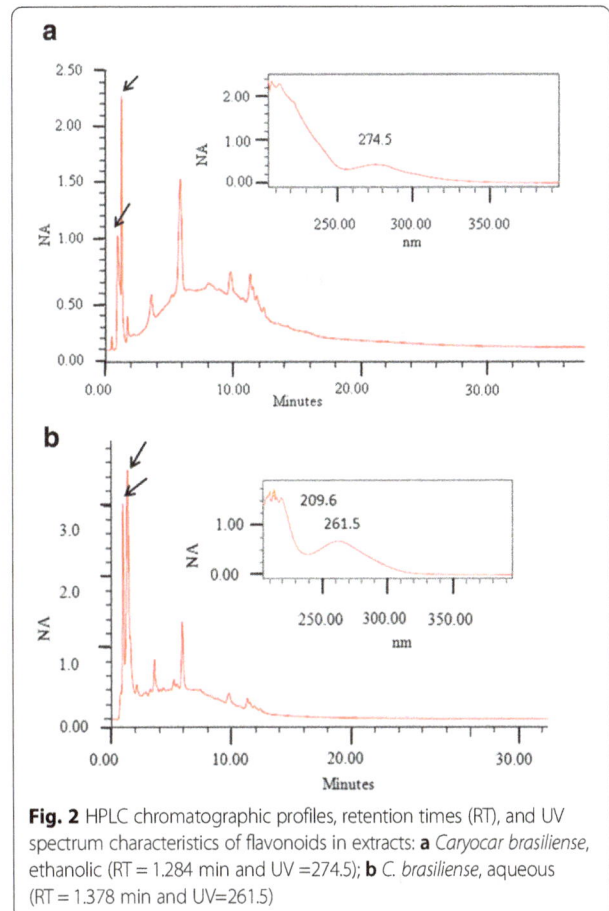

Fig. 2 HPLC chromatographic profiles, retention times (RT), and UV spectrum characteristics of flavonoids in extracts: **a** *Caryocar brasiliense,* ethanolic (RT = 1.284 min and UV =274.5); **b** *C. brasiliense,* aqueous (RT = 1.378 min and UV=261.5)

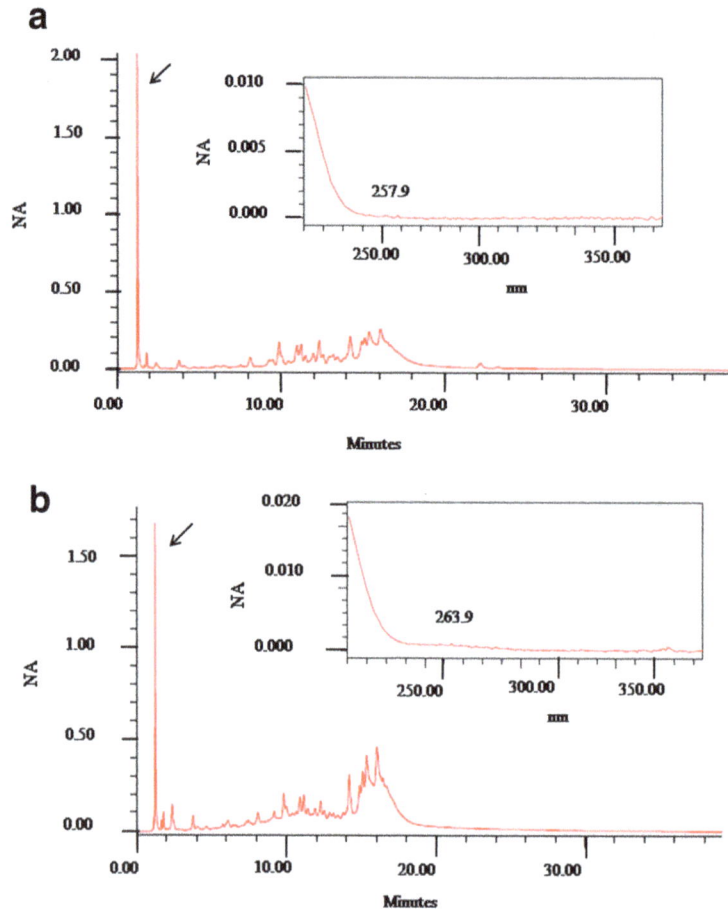

Fig. 3 HPLC chromatographic profiles, retention times (RT), and UV spectrum characteristics of tannins in extracts: **a** *Schinopsis brasiliensis*, ethanolic (RT = 1.053 min and UV = 257.9); **b** *Schinopsis brasiliensis*, aqueous (RT = 1.054 min and UV=263.9)

A specific mechanism of action against *Staphylococcus* spp. or *E. coli* could be explained by differences in cell wall constitutions of Gram-negative and Gram-positive bacteria [27]. Specific constituents in *C. brasiliense* extracts such as saponins and flavonoids could be more toxic to walls with greater lipid contents, as was observed in *E. coli* [28, 29].

We detected that a lower number of the effective extracts against the Gram-negative bacteria. In another antibacterial screening of plants from the Cerrado revealed that the stem bark extracts from *Qualea grandiflora* Mart (Vochysiaceae), *Virola surinamensis* (Rol.) Warb (Myristicaceae), and *Hancornia speciosa* Gome (Apocynaceae) significantly inhibited *S. aureus* but were not effective against *E. coli* [13].

The results obtained in this research corroborate with those of the study by Lima et al. [30] that evaluated the in vitro antimicrobial activity of EEs (1:1) from leaves, fruits, and seeds of *A. crassiflora* against *Staphylococcus* spp. The authors showed that these extracts significantly inhibited bacterial growth with a mean inhibition zone of 10 to 12 mm. Silva et al. [31] also reported inhibition

zones of 10 to 14 mm when analyzing the effects of *A. crassiflora* leaf extracts against multi-resistant human *S. aureus*.

Considering antibacterial effects of *C brasiliensis*, unlike the leaf extracts in our study, bark extracts used at 500 mg/mL showed no inhibition when tested against the reference strains of *S. aureus* and *E. coli* [32]. These results indicate the leaves, which are more available, as sustainable source of the antibacterial metabolites of this plant.

Analyzing the MICs observed in this research, the EEs and AEs from *S. brasiliensis* and *C. brasiliense* were more effective against all bacterial strains than *A. crassiflora* extract. However MBC values were different among the bacterial strains, thus showing intra-specific variations. The EE of *C. brasileiense* showed bacteriostatic effect while other extracts were bactericides.

Although of few scientific studies describe antagonistic effects of plant extracts against Gram-negative bacteria, we detected that the extracts from *C. brasiliense* and *S. brailiensis* had lower MICs against *E. coli* strains than other studies. Paula-Junior et al. [28] reported that the hydroethanolic extract from *C. brasiliense* leaves exhibited MIC values of 4 mg/mL against *E. coli* and *S. aureus,* and Amaral et al. [33] described higher MICs for *C. brasiliense* EE against *E. coli* ATCC 25922 (MIC of 11.25 mg/mL) and *S. aureus* ATCC 6538 (22.5 mg/mL).

In other study, low MICs have also been reported for *S. brasiliensis, E. coli* ATCC 9723, and multi-resistant *S. aureus* isolates, with values varying from 0.025 to 0.100 mg/mL, depending on the fraction of extracts [14].

In this study, the EE of *A. crassiflora* showed MIC and MBC of 6.4 mg/mL for all bacteria evaluated, indicating this extract is safe and can ensure complete inhibition or death of these microorganisms. Silva et al. [31] evaluated the effects of leaf extract of this plant against oxacillin-resistant human *S. aureus* and ATCC 6538, and observed higher MICs (25 mg/mL) for both strains. Furthermore, the authors identified alkaloids and flavonoids as the active compounds.

Considering HPLC analysis, the EE of *A. crassiflora* and both extracts from *C. brasiliense* presented flavonoids while the tannins were main components detected for *S. brasiliensis* extracts. However, these analyses cannot reveal the concentration or chemical structure of these metabolites. The characterization of these extracts by other methods, such as gas chromatography after derivation, could reveal reveal the contents of their specific antibacterial components.

In this study, inhibitory effects were no observed after tannin removal from the selected extracts, indicating that this metabolite represented the main antibacterial agent in these extracts. Future studies should elucidate the mechanism of action of these extracts, and whether such effects can be linked to a group of substances promoting bacterial inhibition.

The antimicrobial role of these plant metabolites has not yet been clearly elucidated. However, there is a consensus of multiple mechanisms of action on bacterial cells. The tannins can interact with the cytoplasmic membrane, inhibiting its function and thereby compromising cellular integrity [11]. These compounds could also inhibit nucleic acid and enzymes synthesis, modify the cellular metabolism via membrane interaction, and complex with metal ions to decrease its availability for the microorganisms [34]. The antibacterial activity of flavonoids has been attributed to inhibition of DNA gyrase, inhibition of cytoplasmic membrane function, and energy metabolism. These compounds represent a novel source that can be utilized to develop pharmacologically acceptable antimicrobial agents [35].

Other studies have characterized the main metabolites of the plants selected in this research. Phytochemical tests of *C. brasiliense* identified condensed tannins, hydrolyzed tannins, flavonoids, terpenoids, and saponins, which could contribute synergistically with antibacterial effects [28, 29]. In the *A. crassiflora* extracts, alkaloids, acetogenins, flavonoids, and phenolic compounds have previously been detected [36].

We used the disk-diffusion assay for the antimicrobial screening of plant extracts which showed simplicity and low cost as reported by Balouiri et al. [37]. However, bactericidal and bacteriostatic effects were not distinguished and special attention should be given to the standardization of bacterial inocula and microbial procedures to reduce the variability of measures of the inhibition zones. The selected leaf extracts of native plants from the Cerrado showed inhibitory effects against three bacterial species related to mastitis or colibacillosis in cattle; both diseases that have caused significant economic loss in cattle production in several continents. The *E. coli* strains used were resistant to tetracycline and the S178 *S. aureus* isolate was multi-resistant. These results could be explained by bovine herds infected with *S. aureus* and *E. coli* being frequently treated with antimicrobials to control mastitis or colibacillosis [2, 3].

The selected extracts from native plants of the Cerrado could prove to be alternative agents for the control of colibacillosis, mastitis, and other diseases associated with these bacteria, after toxicity studies and in vivo tests are performed. The specific metabolites present in these extract could be essential to control resistant or multi-resistant bacterial strains.

Conclusion

In this research, *Staphylococcus* spp. and *E. coli* were sensitive to leaf extracts of native plants from the Cerrado. Inter- and intra-specific bacterial variations were detected with regards to extract susceptibility. Notably, the *A. crassiflora* EE and *S. brasiliensis* extracts show potent bactericidal activity. After removal of the tannins, no antimicrobial effects were observed, indicating these metabolites are the main active antibacterial components.

Abbreviation

AE: Aqueous extract; EE: Ethanolic extract; MBC: Minimum bactericidal concentrations; MIC: Minimum inhibitory concentrations

Acknowledgements

To the Professor Anna Christina de Almeida who provided the bacterial strains of cow mastitis.

Funding

This study was supported by CAPES (Coordenação de Aperfeiçoamento de Pessoal de Nível Superior), Conselho Nacional de Desenvolvimento Científico e Tecnológico (CNPq), Fundação de Amparo à Pesquisa do Estado de Minas Gerias (FAPEMIG) and Pró-reitoria de Pesquisa da Universidade Federal de Minas Gerais.

Authors' contributions

ERD conceived, designed and coordinated the study. ICOR, EGAM, FMC, FMCA, MRS participated in data collection, analyses and drafting the manuscript. ERD and ICOR finalized and submitted the manuscript for publication. All the authors revised and approved the final manuscript.

Consent for publication

Not applicable.

Competing interests

The authors of this manuscript have no financial or personal relationships with other people or organizations that could inappropriately influence or bias the content of the paper.

Author details

[1]Instituto de Ciências Agrárias, Universidade Federal de Minas Gerais, Avenida Universitária, 1000, Bairro Universitário, Montes Claros, Minas Gerais CEP 39401-790, Brazil. [2]Escola de Medicina Veterinária, Universidade Federal de Minas Gerais, Av. Antonio Carlos, 6627 Pampulha, Belo Horizonte, MG CEP 31270-901, Brazil. [3]Instituto de Ciências Agrárias, Universidade Federal de Minas Gerais, Av Universitária 1000, Bairro Universitario, Montes Claros, MG 39400-006, Brazil.

References

1.   Salvadori MR, Valadares GF, Leite DS, Blanco J, Yano T. Virulence factors of *Escherichia coli* isolated from calves with diarrhea in Brazil. Brazil J Microbiol. 2003;34:230–5.

2.   Duse A, Waller KP, Emanuelson U, Unnerstad HE, Persson Y, Bengtsson B. Risk factors for antimicrobial resistance in fecal *Escherichia coli* from preweaned dairy calves. J Dairy Sci. 2015;98:500–16.

3.   Su X, Howell AB, D'Souza DH. Antibacterial effects of plant-derived extracts on methicillin-resistant *Staphylococcus aureus*. Foodborne Pathog Dis. 2012;9:573–8.

4.   Saidi R, Khelef D, Kaidi R. Bovine mastitis: prevalence of bacterial pathogens and evaluation of early screening test. Afr J Microbio Res. 2013;7:777–82.

5.   Toyang NJ, Wanyama J, Nuwanyakpa M, Django S. Ethnoveterinary medicine: a practical approach to the treatment of cattle diseases in sub-Saharan 2 ed Africa Roosendaal. Netherlands: Agromisa Foundation and CTA; 2007.

6.   Sampimon OC, Lam TJGM, Mevius DJ, Schukken YH, Zadoks RN. Antimicrobial susceptibility of coagulase-negative staphylococci isolated from bovine milk samples. Vet Microbiol. 2011;150:173–9.

7.   Samoilova Z, Smirnova G, Muzyka N, Oktyabrskya O. Medicinal plant extracts variously modulate susceptibility of *Escherichia coli* to different antibiotics. Microbiol Res. 2014;169:307–13.

8.   Ammer MR, Zaman S, Khalid M, Bilal M, Erum S, Huang D Che S. Optimization of antibacterial activity of *Eucalyptus tereticornis* leaf extracts against *Escherichia coli* through response surface methodology. J Radiation Res Appl Sc. 2016;9:376–85.

9.   Pereira AV, Silva VA, Freitas AFR, Pereira MSV, Trevisan LFA, Costa MRM. Extratos vegetais: atividade antimicrobiana e genética sobre plasmídios de resistência a antibióticos em microrganismos. Rev Biol Farmac. 2010;4:60–5.

10. Mordmuang A, Voravuthikunchai SP. *Rhodomyrtus tomentosa* (Aiton) Hassk leaf extract: an alternative approach for the treatment of staphylococcal bovine mastitis. Res Vet Sci. 2015;102:242–6.

11. Mello CP, Santos SC. Taninos 4 ed In: Farmacognosia: da planta ao medicamento Porto Alegre: Editora Universitária/UFRGS/Ed da UFSC; 2002.

12. Silva Júnior MCS. 100 Árvores do Cerrado: guia de campo. Brasília: Rede de Sementes do Cerrado; 2005. p. 278.

13. Costa ES, Hiruma-Lima CA, Lima EO, Sucupira GC, Bertolin AO, Lolis SF, Andrade FDP, Vilegas W, Souza-Brito ARM. Antimicrobial activity of some medicinal plants of the cerrado, Brazil. Phytother Res. 2008;22:705–7.

14. Saraiva AM, Saraiva CL, Cordeiro RP, Soares RR, Xavier HS, Caetano N. Atividade antimicrobiana e sinérgica das frações das folhas de *Schinopsis brasiliensis* Engl frente a clones multirresistentes de *Staphylococcus aureus*. Rev Bras Plants Medic. 2013;15:99–207.

15. Chapaval L, Moon DH, Gomes JE, Duarte FR, Tsai SM. An alternative method for *Staphylococcus aureus* DNA isolation. Arq Bras Med Vet Zootec. 2008;60:299–306.

16. Lane DJ. 16S/23S rRNA sequencing. In: Stackebrandt E, Goodfellow M, editors. Nucleic acid techniques in bacterial systematics. Chichester: Wiley; 1991. p. 115–75.

17. Sanger F, Coulson AR. A rapid method for determining sequences in DNA by primed synthesis with DNA polymerase. J Mol Biol. 1975;94:441–8.

18. Reysenbach AL, Longnecker K, Kirshtein J. Novel bacterial and archaeal lineages from an *in situ* growth chamber deployed at a mid-atlantic ridge hydrothermal vent. Appl Environ Microbiol. 2000;66:3798–806.

19. NCCLS. Performance Standards for Antimicrobial Susceptibility Testing. Fifteenth Informational Supplement [Online] CLSI/NCCLS document M100-S15 [ISBN 1–56238–556-9] Clinical and Laboratory Standards Institute, 940 West Valley Road, Suite 1400, Wayne, Pennsylvania 19087–1898 USA. 2005. Available: http://www.anvisagovbr/servicosaude/manuais/clsi/clsi_OPASM100S15.pdf [Accessed 10 May 2015].

20. Alvares CA, Stape JL, Sentelhas PC, JLM G, Sparovek G. Köppen's climate classification map for Brazil. Meteorol Z. 2014;22:711–28.

21. Matos FJA. Introdução a Fitoquímica. UFC: Fortaleza; 2009. p. 150.

22. Nyman U, Joshi P, Madsen LB, Pinstrup M, Rajasekharan S, George V, Pushpangadan P. Ethnomedical information and in vitro screening for angiotensin-converting enzyme inhibition of plants utilized as traditional medicines in Gujarat, Rajasthan and Kerala (India). J Ethnopharmacol. 1998;60:247–63.

23. Hiermann A, Kartnig TH, Azzam S. Ein Beitrag zur quantitativen Bestimmung der Procyanidine in Crataegus. Sci Pharm. 1986;54:331–7.

24. CLSI. Performance Standards for Antimicrobial Susceptibility Testing. Wayne: Twenty-First Informational Supplement CLSI document M100-S21Clinical and Laboratory Standards Institute; 2011.

25. NCCLS. Methods for Dilution Antimicrobial Susceptibility Tests for Bacteria That Grow Aerobically. Approved Standard—Sixth Edition [Online] NCCLS document M7-A6 [ISBN 1–56238–486-4] NCCLS, 940 West Valley Road, Suite 1400, Wayne, Pennsylvania 19087–1898 USA, 2003. Available: (http://www.anvisagovbr/servicosaude/manuais/clsi/clsi_OPASM2-A8.pdf) [Accessed 10 May 2015].

26. Klancnik A, Piskernik S, Jersek B, Mozina SS. Evaluation of diffusion and dilution methods to determine the antibacterial activity of plant extracts. J Microbiol Methods. 2010;81:121–6.

27. Malanovic N, Lohner K. Gram-positive bacterial cell envelopes: the impact on the activity of antimicrobial peptides. Biochim Biophys Acta. 2016;1858:936–46.

28. Paula-Júnior W, Rocha FH, Donatti L, CMT F-p, Weffort-santos AM. Leishmanicidal, antibacterial, and antioxidant activities of Caryocar brasiliense Cambess leaves hydroEE. Rev Bras Farmacogn. 2006;16:625–30.

29. Miranda-Vilela AL, Resck IS, Grisolia CK. Antigenotoxic activity and antioxidant properties of organic and aqueous extracts of pequi fruit (*Caryocar brasiliense* Camb) pulp. Genet Mol Biol. 2008;31:956–63.

30. Lima MRF, Ximenes CPA, Luna JS, Sant'ana AEG. The antibiotic activity of some Brazilian medicinal plants. Braz Med Plants. 2006;16:300–6.

31. Silva JJ, Cerdeira CD, Chavasco JM, Cintra ABP, Silva CBP, Mendonça AN, Ishikawa T, Boriollo MFG, Chavasco JK. *In vitro* screening antibacterial activity of *Bedens pilosa* Linné and *Annona crassiflora* Mart against oxacillin resistant *Staphylococcus aureus* (ORSA) from the aerial environment at the dental clinic. Rev Inst Med Trop Sao Paulo. 2014;56:333–40.

32. Pinho L, Souza PNS, Macedo Sobrinho E, Almeida AC, Martins ER. Atividade antimicrobiana de extratos hidroalcoolicos das folhas de alecrim-pimenta, aroeira, barbatimão, erva baleeira e do farelo da casca de pequi. Cienc Rural. 2012;42:326–31.

33. Amaral LFB, Moriel P, Foglio MA, Mazzola PG. Caryocar brasiliense supercritical CO2 extract possesses antimicrobial and antioxidant properties useful for personal care products. BMC Complement Altern Med. 2014;73:14–73.

34. Scalbert A. Antimicrobial properties of tannins. Phytochemistry. 1991;30: 3875–83.

35. Cushnie TPT, Lamb AJ. Antimicrobial activity of flavonoids. Int J Antimicrobial Agents. 2005;26:343–56.

36. Roesler R, Malta LG, Carrasco LC, Holanda RB, Souza CAS, Pastore GM. Atividade antioxidante de frutas do Cerrado. Cienc Tecnol Aliment. 2007;27:53–60.

6

Lethal chondrodysplasia in a family of Holstein cattle is associated with a *de novo* splice site variant of *COL2A1*

Jørgen S. Agerholm[1*], Fiona Menzi[2], Fintan J. McEvoy[3], Vidhya Jagannathan[2] and Cord Drögemüller[2]

Abstract

Background: Lethal chondrodysplasia (bulldog syndrome) is a well-known congenital syndrome in cattle and occurs sporadically in many breeds. In 2015, it was noticed that about 12 % of the offspring of the phenotypically normal Danish Holstein sire VH Cadiz Captivo showed chondrodysplasia resembling previously reported bulldog calves. Pedigree analysis of affected calves did not display obvious inbreeding to a common ancestor, suggesting the causative allele was not a rare recessive. The normal phenotype of the sire suggested a dominant inheritance with incomplete penetrance or a mosaic mutation.

Results: Three malformed calves were examined by necropsy, histopathology, radiology, and computed tomography scanning. These calves were morphologically similar and displayed severe disproportionate dwarfism and reduced body weight. The syndrome was characterized by shortening and compression of the body due to reduced length of the spine and the long bones of the limbs. The vicerocranium had severe dysplasia and palatoschisis. The bones had small irregular diaphyses and enlarged epiphyses consisting only of chondroid tissue.

The sire and a total of four affected half-sib offspring and their dams were genotyped with the BovineHD SNP array to map the defect in the genome. Significant genetic linkage was obtained for several regions of the bovine genome including chromosome 5 where whole genome sequencing of an affected calf revealed a *COL2A1* point mutation (g. 32473300 G > A). This private sequence variant was predicted to affect splicing as it altered the conserved splice donor sequence GT at the 5′-end of *COL2A1* intron 36, which was changed to AT. All five available cases carried the mutant allele in heterozygous state and all five dams were homozygous wild type. The sire VH Cadiz Captivo was shown to be a gonadal and somatic mosaic as assessed by the presence of the mutant allele at levels of about 5 % in peripheral blood and 15 % in semen.

Conclusions: The phenotypic and genetic findings are comparable to a previously reported *COL2A1* missense mutation underlying lethal chondrodysplasia in the offspring of a mosaic French Holstein sire (Igale Masc). The identified independent spontaneous splice site variant in *COL2A1* most likely caused chondrodysplasia and must have occurred during the early foetal development of the sire. This study provides a first example of a dominant *COL2A1* splice site variant as candidate causal mutation of a severe lethal chondrodysplasia phenotype. Germline mosaicism is a relatively frequent mechanism in the origin of genetic disorders and explains the prevalence of a certain fraction of affected offspring. Paternal dominant *de novo* mutations are a risk in cattle breeding, especially because the ratio of defective offspring may be very high and be associated with significant animal welfare problems.

Keywords: Congenital, Malformation, Rare disease, Type II collagenopathy

* Correspondence: jager@sund.ku.dk
[1]Department of Large Animal Sciences, Faculty of Health and Medical Sciences, University of Copenhagen, Dyrlægevej 68, Frederiksberg C DK-1870, Denmark
Full list of author information is available at the end of the article

Background

Chondrodysplasia is a developmental bone defect occurring due to disturbed endochondral osteogenesis. This leads to a reduced longitudinal growth of bones such as those of the limbs, spine and face. The most severe forms are found in lethal congenital generalized chondrodysplasia, which was originally reported in Dexter cattle in 1904 [1] and usually referred to as "bulldog calves." Congenital chondrodysplasia has been described in a number of cattle breeds with varying phenotype and different modes of inheritance (OMIA 000004-9913, OMIA 000187-9913, OMIA 000311-9913, OMIA 000189-9913).

Development in molecular genetics has revolutionized the research in bovine teratology as the methods to identify genetic causes have improved significantly [2]. The molecular basis of a few bovine chondrodysplasias has been determined. In 2007, two independent mutations affecting the coding region of the aggrecan (*ACAN*) gene causing lethal chondrodysplasia in Dexter cattle (OMIA 001271-9913) were described. These *ACAN* mutations showed incomplete dominant inheritance, leading to a mild form of dwarfism in heterozygotes, while homozygous animals displayed extreme chondrodysplasia and usually died during gestation [3]. For another lethal chondrodysplasia phenotype (OMIA 001926-9913) reported in Holstein cattle in 2004 [4] it was later demonstrated that the sire (Igale Masc) of affected calves was mosaic for a dominant acting collagen type II (*COL2A1*) missense mutation (pG600D) disrupting the Gly-X-Y structural motif essential for the assembly of the collagen triple-helix [5]. In 2015, cases of stillborn bulldog-like calves occurring among the offspring of the Danish Holstein sire VH Cadiz Captivo (DK256588) were reported to the Danish bovine genetic disease programme [6]. This study reports the pathological investigation of the condition and the genetic analyses that led to the identification of a candidate causal mutation.

Methods

Animals

Three Holstein calves were submitted for examination: Case 1: a male delivered at gestation day (GD) 271; Case 2: a female delivered at GD 273; Case 3: a male delivered at GD 269. The cases originated from different herds. Ear tissue from two additional cases, where the herd veterinarian diagnosed the condition, was also submitted. The Holstein bull VH Cadiz Captivo was registered as the sire in all cases. Ethylenediaminetraacetic acid (EDTA) stabilized blood from all five dams was also available as was semen and EDTA stabilized blood of the sire.

Post mortem examinations

Two calves (cases 2 and 3) underwent initial full body computed tomography (CT) scans using a single slice

helical CT machine (Emotion, Siemens, Erlangen, Germany) to obtain a full view of the bone malformations. Slice thickness was 3 mm and surface rendered reconstructions, which help illustrate bone malformations, were made using OsiriX software [7]. The calves were then necropsied, which included longitudinal sectioning of the right limbs and the spine. Specimens of heart, lung, liver, kidneys, adrenal glands, thymus, vertebrae, femur and humerus were fixed in 10 % neutral buffered formalin for histology. The left limbs and the head of all calves were frozen at -20 °C and later examined by radiology. The head was sectioned longitudinally though the midline before radiology to obtain better images. The limbs were sectioned longitudinally after radiology to check for bilateral similarities in gross bone morphology. Tissues for histology were processed by routine methods, embedded in paraffin, sectioned at 2–3 μm and stained with haematoxylin and eosin. Bone specimens were decalcified in a 3.3 % formaldehyde/17 % formic acid solution before processing.

Breeding analysis

Data on the outcomes of inseminations with semen of VH Cadiz Captivo in Denmark were obtained from the Danish Cattle Database. These included offspring recorded as "defective", "stillborn", "dead within 24 hours" or "still alive." A questionnaire was constructed and mailed to owners of offspring recorded as "defective", "stillborn" or "dead within 24 hours." The questionnaire included pictures of two chondrodysplastic calves sired by VH Cadiz Captivo, the ear tack number of the dam of the calf and the date of delivery. The owners were then contacted by phone to assess if the offspring was chondrodysplastic or not.

Data on the gestation period for all calves sired by VH Cadiz Captivo were obtained from the database and the length of the gestation period for chondrodysplastic vs. normal calves was compared using the Welch Two-sample t-test.

Genetic analysis

Genomic DNA was extracted for a total of 11 family members using standard protocols. For one of the affected calves the obtained quality of isolated DNA was very low due to tissue degradation. Therefore only a total of nine samples (four cases with their respective dams and VH Cadiz Captivo) were used for genotyping with the BovineHD BeadChip (Illumina), including 777,962 evenly distributed SNPs, at Geneseek.

MERLIN v 1.1.2 software [8] was used to analyze the dataset and carry out the linkage analysis. The Whittemore and Halpern non-parametric linkage pair statistics Z-mean and Kong and Cox logarithm of the odds (LOD) score were used to test for linkage using allele sharing among affected pedigree members without any assumptions on the mode of inheritance or on the genetic parameters of a specified model.

A polymerase chain reaction (PCR) free fragment library with a 390 base pair (bp) insert was prepared from one affected calf which was sequenced on half of a lane of an Illumina HiSeq3000 instrument using 2 × 150 bp paired-end reads. The genome sequencing data were deposited in the European Nucleotide Archive (ENA) (http://www.ebi.ac.uk/ena) under accession PRJEB12095. The mapping to the UMD3.1 bovine reference genome assembly and variant calling were undertaken as previously described [9]. During the mutation analysis, 118 sequenced genomes of normal cattle from 20 genetically diverse *Bos taurus* breeds were used as a local control cohort. The recent sequence variant database containing 1119 already sequenced genomes of the ongoing 1000 bull genomes project [5] was used as a global control cohort during filtering for private variants of the sequenced affected calf.

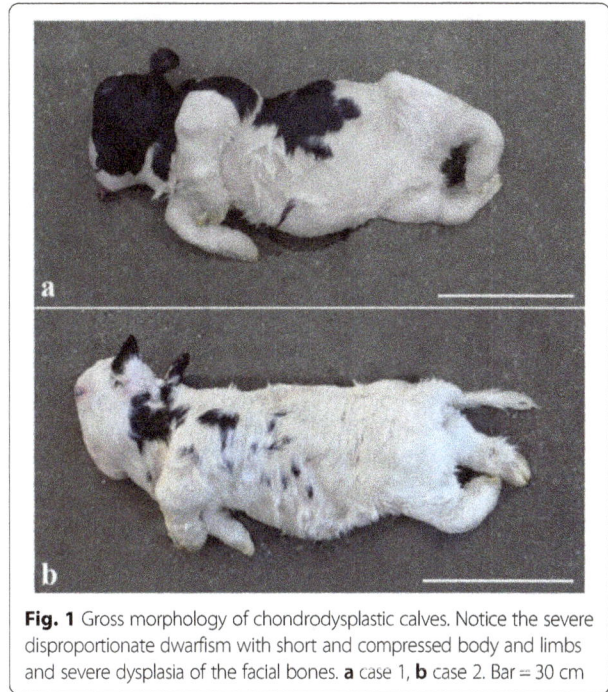

Fig. 1 Gross morphology of chondrodysplastic calves. Notice the severe disproportionate dwarfism with short and compressed body and limbs and severe dysplasia of the facial bones. **a** case 1, **b** case 2. Bar = 30 cm

Results

Phenotype

The three submitted calves were morphologically similar and displayed severe disproportionate dwarfism ("bulldog" syndrome). The body and the limbs were shortened and compressed due to reduced length of the spine and the long bones of the limbs. The distal parts of the limbs were rotated medially and had digits of almost normal size. The limbs were bilateral symmetrically malformed. The face had severe dysplasia with compression and ventral deviation of the facial bones, which had an almost 90° angle to the axis of the brain stem. The dorsal aspect of the cerebellum was compressed (Figs. 1 and 2a). Palatoschisis was present in all cases and the tongue protruded. The body weight was reduced to 26 kg (25–28 kg) compared to full term Danish Holstein calves of approximately 43.5 kg for males and 41.5 kg for females.

Longitudinal sectioning of the spine and long bones of the limbs showed small irregular diaphyses of increased hardness and significantly enlarged epiphyses consisting entirely of a homogenous chondroid rubber-like tissue. Distinct epiphyseal lines could not be identified as the irregular metaphyses opposed the chondroid epiphyses that lacked a center of ossification. The dorsally enlarged epiphyses of the vertebrae caused compression of the spinal cord.

The thorax and abdomen were of reduced volume and the enclosed organs were consequently closely opposed. The heart had biventricular myocardial hypertrophia and dilation of the right ventricle. Slight hydrothorax was present. The lungs had diffuse congenital atelectasis. The liver surface was irregular and the texture of the

parenchyma increased. Males had bilateral abdominal cryptorchidism.

Radiology and CT scanning showed that similar deformities were present in all calves. The neurocranium was relatively normal in size and shape (Fig. 2b). The splanchnocranium was foreshortened and misshapen resulting in severe superior brachygnathism. Mandibular incisors, premolars and molars were present. Maxillary teeth were also present but crowded due to the skull deformity. Bone structure appeared normal, with normal appearing cortical, cancellous and medullary cavities present in the mandible. Nasal and ethmoid turbinate bones were present. The appendicular skeleton was grossly deformed with the scapula, humerus radius, ulna, pelvis, femur, tibia, fibula and phalanges being affected. Only remnants of the diaphyses were identifiable. Bone morphology was so altered that they could only be identified by virtue of their relative location (Fig. 3). Deformities, especially shortening, were most severe for the proximal bones of the limbs while the distal phalanges were almost normal in shape and size. The carpus and tarsus were not ossified.

CT scanning showed lesions similar to those observed at radiology. In addition, there was widespread failure of fusion of the right and left parts of the dorsal spinous processes of the lumbar spine while partial fusion was present in the thoracic spine. Vertebral bodies were only partially formed in the cervical area and appeared as multiple osseous bodies, while vertebral bodies in the thoracic and lumbar spine were better defined but were nonetheless irregular in shape. Ribs

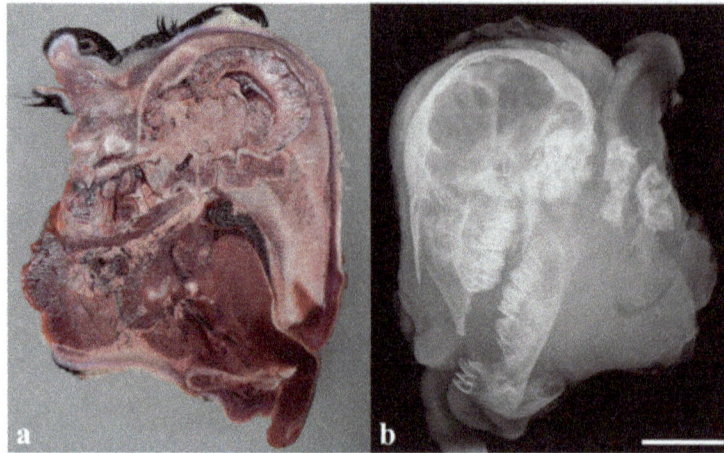

Fig. 2 Lesions in the head of a chondrodysplastic calf. **a** Longitudinal sectioning of the head through the midline showing severe dysplasia and 90° angular ventral deviation of the facial bones to the axis of the brain (Rostral is to the right). The cerebellum appears compressed dorsally. Frozen specimen, case 1. **b** Radiograph of the head shown in "a" (Rostral is to the left). The splanchnocranium is dysplastic resulting in severe superior brachygnathism. Bar = 5 cm

(13 pairs) were present and appeared normal (Fig. 4). A movie showing the malformation in a 360° view is presented in Additional file 1.

Histology showed highly irregular growth zones dominated by hypertrophied chondrocytes and with lack of normal alignment of chondrocytes although some chondrocytes were occasionally arranged normally in columns. Calcification of the intercellular matrix and chondrocyte degeneration occurred as irregular and interrupted areas towards the metaphysis. The ossification process was incomplete and cores of cartilage persisted throughout the meta- and diaphysis (Fig. 5). The epiphyses consisted of hyaline cartilage with chondrocytes located in variably sizes lacunae and with fibrous septae containing dilated vessels. Slight fibrosis and stasis was present in the liver.

Breeding analysis

Analysis of breeding data showed that HV Cadiz Captivo had 521 recorded offspring of which 412 were born alive and had received an ear tack number, i.e. regarded as normal. The owners of the remaining 109 calves were contacted by questionnaire and phone. Sixty-four owners reported that the offspring had been a "bulldog calf" while 39 reported that the calf was of normal proportions. Six calves were excluded as the owner did not respond or could not recall the morphology. HV Cadiz Captivo therefore produced "bulldog calves" at a ratio of 12.4 % (64/515). Pedigree analysis of affected calves did not display obvious inbreeding to a common ancestor.

The length of the gestation period between cows having normal offspring *vs.* "bulldog calves" was compared. The mean among the cows with a normal calving was 280.2 days, which was significantly different ($P < 0.0001$) from the mean among cows giving birth to "bulldog calves" (276.2 days).

Genetic analysis

A multipoint non-parametric linkage analysis detected 12 genomic regions located on nine different chromosomes significantly linked with the chondrodysplasia phenotype at chromosome-wide error probabilities for Z-mean values and LOD scores below 0.05 (Fig. 6; Additional file 2). In light of the few reports of mutations causing chondrodysplasia in livestock, a causative variant affecting one of the known candidate genes was hypothesized. An OMIM database search for osteochondrodysplasia and/or chondrodysplasia revealed a total of 95 candidate genes (Additional file 3) of which nine genes were located in linked genome regions (Fig. 6).

For mutation analysis the whole genome of one affected calf was sequenced to 16.9x coverage using next-generation sequencing technology. Genome-wide filtering for sequence variants in the whole genome that were present only in the affected calf and absent from 118 control cattle genomes (that were sequenced before in the course of other on-going projects) resulted in 6054 private variants (Additional file 4). Subsequent filtering of variants located apart from the linked genome regions allowed the exclusion of 87 % variants remaining with 785 variants including 11 coding. Finally the candidate genes present in the mapped linked regions were screened for possible variants remaining with a single splice site variant in the *COL2A1* gene (Chr 5 g.32473300 G > A). All five available affected calves, their dams and the sire were genotyped by direct

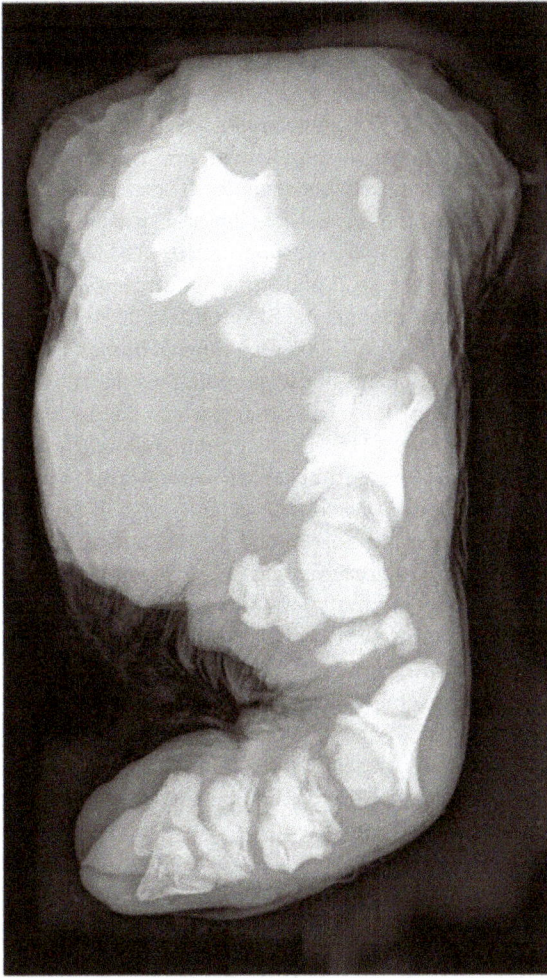

Fig. 3 Radiograph showing abnormally developed bones of a hind limb. Only remnants of the diaphyses can be identified and the individual bones can only be identified by virtue of their relative location. The proximal bones are most deformed, while the distal phalanges are almost normal in shape and size. Case 2

sequencing of a targeted PCR product using standard Sanger sequencing in the pedigree for this *COL2A1* variant revealing a perfect association of the mutant A allele with the chondrodysplasia phenotype (Fig. 7). All affected animals were heterozygous for the mutant allele and none of the dams carried this allele. Interestingly, sequencing showed the presence of the identified mutation in the sire VH Cadiz Captivo, who clearly carried the mutant allele at a low level in comparison to the wild type allele. The presence of both the wild type G allele and a smaller sized peak for the mutant A allele was detected in both available samples. The manually estimated relative area ratio of the mutant A allele was about 5 % in peripheral blood and about 15 % in semen (Fig. 7).

Discussion

The gross, microscopic and radiographic lesions in the examined offspring of the sire VH Cadiz Captivo are consistent with generalized congenital chondrodysplasia, a condition usually referred to as "bulldog calf syndrome." Pedigree analysis of affected calves did not display obvious inbreeding to a common ancestor, suggesting the causative allele is not a rare recessive. The normal phenotypes of the sire and the dams suggest a dominant inheritance with incomplete penetrance or a mosaic mutation. Due to the lethal effect of the mutation we concentrated on variants that were located within the coding sequences or within the splice sites of the candidate genes in the identified linked region of the bovine genome. Finally, mutation analysis for chondrodysplasia detected a perfect association between the splice site mutation of *COL2A1* and the disease phenotype. The *COL2A1* variant (c.2463 + 1G > A) is predicted to affect splicing because it alters the conserved splice donor sequence GT at the 5'-end of intron 36, which was changed to AT. The predicted consequences are either skipping of exon 36 or retention of intron 36. Both possible scenarios will lead to aberrant splice variants encoding truncated proteins, but nonsense-

Fig. 4 Surface rendered computed tomography images of a chondrodysplastic calf. The scanning data are rendered with bone and soft tissue surfaces. The latter has been set to have a degree of transparency in the reconstruction thus allowing visualisation of the calf's overall morphology and its relation to the underlying skeletal abnormalities. Case 3

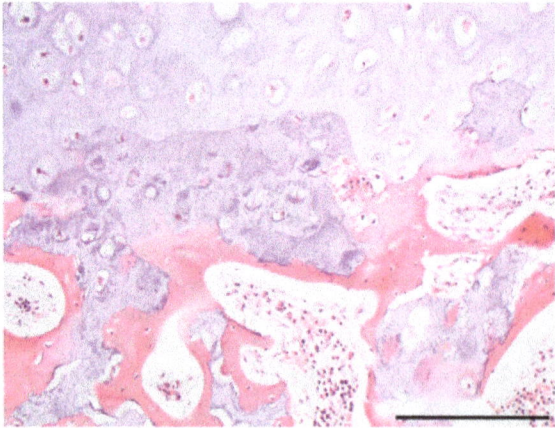

Fig. 5 Photomicrograph of the growth zone. The growth zone is characterized by irregular areas a granular basophilic appearance and degeneration of chondrocytes. Notice absence of the alignment of chondrocytes normally seen in a growth zone. Large cores of chondroid matrix persists into the metaphysis. Vertebra, case 1, haematoxylin and eosin, bar = 200 µm

mediated decay might be the most likely major consequence of the splice site mutation. Unfortunately, the RNA was too degraded by the time the calves were available for examination so reverse transcription PCR failed to verify the consequence of the sequence variant experimentally.

Heterozygous mutations of *COL2A1* (OMIM 120140) have been identified in human patients with various rare autosomal dominant conditions characterized by skeletal dysplasia, short stature, and sensorial defects collectively termed as type II collagenopathies [10, 11]. These disorders not only impair skeletal growth but also cause ocular and otolaryngological abnormalities. The classical phenotypes include the spondyloepiphyseal dysplasia (SED) spectrum with variable severity from lethal SED, including achondrogenesis type II and hypochondrogenesis (OMIM 200610), through congenital SED (OMIM 200160) to late-onset SED (OMIM 271700). One third of human *COL2A1* defects are dominant-negative mutations in the triple-helical region of alpha 1 (II) chains which disrupt the collagen triple helix [11]. These predominantly glycine to nonserine residue substitutions exclusively create more severe phenotypes like achondrogenesis type II and hypochondrogenesis and correspond to the bovine *COL2A1* allele causing chondrodysplasia in Holstein cattle [5]. In humans, splice-site mutations are assumed to cause haploinsufficiency through nonsense-mediated mRNA decay [10]. The *COL2A1* mutations resulting in a premature stop codon are found in less severe phenotypes such as Stickler dysplasia type I (OMIM 108300) characterized by ocular, auditory, skeletal, and orofacial abnormalities or Kniest dysplasia (OMIM 156550) characterized with short stature, restricted joint mobility, and blindness [10, 11].

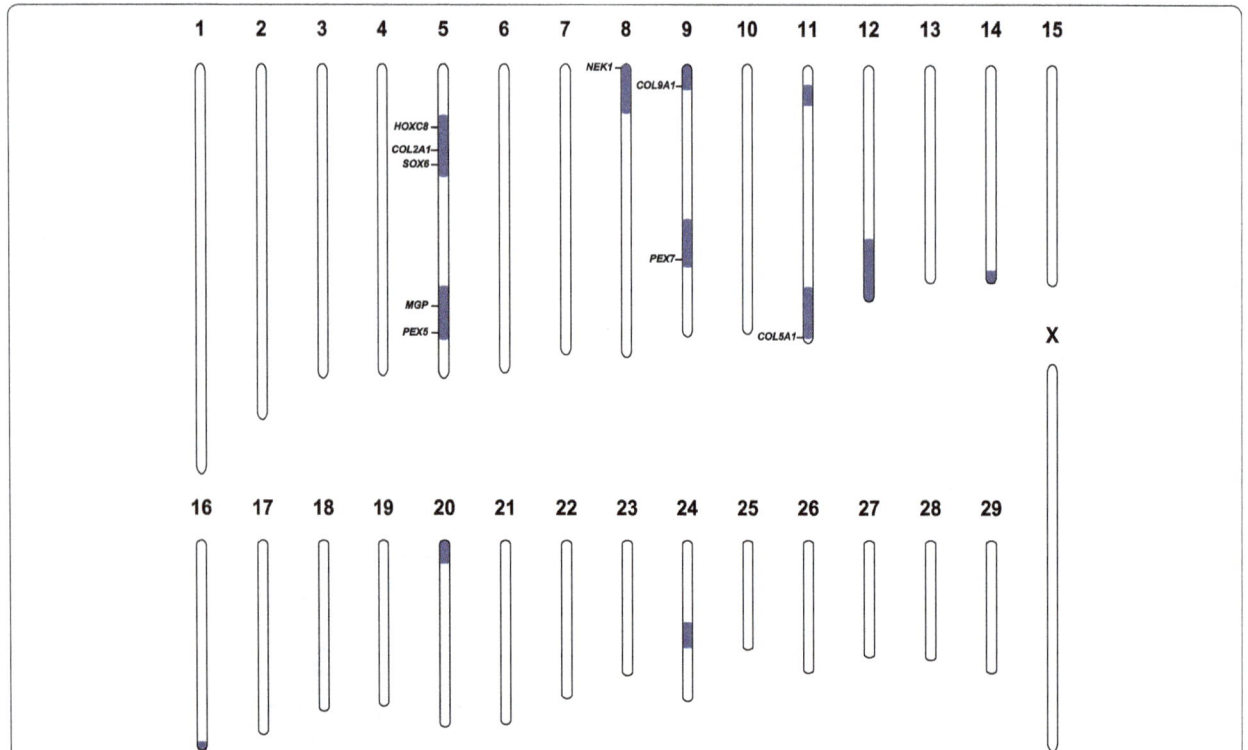

Fig. 6 Non-parametric multipoint linkage analysis for chondrodysplasia. A total of 12 significantly linked genome regions are shown in blue. Note, that 9 out of 95 candidate genes for chondrodysplasia including *COL2A1* are located in linked regions

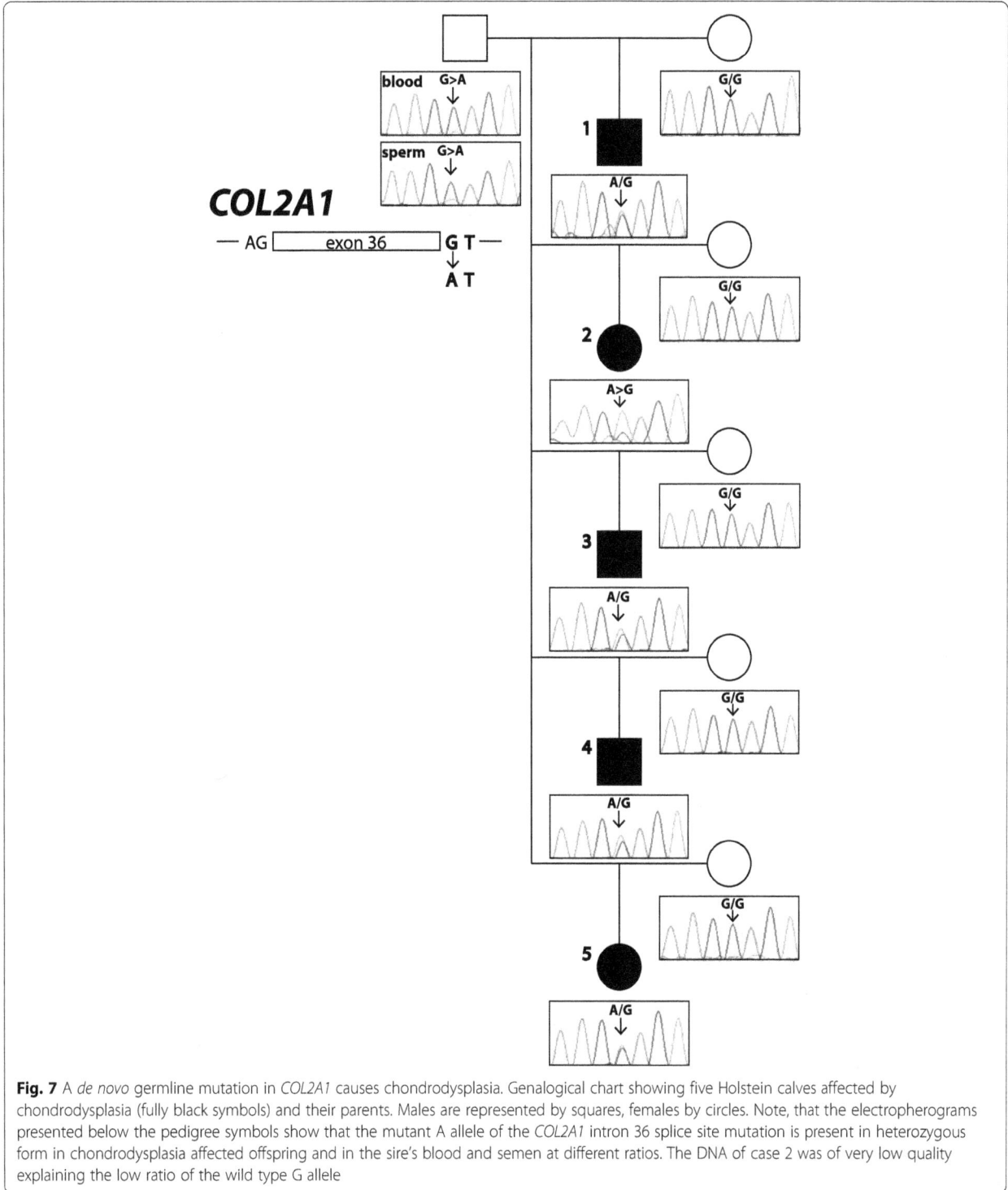

Fig. 7 A *de novo* germline mutation in *COL2A1* causes chondrodysplasia. Genalogical chart showing five Holstein calves affected by chondrodysplasia (fully black symbols) and their parents. Males are represented by squares, females by circles. Note, that the electropherograms presented below the pedigree symbols show that the mutant A allele of the *COL2A1* intron 36 splice site mutation is present in heterozygous form in chondrodysplasia affected offspring and in the sire's blood and semen at different ratios. The DNA of case 2 was of very low quality explaining the low ratio of the wild type G allele

Only in two cases, splice site mutations, similar to the mutation identified in this study, are shown to produce phenotypes at the lethal end of the type II collagenopathy spectrum [12, 13]. Therefore, it seems to be highly likely that the detected bovine *COL2A1* splice site variant explains the phenotype in the affected calves.

Probably a functional haploinsufficiency of collagen II, resulting from degradation of mutant transcripts due to nonsense-mediated mRNA decay, leads to a quantitative deficit of structurally normal collagen II causing the major malformations observed in the affected calves.

This lethal syndrome was present in 12.4 % of VH Cadiz Captivo offspring and was associated with the *COL2A1* splice site variant being present in his germ cells, presumable in around 15 % corresponding to the prevalence of malformed offspring. Therefore this sire is confirmed as a founder mosaic as shown before for "Solid Gold" for ovine callipyge [14] and "Campus" for porcine myopathy [15]. On the other hand, the genetic analysis of blood from the sire showed that the mutant *COL2A1* allele was present in peripheral blood cells at a low level of about 5 %. These results reveal that VH Cadiz Captivo is a gonadal and somatic mosaic. Germline mosaicism is a relatively frequent mechanism in the origin of genetic disorders [16]. Depending on various factors, such as the gene involved and/or the degree of mosaicism, the carrier of somatic and germline mosaicism may be asymptomatic or may present with various symptoms. There are two possibilities for the existence of such a mosaicism: one is that the mutation occurs in a germ cell that continues to divide. The other possibility is that the mutation occurs very early in a somatic cell before the separation to germinal cells and is therefore present both in somatic and germinal cells [16]. This seems to be the most likely explanation for VH Cadiz Captivo as he carried the mutant allele in both somatic and germinal cells, although he appeared normal.

A similar chondrodysplastic phenotype was observed in the offspring of the Holstein sire Igale Masc but affected only 1–2 % of his offspring [4, 5], which indicates that the mutation probably occurred later in the development of the testicular germ cells than in VH Cadiz Captivo. These findings show that the proportion of malformed calves may vary considerably between offspring of sires carrying germ cell mutations.

Paternal dominant germ cell mutations is a risk in cattle breeding, especially because the ratio of defective offspring may be very high. Presence of a dominant mutation in a certain proportion of spermatozoa has been hypothesized in cases of osteogenesis imperfecta. Denholm and Cole [17] found this lethal bone disorder in 44 % of a clinically normal Friesian bull mated to unrelated females, while Agerholm et al. [18] diagnosed osteogenesis imperfecta in 9 % of the offspring of a normal Holstein sire. Paternal dominant germ cell mutations are a challenge to cattle breeding as diseased progeny may occur without inbreeding, which in other ways is the most significant challenge regarding bovine genetic diseases. Reporting systems and disease programs that allow proper diagnostic of defective calves and targeted genetic investigations are highly valuable to limit the occurrence of severe disorders as lethal chondrodysplasia and osteogenesis imperfecta. The sire VH Cadiz Captivo was omitted for breeding by the breeding association as soon as the condition was diagnosed in his offspring.

It is important to recognize that even within a single breed, cases of lethal generalized chondrodysplasia show allelic heterogeneity. The cases reported here had a morphology indistinguishable from the Igale Masc bulldog calves [4] even by detailed post mortem examinations. A mutation in the same gene was therefore possible to be the cause. However, lethal congenital chondrodysplasia in Holsteins has been reported previously. Although such cases may share the basic lesion, i.e. chondrodysplasia, during examination in the field, significant variation in the phenotype may be found during post mortem examination, e.g. the syndrome reported by Berger and Innes [19] had severe hydrocephalus, which distinguishes it from the *COL2A1* associated "Igale Masc" and "VH Cadiz Captivo" types. Investigations and reports detailing characteristics of chondrodysplasia phenotypes are needed.

Conclusions

Genomic analyses identified a causative spontaneous dominant acting candidate causal mutation in *COL2A1*. This study provides a first example of a dominant *COL2A1* splice site variant associated with a severe lethal chondrodysplasia phenotype. This mutation must have occurred during the early development of the asymptomatic sire VH Cadiz Captivo. Germline mosaicism as a relatively frequent mechanism in the origin of genetic disorders explains the prevalence of a certain fraction of affected offspring. Paternal dominant *de novo* mutations are a risk in cattle breeding, especially because the ratio of defective offspring may be very high and be associated with significant animal welfare problems. It is therefore important to have surveillance programmes for congenital syndromes in cattle.

Additional files

Additional file 1: Surface rendered computed tomography scanning movie of a chondrodysplastic calf.

Additional file 2: Non-parametric multipoint linkage analysis output data. Z-mean values and LOD scores and their chromosome-wide error probabilities (*P*) along a grid of equally spaced locations (2 Mb) of all 29 autosomes. The minimum and maximum achievable values in the present linkage analyses are given in the first two rows. Chromosome-wide significant Z-mean values and LOD scores and their *P*-values are highlighted in green.

Additional file 3: Candidate genes for chondrodysplasia. OMIM database search for genes associated with osteochondrodysplasia/chondrodysplasia and their genomic position in the cattle genome.

Additional file 4: Private sequence variants of the sequenced chondrodysplasia affected calf. A total of 6054 sequence variants which were absent in 118 control genomes. Variants (*n* = 785) located in the linked genome regions are highlighted in green.

Abbreviations

ACAN, aggrecan gene; bp, base pair; CT, computed tomography; EDTA, ethylenediaminetraacetic acid; ENA, European Nucleotide Archive; GD, gestation day; LOD, logarithm of the odds; PCR, polymerase chain reaction; SED, spondyloepiphyseal dysplasia

Acknowledgements

The breeders and their veterinarians are acknowledged for referring the cases. Ms L Pedersen and Ms AB Brandt are thanked for performing the telephone interviews. Mrs H Holm is thanked for preparation of tissues sections and Mrs CS Due and Mrs MK Carlsen for technical assistance in radiology. We acknowledge the Next Generation Sequencing Platform of the University of Bern for performing the sequencing analyses. The study was supported by Seges Cattle, Aarhus, Denmark through funding of the Danish Bovine Genetic Disease Programme.

Authors' contributions

JSA was study leader, examined the defective calves, designed the questionnaire and analysed the breeding data. FJM did the CT scanning and radiology. FM performed linkage analysis, did the COL2A1 genotyping and prepared the genetic figures. VJ did the bioinformatics of the whole genome sequencing experiment. CD supervised the genetic investigations. All authors participated in writing the manuscript and have read and approved the final version.

Competing interests

The authors declare that they have no competing interests.

Consent for publication

Not applicable.

Author details

[1]Department of Large Animal Sciences, Faculty of Health and Medical Sciences, University of Copenhagen, Dyrlægevej 68, Frederiksberg C DK-1870, Denmark. [2]Institute of Genetics, Vetsuisse Faculty, University of Bern, Bremgartenstrasse 109a, Bern CH-3001, Switzerland. [3]Department of Veterinary Clinical and Animal Sciences, Faculty of Health and Medical Sciences, University of Copenhagen, Dyrlægevej 16, Frederiksberg C DK-1870, Denmark.

References

1. Seligmann MB. Cretinism in calves. J Pathol Bacteriol. 1904;9:311–22.
2. Nicholas FW, Hobbs M. Mutation discovery for Mendelian traits in non-laboratory animals: a review of achievements up to 2012. Anim Genet. 2014; 45:157–70.
3. Cavanagh JA, Tammen I, Windsor PA, Bateman JF, Savarirayan R, Nicholas FW, Raadsma HW. Bulldog dwarfism in Dexter cattle is caused by mutations in ACAN. Mamm Genome. 2007;18:808–14.
4. Agerholm JS, Arnbjerg J, Andersen O. Familial chondrodysplasia in Holstein calves. J Vet Diagn Invest. 2004;16:293–8.
5. Daetwyler HD, Capitan A, Pausch H, Stothard P, van Binsbergen R, Brøndum RF, et al. Whole-genome sequencing of 234 bulls facilitates mapping of monogenic and complex traits in cattle. Nat Genet. 2014;46:858–65. doi:10.1038/ng.3034.
6. Agerholm JS, Basse A, Christensen K. Investigations on the occurrence of hereditary diseases in the Danish cattle population 1989-1991. Acta Vet Scand. 1993;34:245–53.
7. Rosset A, Spadola L, Ratib O. OsiriX: An open-source software for navigating in multidimensional DICOM images. J Digit Imaging. 2004;17:205–16.
8. Abecasis GR, Cherny SS, Cookson WO, Cardon LR. Merlin-rapid analysis of dense genetic maps using sparse gene flow trees. Nat Genet. 2002;30:97–101.
9. Murgiano L, Shirokova V, Welle MM, Jagannathan V, Plattet P, Oevermann A, et al. Hairless streaks in cattle implicate TSR2 in early hair follicle formation. PLoS Genet. 2015;11:e1005427.
10. Nishimura G, Haga N, Kitoh H, Tanaka Y, Sonoda T, Kitamura M, et al. The phenotypic spectrum of COL2A1 mutations. Hum Mutat. 2005;26:36–43.
11. Barat-Houari M, Sarrabay G, Gatinois V, Fabre A, Dumont B, Genevieve D, et al. Mutation update for COL2A1 gene variants associated with type II collagenopathies. Hum Mutat. 2016;37:7–15.
12. Körkkö J, Cohn DH, Ala-Kokko L, Krakow D, Prockop DJ. Widely distributed mutations in the COL2A1 gene produce achondrogenesis type II/ hypochondrogenesis. Am J Med Genet. 2000;92:95–100.
13. Mortier GR, Weis M, Nuytinck L, King LM, Wilkin DJ, De Paepe A, et al. Report of five novel and one recurrent COL2A1 mutations with analysis of genotype-phenotype correlation in patients with a lethal type II collagen disorder. J Med Genet. 2000;37:263–71.
14. Smit M, Segers K, Carrascosa LG, Shay T, Baraldi F, Gyapay G, et al. Mosaicism of Solid Gold supports the causality of a noncoding A-to-G transition in the determinism of the callipyge phenotype. Genetics. 2003; 163:453–6.
15. Murgiano L, Tammen I, Harlizius B, Drögemüller C. A de novo germline mutation in MYH7 causes a progressive dominant myopathy in pigs. BMC Genet. 2012;13:99.
16. Zlotogora J. Germ line mosaicism. Hum Genet. 1998;102:381–6.
17. Denholm LJ, Cole WG. Heritable bone fragility, joint laxity and dysplastic dentin in Friesian calves: a bovine syndrome of osteogenesis imperfecta. Aust Vet J. 1983;60:9–17.
18. Agerholm JS, Lund AM, Bloch B, Reibel J, Basse A, Arnbjerg J. Osteogenesis imperfecta in Holstein-Friesian calves. Zentralbl Veterinarmed A. 1994;41: 128–38.
19. Berger J, Innes JR. Bull-dog calves (chondrodystrophy, achondroplasia) in a Friesian herd. Vet Rec. 1948;60:57.

7

Spatial and temporal distribution of lumpy skin disease outbreaks in Uganda (2002–2016)

Sylvester Ochwo[1], Kimberly VanderWaal[2*], Anna Munsey[2], Christian Ndekezi[1], Robert Mwebe[3], Anna Rose Ademun Okurut[3], Noelina Nantima[3] and Frank Norbert Mwiine[1]

Abstract

Background: Lumpy skin disease (LSD) is a devastating transboundary viral disease of cattle which causes significant loss in production. Although this disease has been reported in Uganda and throughout East Africa, there is almost no information about its epidemiology, spatial or spatio-temporal distribution. We carried out a retrospective study on the epidemiology of LSD in Uganda between the years 2002 and 2016, using data on reported outbreaks collected monthly by the central government veterinary administration. Descriptive statistics were computed on frequency of outbreaks, number of cases, vaccinations and deaths. We evaluated differences in the number of reported outbreaks across different regions (agro-ecological zones), districts, months and years. Spatial, temporal and space-time scan statistics were used to identify possible epidemiological clusters of LSD outbreaks.

Results: A total of 1161 outbreaks and 319,355 cases of LSD were reported from 55 out of 56 districts of Uganda. There was a significant difference in incidence between years ($P = 0.007$) and across different regions. However, there was no significant difference in the number of outbreaks per month ($P = 0.443$). The Central region reported the highest number of outbreaks ($n = 418$, 36%) followed by Eastern ($n = 372$, 32%), Southwestern ($n = 140$, 12%), Northern ($n = 131$, 11%), Northeastern ($n = 37$, 3%), Western ($n = 41$, 4%) and Northwestern ($n = 22$, 2%) regions. Several endemic hotspots for the circulation of LSD were identified in the Central and Eastern regions using spatial cluster analyses. Outbreaks in endemic hotspots were less seasonal and had strikingly lower mortality and case-fatality rates than the other regions, suggesting an underlying difference in the epidemiology and impact of LSD in these different zones.

Conclusion: Lumpy Skin disease is endemic in Uganda, with outbreaks occurring annually in all regions of the country. We identified potential spatial hotspots for LSD outbreaks, underlining the need for risk-based surveillance to establish the actual disease prevalence and risk factors for disease maintenance. Space-time analysis revealed that sporadic LSD outbreaks tend to occur both within and outside of endemic areas. The findings from this study will be used as a baseline for further epidemiological studies for the development of sustainable programmes towards the control of LSD in Uganda.

Keywords: Lumpy skin disease, Epidemiology, Agro-ecological zones, Spatio-temporal epidemiology, Uganda

* Correspondence: kvw@umn.edu
[2]College of Veterinary Medicine, University of Minnesota, 1365 Gortner Avenue, St. Paul, MN 55108, USA
Full list of author information is available at the end of the article

Background

Lumpy skin disease (LSD) is an acute to sub-acute viral disease of cattle that is defined by fever, increased nasal secretions, enlarged lymph nodes, formation of nodules on the skin, mucous membranes and internal organs, edema of the skin and sometimes death [1–3]. Lumpy skin disease virus (LSDV) is classified within the genus Capripoxvirus in the family Poxviridae, which includes the closely related viruses of sheep pox and goat pox [2]. Although cattle are the natural host of LSDV, clinical infection has been observed in the Asian water buffalo from Egypt [4] and antibodies have been reported in black and blue wildebeest, eland, giraffe, greater kudu, African buffalo, and other animal species [5, 6]. LSDV neither infects nor is it transmitted between sheep and goats [7].

LSD has been reported in a number of regions of Africa, where it is endemic, in the Middle East, and more recently in parts of Europe. There is a potential risk that LSDV could spread further into Europe and eventually worldwide [8–10]. In Eastern Africa, LSD was first reported in Kenya in 1957 [11], Sudan in 1972, and in Somalia in 1983 [12, 13]. There is no published literature about when LSD was first identified in Uganda, however the disease is thought to have spread from Southern Africa into Uganda between 1955 and 1960 [12]. The disease is currently present in all geographical regions of the country, with several outbreaks reported annually.

Outbreaks of LSD tend to be sporadic, and are likely dependent on animal movements, immune status of animals, and changes in weather patterns which affect vector populations. The main mode of transmission of LSDV is by mechanical arthropod vectors, such as biting flies, *Aedes aegypti* mosquitoes and three tick species belonging to the family Ixodidae (*Rhipicephalus (Boophilus) decoloratus, R. appendiculatus and Amblyomma hebraeum*) [14]. Predators, vermin and wild birds might also act as mechanical carriers of the virus [15, 16]. The virus can also be transmitted by fomites, such as equipment, clothing, and personnel [16]. Epidemics of LSD in non-endemic regions are reported to be associated with hot and wet seasons, as well as areas close to water bodies, swamps and marshlands that are conducive for breeding and multiplication of insects [17]. Spread of LSDV between farms and districts might be due to the lack of complete restriction of animal movements [18, 19].

LSD is known to cause substantial economic losses in the form of severe emaciation, lowered milk production, abortion, secondary mastitis, loss of fertility, extensive damage to hides leading to low quality of leather and loss of draught power from lameness [13, 20]. The morbidity rate in cattle can vary from 3 to 85% depending on the presence of insect vectors and host susceptibility. Mortality usually ranges between 1 to 5% [2], but can occasionally be as high as 20 to 85%. The disease is therefore a serious threat to the cattle farming community in endemic areas and is associated with trade restrictions following outbreaks. The livestock sector is one of Uganda's important growth sectors contributing about US $290 million to the total GDP. Livestock constitutes 17% of the agricultural GDP and is a source of livelihood to about 4.5 million people in the country [21]. Trade in cattle hides generates about US $17 million annually and has potential for continued growth if conditions like LSD, which lower the quality of hides and skins, can be managed [22].

The success of any disease control program depends on a clear understanding of the epidemiology of the disease [23]. This requires analysis of available data to understand the distribution and patterns of spread of the disease [24]. Little has been studied about the epidemiology of LSD in Uganda, yet cattle farmers, district veterinary authorities, and monthly surveillance reports indicate the presence and impact of the disease in the country. Therefore, this study was conducted to describe the temporal and spatial distribution of reported outbreaks of LSD from 2002 to 2016 and to generate baseline epidemiological information on LSD in Uganda, which will facilitate further studies on disease prevalence and risk factors.

Methods

Study area

Uganda is a landlocked country located on the East African Plateau; it lies between latitudes 4°N and , and longitudes 29°35'E, with an area of about 241,038 km². It is bordered by Kenya to the east, the Democratic Republic of the Congo to the west, South Sudan to the north, and Rwanda and Tanzania to the south. The southern part of the country includes a considerable portion of Lake Victoria, which is shared with Kenya and Tanzania (Fig. 1). Uganda lies within the Nile basin as well as the African Great lakes region, and has a diverse but generally equatorial climate. It is on average about 1100 m (3,609 ft) above sea level. Currently, the country is divided into 112 districts; each district is sub-divided into counties and sub-counties; each sub-county consists of several parishes and villages. Uganda has a number of national parks, however the major parks considered in this study are: Bwindi Impenetrable National Park (BINP), Kibale National Park (KINP), Kidepo Valley National Park (KVNP), Lake Mburo National Park (LMNP), Mount Elgon National Park (MENP), Murchison Falls National Park (MFNP), and Queen Elizabeth National Park (QENP).

Data source and collection

Retrospective data on LSD outbreaks in Uganda during 2002–2016 were retrieved from the Ministry of Agriculture

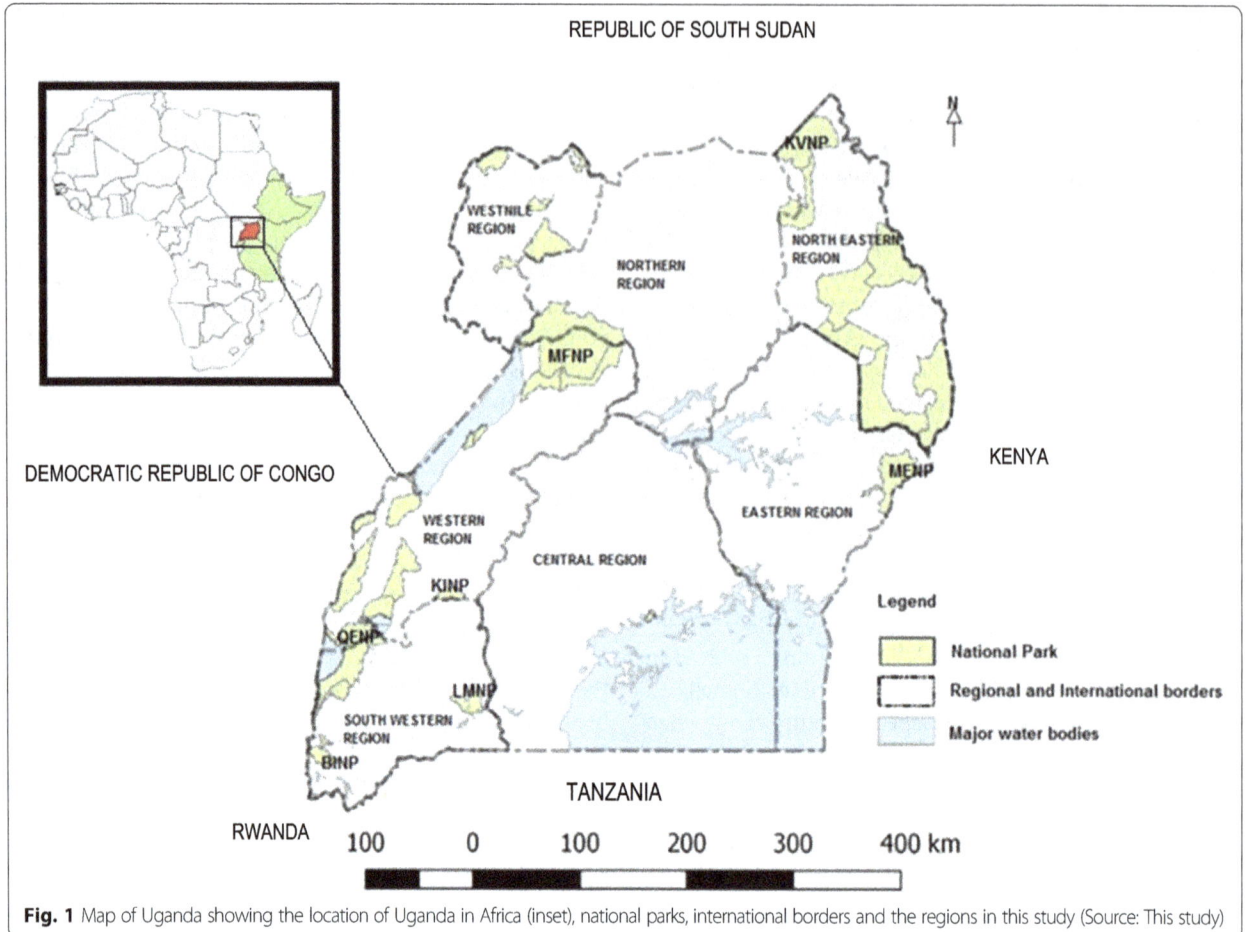

Fig. 1 Map of Uganda showing the location of Uganda in Africa (inset), national parks, international borders and the regions in this study (Source: This study)

Animal Industry and Fisheries (MAAIF), Uganda. This information is based on the monthly disease surveillance reports submitted to MAAIF by District Veterinary Officers (DVOs). For the period considered in this study, Uganda had a varying number of districts (56–112). Reporting of LSD outbreaks was at the district level, thus making it impossible to disaggregate data from earlier years (56 districts) into the present 112 districts. Thus, all the data analysed here were aggregated into 56 districts consistent with 2002 boundaries. Districts were classified as adjacent to international borders and national parks. For data handling and presentation, districts were also grouped into seven geographical agro-ecological regions, which typically vary by rainfall and farming production systems (Fig. 1) [25]. Data on livestock numbers were obtained from the Uganda National livestock census report 2009 [26], and used to calculate cattle density.

In this paper, a case was defined as an animal with clinical signs or nodular lesions characteristic of LSD (with or without laboratory confirmatory diagnosis). An outbreak was defined as the occurrence of one or more cases of LSD in a particular herd. However, cases from nearby herds with frequent animal contact or shared grazing areas were also considered part of the same outbreak. New outbreaks were defined as those occurring in a herd separated from other herds by a fence or physical barrier such as hills, water bodies, forests or mountains. Each outbreak report contained data on the number of affected animals, susceptible animals, vaccinated animals, and deaths. Since the exact locations of the affected herds were not recorded, geographic coordinates of each district were defined as the centroid of the district.

Data analysis

Descriptive analysis was performed on outbreaks, cases, and vaccination data. Count data on number of outbreaks and cases were tested for normal distribution using Shapiro-Wilk test and qqplots and found to be over-dispersed and positively skewed, with variance much larger than the mean. Therefore, Kruskal-wallis chi-squared tests and post-hoc Dunn's tests with Bonferroni corrections for p-values were carried out to assess whether the differences in number of outbreaks between regions, months and years were statistically significant. Spearman's correlation was used to investigate the relationship between cattle density and number of outbreaks and cases. QGIS version 2.18.9 with GRASS 7.2.1 [27] was used to plot the distribution of LSD outbreaks per district (2002–2016) and to create maps of the spatial and temporal distribution of LSD in Uganda. Software used for data analysis were Microsoft Office Excel, 2013 and R version 3.4.2.

Purely spatial, purely temporal, and space-time scan statistical analyses were performed using SaTScan™ v9.4.4 [28, 29]. The purely spatial scan statistic imposes a circular window of varying size upon the locations of possible outbreaks. The space-time scan statistic utilizes a dynamic cylindrical window, with a circular geographic base and with height corresponding to time. The purely temporal scan statistic uses a window with varying height corresponding to time, in the same way the height of the cylinder is used in the space-time scan. For each scan, the number of outbreaks in the window is recorded and compared to the null hypothesis of a random Poisson distribution, accounting for population size. A relative risk is calculated as the number of observed outbreaks within a window divided by the number of expected outbreaks across the study area. The window with the maximum log likelihood ratio (LLR) is defined as the most likely cluster. LLR is calculated by

$$LLR = \log\left(\frac{n}{E(n)}\right)^2\left(\frac{N-n}{N-E(n)}\right)^{(N-n)}I'$$

where N is the total number of cases; n is the observed number of cases within the scan window; $E(n)$ and $N - E(n)$ are the expected number of cases within and outside the window under the null hypothesis, respectively, and I is an indicator function (equal to 1 when the window has more cases than expected under the null hypothesis and 0 otherwise). Here, scans were conducted for areas of high rates, testing for elevated risk within a window as compared to outside. District centroids were tested as potential outbreak locations, and the maximum possible spatial and/or temporal cluster size was set to 50% of the total population at risk. Monte Carlo simulation ($n = 999$ permutations) was used to determine the significance of detected clusters [30].

Results

A total of 1161 LSD outbreaks were reported at the district level from January 1, 2002 to December 31, 2016, with an average of 77 (± 51.4 SD) outbreaks per year and a median of 70 outbreaks per year. During this 15-year period, 319,552 cases were recorded, with an average of 21,303 ± 4121 SD cases per year, and 2169 recorded deaths (average of 146 ± 17 SD deaths per year) attributed to LSD. Morbidity, mortality and case fatality rates were 4.77, 0.03 and 0.72%, respectively (Table 1).

Spatial distribution of LSD

The distribution of LSD at the district and regional level was mapped (Fig. 2) to represent the spatial pattern of outbreaks (2002–2016). The disease was reported in 55 out of 56 districts during this period. Lira ($n = 84$, 6.3%) and Tororo ($n = 83$, 6.3%) had the highest number of outbreaks (2002–2016) while Kisoro ($n = 1$, 0.075%), Mayuge ($n = 1$, 0.075%) and Ntungamo ($n = 1$, 0.075%) had the lowest numbers of outbreaks during the period studied. No LSD outbreaks were reported in Yumbe district. There was a significant difference between the numbers of outbreaks by region ($P < 0.002$), with the Central region ($n = 418$, 36%) reporting the highest number of outbreaks followed by the Eastern region ($n = 372$, 32%), Southwestern region ($n = 140$, 12%), Northern region ($n = 131$, 11%), Western region ($n = 41$, 4%), Northeastern region ($n = 37$, 3%), and Northwestern ($n = 22$, 2%) region. An additional file shows this in more detail [see additional file 1]. We found significant differences in number of outbreaks by region for the following pairs of regions; Central-West (Dunn's test, $p = 0.004$), Central-West Nile (Dunn's test, $p = 0.013$), North-West (Dunn's test, $p = 0.02$), and North-West Nile (Dunn's test, p = 0.02). Spearman's correlation showed a significant correlation between cattle density and number of outbreaks, and no significant correlation between cattle density and cases respectively ($r_s = 0.27$, p-value = 0.04; $r_s = 0.12$, p-value = 0.37 respectively). Seventeen (17) out of 56 districts adjacent to national parks reported only 45 (3.9%) outbreaks, while 332 (28.6%) outbreaks were reported in districts adjacent to international borders. When outbreaks in districts adjacent to national parks were compared according to which national park they bordered, we observed that districts bordering Queen Elizabeth National Park (QENP) reported a higher number of outbreaks than those reported by districts bordering the other six national parks (an additional file shows this in more detail [see additional file 2]).

Incidence of LSD outbreaks adjacent to the international Borders

The 22 of 56 districts adjacent to the international borders reported 332 (28.6%) LSD outbreaks as compared to 829 (71.4%) outbreaks from districts with no international border. The number of LSD outbreaks varied between the different international borders, the highest being adjacent with Kenya (157 outbreaks in 6 districts) and DRC borders (87 outbreaks in 7 districts), while 55 outbreaks were reported in 2 districts bordering Tanzania and 21 outbreaks were reported in 4 districts bordering South Sudan. The lowest number of LSD outbreaks was reported among the districts bordering Rwanda (12 outbreaks in 3 districts). Analysis of these differences by Kruskal Wallis test however revealed no

Table 1 Average annual number of outbreaks, morbidity, mortality and case fatality rates in different regions of Uganda. Population at risk refers to the number of susceptible cattle in herds where at least one case was reported

Region	No. of Outbreaks	Population at Risk	No. of Sick	No. of Dead	Morbidity rate (%)	Mortality rate (%)	Case fatality rate (%)
Central	28	306,452	5746	49	1.88	0.02	0.86
East	25	75,323	12,112	16	16.08	0.02	0.13
North	9	18,878	1440	15	7.63	0.08	1.02
North East	2	268	61	2	22.90	0.87	3.80
South West	9	27,786	1228	24	4.42	0.09	1.95
West	3	17,222	639	43	3.71	0.25	6.69
West Nile	1	647	77	4	11.97	0.64	5.34
Total	77	446,575	21,303	153	4.77	0.03	0.72

significant difference in the numbers of outbreaks per international border (p-value = 0.41).

Temporal distribution of LSD

On average, 22 districts (± 9.8 SD) experienced outbreaks of LSD each year. High annual incidences of LSD outbreaks were reported in 2002 (n = 182 outbreaks), 2003 (n = 153), 2004 (n = 117), 2011 (n = 110) and 2012

(n = 121) while the lowest annual incidence was reported in 2009 (n = 9) (Fig. 3). The highest incidence was reported in the month of January (n = 117 across all years), which accounted for 10% of all outbreaks reported, and the lowest in November (n = 80), accounting for 6.9% of all reported outbreaks. There was no significant difference in the incidence of outbreaks between months (p = 0.443). When the overall data were grouped into

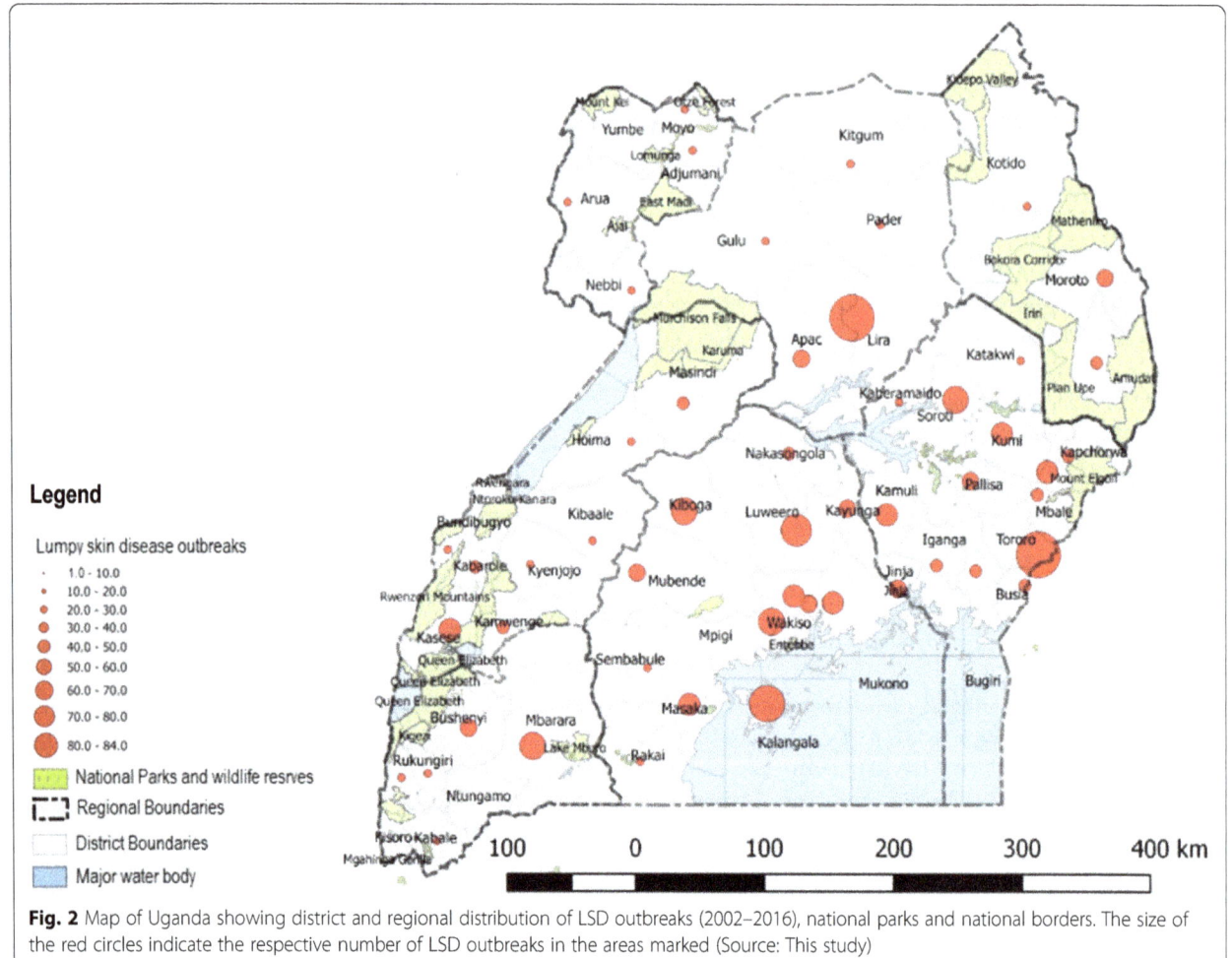

Fig. 2 Map of Uganda showing district and regional distribution of LSD outbreaks (2002–2016), national parks and national borders. The size of the red circles indicate the respective number of LSD outbreaks in the areas marked (Source: This study)

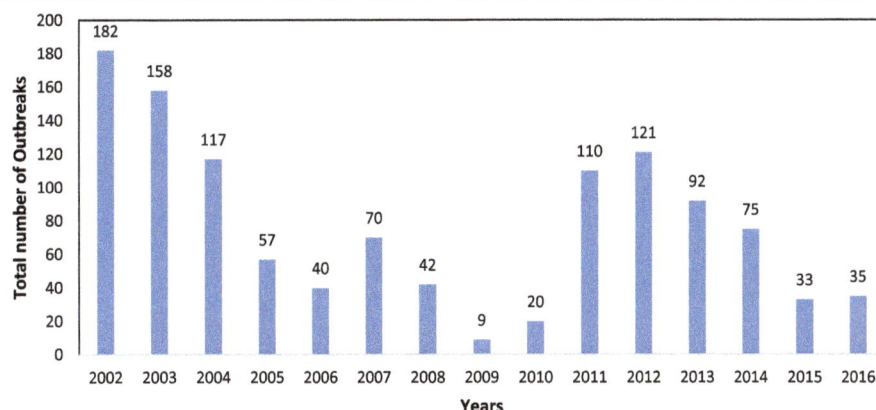

Fig. 3 Total yearly Lumpy skin disease outbreaks in Uganda from 2002 to 2016

four seasons, two wet seasons and two dry seasons, the highest incidence was reported in the first dry season (Dec–Feb, $n = 312$, 26.9%) followed by second dry season (Jun–Aug, $n = 300$, 25.8%), first wet season (Mar–May, $n = 286$, 24.6%) and second wet season which had the lowest incidence (Sep-Nov, $n = 263$, 22.7%). More marked intra-annual variation was observed when subdividing the analysis by region, with the northeastern (Fig. 4d), western (Fig. 4f), and West Nile region showing more profound seasonal patterns (Fig. 4g).

Purely spatial clusters of lumpy skin disease

The spatial pattern of LSD was found to be nonrandom. A total of 7 clusters were identified, two (2) of which were located in Central region, three (3) in Eastern region, one (1) in Southwestern region and one (1) in Northern region (Fig. 5). The most likely cluster was observed in the Kalangala district in Central Uganda. The radius of the cluster was 0 km, indicating that the cluster only included Kalangala district. The relative risk (RR) was 156.17, indicating that cattle within this district were around 156 times more likely to be affected by LSD than in areas outside the cluster (Table 2). The observed number of outbreaks for this cluster was 66 compared with a calculated 0.45 expected outbreaks. Secondary clusters were located in (Luwero, Kayunga, Wakiso, and Kampala), found in central Uganda; (Busia, Tororo), Jinja, (Kapchorwa, Sironko, Mbale and Kumi) in Eastern Uganda; Kasese in Southwestern Uganda; and Lira in Northern Uganda; (Table 2 and Fig. 5). RR for these clusters ranged from just over 1.8 to over 9.

Space-time clusters of lumpy skin disease

One space-time cluster was identified and it persisted for a duration of 3 years. This space-time cluster was located in 24 districts found in Eastern and Central region, and in 2 districts found in Northern region. This space-time cluster was from January 1, 2002 - December 31, 2005, with 383 observed outbreaks, compared to a calculated 137.97 expected outbreaks. The space-time cluster is shown in Table 3 and Fig. 6.

Purely temporal clusters of LSD

Temporal cluster analysis of LSD outbreaks in Uganda showed one peak period with only one cluster identified during January 1, 2002 to December 31, 2004. The overall RR within the cluster was 2.34 (LLR =85.92, $P = 0.001$) with 417 observed outbreaks compared to 226.24 expected outbreaks.

Discussion

To understand the spatial epidemiology of lumpy skin disease (LSD) outbreaks in Uganda, we described the geographic and temporal occurrence of LSD and analyzed the data for spatial and temporal clusters using retrospective data collected between 2002 and 2016. During this period, an average of 77 LSD outbreaks were reported across 22 (±9.8 SD) districts each year, demonstrating that LSD is endemic in Uganda.

Incidence of reported LSD outbreaks differed between regions, with more outbreaks reported in the Central and Eastern regions as compared to the rest of the regions. The Central and Eastern regions represented more than half of the reported LSD outbreaks during this time-frame. This marked difference could be due to a number of factors including animal husbandry practices, presence of high numbers of insect vectors, higher frequency of exotic cattle breeds, awareness of disease control, uncontrolled animal movements, and potentially biases in disease reporting related to proximity to the central administrative center of MAAIF in Kampala [24, 31, 32]. However, the primary contributor to the high rates of reported outbreaks in this region of Uganda may be climate, given that the Central and Eastern regions of Uganda form part of the Lake Victoria basin; these two regions also have other lakes (Kyoga, Opeta, and Bisina), rivers (Nile, Manafwa, Mpologoma,

a Central

b East

c North

d North East

e South Western

f Western

g West Nile

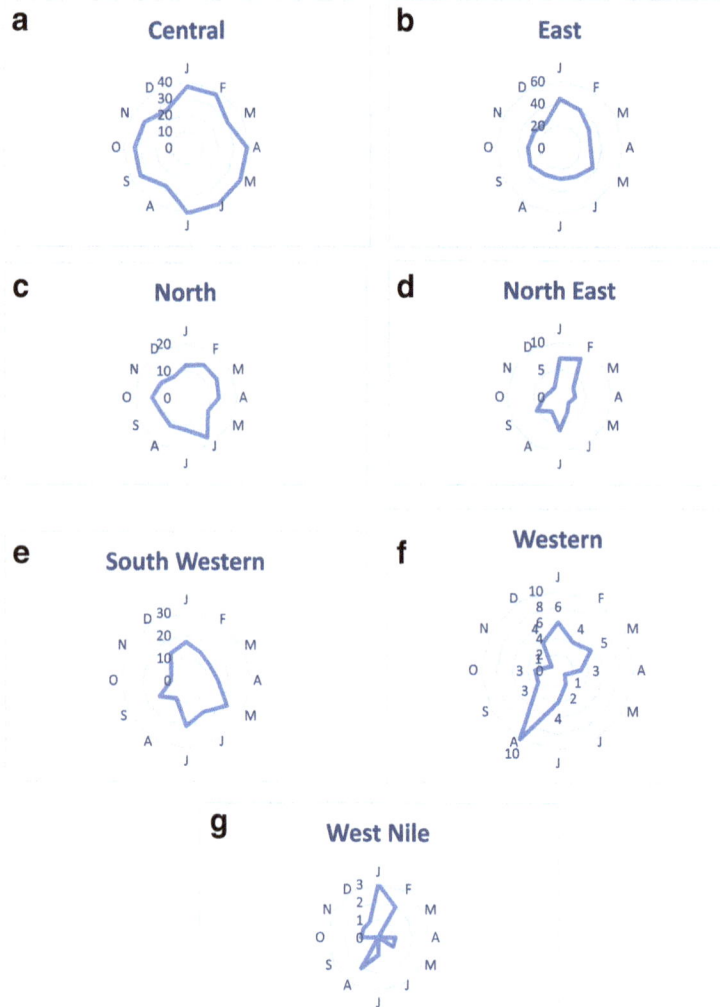

Fig. 4 Spider plots showing the monthly distribution of LSD outbreaks per region from 2002 to 2016

Malaba) and wetlands which provide wet and humid micro-climates [33, 34]. This, coupled with an average monthly temperature range of 22 °C–29 °C, provides suitable conditions for multiplication of arthropod vectors for LSD [14]. The only published studies about arthropod vectors in this region are studies reporting distribution of *Glossina spp*, [35, 36] which are known to transmit LSD mechanically [37]. These studies have found high density of these flies in the Central and Eastern regions of Uganda, thus suggesting that the climatic conditions are suitable for arthropod vector multiplication and survival.

Spatial, temporal and space-time scan statistics are tools used to detect aggregations of disease outbreaks or cases and identify whether these outbreaks or cases of disease in space or time can be explained by chance alone or are statistically significant. Clusters may occur due to local transmission of the disease or due to shared risk factors within an area. We investigated the spatial

distribution of LSD outbreaks and identified areas with high endemicity of LSD and clustering patterns using spatial scan statistics. We showed that in the period from 2002 to 2016 as a whole, the geographic distribution patterns of LSD outbreaks in Uganda were not random. Spatial cluster analysis identified 7 clusters, which were primarily located in the Central and Eastern regions. The most likely spatial cluster was observed in Kalangala district in Central Uganda. High incidence in this region is likely driven by climate and presence of wetlands in this district. Kalangala is a district made up of 84 islands surrounded by Lake Victoria, with 95% of the district area covered by water bodies, and mean annual rainfall ranging from 1125 to 2250 mm [38]. The climate of this district is generally moist and humid all through the year with moderately small seasonal variations of temperature, humidity, and wind throughout the year [38]. These conditions are known to maintain arthropod vectors

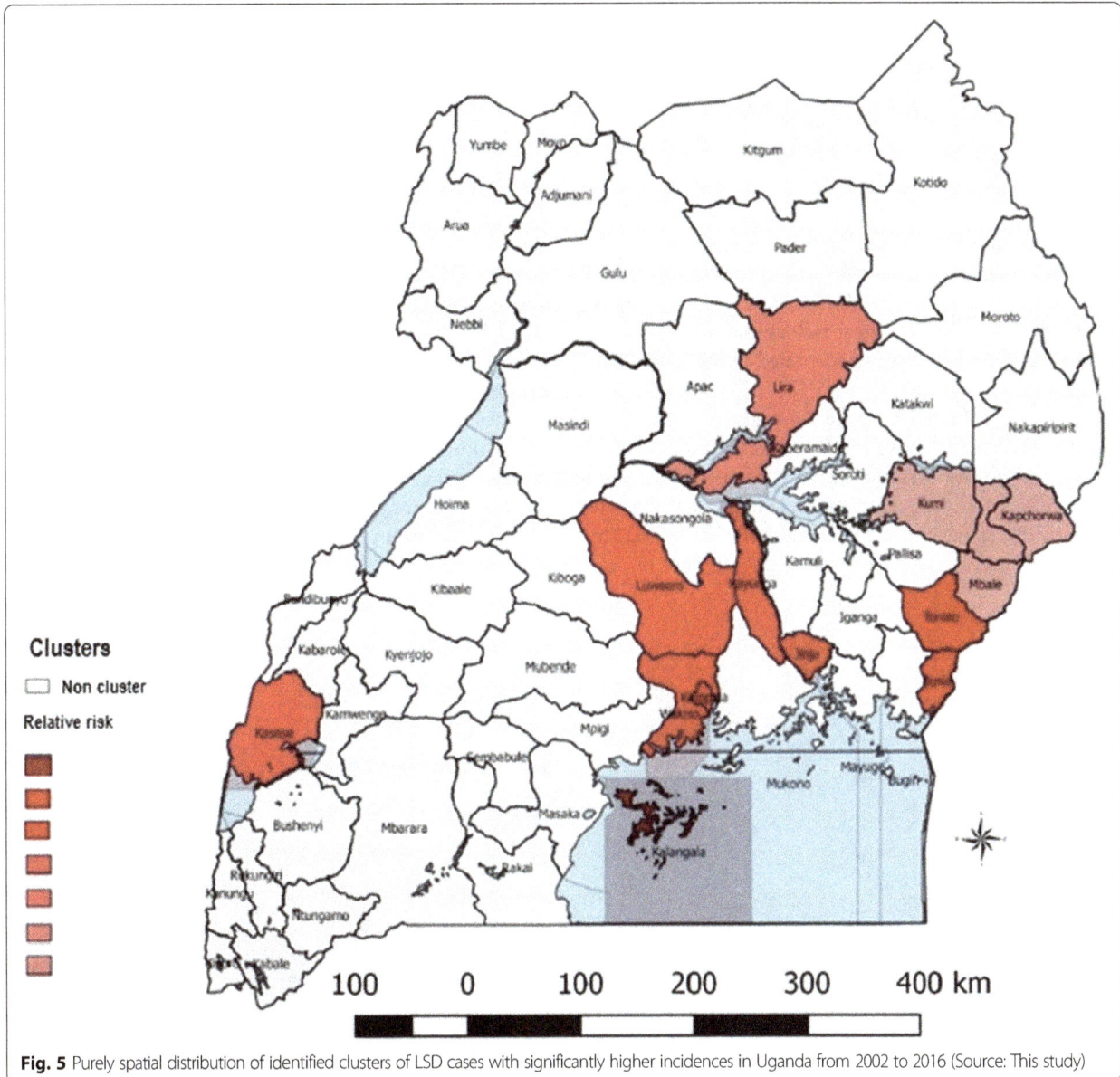

Fig. 5 Purely spatial distribution of identified clusters of LSD cases with significantly higher incidences in Uganda from 2002 to 2016 (Source: This study)

Table 2 SaTScan statistics for purely spatial clusters with significantly higher incidence of LSD in Uganda from 2002 to 2016

District/location	Coordinates/radius	Number of outbreaks	Expected outbreaks	Relative risk	Log likelihood ratio	P-value
Kalangala	(0.320837 S, 32.293743 E) / 0 km	66	0.45	56.17	265.81	$< 10^{-17}$
Busia, Tororo	(0.470669 N, 34.091980 E) / 25.35 km	103	11.30	9.93	139.83	$< 10^{-17}$
Luwero, Kayunga, Wakiso, Kampala	(0.840409 N, 32.497668 E) / 55.57 km	135	30.46	4.90	101.58	$< 10^{-17}$
Kasese	(0.169899 N, 30.078078 E) / 0 km	34	7.50	4.64	25.19	$< 2.3 \times 10^{-10}$
Jinja	(0.447857 N, 33.202612 E) / 0 km	20	3.11	6.54	20.48	$< 2.5 \times 10^{-8}$
Lira	(2.258083 N, 32.887407 E) / 0 km	95	49.24	2.01	17.65	$< 4.2 \times 10^{-6}$
Kapchorwa, Sironko, Mbale, Kumi	(1.335021 N, 34.397636 E) / 54.24 km	94	52.18	1.87	14.33	$< 1.1 \times 10^{-5}$

Table 3 SaTScan statistics for a space-time cluster with a significantly higher incidence of LSD in Uganda from 2002 to 2016

District/location	Coordinates/radius	Timeframe	Number of outbreaks	Expected outbreaks	Relative risk	Log likelihood ratio	P-value
Kamuli, Kayunga, Iganga, Jinja, Pallisa, Luwero, Mukono, Bugiri, Kaberamaido, Nakasongola, Kampala, Mayuge, Wakiso, Soroti, Kumi, Mbale, Busia, Mpigi, Tororo, Sironko, Apac, Kapchorwa, Lira, Katakwi, Kiboga, Kalangala	(0.944785 N, 33.126717 E) / 168.37 km	2002.1.1 to 2005.12.31	383	137.97	3.69	179.08	$< 10^{-18}$

which transmit LSD. District local government reports list Lumpy skin disease among the most economically important Livestock diseases in the district [38], which is in agreement with our findings. Similar factors may also play a role in creating the other hotspots identified in the spatial cluster analysis.

We also conducted a space-time cluster analysis in addition to the purely spatial cluster analysis. A single space-time cluster was identified. When we compared the results of the purely spatial cluster analysis with those of the space-time cluster analysis, we found that thirteen of fourteen districts, identified as purely spatial clusters, were also identified as part of the space-time cluster. However, one district (Kasese) identified by the purely spatial analysis did not appear in the space-time cluster. The districts in the space-time cluster were found in Central, East and Northern parts of the country, and the duration of the associated space-time cluster was for four years. These areas thus appear to have experienced an epidemic wave of LSD for this four year period, occurring mainly within the endemic hotspots.

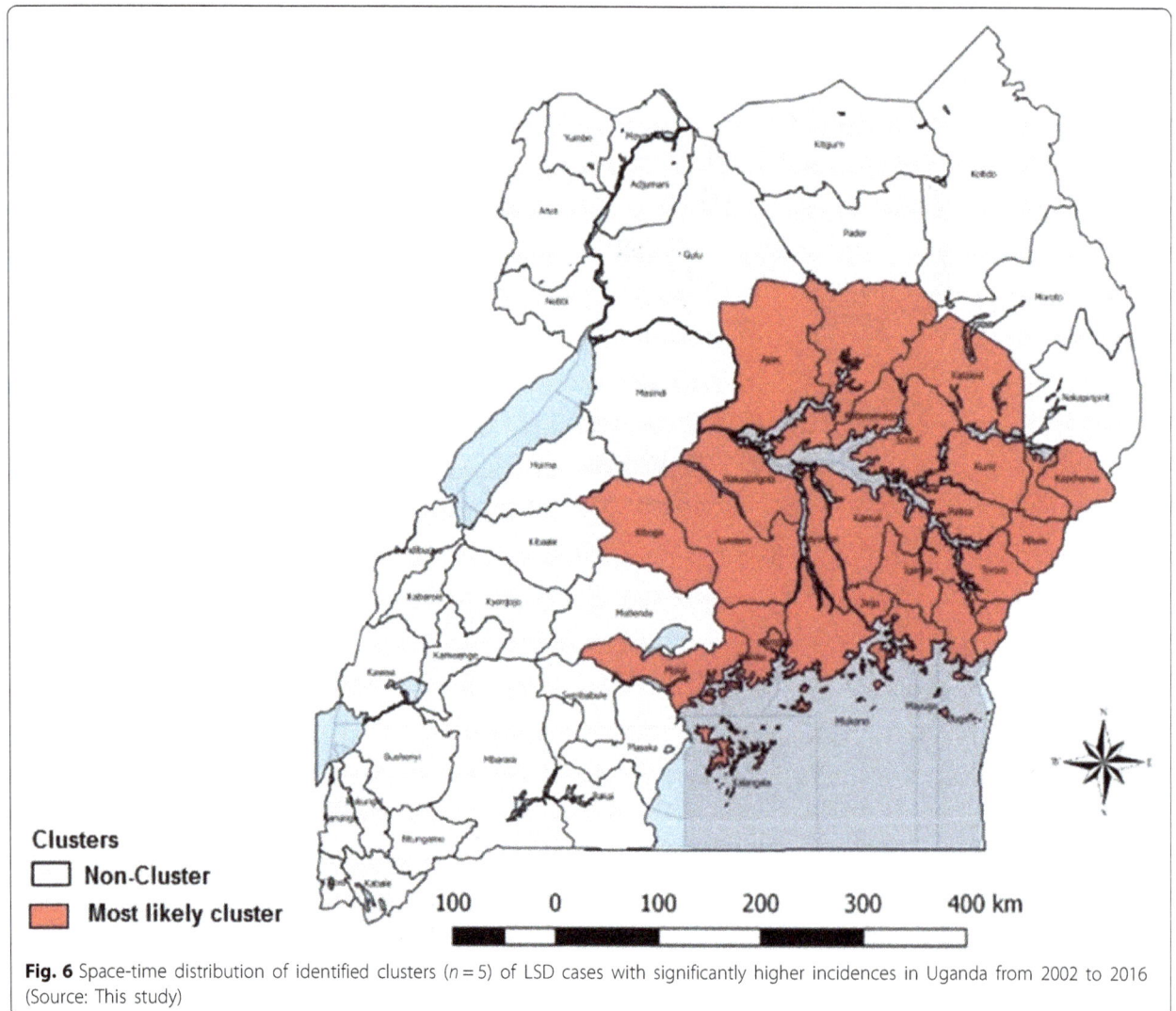

Fig. 6 Space-time distribution of identified clusters ($n = 5$) of LSD cases with significantly higher incidences in Uganda from 2002 to 2016 (Source: This study)

We found that LSD occurs throughout the year with outbreaks reported every month. In the more endemic areas around the Lake Victoria and Lake Kyoga basin of Uganda, there is rainfall throughout most of the year, providing hot and wet weather conditions which are conducive for breeding of biting flies which are known to transmit LSD. During the dry season (December to February), there is reduced availability of pasture and water, so cattle are moved to swampy marsh lands which are common in Central and Eastern regions of the country and present in the other regions as well. These swampy areas maintain a hot and wet micro-climate which support large populations of biting insects; this together with a surge in cattle herds competing for limited grazing areas may lead to spread of LSD and therefore an increase in the number of new outbreaks reported. The Central and Eastern regions showed no seasonal pattern of LSD outbreaks, however the Northeast, West Nile and West showed more seasonal patterns of outbreaks. A slight increase in the number of outbreaks was observed around the month of August for the West and West Nile regions, and an increase in outbreaks was observed in January and February for the Northeast and West Nile regions (Fig. 4). This suggests that seasonal factors have greater effects on incidence for these regions. These results further substantiate the suggestion that the epidemiology of LSD may differ in the endemic hotspots of Central and Eastern Uganda, characterized by less seasonality, presence of spatial clusters as well as space-time clusters of outbreaks, and non-endemic zones that experience sporadic outbreaks but no persistent circulation. Interestingly, endemic hotspots (Central and Eastern) had strikingly lower mortality and case-fatality rates than the other regions, which further suggests an underlying difference in the disease's epidemiology and impact in these different zones. However, management and ecological factors could also impact the fatality rate of the disease.

There were two evident temporal waves of LSD spread during which high number of outbreaks were reported, spaced about ten years apart (Fig. 3). Temporal cluster analysis also identified the first of these two temporal waves, January 2002 to December 2004 as a period with heightened occurrence of LSD in Uganda. Low numbers of outbreaks were reported in 2009, but we do not have sufficient data to propose factors responsible for this occurrence.

Uganda has seven (7) major game parks; these parks are not fenced and it is therefore common for livestock to graze with wildlife. While there have been few studies elsewhere in Africa investigating the role of wildlife in the transmission of LSD, the 17 districts bordering national parks accounted for only 3.9% ($n = 45$) of the total LSD cases reported. However, parks may vary in terms of the types of wildlife species present and the extent to which wildlife and livestock interact. It is notable that 20 out of the total 45 outbreaks bordering national parks (44.4%) were reported in districts bordering Queen Elizabeth National Park (QENP). QENP holds populations of African Buffalo (*Syncerus caffer*), kudu and waterbuck, which have previously been shown to have antibodies against LSDV [6, 39, 40] and therefore could be potential hosts for the virus. It must however be noted that three (3) other parks are inhabited by buffaloes, and the extent of wildlife-livestock interaction may vary in these parks thus limiting cross species transmission of LSD. Neutralizing antibodies have previously been detected in African Buffalo sera from QENP [39]. Though this was in the 1980s, these findings suggest that wildlife may play a role in the maintenance cycle of LSD. More research is needed to clarify the role of buffalo. Genotyping of LSD at wildlife-livestock interfaces, as well as at international borders, should be performed to determine the molecular epidemiology of the disease and shed more light on the effect of wildlife and cross-border animal movements. When we compared outbreaks in districts adjacent to international borders, we found that even if 47.3% of these 332 outbreaks were reported at the Kenya-Uganda border, this difference was found not statistically significant when compared with outbreaks from the four other international borders of Uganda.

The findings of this study should be interpreted with caution because of the potential bias related to underreporting of outbreaks and cases [41]. In addition, the cases were determined based on clinical signs with no confirmatory diagnostic tests, which may have led to biases occurring from nonreporting of sub-clinical cases. Outbreak location information was at the district level, which therefore prevented more elegant spatial analyses and made it difficult to more precisely assess the role of spatial proximity to international boundaries or national parks. More purposeful sampling schemes based on active surveillance and molecular epidemiology are needed to better resolve risk factors and dynamics of LSD spread in Uganda.

Conclusions

Uganda's hot and wet climate provides a conducive environment for biting arthropods which are known to transmit LSD. In this study, we demonstrate that LSD is endemic in Uganda, with annual outbreaks in all regions of the country, albeit in varying incidence. We identified potential endemic hotspots for LSD outbreaks, highlighting the need for risk-based surveillance in these areas to establish the actual disease prevalence and risk factors for maintenance of the disease. Our space-time analysis also revealed that sporadic LSD outbreaks tend to occur within endemic hotspot areas. Interestingly, endemic hotspots had less seasonality in incidence and strikingly lower mortality and case-fatality rates than the other regions, suggesting that epidemiology and impact of LSD may vary within and outside these hotspots. Based on our findings, we suggest that true prevalence of the disease, and viral genotypes, should be determined in order to inform appropriate control measures in these endemic hotspots, such as vaccination, to prevent further spread of the disease. LSD should be included amongst the priority cattle diseases in Uganda, where regular surveillance and vaccination are done by the government. Our findings provide a baseline for further studies into the epidemiology of LSD in Uganda and East Africa.

Additional files

Additional file 1: Mean annual Lumpy skin disease outbreaks across different regions (agro-ecological zones) from 2002 to 2016. The mean annual Lumpy skin disease outbreaks reported in the Central, East, North, Northeast, Southwest, West and Westnile regions of Uganda from 2002 to 2016.

Additional file 2: Occurrence of LSD outbreaks in districts adjacent to national parks in Uganda 2002–2016. This table shows the yearly number of Lumpy skin disease outbreaks reported in districts bordering each of the seven major national parks in Uganda. A total of forty five outbreaks were reported, notably twenty out of these forty five outbreaks are from districts bordering Queen Elizabeth national park.

Abbreviations

BINP: Bwindi Impenetrable National Park; DVO: District Veterinary Officer; KINP: Kibale National Park; KVNP: Kidepo Valley National Park; LMNP: Lake Mburo National Park; LSD: Lumpy Skin Disease; MAAIF: Ministry of Agriculture Animal Industry and Fisheries; MENP: Mount Elgon National Park; MFNP: Murchison Falls National Park; QENP: Queen Elizabeth National Park

Acknowledgements

The authors would like to acknowledge the staff of the Ministry of Agriculture, Animal Industry and Fisheries for the valuable information on Lumpy skin disease especially Ms. Esther Nambo for valued support offered while accessing the data archives at the department of epidemiology.

Funding

This research was funded by the University Of Minnesota Academic Health Center and they had no role in the design and execution of the study as well as the decision to publish this manuscript.

Authors' contributions

SO contributed to the conception of the idea, design, and data collection, drafting and writing of the manuscript. KVW contributed to statistical analysis, interpretation of results and manuscript preparation. CN contributed to statistical analysis and drafting of the manuscript. AM contributed to design and writing of the manuscript, ARAO contributed to data collection and writing of the manuscript. NN contributed to data collection and writing of the manuscript. RM contributed to data collection and writing of the manuscript. FNM contributed to conception of the idea, design and writing of the manuscript. All authors read and approved the manuscript.

Authors' information

SO is a PhD candidate with interest in epidemiology and diagnostics for Animal and zoonotic viruses. KVW is an assistant professor at the Department of Veterinary Population Medicine, University of Minnesota, CN is an MSc candidate with interest in epidemiology and molecular biology, AM is a PhD candidate studying epidemiology of animal viruses in Uganda, ARAO is the assistant commissioner diagnostics and epidemiology at the Ministry of Agriculture Animal Industry and Fisheries Uganda, NN is the assistant commissioner disease control at the Ministry of Agriculture Animal Industry and Fisheries Uganda, RM is a Senior Veterinary Officer-epidemiology at the Ministry of Agriculture Animal Industry and Fisheries Uganda, FNM is a professor of Veterinary virology at Makerere University with vast experience in epidemiology and diagnosis of livestock diseases.

Competing interests

The authors of this paper do not have any financial or personal relationship with other people or organisations that could inappropriately influence or bias the content of the paper. The authors therefore declare that they have no competing interests in the publication of this paper.

Author details

[1]College of Veterinary Medicine, Animal resources and Biosecurity, Makerere University, P.O.BOX 7062 Kampala, Uganda. [2]College of Veterinary Medicine, University of Minnesota, 1365 Gortner Avenue, St. Paul, MN 55108, USA. [3]Ministry of Agriculture Animal Industry & Fisheries, Berkley Ln, Entebbe, Uganda.

References

1. Babiuk S, Bowden TR, Parkyn G, Dalman B, Manning L, Neufeld J, et al. Quantification of lumpy skin disease virus following experimental infection in cattle. Transbound Emerg Dis. 2008;55:299–307.
2. OIE Terrestrial Manual. Aetiology epidemiology diagnosis prevention and control references. Oie. 2012:1–5.
3. Abutarbush SM, Ababneh MM, Al Zoubi IG, Al Sheyab OM, Al Zoubi MG, Alekish MO, et al. Lumpy skin disease in Jordan: disease emergence, clinical signs, complications and preliminary-associated economic losses. Transbound Emerg Dis. 2015;62:549–54.
4. Ali AA, Esmat M, Attia H, Selim A, Abdel-Hamid YM. Clinical and pathological studies of lumpy skin disease in Egypt. Vet Rec. 1990;127:549–50.
5. Hedger RSHC. Neutralising antibodies to lumpy skin disease virus in African wildlife. Comp Immunol Microbiol Infect Dis Microbiol Infect Dis. 1983;6: 209–13.
6. Fagbo S, JAW C, Venter EH. Seroprevalence of Rift Valley fever and lumpy skin disease in African buffalo (<i>Syncerus caffer</i>) in the Kruger National Park and Hluhluwe-iMfolozi Park, South Africa. J S Afr Vet Assoc. 2014;85:1–8. Available from: http://www.jsava.co.za/index.php/jsava/article/view/1075
7. OIE. Lumpy skin disease. OIE Terr. Anim. Heal. Code [Internet]. 2016;1–4. Available from: http://www.oie.int/fileadmin/Home/eng/Animal_Health_in_the_World/docs/pdf/Disease_cards/SHEEP_GOAT_POX.pdf.
8. FG D. Lumpy skin disease. Virus Dis. Food Anim. Gibbs EPJ (ed), Acad. Press. London 1981;2:751–764.
9. Al-Salihi KA, Hassan IQ. Lumpy skin disease in Iraq: study of the disease emergence. Transbound Emerg Dis. 2015;62:457–62.
10. Tuppurainen ESM, Venter EH, Shisler JL, Gari G, Mekonnen GA, Juleff N, et al. Review: Capripoxvirus diseases: current status and opportunities for control. Transbound Emerg Dis. 2017;64:729–45.
11. Burdin ML, Prydie J. Lumpy Skin Disease of Cattle in Kenya. Nature. 1959; 183:55–6. Available from: https://doi.org/10.1038/183055a0
12. Nawathe, Paden J, Confl R. In Nigeria. Pieleg Polozna 1982;36:19, 25.
13. FG D. Lumpy skin disease of cattle: a growing problem in Africa and the near east [internet]. 1991. Available from: http://www.fao.org/ag/aGa/agap/frg/feedback/war/u4900b/u4900b0d.htm.
14. Lubinga JC, Clift SJ, Tuppurainen ESM, Stoltsz WH, Babiuk S, Coetzer JAW, et al. Demonstration of lumpy skin disease virus infection in Amblyomma hebraeum and Rhipicephalus appendiculatus ticks using immunohistochemistry. Ticks Tick Borne Dis. 2014;5:113–20. Elsevier GmbH. Available from: https://doi.org/10.1016/j.ttbdis.2013.09.010

15. Kitching RP, Mellor PS. Insect transmission of capripoxvirus. Res Vet Sci. 1986;40:255–8.

16. Australian Veterinary Emergency Plan (AUSVETPLAN). Animal health Australia. Disease strategy: Lumpy skin disease (Version 3.0). 2009.

17. European Food Safety Authority. Lumpy skin disease: I. Data collection and analysis. EFSA J. 2017;15:54.

18. Tuppurainbwe ESM, Oura CAL. Review: lumpy skin disease: an emerging threat to Europe, the Middle East and Asia. Transbound Emerg Dis. 2012;59:40–8.

19. Tuppurainen ESM, Venter EH, Coetzer JAW. The detection of lumpy skin disease virus in samples of experimentally infected cattle using different diagnostic techniques. Onderstepoort J Vet Res. 2005;72:153–64.

20. Molla W, MCM d J, Gari G, Frankena K. Economic impact of lumpy skin disease and cost effectiveness of vaccination for the control of outbreaks in Ethiopia. Prev Vet Med. 2017;147:100–7. Elsevier. Available from: https://www.sciencedirect.com/science/article/pii/S0167587717303999?via%3Dihub

21. Behnke R, Nakirya M. The contribution of livestock to the Ugandan economy. IGAD Livest. Policy Initiat. Work. Pap. [Internet]. 2012;1–37. Available from: http://citeseerx.ist.psu.edu/viewdoc/download?doi=10.1.1.366.4644&rep=rep1&type=pdf.

22. Uganda Investment Authority. Uganda livestock sector profile. In: Uganda Investiment Auth; 2014.

23. Dukpa K, Robertson ID, Edwards JR, Ellis TM. A retrospective study on the epidemiology of foot-and-mouth disease in Bhutan. Trop Anim Health Prod. 2011;43:495–502.

24. Ayebazibwe C, Tjørnehøj K, Mwiine FN, Muwanika VB, Ademun Okurut AR, Siegismund HR, et al. Patterns, risk factors and characteristics of reported and perceived foot-and-mouth disease (FMD) in Uganda. Trop Anim Health Prod. 2010;42:1547–59.

25. Basalirwa CPK. Delineation of Uganda into climatological rainfall zones using the method of principal component analysis. Int J Climatol. 1995;15:1161–77.

26. The Ministry Of Agriculture AIAF. The National Livestock Census a Summary Report of the National Livestock Census. 2009.

27. QGIS Development Team. QGIS Geographic Information System. v 2.18.7-Las Palmas. Open Source Geospatial Found. Proj. 2015.

28. Kulldorff M. A spatial scan statistic. Commun. Stat - Theory Methods. 1997;26:1481–96.

29. Kulldorff M, Heffernan R, Hartman J, Assunção R, Mostashari F. A space-time permutation scan statistic for disease outbreak detection. PLoS Med. 2005;2:0216–24.

30. Meyer D. Modified Randomization Tests for Nonparametric Hypotheses. Ann Math Stat. 1957;28:181–7. Available from: http://www.jstor.org/stable/2237031

31. Alexandersen S, Mowat N. Foot-and-mouth disease: host range and pathogenesis. Curr Top Microbiol Immunol. 2005;288:9–42. Available from: http://www.ncbi.nlm.nih.gov/pubmed/15648173

32. Kalenzi Atuhaire D, Ochwo S, Afayoa M, Norbert Mwiine F, Kokas I, Arinaitwe E, et al. Epidemiological overview of African swine fever in Uganda (2001–2012). J Vet Med. 2013;2013:1–9. Available from: http://www.hindawi.com/journals/jvm/2013/949638/. http://downloads.hindawi.com/journals/jvm/2013/949638.pdf%5Cn.

33. UBOS. 2004 Statistical Abstract 2004;256.

34. Nsubuga FNW, Namutebi EN, Nsubuga-Ssenfuma M. Water Resources of Uganda: An Assessment and Review. J. Water Resour. Prot. Water Resour. Uganda An Assess. Rev. J. Water Resour Prot. [Internet]. 2014;6:1297–315. Available from: http://www.scirp.org/journal/jwarp. http://dx.doi.org/10.4236/jwarp.2014.614120.

35. Nakato T, Jegede OO, Ayansina A, Olaleye VF, Olufemi B. Mapping the Distribution of Tsetse Flies in Eastern Uganda. Geogr. Inf Syst [Internet]. 2013;938–51. Available from: http://sci-hub.tw/https://www.igi-global.com/journal/international-journal-ictresearch-development/1172.

36. Albert M, Wardrop NA, Atkinson PM, Torr SJ. Tsetse Fly (G . f . fuscipes) Distribution in the Lake Victoria Basin of Uganda. 2015;1–14. Available from: https://doi.org/10.1371/journal.pntd.0003705.

37. Smith J. Lumpy skin disease in Bulgaria and Greece. 2016.

38. Kalangala District Local Government. State of Environment Report [Internet]. 2005. Available from: http://www.nemaug.org/district_reports/Kalangala_DSOER_2004.pdf.

39. Davies FG. Observations on the epidemiology of lumpy skin disease in Kenya. J Hyg (Lond). 1982;88:95–102. Available from: https://www.ncbi.nlm.nih.gov/pmc/articles/PMC2134151/.

40. Hedger RS, Hamblin C. Neutralising antibodies to lumpy skin disease virus in african wildlife. Comp Immun Microbiol Infect Dis. 1983;6(3):209–13.

41. Thrusfield M. Veterinary Epidemiology. Third Edit. Blackwell Sci. Ltd. Wiley; 2005. p. 170–171.

8

First molecular characterization of *Echinococcus granulosus* (sensu stricto) genotype 1 among cattle in Sudan

Mohamed E. Ahmed[1], Bashir Salim[2], Martin P. Grobusch[3] and Imadeldin E. Aradaib[1,4*]

Abstract

Background: *Echinococcus granulosus* sensu *lato* (*s.l.*) is the causative agent of cystic echinococcosis (CE), which is a cosmopolitan zoonotic parasitic disease infecting humans and a wide range of mammalian species including cattle. Currently, little information is available on the genetic diversity of *Echinococcus* species among livestock in Sudan. In the present study, fifty (*n* = 50) hydatid cysts were collected from cattle carcasses (one cyst sample per animal) at Al-kadarou slaughterhouse, Khartoum North, Sudan. DNA was extracted from protoscolices and the germinal layer of each cyst and subsequently amplified by PCR targeting the mitochondrial NADH dehydrogenase subunit 1 (NADH-1) gene. The amplified PCR products were purified and subjected to direct sequencing for subsequent construction of phylogenetic tree and net work analysis.

Results: The phylogenetic tree revealed the presence of *Echinococcus canadenesis* genotype 6 (G6) in 44 cysts (88.0%), *Echinococcus ortleppi* genotype 5 (G5) in 4 cysts (8.0%) and *Echinococcus granulosus* sensu stricto (*s.s*) genotype 1 (G1) in 2 cysts (4.0%). The phylogenetic network analysis revealed genetic variation among the different haplotypes/genotypes. This report has provided, for the first time, an insight of the role of cattle in the transmission of the zoonotic G1 echinococosis.

Conclusions: The results of the study illustrate that Sudanese breeds of cattle may play an important role in the transmission dynamics and the epidemiology of cystic echinococcosis in Sudan. This study reports the first molecular identification of *E. granulosus s.s.* in cattle in Central Sudan.

Background

Cystic echinococcosis (CE) is a significant public health problem with high endemicity in east and central Africa including Sudan [1–4]. The larval stage of *Echinococcus granulosus* sensu *lato* (*s.l.*) causes CE in humans and a wide range of mammalian species. The life cycle involves the ingestion of parasite eggs by an intermediate host belonging to wildlife and domestic livestock species, including cattle. The dog is considered as the definitive host for this parasitic infection [5]. Humans are accidental dead end hosts. It is estimated that CE results in economic losses in the livestock sector due to morbidity.

In addition, partial or total condemnations of infected organs of slaughtered animals are frequently encountered in endemic areas [5–9]. Echinococcosis has recently been included by the World Health Organization (WHO) as a neglected tropical disease [10]. CE may significantly affect the overall development and work productivity in endemic areas. In pastoral Sudanese communities, CE remains highly endemic with higher prevalence compared to agricultural communities. CE is endemic in most parts of the world, including regions of South America, the Mediterranean, Eastern Europe, East Africa, the Near and Middle East, Central Asia, China and Russia [7, 10–13]. Currently, ten distinct genotypes of *E. granulosus s.l.*, designated as G1-G10, have been described worldwide on the basis of genetic diversity related to nucleotide sequences of the mitochondrial NADH dehydrogenase subunit 1 (NADH 1) and cytochrome C oxidase subunit 1 (COX1) genes.

* Correspondence: aradaib@uofk.edu; aradaib@yahoo.com
[1]EBH Research Center, Zamzam University College (ZUC), Khartoum, Sudan
[4]Molecular Biology Laboratory (MBL), Department of Clinical Medicine, Faculty of Veterinary Medicine, University of Khartoum, P.O. Box 32, Khartoum North, Sudan
Full list of author information is available at the end of the article

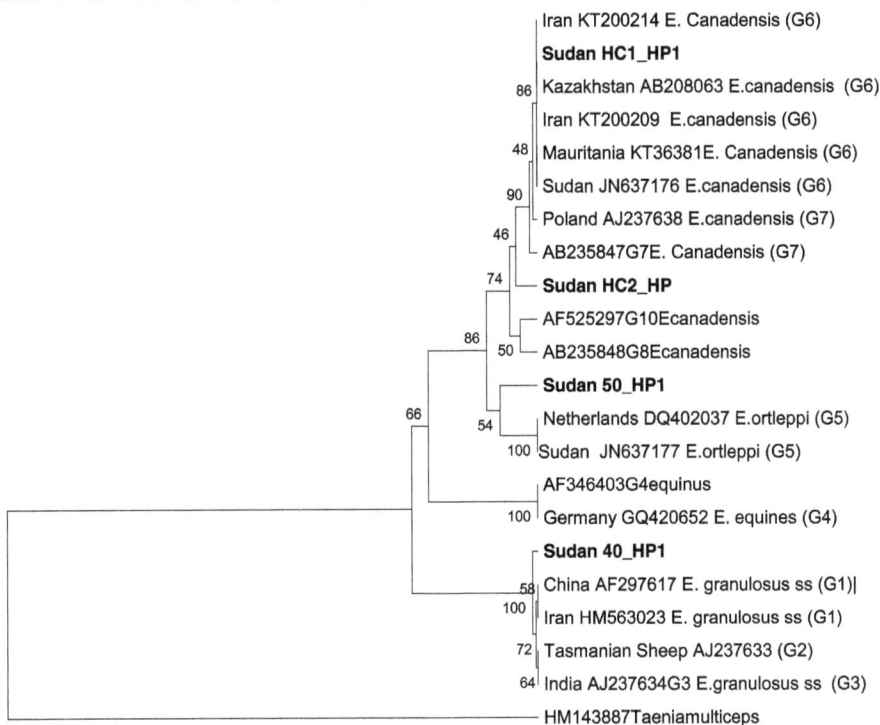

Fig. 1 Phylogenetic relationship of hydatid cysts of *Echinococcus* granulosus sensu lato recovered from Sudanese cattle and other genotypes identified globally. NADH dehydrogenase subunit 1 (NADH-1) partial sequences generated from this study were aligned with sequences of other strains from different parts of the world. Sequences were analyzed with the BioEdit software (Ibis Biosciences, Carlsbad, CA, USA). The phylogenetic tree was constructed using unweighted pair group method with arithmetic mean (UPGMA) implemented in MEGA software version 6.0 [31]. Bootstrap values were calculated from analysis of 500 replicates of the data set, and values greater than 50% are indicated at the appropriate nodes. Each genotype was designated by its GenBank accession number and the country of origin when available. The GenBank accession numbers (LC167080, LC167081) were given for *Echinococcus granulossus* sensu stricto (G1) and *Echinococcus orteleppi* (G5), respectively. *Echinococcus canadensis* genotypes (G6) were given accession numbers LC167082 and LC167083). Corresponding nucleotide sequence of NADH 1 of *Taenia multiceps*, GenBank accession number HM143887, was used as an out group. The partial NADH-1gene sequences identified in this study were highlighted in red color for clarity of the constructed phylogenetic tree

These different genotypes are associated with distinct intermediate hosts including sheep, goats, horses, cattles, pigs, camels and and members of the cervid family [14–19]. Of the ten genotypes of *E. granulosus s.l.*, the cattle (G5) and the camel (G6) strains have already been reported among humans and livestock in Sudan [2, 20, 21]. Recent epidemiological studies indicated that the camel genotype (G6) was the most prevalent strain in Sudan [4, 22]. The extensive intra-specific genetic variation of *E. granulosus s.l.* could be better understood within the context of variations in the life cycle pattern [23, 24]. It is suggested that, different genotypes would probably exhibit different antigenicity, transmission profiles, pathological consequences, and different sensitivity to chemotherapeutic agents [25]. A lot of research efforts have been directed towards the epidemiology of CE in Sudan [26–29]. However, only few reports of the genetic diversity of the parasite among the cattle in Sudan employed sequence analysis of mitochondrial markers [2, 4, 21]. It is, therefore, becoming increasingly obvious that expanding the existing sequence data on the genetic diversity of *E. granulosus s.l.* is necessary to better understand the biology, ecology and molecular epidemiology of this parasite. In this investigation, a molecular characterization was conducted to identify hydatid cysts recovered from local cattle breed in Central Sudan.

Methods

Collection of samples and processing

Fifty hydatid cysts ($n = 50$) were collected over a period of 6 months from cattle during April–October, 2016, at Al-kadarou slaughterhouse, Khartoum North, Central Sudan. This slaughter house is the major cattle battoir in Khartoum North, Sudan. Hydatid cysts were obtained from cattle instantly after slaughtering and transferred in thermo-flasks to the Molecular Biology Laboratory at the Faculty of Veterinary Medicine, University of Khartoum, for processing and molecular characterization.

DNA extraction from hydatid cysts

Parasite genomic DNA was extracted from hydatid cysts as described by Ahmed and his coworkers [4]. Maximum DNA yield was obtained by spinning at 12,000 rpm for 1 min at room temperature. From the suspended nucleic acid, 5 μl was used in the PCR amplification.

Primers design and PCR assays

The primers were designed based on the published sequences of NADH dehydrogenase subunit 1 (NADH-1) gene of *E. granulosus* genotype 6 (G6) reported by Bowles and McManus [15]. Briefly, primer EGL1: 5′TGA AGT TAG TAA TTA AGT TTA A′3 and primer EGR2: 5′AAT CAA ATG GAG TAC GAT TA′3 were designed to amplify a fragment of 435 bp of *E. granulosus s.l.* by PCR. The details of PCR amplification, visualization and of results were described previously [4].

Sequence processing and phylogenetic analysis

The PCR products were purified using QIAquick PCR purification kit (Felden, Germany) and submitted for sequencing to a commercial company (Macrogen, Seoul, Korea). Bidirectional sequence fragments of the forward and reverse primers were generated for each sample. These were edited manually to correct possible base calling errors using BIOEDIT 7.0 and were subsequently joined to reconstruct a fragment of 344 bp of the parasite (NADH-1) gene. The consensus sequences were aligned with the corresponding region of NADH-1 gene of known genotypes circulating globally using CLUSTAL-X 2.1 [30]. The phylogenetic tree was constructed using the unweighted pair group method with arithmetic mean (UPGAM) implemented in MEGA software version 6.0 with 1000 bootstrap replicates [31]. Corresponding nucleotide sequences of NADH-1 of *Taenia multiceps* with GenBank accession number HM143887 were used as out groups in the constructed phylogenetic trees.

Phylogenetic network analysis

To measure the genetic variability, the number of haplotypes was determined using DNASP v5 [31] with insertions and deletions considered as variable sites. We used the median-joining (MJ) network algorithm [32] implemented in NETWORK 4.6 (www.fluxus-engineering.com).

Results and discussion

Microscopic examination revealed that all hydatid cysts were fertile and measured 2–10 cm in diameter. The predilection sites of the cysts were found to be the lung and the liver. All fifty DNA samples were amplified by PCR and generated a fragment of 435 bp of the NADH-1 gene. The partial sequences of the NADH-1 gene representing genotypes G1 (accession number LC167080), G5 (accession number LC167081) and G6 (accession numbers LC167082 and LC167083) were submitted to GenBank, DNA Data Base of Japan (DDBJ). The sequence analysis indicated a prevalence of (88.0%, $n = 44$), (8.0%, n = 4), (4.0%, $n = 2$) for *Echinococcus canadenesis* (G6), *Echinococcus ortleppi* (G5), and *E.granulosus* sensu stricto s.s (G1), respectively. The phylogenetic network analysis revealed clear genetic variation between the different genotypes and haplotypes. The present investigation indicated that at least three different genotypes of *E. granulosus s.l.* are actively circulating in cattle in Sudan as illustrated by the phylogenetic tree (Fig. 1) and phylogenetic network analyses (Fig. 2). The sample Sudan HC1_HP was well grouped with haplotype 6 and samples

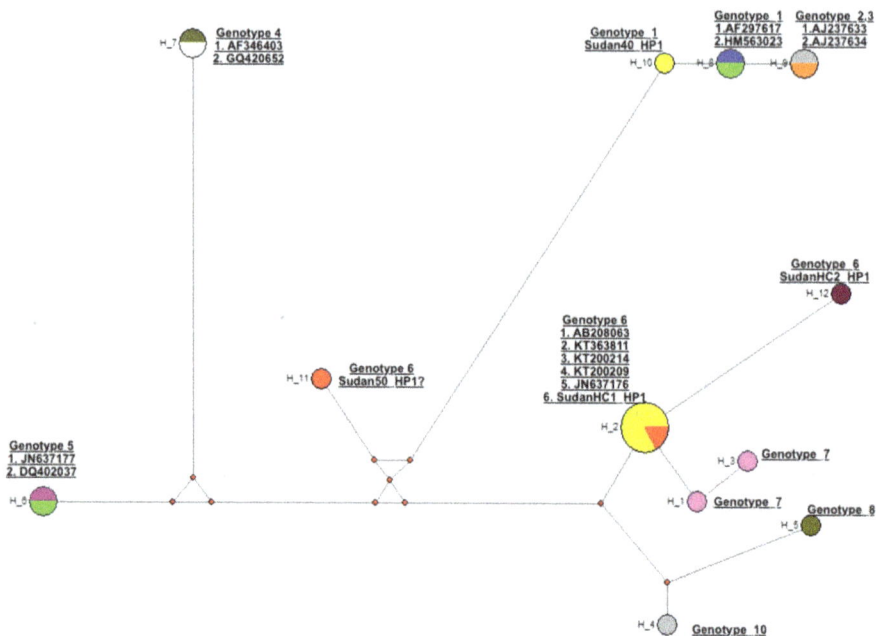

Fig. 2 Phylogenetic network analysis of haplotypes. The number of haplotypes was determined with insertions and deletions considered as variable sites. Median-joining (MJ) network algorithm [32] implemented in NETWORK 4.6 was used to construct the phylogenetic network. The GenBank accession numbers were the same as indicated for the phylogenetic tree

SudanHC2_HP1 that was clustered with genotype 6 in the phylogenetic tree was six SNPs different from the haplotype 6. Sudan 40_HP was grouped with genotype 1 and differed with only 2 SNPs from the previously known genotype 1.The three *Echinococcus* genotypes (G1, G5 and G6) reported in this study are all known human pathogens of significant public health concern [33]. The exclusive occurrence and a predominant circulation of the camel genotype (G6) in the bovine species suggested that cattle can play an important role in the transmission dynamic and the epidemiology of the disease [4]. The present study indicated that *E. granulosus s.s.*, the sheep strain (G1), should equally be considered as an important infectious form of CE among cattle in Central Sudan.

Conclusions

The present study represents the first molecular record of *E. granulosus s.s* G1, thus reinforcing its role as a source of infection among Sudanese cattle breeds. In addition, this investigation provides additional information on the existing data indicating that *Echinococcus granulosus s.s.* G1, which was previously restricted to other region in the African continent, is now becoming broadly distributed in the country. Active surveillance is required to determine the distribution and prevalence of CE and to identify the genotypes/strains circulating in different regions of Sudan.

Abbreviations

CE: Cystic echinococcosis; cox1: Cytochrome C oxidase subunit 1; DDBJ: DNA Data Base of Japan; *E.granulosus s.l.*: *Echinococcus granulosus* sensu *lato*; *E.granulosus s.s.*: *Echinococcus granulosus* sensu *strict*: s.s; G1: Genotype 1; G5: Genotype 5; G6: Genotype 6; N: Number; NADH-1: NADH dehydrogenase subunit 1

Acknowledgments

We thank the Veterinary Officers at Al-kadarou slaughterhouse for many assistance to this study. The authors would also like to thank Mr. Abdalla M. Fadlemoula for technical assistance. The findings and conclusions in this report are those derived by the authors and do not necessarily represent the views of the funding source. The authors would also like to thank Professor M.O. Khedir for English editing of the manuscript.

Funding

This study has no available funding.

Authors' contributions

MEA help with collection of hydatid cyst samples, extracted the DNA, optimized the polymerase chain reaction-based detection assay, editing of sequences and helped with the manuscript writting; BS collected hydatid cyst samples, edited and analyzed the sequence data; MPG designed the experiment and helped with preparation of the final manuscript; IEA designed the experiment, helped with collection of hydatid cyst samples and prepared the final manuscript. All authors read and approved the final version of the manuscript.

Ethics approval and consent to participate

The study protocol was approved by the Institutional Research Board (IRB), Deanship of Scientific Research, Al-Neelain University, Khartoum, Sudan.

Hydatid cysts were collected from slaughtered cattle during post-mortem inspection by qualified veterinary officers at Al-kadarou slaughter house, Khartoum North, Sudan. Formal consent and permission for research use of hydatid cysts were obtained from both the university and abattoir veterinarians. In this study, no experiment was conducted on live animals.

Consent for publication

Not applicable.

Competing interests

The authors declare that they have no competing interests. All authors have read and approved the final version of this manuscript.

Author details

[1]EBH Research Center, Zamzam University College (ZUC), Khartoum, Sudan. [2]Department of Parasitology, Faculty of Veterinary Medicine, University of Khartoum, Khartoum, Sudan. [3]Center of Tropical Medicine and Travel Medicine, Department of Infectious Diseases, Division of Internal Medicine, Amsterdam Medical Center, University of Amsterdam, Amsterdam, The Netherlands. [4]Molecular Biology Laboratory (MBL), Department of Clinical Medicine, Faculty of Veterinary Medicine, University of Khartoum, P.O. Box 32, Khartoum North, Sudan.

References

1. Saad MB, Magzoub M. Echinococcus granulosus infection in Tamboul. Sudan J Helminthol. 1986;60:299–300.
2. Omer RA, Dinkle A, Romig T, Mackenstedt U, Elnahas AA, Aradaib IE, Ahmed ME, Elmalik KH, Adam A. A molecular survey of cystic echinococcosis in Sudan. Vet Parasitol. 2010;169:340–6.
3. Wahlers K, Menezes CN, Wong ML, Zeyhle E, Ahmed ME, Ocaido M, Stijnis C, Romig T, Kern P, Grobusch MP. Cystic echinococcosis in sub-Saharan Africa. Lancet Infect Dis. 2012;12:871–80.
4. Ahmed ME, Eltom KH, Musa NO, Ali IA, Elamin FM, Grobusch MP, Aradaib IE. First report on circulation of Echinococcus ortleppi in the one humped camel (Camelus dromedaries), Sudan. BMC Vet Res. 2013;9:127.
5. Eckert J, Deplazes P. Biological, epidemiological, and clinical aspects of echinococcosis, a zoonosis of increasing concern. Am Soc Mic. 2004;17:107–35.
6. Torgerson PR. Economic effect of echinococcosis. Acta Trop. 2003;85:113–8.
7. Musa NO, Eltom K, Awad S, Gameel AA. Causes of condemnation of sheep carcasses in abattoirs in Khartoum. In: Tielkes E, editor. Tropentag book of abstracts: Tropentag 19–21 Sept. 2012. Göttingen: Cuvillier; 2012. p. 54.
8. Budke C, Deplazes P, Torgerson P. Global socioeconomic impact of cystic echinococcosis. Emerg Infect Dis. 2006;12:296–303.
9. Deplazes P, Rinaidi L, Rojas A, Harandi MF, Romig T, Antolova D, Jm S, Lahmar S, Cringol G, Magambo J, Thompson RCA, Jenkins EJ. Global distribution of alveolar and cystic echinococcosis. Adv Parasitol. 2017; 95:315–493.
10. Thompson RC. The taxonomy, phylogeny and transmission of Echinococcus. Exp Parasitol. 2008;119:439–46.
11. Brunetti E, Kern P, Vuitton D. (writing panel for the WHO-IWGE). Expert consensus for the diagnosis and treatment of cystic and alveolar echinococcosis in humans. Acta Trop. 2010;114:1–16.
12. Sadjjadi S. Present situation of echinococcosis in the Middle East and Arabic North Africa. Parasitol Int. 2006;55:197–202.
13. Taha H. Genetic variations among *Echinococcus granulosus* isolates in Egypt using RAPD-PCR. Parasitol Res. 2012;111:1993–2000.
14. Zhang W, Zhang Z, Wu W, Shi B, Li J, Zhou X, et al. Epidemiology and control of echinococcosis in central Asia, with particular reference to the People's Republic of China. Acta Trop. 2014;141:235–43.
15. Bowles J, Blair D, McManus DP. Genetic variants within the genus Echinococcus identified by mitochondrial DNA sequencing. Mol Biochem Parasitol. 1992;54:165–73.

16. Bowles J, Blair D, McManus DP. A molecular phylogeny of the genus Echinococcus. Parasitology. 1995;110:317–28.

17. Bowles J, McManus DP. Rapid discrimination of Echinococcus species and strains using a polymerase chain reaction-based RFLP method. Mol Biochem Parasitol. 1993;57:231–9.

18. Bowles J, McManus DP. NADH dehydrogenase 1 gene sequences compared for species and strains of the genus Echinococcus. Int J Parasitol. 1993;23:969–72.

19. Lavikainen A, Lehtinen MJ, Meri T, Hirvila-Koski V, Meri S. Molecular genetic characterization of the Fennoscandien cervid strain, a new genotypic group (G10) of Echinococcus granulosus. Parasitology. 2003;127:207–15.

20. McManus DP. The molecular epidemiology of Echinococcus granulosus and cystic hydatid disease. Trans R Soc Trop Med Hyg. 2002;96:51–157.

21. Elmahdi IE, Ali QM, Magzoub MM, Ibrahim AM, Saad MB, Romig T. Cystic echinococcosis of livestock and humans in central Sudan. Ann Trop Med Parasitol. 2004;98:473–9.

22. Ibrahim K, Romig T, Peter K, Omer RA. A molecular survey on cystic echinococcosis in Sinnar area, Blue Nile state (Sudan). Chinese Med J. 2011; 124:2829–33.

23. Ahmed ME, Abdelrahim MI, Ahmed FM. Hydatid disease, a morbid drop needs awareness. Sudan Med J. 2011;47:4–8.

24. Nakao M, Li T, Han X, Ma X, Xiao N, Qiu J, Wang H, Yanagida T, Mamuti W, Wen H, Moro PL, Giraudoux P, Craig PS, Ito A. Genetic polymorphisms of Echinococcus tapeworms in China as determined by mitochondrial and nuclear DNA sequences. Int J Parasitol. 2010;40:7.

25. Thompson RC, Lymbery AJ, Constantine CC. Variation in Echinococcus: towards a taxonomic revision of the genus. Adv Parasitol. 1995;35:145–76.

26. Thompson RC, Lymbery AJ. The nature, extent and significance ofvariation within the genus Echinococcus. Adv Parasitol. 1988;27:209–58.

27. Abushhewa M, Abushhiwa M, Nolan M, Jex A, Campbell B, Jabbar A, et al. Genetic classification of Echinococcus granulosus cysts from humans, cattle and camels in Libya using mutation scanning-based analysis of mitochondrial loci. Mol Cell Probes. 2010;24:346–51.

28. Eisa AM, Mustfa AA, Soliman KN. Preliminary report on cysticercosis and hydatidosis in the southern Sudan. Sud J Vet Sc Anim Husb. 1962;3:97–102.

29. El Khawat SE, Eisa AM, Slepnev NK, Saad MB. Hydatidosis of domestic animals in the central region of the Sudan. Bull Anim Hlth Prod Afr. 1979;27:249–51.

30. Hall TA. BioEdit: a user-friendly biological sequence alignment editor and analysis program for windows 95/98/. Nt Nucl Acids Symp Ser. 1999;41:95–8.

31. Tamura K, Stecher G, Peterson D, Filipski A, Kumar S. MEGA6: Molecular Evolutionary Genetics Analysis version 6.0. Mol Biol Evol. 2013;30:2725–9.

32. Bandelt HJ, Forster P, Rohl A. Median-joining networks for inferring intraspecific phylogenies. Mol Biol Evol. 1999;16:37–48.

33. Alvarez, Rojas CA, Romig T, Lightowlers. M.W. Echinococcus granulosus sensu lato genotypes infecting humans—review of current knowledge. Int J Parasitol. 2014;44:9–18.

9

Cattle transhumance and agropastoral nomadic herding practices in Central Cameroon

Paolo Motta[1,2]* (iD), Thibaud Porphyre[1], Saidou M. Hamman[3], Kenton L. Morgan[4], Victor Ngu Ngwa[5], Vincent N. Tanya[6], Eran Raizman[7], Ian G. Handel[8] and Barend Mark Bronsvoort[1]

Abstract

Background: In sub-Saharan Africa, livestock transhumance represents a key adaptation strategy to environmental variability. In this context, seasonal livestock transhumance also plays an important role in driving the dynamics of multiple livestock infectious diseases. In Cameroon, cattle transhumance is a common practice during the dry season across all the main livestock production zones. Currently, the little recorded information of the migratory routes, grazing locations and nomadic herding practices adopted by pastoralists, limits our understanding of pastoral cattle movements in the country. GPS-tracking technology in combination with a questionnaire based-survey were used to study a limited pool of 10 cattle herds from the Adamawa Region of Cameroon during their seasonal migration, between October 2014 and May 2015. The data were used to analyse the trajectories and movement patterns, and to characterize the key animal health aspects related to this seasonal migration in Cameroon.

Results: Several administrative Regions of the country were visited by the transhumant herds over more than 6 months. Herds travelled between 53 and 170 km to their transhumance grazing areas adopting different strategies, some travelling directly to their destination areas while others having multiple resting periods and grazing areas. Despite their limitations, these are among the first detailed data available on transhumance in Cameroon. These reports highlight key livestock health issues and the potential for multiple types of interactions between transhumant herds and other domestic and wild animals, as well as with the formal livestock trading system.

Conclusion: Overall, these findings provide useful insights into transhumance patterns and into the related animal health implications recorded in Cameroon. This knowledge could better inform evidence-based approaches for designing infectious diseases surveillance and control measures and help driving further studies to improve the understanding of risks associated with livestock movements in the region.

Keywords: Transhumance, Cameroon, GPS, Cattle, Livestock movements

Background

In sub-Saharan Africa (SSA), transhumance of livestock is a common practice for pastoralist communities to cope with local environmental constraints, and fully exploit seasonal availabilities of grazing and water resources [1–3]. Transhumance, therefore, describes the movement of pastoralists and their livestock in response to the variability of environmental and ecological resources [1, 2]. Usually, these migrations are towards regions of different climate and tend to be to remoter riverine areas with poorer veterinary or medical facilities [2]. Long-distance livestock movements can contribute to the dissemination of endemic diseases, or to the introduction and spread of exotic animal diseases [4]. In particular, increased movements and mixing of stock during transhumance are common risks factors for the dissemination of a number of diseases in SSA [5, 6].

* Correspondence: motta.paolo@outlook.com
[1]The Roslin Institute, Royal (Dick) School of Veterinary Studies, University of Edinburgh, Edinburgh, Easter Bush, Midlothian EH25 9RG, UK
[2]The European Commission for the Control of Foot-and-Mouth Disease (EuFMD) - Food and Agricolture Organization (FAO), Viale delle Terme di Caracalla, 00153 Rome, Italy

In Cameroon, transhumance is an established practice among cattle herders to overcome the constrains of the dry season [7, 8], which usually extends from September/October through to April/May of the following year [3]. During this period, a large proportion of cattle herds from the main livestock production areas of the country migrate as a coping mechanism to the ecological and environmental constraints. In particular, transhumance represents an integral component of the livestock production system in the Adamawa Region, with around 50% of the cattle herders implementing such a management practice [7, 8]. While herds in the North and Adamawa Regions usually migrate extensively covering long distances from their Region of origin, in the North-West and West Regions of the country most cattle herds tend to undertake a more local migration, largely within the Region [9]. However, knowledge of these migratory routes and trajectories in Central Cameroon is limited to anecdotal and informal reporting. Characterizing the seasonal transhumance trajectories and the nomadic herding patterns is, therefore, of importance for better understanding interactions within the livestock population and, hence, their potential implications for infectious diseases epidemiology and prevention.

Since 1997, several studies have investigated the movement behaviour of wildlife and livestock animals in Africa using global positioning system (GPS) technology [10–12] and, more recently, mobile phone systems [13]. Notably, GPS-tracking technology has been used in SSA to study grazing behaviour of free-ranging cattle and their response to the spatio-temporal variability of vegetation resources [14–21], to characterize the movements of nomadic pastoralist communities [22] and to collect data on the movements of both traders and traded herds [23]. However, to date, the formal application of GPS-tracking devices on transhumant cattle herds for the entire duration of the migration is still limited, and the understanding of transhumance routes, and associated migratory patterns, is particularly poor in the Central African region. Despite this paucity of specific information on the livestock transhumance patterns in Central Africa, previous investigations of grazing behavior of free-ranging cattle showed that the trajectory of a single animal is representative of the daily grazing orbit and movement patterns of the rest of the herd [16, 18, 21]. Tracking one animal from a migrating herd with GPS technology provides, therefore, a suitable framework for studying transhumance routes and migration patterns during long-distance movements.

Movements and contact patterns within and between animal populations are known to be central drivers of livestock disease dynamics [4] and empirical information, including precise seasonal cattle transhumance trajectories, would help informing a more evidence-based

approach to animal health management in Cameroon. This builds on recent work for improving the understanding of cattle trade-related movements [24] and for identifying constraints for disease controls in pastoral and small-scale livestock husbandry and production systems in Cameroon [25]. An increased understanding of the common patterns and practices during this seasonal migration would help informing the veterinary authorities in designing interventions aiming at enhancing disease surveillance and improving disease control in the study areas.

Here, we present the first formal study of transhumance patterns in Cameroon, while assessing the feasibility of applying GPS collar devices on cattle herds for the entire duration of the migration. Cattle herds originally located in the Adamawa Region of Cameroon were tracked for a period of over six months. Upon their return from seasonal pastoral movements, a questionnaire-based survey was used to collect further information on tracked herds' experience during their migration. The objectives of this study were (1) to characterise the seasonal transhumance routes and daily movement patterns of a restricted pool of cattle herds normally grazing in the Adamawa Region, and (2) to describe the main animal health related issues and interaction patterns during this long-distance migration.

Methods
Study area and herd identification
The Adamawa Region is mainly an open woodland Guinea savannah ecotype above 1000 m, covering an area of approximately 64,000 km2. It is considered to be the main cattle production area of Cameroon with a reported cattle population of about 1.25 million head of cattle [26].

Between October and November 2014, a convenience sample of ten cattle herds whose owner/herdsman were prepared to participate in the study were identified. It was possible to select herds originating/normally grazing in three different Divisions (one administrative level below Region) of the Adamawa Region. This convenience sample was the only applicable due to the nature of the context and the complexity of identifying suitable and available candidates willing to participate in the study. In similar settings, cattle herds have been observed to synchronise their behaviour, to a large extent, to more socially dominant individuals, particularly during travelling and grazing activities [27–29]. In each herd, one animal was selected to carry the GPS device. Selection was first based on discussions with the herdsman to identify socially dominant animals within their herd. Each identified animal was then inspected clinically to ensure that they were robust and healthy. Details on the animals chosen to carry the GPS device in each selected herd are shown in Table 1.

Table 1 Cattle herds identified in the Adamawa Region of Cameroon in October and November 2014

Herd/Collar number	Location (village of origin)	Administrative Division	Collar deployment date	Herd size	Tracked Animal and age (years)	Transhumance completed and survey carried out	Complete GPS data retrieved
1299	Likok	Vina	18/11/2014	45	cow (4y)	Yes	Yes
1300	Belel	Vina	25/10/2014	57	cow (4y)	Yes	Yes
1301	Nyambaka	Vina	24/10/2014	35	bull (4y)	Yes	Yes
1302	Likok	Vina	03/11/2014	40	bull (4y)	Yes	Yes
1303	Margol	Vina	03/11/2014	71	cow (5y)	Yes	No
1304	Mbe	Vina	03/11/2014	71	cow (4y)	No	No
1305	Dir	Mbere	05/11/2014	93	bull (4y)	Yes	Yes
1307	Lougga	Vina	19/11/2014	50	cow (6y)	Yes	No
1308	Martap	Vina	08/11/2014	45	cow (4y)	Yes	Yes
1350	Banyo	Mayo Banyo	30/11/2014	33	bull (4y)	Yes	No

Date of deployment of the GPS device depended on the date of start of the transhumance, the availability of the herds' owners and locations of these herds. In brackets are reported the ages of the cattle (in year) that were selected to be tracked

Data collection

Lightweight (320 g) GPS collars (Savannah Tracking Ltd., Kenya) with global system for mobile (GSM) communications network access and on-board backup data storage were fitted to each selected animal. The data collection schedule and data recording parameters were set through an on-line software interface. The GPS sampling frequency was set to every 2 h and transmission over the GSM network was set for once daily in accordance with other studies tracking cattle in similar settings [11, 15, 30]. More frequent GPS sampling would have provided the opportunity to increase the accuracy of the estimates of distances travelled between GPS locations over the observation period [31, 32]. However, the trade-off between data storage capacity, the expected duration of the transhumance and the GSM network coverage, led to identify this as the optimal sampling frequency for the objectives of the current study.

The GPS collars were retrieved upon return from transhumance in May 2015 and sent back to Savannah Tracking Ltd. in Kenya to retrieve the stored data that could not be downloaded through the GSM network. The lightweight (320 g) GPS collars were easily retrieved from the tracked animals of each herd that after the study continued to be part of their herds. The datasets from each collar unit were downloaded as csv files and included the complete record of the latitude and longitude and the travelling speed (km/hour) at each recording point and an estimation of the accuracy of the location. In addition, at the time of recovering the collar, a structured interview was carried out with the returning herdsmen. The questionnaire took 15–20 min to administer and aimed at collecting information on the daily routines of the herd and herd management practices during the transhumance period, including animal health conditions, interactions with livestock or wildlife populations and on trading activities. The hard copies of the

questionnaires were manually transcribed to pre-designed Excel 2007 (Microsoft) spreadsheet and stored as a csv file.

Data analysis

A combination of descriptive analytical approaches was applied to characterize the transhumance trajectories of the tracked cattle and to assess the general characteristics of the period of movements of the tracked herds. The recorded GPS coordinates and speed of travel of the tracked cattle were assessed using simple descriptive tools such as histograms and interquartile box plots.

The distance traveled between any two consecutive recorded GPS locations was estimated in kilometers using latitude/longitude (degree) georeferences and calculating the Euclidean distance between these GPS locations. The distances travelled were also aggregated at daily and weekly intervals in order to assess the variability of the distances travelled by the different tracked herds over different intervals. This was then used to characterize the range of daily distances traveled and to estimate the daily mean distances travelled during each of the weeks of observation.

A hot spot analysis [33] was carried out to assist identification of locations with unusual high concentrations of data points, identifying spatial clusters. In order to identify and count these hot spots, or activity locations, 2D kernel density were used [34]. Kernel density estimation is a non-parametric method where a symmetrical kernel function is superimposed over each GPS location and requires the definition of a spatial and temporal parameter [35]. The spatial parameter, or bandwidth value, corresponds to the roaming radius while the temporal parameter defines the minimal duration of stay at a given location to qualify as a significant stop. For the purpose of this study, the temporal parameter was given by the time between any two GPS recordings (2 h) and the spatial parameter (bandwidth) was set to a cell size likely to host all of the cattle of the tracked herd according to field observations (500 m) [34].

All analyses and graphics were performed using the raster [36], rgdal [37], ks [38] and ggplot2 [39] packages in the statistical software R version 3.2.3 [40].

Ethical statement

This research was authorised by the Ministry of Livestock, Fisheries and Animal Industries (MINEPIA) (Research permit number: 0119/MINRESI/B00/C00/C010/nye), and approved by the Cameroon Academy of Sciences (approval number 0371/CAS/PR/ES/PO). In the United Kingdom approval was given by the Veterinary Ethical Review Committee (VERC) of the Royal (Dick) Veterinary School of the University of Edinburgh (approval number 28/14).

All methods were performed in accordance with the relevant guidelines and regulations and informed consent was obtained from all subjects. Interviewers were trained to provide the information regarding the consent process to be communicated to the participants and the informed consent was obtained from all subjects. Oral consent was obtained due to the variable level of literacy of the respondents. Prior to interviewing, the study objectives, procedures and the content of the questionnaires were also explained to the participants who were made aware that they were under no obligation to participate if they did not want to.

Results
Collar and data retrieval

Out of the 10 deployed GPS collar units, all of them were successfully retrieved from the animals in May 2015, including seven with complete records of spatial locations and three collars with partial records. Among the seven collars with complete records, one collar belonged to a herd whose herder finally decided not to go on transhumance. Partial recordings from the three collars were due to the memory being overwritten with later locations as a consequence of poor GSM network coverage which resulted in excessive use of the on-board memory storage. In one case (collar 1350) the recordings were almost entirely unavailable, likely due to a concomitant technical failure of the GPS device. Data retrieved from collars 1303 and 1307 were only partially complete, 42 and 44% of the transhumance days, respectively. As a consequence, these three collars were excluded and, along with the herd that failed to leave on transhumance, left data from 6 collars for analysis.

Spatial movements of the tracked herds

The six herds that went on transhumance showed different migratory patterns (Fig. 1 and Table 2). In most cases (5/6) the seasonal migration was towards the southern Regions of the country (herd 1299, 1302 and 1308 to the Centre Region; herd 1301 and 1305 to the East Region), while in one case it was towards the north, to the North Region (herd 1300). The three herds migrating to the Centre Region were directed towards areas of the Mbam and Djerem National Park (about 170 km of distance from their origin), while the two herds migrating to the East Region were directed towards the Pangar and Djerem Reserve (about 150 and 120 km from their origin, respectively). The herd migrating towards north was directed to the Mayo-Rey Division of the North Region of Cameroon, 53 km from its origin. The duration of the transhumance was relatively similar among the 6 herds and varied between 26 and 32 weeks. Although the straight line distances between the origin and the final destination of the transhumance ranged between 53 and 170 km, the overall estimated distance covered by the herds during the whole duration of the transhumance was relatively similar, ranging between 633 and 763 km.

Speed and daily movements

The speed of movement of the tracked herds, as recorded every 2 h, ranged between 0.1 and 7.8 km/hour (Fig. 2a). The overall median speed of each of the 6 herds during the whole period of observation ranged between 0.48 and 1.02 km/hour (Fig. 2b). Although the absolute range of the recorded speed was approximately similar between the herds, herds 1305 and 1308 displayed a wider interquartile range of speeds compared to the other herds (0.48–1.82 km/hour and 0.51–1.91 km/hour, respectively) (Fig. 2b).

Overall, herds showed consistent patterns of movements across the 24 h cycle. The mean speed of the herds tended to increase between 06:00 and 18:00 h. Similarly, the absolute peaks of speed were recorded within this time window (Additional file 1: Figure S1). Nevertheless, across the whole study period, all the 6 herds were recorded to have moved at least 4 km/hour, during all of the recorded time points throughout the 24 h daily cycle. In other words, during the study period herds were recorded making significant movements even during the night.

The daily distance travelled by a herd ranged between 0.3 and 22.9 km/day, with 86% of herd-days below 5 km/day (Fig. 3a). The median distance covered per day by each herd over the whole transhumance period ranged between 3.2 and 4.1 km/day (Fig. 3b). Only during relative short periods greater daily distances were travelled and these were mainly at the start and end of the migration, reflecting movements from and to the main transhumance grazing zone (Fig. 4). However, in 3 cases (herds 1299, 1305 and 1308) greater daily distances were also travelled during other weeks of the transhumance. Overall, these

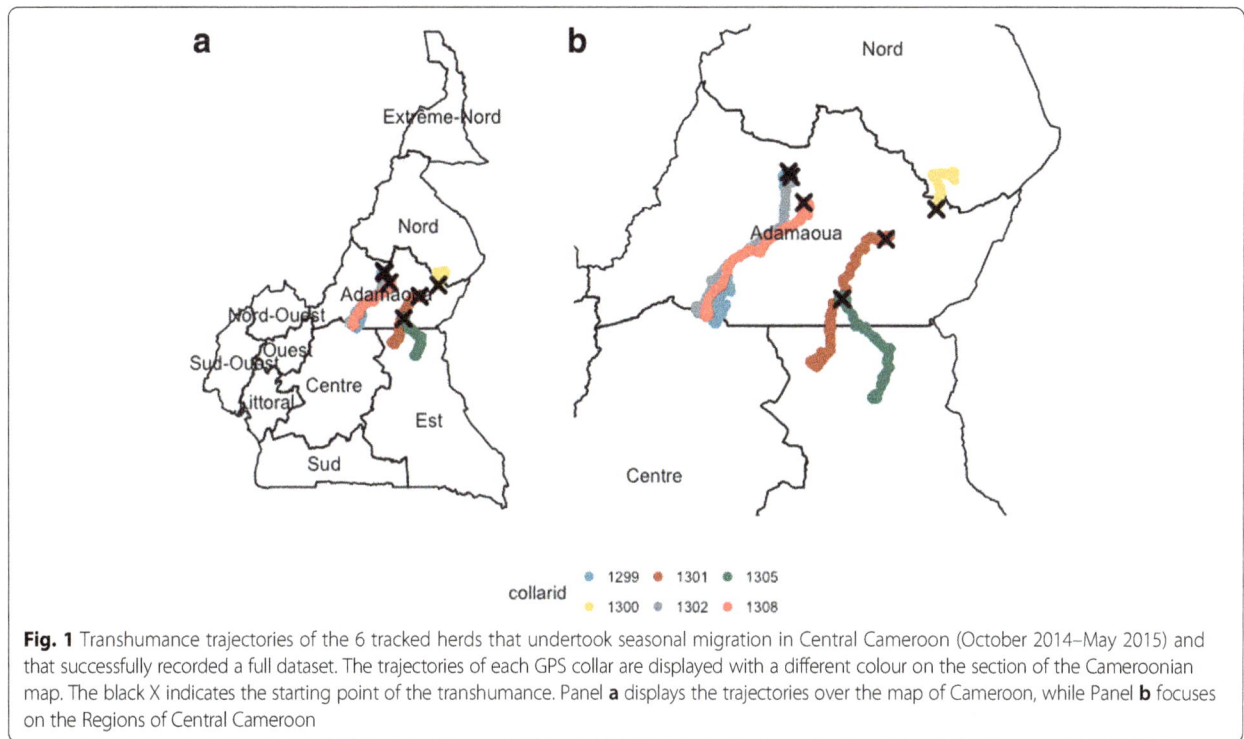

Fig. 1 Transhumance trajectories of the 6 tracked herds that undertook seasonal migration in Central Cameroon (October 2014–May 2015) and that successfully recorded a full dataset. The trajectories of each GPS collar are displayed with a different colour on the section of the Cameroonian map. The black X indicates the starting point of the transhumance. Panel **a** displays the trajectories over the map of Cameroon, while Panel **b** focuses on the Regions of Central Cameroon

periods of higher median daily travelled distance tended to last between 1 and 3 weeks (Fig. 4). Herd 1300 was an exception, with an overall shorter weekly median daily distance travelled across the whole period of observation.

The hot spot, or activity locations, analysis, also showed geographical areas where the herds spent longer periods of activity and areas where, by the contrary, the herds were only transiting (Fig. 5). For all of the tracked herds the origin and destination of the migration represented hot spots of activity. Nevertheless, herds 1299, 1305 and 1308 spent longer periods also in other zones along their transhumance routes (Fig. 5).

Contacts and interactions with other cattle herds during transhumance

During periods of active trekking and movement towards transhumance destinations or returning back from these locations, 5/9 herdsmen reported that > 15 cattle herds were usually encountered each day (Fig. 6a). The four other herdsmen reported routinely encountering 1–3, 4–5, 6–10 and 11–15 other cattle herds per day, respectively. In contrast, in grazing areas such as the transhumance destinations, herds tended to meet fewer other herds per day: only one herdsman reported that more than 15 cattle herds were encountered on an average day at this location, 2 herdsmen reported a mean of 11 to 15 and all the other reported fewer contacts (Fig. 6b).

The typical duration of these encounters was estimated by 6/9 herdsmen to last less than 1 h, while 2/9 herdsmen reported a duration of interaction between 4 and 6 h and 1/9 between 13 and 24 h (Additional file 1: Figure S2). Interestingly, two of the tracked herds (1301 and 1305) physically met during the period of observation

Table 2 Distances travelled by the cattle herds during transhumance in Central Cameroon (October 2014–May 2015)

Herd/Collar number	Total distance covered (km)	Median distance per day (km)	Shortest distance between origin and destination (Km)	Transhumance duration (weeks)	Final num. Transhumance destination
1299	746	4.14 (SD 2.21)	170.9	26	Centre Region
1300	730	3.32 (SD1.74)	53.3	32	North Region
1301	763	3.97 (SD 2.03)	154.0	29	East Region
1302	633	3.23 (SD2.01)	172.7	28	Centre Region
1305	649	3.55 (SD 1.91)	115.5	27	East Region
1308	726	3.63 (SD 1.56)	157.1	29	Centre Region

The total distance covered during the entire period of observation, the median distance travelled per day (standard deviation in brackets) and the distance between the two most far apart locations are all reported in kilometers. The duration of the transhumance, in weeks, is estimated from the week the herds left the grazing location in the Adamawa Region until the week they returned to same location

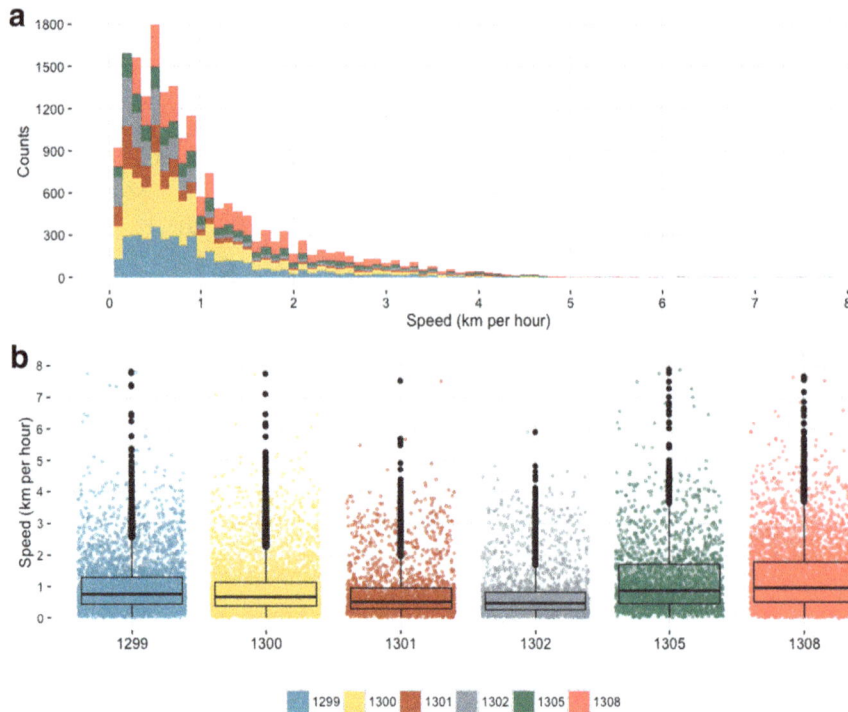

Fig. 2 Speed of movement of the transhumant herds tracked between October 2014 and May 2015 in Central Cameroon (in km/hour). **a**: Distribution of the recorded speed of movements at each GPS captured location for the tracked cattle (km/hour on the x-axis and counts on the y-axis). **b**: Boxplot of the recorded speed of movements at each GPS captured location for each tracked cattle (x axis). For each box the dots refer to the recorded speed at each GPS captured location, the upper and lower hinges correspond to the 1st and 3rd quartiles (the 25th and 75th percentiles) and the horizontal line to the median value

(Additional file 1: Figure S3). The encounter lasted, approximately, between 4 and 6 h, while herds were returning from the seasonal migration, and was in a usual grazing location for the herd 1305, but about 59 km from the migration origin of herd 1301.

Contacts with other animal species and health issues reported during transhumance

Herders were asked what other animal species had been encountered during the transhumance. Among domestic species, sheep were the most frequently encountered and reported by all the 9 interviewees, followed by poultry (5/9 herdsmen), goats (4/9 herdsmen), horses and dogs (3/9 herdsmen) and pigs (1/9 herdsmen). The most frequently encountered wildlife species were the antelopes (4/9 herdsmen), followed by warthogs (2/9 herdsmen) and buffaloes (1/9 herdsmen) and another unspecified animal (1/9 herdsmen) (Fig. 6c).

Among the reported health problems faced by the cattle herds during the transhu- mance period, the most commonly reported was trypanosomiasis (7/9 herdsmen) followed by liver fluke (4/9 herdsmen), foot and mouth disease (FMD) (3/9 herdsmen), dermatophilosis (2/9 herdsmen) and plant intoxication (1/9 herdsmen) (Fig. 6d).

When assessing the causes of death of cattle during transhumance, accidents were reported as having caused the loss of at least one animal (5/9 herdsmen), followed by disease (4/9 herdsmen) and by plant intoxication (3/9 herdsmen) (Table 3).

Trading activities during transhumance

None of the herdsmen (0/9) reported having acquired cattle during the transhumance period, either from livestock markets or from outside the trading system. However, four (4/9) herdsmen reported having sold cattle at livestock markets during the transhumance, while two (2/9) herdsmen reported having sold cattle outside the trading system (Table 3). When traded at the market place, respectively, 2/45, 3/35, 3/40 and 5/93 cattle were sold, while 1/45 and 3/57 cattle were sold outside the market place (Table 3).

Discussion

In SSA, seasonal livestock mobility is an important adaptation mechanism for pastoralist communities, and a key strategy to manage the variability of the natural resources in the ecosystem [1]. In Cameroon, transhumance is a common practice for many pastoralists and their cattle herds to cope with the ecological and

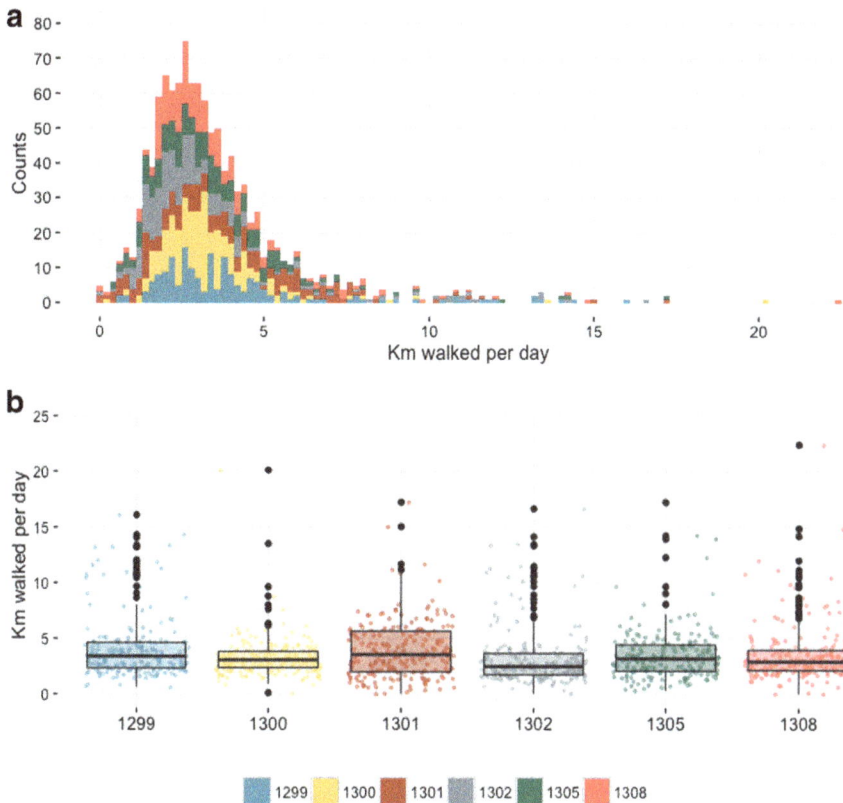

Fig. 3 Daily distances covered by the transhumant herds in Central Cameroon between October 2014 and May 2015. **a**: Distribution of the daily distances walked by the tracked cattle (distance in km on the x-axis and counts on the y-axis). **b**: Boxplot of the daily distance travelled by each tracked cattle (x-axis). For each box the dots refer to the the daily distance for each day of observation, the upper and lower hinges correspond to the 1st and 3rd quartiles (the 25th and 75th percentiles) and the horizontal line to the median value

environmental constraints of the dry season. However, knowledge of migratory routes and patterns in Central Cameroon, and their potential implications for infectious diseases epidemiology and prevention, is still limited. Here, we characterized migrating patterns of a few, but representative, GPS-tracked cattle herds and described key activities and experiences along their transhumance across Central Cameroon.

Long daily distances were relatively rarely traveled (approximately 15% of the recorded days), typically at the beginning and the end of the transhumance. During these periods of more active mobility, an average speed of movement compatible with traveling behavior (between 3 and 4 km/hour) was mostly recorded during the daylight, consistently to reports in the East African rangelands [41]. However, herds were traveling at any time during transhumance, irrespective of being day or night. Overall, the proportion of daily distances traveled and the variability of walking speed are in line with previous findings in East and West Africa [31, 41, 42].

Over the six GPS-tracked herds we found three cattle herds having multiple repeated traveling and grazing periods through different temporary transhumance locations.

Similarly to previous findings in SSA [41, 43], hence, the seasonal cattle transhumance in Central Cameroon, rather than a simple transit between two locations, tended to be a more complex journey through multiple grazing areas. This migration lasted a significant length of time (even greater than half of the year). Furthermore, the trajectories of all of the three tracked herds moving from the Central-Eastern part of the Adamawa Region towards the Centre Region highlighted the presence of a common migratory route, or transhumance corridor. In Northern Cameroon, and the larger Chad Basin, pastoralists and their herds move through established transhumance corridors connecting seasonal grazing lands [44]. Further confirmation of the observed migratory routes between the Adamawa and the Central Regions can provide evidence of common transhumance corridors and important indications for designing strategic and efficient veterinary interventions. For example, surveillance posts could be established along these corridors for providing animal health services to the migrating herds (e.g. free dipping or spraying points), and potentially using these locations for control measures (e.g. vaccination points). In addition, if the common transhumance destinations in the Central Region can be further confirmed for a

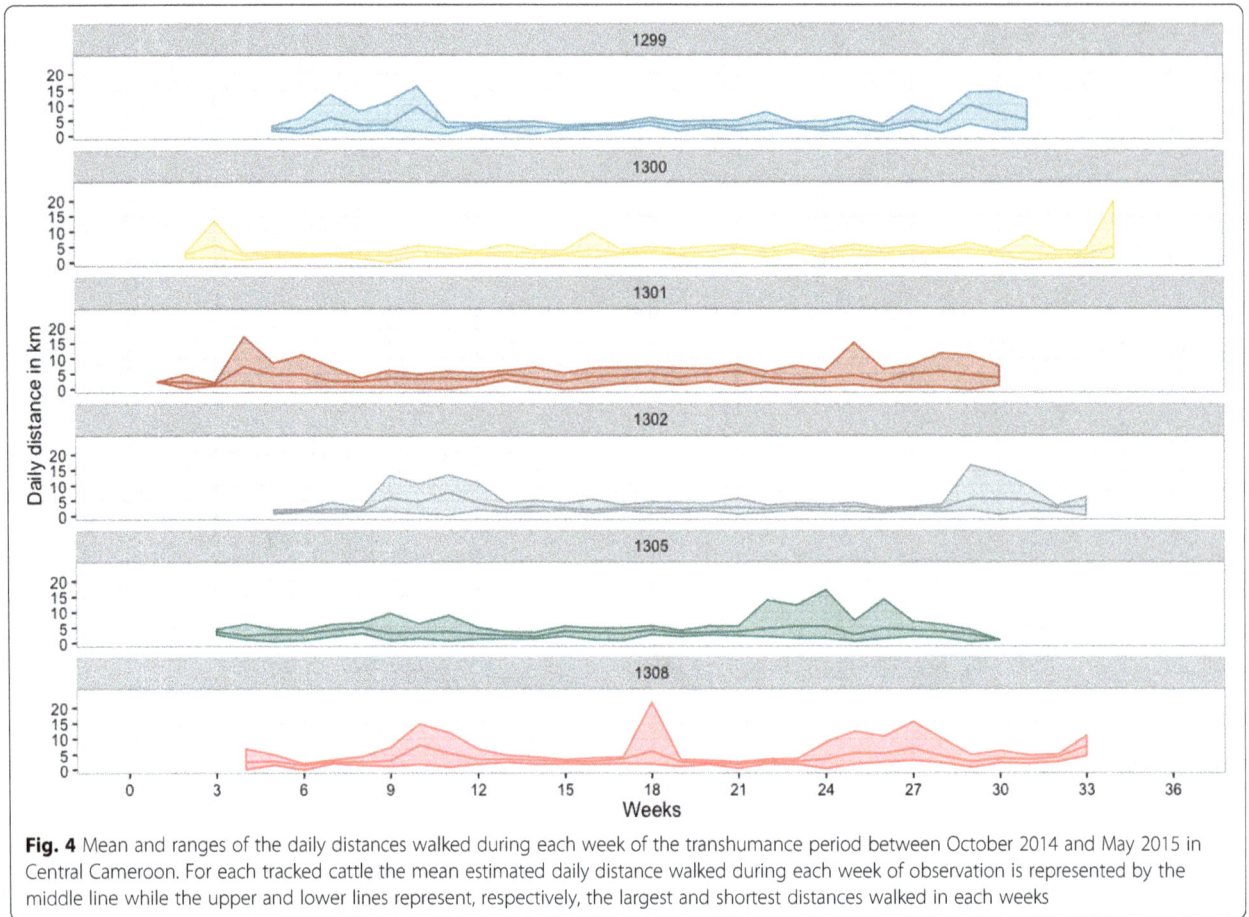

Fig. 4 Mean and ranges of the daily distances walked during each week of the transhumance period between October 2014 and May 2015 in Central Cameroon. For each tracked cattle the mean estimated daily distance walked during each week of observation is represented by the middle line while the upper and lower lines represent, respectively, the largest and shortest distances walked in each weeks

high number of cattle herds, and over multiple years, it would suggest these zones should be considered as high cattle density areas in Cameroon, at least for a significant part of the year. This information would provide evidence for appropriately considering these areas within infectious diseases surveillance and control strategies.

Multiple grazing locations during transhumance increase exposure of herds to geographically limited or seasonally abundant diseases [5]. In our survey, most herdsmen reported trypanosomiasis and liver fluke as health issues for their herds. This finding suggests that, while grazing areas provide greener pastures and greater water resources (e.g. natural water points) for transhumant herds, they also offer ideal habitats for vector and parasites proliferation [45]. As such, transhumant herds potentially contribute to the persistence and circulation of vector-borne diseases and parasites in Cameroon [45].

Because of the complexity of engaging stakeholders in this study, it was not possible to use a statistically robust sampling approach. Instead, we used a convenient sampling of a limited number of cattle herds. Clearly, this is a major shortcoming of this survey and additional studies, with increased number of herds, are required to confirm our findings. In particular, our survey provides

information of only a few diseases affecting cattle herds through anecdotal reporting and without any supporting laboratory evidence. Although this list is certainly not exhaustive, we believe that it represents the list of infectious diseases which are perceived by herdsmen as the most important for their herds. It is also worth noting that the reported health conditions included infectious diseases for which direct contacts between animals are key transmission mechanisms (e.g. FMD and dermatophilosis), and which are well recognized by livestock stakeholders in the study areas [46].

During transhumance, particularly while traveling towards grazing areas, cattle herds tended to have more frequent contacts with other herds and with wildlife, compared to when they are sedentary at grazing locations. Despite this variability in contact rates, the interactions between cattle herds, both during traveling and at grazing areas, were reported to have relatively short durations (< 1 h). Nevertheless, a short contact time, particularly if at close proximity, could be sufficient for the transmission of highly infectious diseases, especially during the peak of the infectious periods [47, 48].

Over the eight months of study, two tracked herds, which originated from very distant areas, were recorded

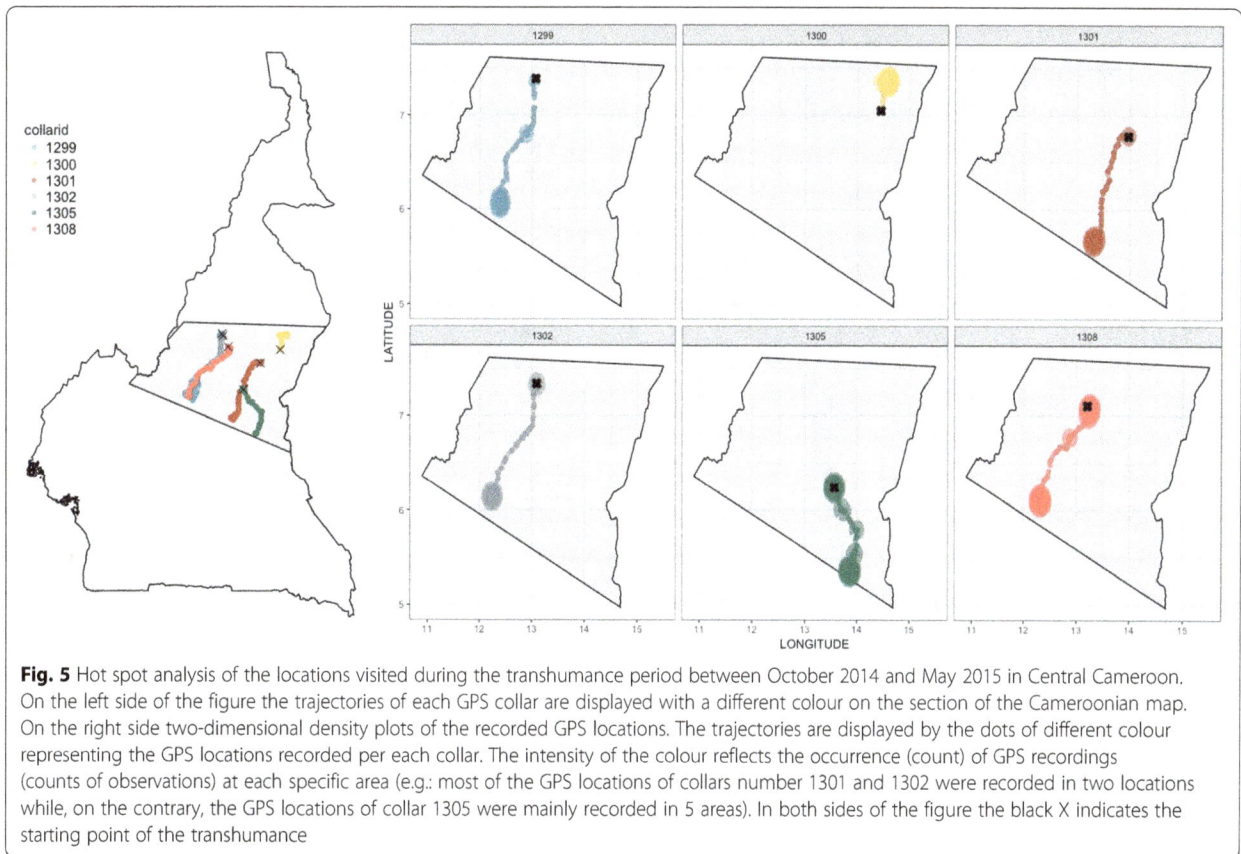

Fig. 5 Hot spot analysis of the locations visited during the transhumance period between October 2014 and May 2015 in Central Cameroon. On the left side of the figure the trajectories of each GPS collar are displayed with a different colour on the section of the Cameroonian map. On the right side two-dimensional density plots of the recorded GPS locations. The trajectories are displayed by the dots of different colour representing the GPS locations recorded per each collar. The intensity of the colour reflects the occurrence (count) of GPS recordings (counts of observations) at each specific area (e.g.: most of the GPS locations of collars number 1301 and 1302 were recorded in two locations while, on the contrary, the GPS locations of collar 1305 were mainly recorded in 5 areas). In both sides of the figure the black X indicates the starting point of the transhumance

in contact at the same grazing location for 4 to 6 h. This observation confirms that opportunities for close interactions occur not just locally, at transhumance destinations with communal grazing, but also through the transhumance migration routes. This finding further reinforces the potential strategic role of veterinary surveillance and control points along the migratory routes, or transhumance corridors. Veterinary check points would represent key locations for designing and implementing efficient surveillance and control measures against infectious and parasitic diseases, including other priority livestock diseases in Cameroon other than the ones highlighted in this study, such as pasteurellosis, Contagious Bovine Pleuropneumonia (CBPP) and tick-borne diseases [45].

During the transhumance period, herdsmen of six of the nine herds under study reported to have sold cattle, either within or outside the formal trading system. Although livestock markets are known to greatly influence the spread of multiple infectious diseases throughout livestock industries [49–51], they are also places where social and cultural interactions occur. Such a report highlights the role of markets as an interface between the pastoral and the trading systems in the country. It also underlines their potential complementary role for

risk-based approaches to surveillance, control and communication strategies for pastoral communities.

Although infectious diseases are the major animal health problem for the livestock sector in Cameroon [3, 45], the most commonly reported causes of death for cattle during transhumance included accidents and plant intoxications. Despite the small sample size available for this study, these findings show that transhumance presents specific challenges. Trekking for long distances poses specific risks and higher exposures for various types of physical accidents and environmental hazards, including plant intoxication for which very little knowledge is currently available. Furthermore, long distance livestock migrations may generate conflicts between pastoralists and local farming communities over limited natural resources and damages to crops [52, 53].

The potential tension then arising could pose additional security challenges to herds, as informally reported during the data collection of the current study.

Conclusion

This limited study provides a general characterization of cattle transhumance pat- terns and of the key associated issues in Central Cameroon. The spatial and temporal

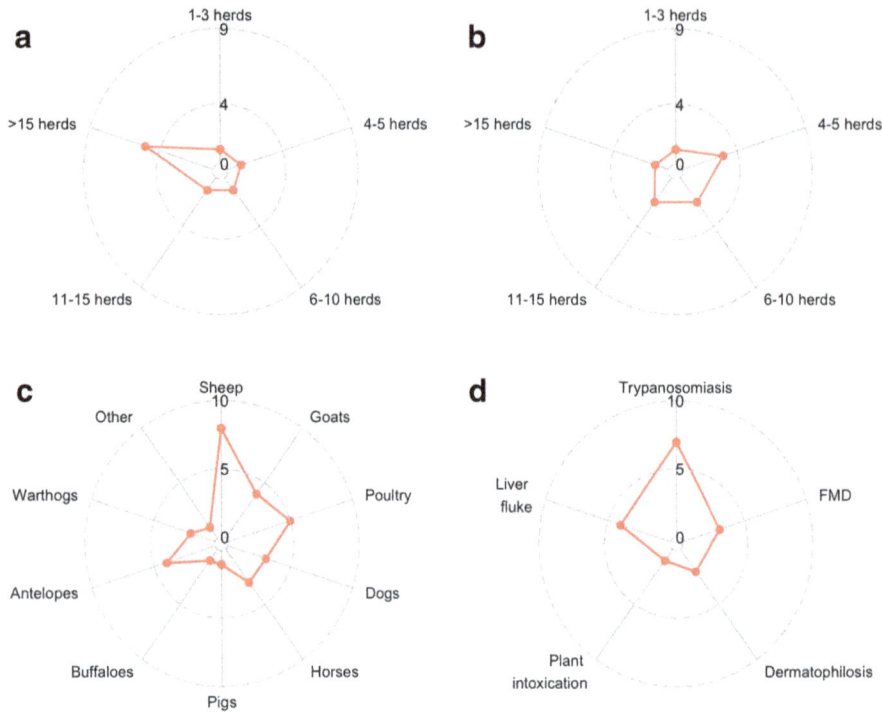

Fig. 6 Encounters with other animal species and common health problems reported during transhumance in Central Cameroon (October 2014–May 2015). **a** and **b** report the frequency of specific answers from each of the 9 interviewees. **c** and **d** report the sum of the times animals species and diseases were mentioned by the 9 interviewees. **a**: Reported number of cattle herds encountered on average every day during the trekking towards transhumance destinations or returning towards the usual grazing locations. The plot displays the answers of the interviewees and the red line refers to the number of reports by the 9 interviewees. **b**: Reported number of cattle herds encountered on average every day during grazing activities at the transhumance destination. The plot displays the answers of the interviewees and the red line refers to the number of reports by the 9 interviewees. **c**: Reported species of domestic and wild animals encountered during the transhumance. The plot displays the animal species and the red line refers to the number of reports by the interviewees. **d**: Reported health problems faced by the herds during the transhumance. The plot displays the reported health problems and the red line refers to the number of reports by the interviewees

overlapping between tracked cattle herds highlighted the opportunity for direct contacts and interactions between herds of distant origins, as well as with other domestic and wildlife species, both during traveling and grazing periods. The recorded speed of movement and interaction

Table 3 Number of herdsmen reporting cattle death and trade during the transhumance period (October 2014–May 2015). NH: Number of herdsmen reporting events; NC: Number of cattle involved in each reported event

Reported events		N_H	N_C
Causes of cattle losses	Accident	5	7
	Diseases	4	6
	Plant Intoxication	3	10
Cattle trade	Sale (within market)	4	13
	Sale (outside market)	2	4
	Purchase (within market)	0	0
	Purchase (outside market)	0	0

Reported cattle losses during transhumance were stratified by the cause of death as diagnosed by herdsmen. Reported trade events during transhumance were stratified by whether animals were sold (or purchased) within or outside the trading system (livestock markets)

frequencies, and durations, hence, could potentially inform parametrization of further epidemiological studies.

Specific infectious and parasitic diseases were reported affecting the migrating herds, however physical accidents and environmental hazards (e.g. plant intoxications) were also reported as key factors impacting these herds. Importantly, the transhumant herds have also been shown to connect to the formal cattle trading system, highlighting the complexity of the pastoral, but increasingly market-orientated, livestock system in the country.

The overall characterization of transhumnce patterns in the study areas, and the related key aspects, represent a preliminary step for better understanding their implications in the epidemiology of livestock infectious diseases, and for their potential applications to inform surveillance and control strategies.

The further confirmation of some of the characterized migratory routes as common transhumance corridors, would provide the evidence to strategically design robust and efficient surveillance and control interventions at key locations.

Increased knowledge and understanding of pastoral movements and contacts between livestock populations at local and long-distance levels is essential for supporting the veterinary services in designing and planning more effective surveillance and control strategies of infectious diseases. Considering the key role of livestock transhumance as a key adaptation and ecological management mechanism for pastoralist communities in the region, this seasonal migration should be increasingly addressed in animal health management.

Additional file

Additional file 1: Figure S1. Mean and ranges of the speed over the 24 h period. For each tracked cattle the speed of movements was recorded every two hours of the observation period. The middle line represents the mean speed at that hour of the day while the upper and lower lines represent, respectively, the fastest and slowest speed recorded at that specific time during the observation period. **Figure S2.** Reported duration of interaction with other cattle herds. On the x axis the reported usual duration of interaction with other cattle herds during transhumance and on the y axis the number of interviewees. **Figure S3.** Recorded encounter between 2 tracked herds. The herdsmen of these 2 herds (1301 and 1305) reported having met each other at the time of interview. The analysis of the GPS recordings enabled to identify the exact time and location of this encounter. The herds were recorded interacting for about 4 h between 8 am and 12 am of the 23rd April 2015, while returning to their respective grazing locations for the rainy season.

Abbreviations
CBPP: Contagious Bovine Pleuropneumonia; FMD: Foot and mouth disease; GPS: Global positioning system; GSM: Global system for mobile; MINEPIA: Ministry of Livestock, Fisheries and Animal Industries; SSA: Sub-Saharan Africa

Acknowledgments
The authors would like to thank all the herds owners and keepers for their willingness to be involved in the research. In addition, the authors gratefully acknowledge all the Delegates and veterinarians of the Ministry of Livestock, Fisheries and Animal Industries (MINEPIA) for their cooperation in the study. Florian Druke for his support during the trekking for reaching some of these cattle herds and Henrik Rasmussen of Savannah Tracking Ltd (http://www.savannahtracking.com/) for technical support with the GPS collars.

Funding
P.M. was supported by the University of Edinburgh through a Principal's Career Development Scholarship. B.M.B. receives core strategic funding from the BBSRC (BB/J004235/1). The funding body was not involved in the design of the study and collection, analysis, and interpretation of data and writing of the manuscript.

Authors' contributions
PM, IH, VT, ER, KLM and BMB contributed to the design of the study; PM, VNN and SMH performed the field work. PM conducted the analyses, interpreted the results and wrote the manuscript. TP provided statistical supports. TP, IH, ER, KLM and BMB revised and reviewed the manuscript. All authors read and approved the final manuscript.

Ethics approval
This research was authorised by the Ministry of Livestock, Fisheries and Animal Industries (MINEPIA) (Research permit number: 0119/MINRESI/B00/C00/C010/nye), and approved by the Cameroon Academy of Sciences (approval number 0371/CAS/PR/ES/PO). In the United Kingdom approval was given by the Veterinary Ethical Review Committee (VERC) of the Royal (Dick) Veterinary School of the University of Edinburgh (approval number 28/14).
All methods were performed in accordance with the relevant guidelines and regulations and informed consent was obtained from all subjects. Interviewers were trained to provide the information regarding the consent process to be communicated to the participants and the informed consent was obtained from all subjects. Oral consent was obtained because of the variable level of literacy of the respondents. Prior to interviewing, the study objectives, procedures and the content of the questionnaires were also explained to the participants who were made aware that they were under no obligation to participate if they did not want to.

Competing interests
The authors declare that the research was conducted in the absence of any commercial or financial relationships that could be construed as a potential conflict of interest.

Author details
[1]The Roslin Institute, Royal (Dick) School of Veterinary Studies, University of Edinburgh, Edinburgh, Easter Bush, Midlothian EH25 9RG, UK. [2]The European Commission for the Control of Foot-and-Mouth Disease (EuFMD) - Food and Agricolture Organization (FAO), Viale delle Terme di Caracalla, 00153 Rome, Italy. [3]Institute of Agricultural Research for Development, Regional Centre of Wakwa, NgaoundereP.O. Box 454Cameroon. [4]Institute of Ageing and Chronic Disease and School of Veterinary Science, University of Liverpool, Leahurst Campus, Neston, Liverpool, Wirral CH64 7TE, UK. [5]School of Veterinary Medicine and Sciences, University of Ngaoundere, NgaoundereP.O. Box 454Cameroon. [6]Cameroon Academy of Sciences, Yaound'eP.O. Box 1457Cameroon. [7]Food and Agriculture Organization (FAO), Animal Production and Health Division, Viale delle Terme di Caracalla, 00153 Rome, Italy. [8]Royal (Dick) School of Veterinary Studies, University of Edinburgh, Easter Bush, Edinburgh, Midlothian EH25 9RG, UK.

References
1. Daniel MG. Livestock mobility and animal health policy in southern Africa: the impact of veterinary cordon fences on pastoralists. Hum Ecol. 2002; 30(2):215–26. https://doi.org/10.1023/A:1015692730088.
2. Scoones I, Wolmer W. Livestock, disease, trade and markets: policy choices for the livestock sector in Africa. Technical report, Institute for Development Studies. In: University of Sussex; 2006.
3. Pamo ET. Country pasture/forage resource profiles - Cameroon. Technical report, food and agriculture Organization of the United Nations. Rome. 2008;
4. F'evre EM, Bronsvoort MC, Hamilton KA, Cleaveland S. Animal movements and the spread of infectious diseases. Trends Microbiol. 2006;14(3) https://doi.org/10.1016/j.tim.2006.01.004.
5. Macpherson CNL. The effect of transhumance on the epidemiology of animal diseases. Preventive Veterinary Medicine. 1995;25(2):213–24. https://doi.org/10.1016/0167-5877(95)00539-0.
6. Bronsvoort MC, Nfon C, Hamman SM, Tanya VN, Kitching RP, Morgan KL. Risk factors for herdsman-reported foot-and-mouth disease in the Adamawa Province of Cameroon. Preventive Veterinary Medicine. 2004;66:127–39. https://doi.org/10.1016/j.prevetmed.2004.09.010.
7. Bronsvoort M, Tanya VN, Kitching RP, Nfon C, Hamman SM, Morgan KL. Foot and mouth disease and livestock husbandry practices in the Adamawa Province of Cameroon. Trop Anim Health Prod. 2003;35(6):491–507. https://doi.org/10.1023/A:1027302525301.
8. Kelly RF, Hamman SM, Morgan KL, Nkongho EF, Ngwa VN, Tanya V, Andu WN, Sander M, Ndip L, Handel IG, Mazeri S, Muwonge A. Bronsvoort, B.M.D. C.: knowledge of bovine tuberculosis, cattle husbandry and dairy practices amongst pastoralists and small-scale dairy farmers in Cameroon. PLoS One. 2016;11(1):0146538. https://doi.org/10.1371/journal.pone.0146538.
9. Xiao N, Cai S, Moritz M, Garabed R, Pomeroy LW. Spatial and temporal characteristics of pastoral mobility in the far north region, Cameroon:

data analysis and modeling. PLoS One. 2015;10(7):131–9. https://doi.org/10.1371/journal.pone.0131697.

10. Handcock RN, Swain DL, Bishop-Hurley GJ, Patison KP, Wark T, Valencia P, Corke P, O'Neill CJ. Monitoring animal behaviour and environmental interactions using wireless sensor networks, GPS collars and satellite remote sensing. Sensors. 2009;9(5):3586–603. https://doi.org/10.3390/s90503586.

11. Anderson DM, Estell RE, Cibils AF. Spatiotemporal cattle data Plea for protocol standardization. Positioning. 2013;4:115–36. https://doi.org/10.4236/pos.2013.41012.

12. Thomas LF, De Glanville WA, Cook EA, F'evre EM. The spatial ecology of free-ranging domestic pigs (Sus scrofa) in western Kenya. BMC Vet Res. 2013;9(46) https://doi.org/10.1186/1746-6148-9-46.

13. Jean-Richard V, Crump L, Moto Daugla D, Hattendorf J, Schelling E, Zinsstag J. The use of mobile phones for demographic surveillance of mobile pastoralists and their animals in Chad: proof of principle. Glob Health Action. 2014;7:23209.

14. Schlecht E, Hu¨lsebusch C, Mahler F, Becker K. The use of differentially corrected global positioning system to monitor activities of cattle at pasture. Appl Anim Behav Sci. 2004;85(3):185–202. https://doi.org/10.1016/j.applanim.2003.11.003.

15. Schlecht E, Hiernaux P, Kadaour'e I, Hulsebusc¨h C, Mahler F. A spatio-temporal analysis of forage availability and grazing excretion behaviour of herded and free grazing cattle, sheep and goats in western Niger. Agric Ecosyst Environ. 2006;113(1):226–42. https://doi.org/10.1016/j.agee.2005.09.008.

16. Butt B, Shortridge A, WinklerPrins AMGA. Pastoral herd management, drought coping strategies, and cattle mobility in southern Kenya. Ann Assoc Am Geogr. 2009;99(2):309–34. https://doi.org/10.1080/00045600802685895.

17. Butt B. Seasonal space-time dynamics of cattle behavior and mobility among Maasai pastoralists in semi-arid Kenya. J Arid Environ. 2010;74(3):403–13. https://doi.org/10.1016/J.JARIDENV.2009.09.025.

18. Moritz M, Galehouse Z, Hao Q, Garabed RB. Can one animal represent an entire herd? Modeling pastoral mobility using GPS/GIS technology. Hum Ecol. 2012;40(4):623–30. https://doi.org/10.1007/s10745-012-9483-6.

19. Augustine D, Derner J. Assessing herbivore foraging behavior with GPS collars in a semiarid grassland. Sensors. 2013;13(3):3711–23. https://doi.org/10.3390/s130303711.

20. Raizman EA, Rasmussen HB, King LE, Ihwagi FW, Douglas-Hamilton I. Feasibility study on the spatial and temporal movement of Samburu's cattle and wildlife in Kenya using GPS radio-tracking, remote sensing and GIS. Preventive Veterinary Medicine. 2013;111(1):76–80. https://doi.org/10.1016/j.prevetmed.2013.04.001.

21. Feldt T, Schlecht E. Analysis of GPS trajectories to assess spatio-temporal differences in grazing patterns and land use preferences of domestic livestock in southwestern Madagascar. Pastoralism. 2016;6(1):5. https://doi.org/10.1186/s13570-016-0052-2.

22. Sonneveld BGJS, Keyzer MA, Georgis K, Pande S, Seid Ali A, Takele A. Following the afar: using remote tracking systems to analyze pastoralists' trekking routes. J Arid Environ. 2009;73(11):1046–50. https://doi.org/10.1016/j.jaridenv.2009.05.001.

23. Tempia S, Braidotti F, Aden HH, Abdulle MN, Costagli R, Otieno FT. Mapping cattle trade routes in southern Somalia: a method for mobile livestock keeping systems. Revue Scientifique Et Technique-Office International Des Epizooties. 2010;29:485–95.

24. Motta P, Porphyre T, Handel I, Hamman SM, Ngu Ngwa V, Tanya V, Morgan K, Christley R, Bronsvoort BM. Implications of the cattle trade network in Cameroon for regional disease prevention and control. Sci Rep. 2017;7:43932. https://doi.org/10.1038/srep43932.

25. Kelly RF, Hamman SM, Morgan KL, Nkongho EF, Ngwa VN, Tanya V, Andu WN, Sander M, Ndip L, Handel IG, Mazeri S, Muwonge A, Bronsvoort BMDC. Knowledge of bovine tuberculosis, cattle husbandry and dairy practices amongst pastoralists and small-scale dairy farmers in Cameroon. PLoS One. 2016; https://doi.org/10.1371/journal.pone.0146538.

26. MINEPIA: National and Regional Reports: Annual Livestock Productions. Technical report, Ministry of Livestock, Fisheries and Animal Industries of Cameroon (2014).

27. Rout PK, Mandal A, Singh LB, Roy R. Studies on behavioral patterns in Jamunapari goats. Small Rumin Res. 2002;43(2):185–8. https://doi.org/10.1016/S0921-4488(02)00011-1.

28. Dumont B, Boissy A, Achard C, Sibbald AM, Erhard HW. Consistency of animal order in spontaneous group movements allows the measurement of leadership in a group of grazing heifers. Appl Anim Behav Sci. 2005; 95(1):55–66. https://doi.org/10.1016/j.applanim.2005.04.005.

29. S''arov'a R, S'pinka M, Panam'a JLA. Synchronization and leadership in switches between resting and activity in a beef cattle herd: a case study. Appl Anim Behav Sci. 2007;108(3):327–31. https://doi.org/10.1016/j.applanim.2007.01.009.

30. Perotto-Baldivieso HL, Cooper SM, Cibils AF, Figueroa-Pag'an M, Udaeta K, Black-Rubio CM. Detecting autocorrelation problems from GPS collar data in livestock studies. Appl Anim Behav Sci. 2012;136(2):117–25. https://doi.org/10.1016/j.applanim.2011.11.009.

31. Ungar ED, Henkin Z, Gutman M, Dolev A, Genizi A, Ganskopp D. Inference of animal activity from GPS collar data on free-ranging cattle. Rangeland Ecology and Management. 2005;58:256–66. https://doi.org/10.2111/1551-5028.

32. Johnson DD, Ganskopp DC. GPS collar sampling frequency: effects on measures of resource use. Rangeland Ecology Management. 2008;61(61):226–31.

33. Lu Y. Spatial cluster analysis of point data: location quotients versus kernel density. In: University consortium of geographic information science summer assembly. Portland: University of Oregon; 2000.

34. Venables, W.N., Ripley, B.D.: Modern Appl Stat with S Fourth Edition, 4th EDN, pp. 125–134. Springer, ??? (2002).

35. Carlos HA, Shi X, Sargent J, Tanski S, Berke EM. Density estimation and adaptive bandwidths: a primer for public health practitioners. Int J Health Geogr. 2010;9:39.

36. Hijmans, R., van Etten Jacob, Cheng, J., Mattiuzzi, M.: package raster (2106).

37. Bivand, R., Keitt, T., Pebesma, E., Rouault, E.: Package rgdal (2017).

38. Duong, T.: Package 'ks' title kernel smoothing (2017).

39. Wickham, H., Chang, W.: Package ggplot2 (2016).

40. R Core Team. R: a language and environment for statistical computing. Vienna, Austria: R Foundation for Statistical Computing; 2013.

41. Chuan L, Patrick EC, Mohamed S, DeGloria SD. Spatiotemporal dynamics of cattle behavior and resource selection patterns on east African rangelands: evidence from GPS-tracking. Int J Geogr Inf Sci. 2018;10 https://doi.org/10.1080/13658816.2018.1424856.

42. Homburger H, Schneider MK, Hilfiker S, Lu, Scher A. Inferring behavioral states of grazing livestock from high-frequency position data alone. PLoS One. 2014;9(12) https://doi.org/10.1371/journal.pone.0114522.

43. Turner M, Kitchell E, Mcpeak J, Bourgoin J. Digital wiki map of pastoral geographies in eastern Senegal. Pastoralism: research, policy and Practice. 2017;7(31) https://doi.org/10.1186/s13570-017-0104-2.

44. Moritz M, Catherine BL, Drent AK, Kari S, Mouhaman A, Scholte P. Rangeland governance in an open system: protecting transhumance corridors in the far North Province of Cameroon. Pastoralism: Research, Policy and Practice. 2013;3:26. https://doi.org/10.1186/2041-7136-3-26.

45. Awa DN, Achukwu MD. Livestock pathology in the central African region: some epidemiological considerations and control strategies. Anim Health Res Rev. 2010;11(02):235–44. https://doi.org/10.1017/S1466252309990077.

46. Morgan KL, Handel IG, Tanya VN, Hamman SM, Nfon C, Bergman IE, Malirat V, Sorensen KJ, de C Bronsvoort BM. Accuracy of herdsmen reporting versus serologic testing for estimating foot-and-mouth disease prevalence. Emerg Infect Dis. 2014;20(12):2048–54. https://doi.org/10.3201/eid2012.140931.

47. Sellers RF, Parker J. Airborne excretion of foot-and-mouth disease virus. J Hyg, Camb. 1969:67.

48. Mardones F, Perez A, Sanchez J, Alkhamis M, Carpenter T. Parameterization of the duration of infection stages of serotype O foot-and-mouth disease virus: an analytical review and meta-analysis with application to simulation models. Vet Res. 2010;41(4):45. https://doi.org/10.1051/vetres/2010017.

49. Keeling MJ, Eames KTD. Networks and epidemic models. J R Soc Interface. 2005;2(4):295–307. https://doi.org/10.1098/rsif.2005.0051.

50. Robinson SE, Christley RM. Exploring the role of auction markets in cattle movements within great Britain. Preventive Veterinary Medicine. 2007;81(1):21–37. https://doi.org/10.1016/j.prevetmed.2007.04.011.

51. Dean AS, Fourni'e G, Kulo AE, Boukaya GA, Schelling E, Bonfoh B. Potential risk of regional disease spread in West Africa through cross-border cattle trade. PLoS One. 2013;8(10):75570. https://doi.org/10.1371/journal.pone.0075570.

52. Mahmoud, H.A.: Livestock Trade in the Kenyan, Somali and Ethiopian Borderlands. Technical report, The Royal Institute of International Affairs, London (2010).

53. crisisgroup.org: The Security Challenges of Pastoralism in Central Africa Technical report, International Crisis Group, Nairobi/Brussels (2014). https://d2071andvip0wj.cloudfront.net/the-security-challenges-of-pastoralism-in-central-africa.pdf

10

Polysplenia syndrome with duodenal and pancreatic dysplasia in a Holstein calf

Daisuke Kondoh[1]* , Tomomi Kawano[2], Tomoaki Kikuchi[1], Kaoru Hatate[2], Kenichi Watanabe[1], Motoki Sasaki[1], Norio Yamagishi[2], Hisashi Inokuma[2] and Nobuo Kitamura[1]

Abstract

Background: Laterality disorders of the abdominal organs include situs inversus totalis that mirrors the arrangements of all internal organs and heterotaxy syndrome (situs ambiguus) in which the thoracic or abdominal organs are abnormally arranged. Heterotaxy is often accompanied by multiple congenital malformations, and it generally comprises asplenia and polysplenia syndromes. To our knowledge, polysplenia syndrome has been reported in only three cattle, and computerized tomographic (CT) images of these animals were not obtained.

Case presentation: A six-month-old Holstein heifer had ruminal tympani and right abdominal distension. CT imaging showed that the rumen occupied the right side of the abdominal cavity, the omasum and abomasum occupied the left ventral side and the liver was positioned on the left. The colon and cecum were located at the left dorsum of the cavity, and the left kidney was located more cranially than the right. Postmortem findings revealed two spleens attached to the rumen. Significantly, the duodenum was too short to be divided into segments, except the cranial and descending parts, or flexures, except the cranial flexure, and the pancreas, which lacked a left lobe, was covered with mesojejunum. The liver comprised a relatively large right lobe and a small left lobe without quadrate and caudate lobes. The caudal vena cava that connected to the left azygous vein passed irregularly through the aortic hiatus of the diaphragm, and the common hepatic vein without the caudal vena cava passed through the caval foramen. Although the lungs and heart were morphologically normal, the right atrium received three major systemic veins. Polysplenia syndrome was diagnosed based on the CT and postmortem findings.

Conclusion: We defined the positions of the abdominal organs and morphological abnormalities in various organs of a calf with polysplenia syndrome based on CT and postmortem findings. These findings will improve understanding of the malpositioning and malformations that can occur in the organs of cattle with polysplenia syndrome.

Keywords: Cattle, Duodenum, Dysplasia, Heterotaxy, Laterality disorder, Malposition, Pancreas, Polysplenia

* Correspondence: kondoh-d@obihiro.ac.jp
[1]Division of Basic Veterinary Medicine, Obihiro University of Agriculture and Veterinary Medicine, Obihiro, Hokkaido, Japan
Full list of author information is available at the end of the article

Background

Laterality disorders are atypical arrangements of internal organs, including situs inversus totalis and heterotaxy (situs ambiguus). Situs inversus totalis is a condition in which all thoracic and abdominal organs mirror the normal arrangement, and in humans it is closely associated with primary ciliary dyskinesia, also known as immotile cilia syndrome. Heterotaxy is an abnormal arrangement of the thoracic or abdominal organs that is often accompanied by multiple congenital malformations, especially cardiovascular malformations that are associated with high morbidity rates in humans [1, 2]. Heterotaxy generally comprises asplenia (right isomerism) and polysplenia (left isomerism) syndromes but exceptions are numerous, and the spectrum of abnormalities seems to overlap [2]. The estimated prevalence of situs inversus totalis and of heterotaxy with cardiovascular malformation in humans is 1 per 8000–25,000 individuals and 1 per 10,000 live births, respectively [1]. Situs inversus totalis has been found in several dogs [3], cats [4], horses [5] and pigs [6], and heterotaxy in two dogs [7, 8] and a sheep [9] has also been reported.

A few reports have described laterality disorders of the abdominal organs in cattle [10–13]. As far as we can ascertain, only one of six reported cattle with laterality disorders of the abdominal organs had situs inversus totalis [12]. The other five had heterotaxy, consisting of polysplenia and asplenia syndromes in three and two, respectively [10, 11, 13], accompanied by cardiovascular malformations (irregular continuation of the caudal vena cava). These findings indicated that the features of heterotaxy in cattle are at least partly similar to those in humans. However, the location of abdominal organs in live cattle with laterality disorders has not been described in detail.

Here, we show the first computerized tomographic (CT) images of a calf with polysplenia syndrome and detailed postmortem findings of complicating duodenal and pancreatic dysplasia.

Case presentation

Animal and clinical findings

A six-month-old Holstein heifer with ruminal tympani was initially examined by a local veterinarian (day 1) who found right abdominal distension (Fig. 1a), and ruminal sounds from the right flank upon auscultation. Rectal palpation revealed the rumen and kidney on the right side. Ultrasound showed that the liver was located on the left side between the 5th and 8th intercostal spaces. Situs inversus was suspected, and the calf was transferred to the Animal Teaching Hospital, Obihiro University of

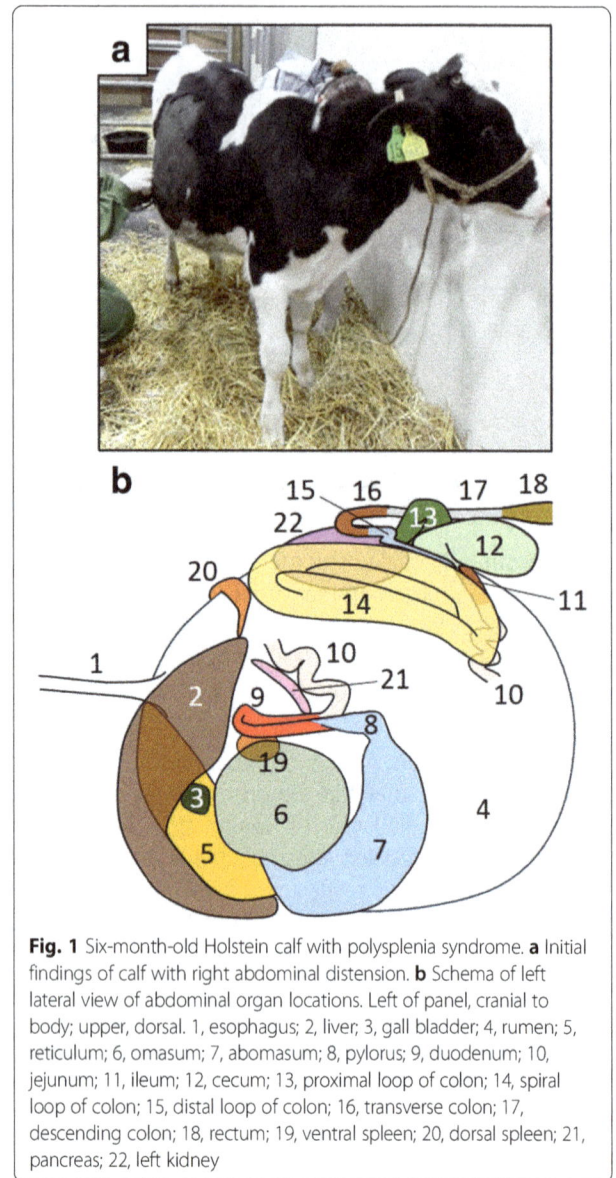

Fig. 1 Six-month-old Holstein calf with polysplenia syndrome. **a** Initial findings of calf with right abdominal distension. **b** Schema of left lateral view of abdominal organ locations. Left of panel, cranial to body; upper, dorsal. 1, esophagus; 2, liver; 3, gall bladder; 4, rumen; 5, reticulum; 6, omasum; 7, abomasum; 8, pylorus; 9, duodenum; 10, jejunum; 11, ileum; 12, cecum; 13, proximal loop of colon; 14, spiral loop of colon; 15, distal loop of colon; 16, transverse colon; 17, descending colon; 18, rectum; 19, ventral spleen; 20, dorsal spleen; 21, pancreas; 22, left kidney

Agriculture and Veterinary Medicine for further examination on day 9. At that time, the general health status of the calf appeared normal. Rectal temperature, heart rate and respiratory rate were 39.2 °C, 96 bpm and 48 breaths/min, respectively. Abdominal auscultation identified three ruminal movements every 2 min and a "ping" sound on the right flank. The findings of rectal palpation were similar to those on day 1.

Hematological findings of the red blood cell count (738 × 10^4/μL), hemoglobin concentration (9.1 g/dL), packed cell volume (27.7%) and white blood cell count (10,800/μL) were normal. Blood chemistry showed normal values for aspartate aminotransferase activity (91 U/L), blood urea nitrogen (6.8 mg/dL), sodium (142 mEq/L), potassium (4.5 mEq/L), chloride (101 mEq/L) and total protein (6.6 g/dL). The general health status of the calf did not change until euthanasia on day 35.

Figure 1b and Additional file 1: Figure S1 summarize a conceivable arrangement of the major abdominal organs in this calf based on the CT and postmortem findings described below.

CT findings

CT images showed that the rumen occupied the right side of the abdominal cavity (Fig. 2, Additional file 1: Figure S2). The saccus cranialis was undeveloped (Fig. 2b–j), and other compartments including saccus dorsalis and saccus ventralis were not detected in the rumen (Fig. 2d, e). The reticulum was located anterior to the rumen and contacted the diaphragm, and the omasum and abomasum occupied the left ventral side of the abdominal cavity (Fig. 2b–d, h–j). The colon and cecum were positioned at the most dorsal part of the abdominal cavity and to the left side of the left kidney (Fig. 2c–f). The liver was located at the most cranial position in the left side of the abdominal cavity and was surrounded by the diaphragm, reticulum and omasum (Fig. 2b, g–j). The left kidney was located more cranial than the right (Fig. 2c–f).

Postmortem findings

The calf was euthanized under anesthesia via an intravenous injection of xylazine followed by thiamylal sodium. The abdominal and thoracic organs were sequentially examined in detail throughout necropsy and isolated to evaluate morphological abnormalities. Various organs were fixed in Bouin's fluid, and paraffin-embedded sections were processed for histopathological analysis.

Organs in the abdominal cavity

The esophagus joined the rumen at the relative left cranial position and opened immediately dorsal to the reticulum (Fig. 3a). Sulci of the rumen were not significant, and only two longitudinal pillars were detected at the mucosal side (Fig. 3b). The reticulum

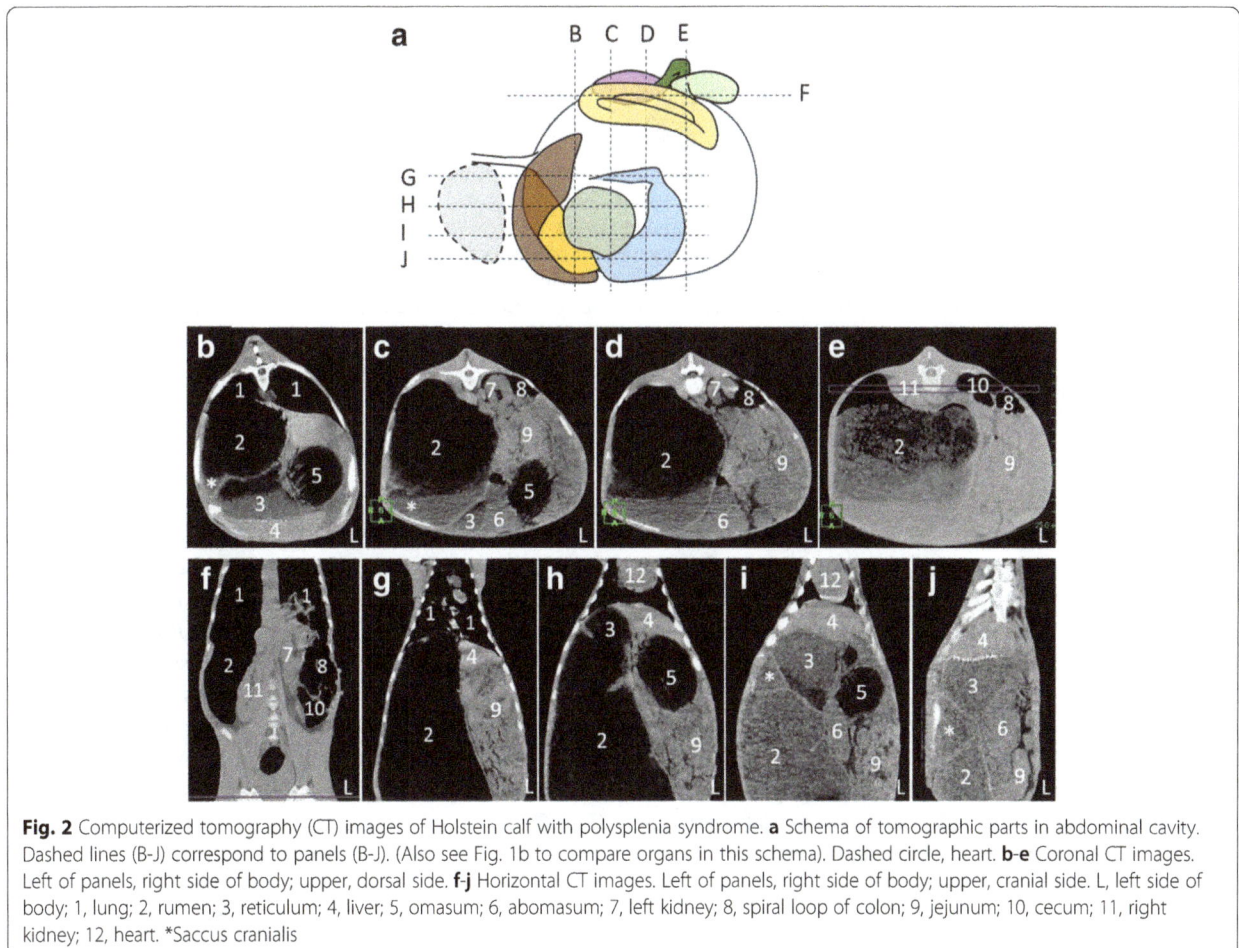

Fig. 2 Computerized tomography (CT) images of Holstein calf with polysplenia syndrome. **a** Schema of tomographic parts in abdominal cavity. Dashed lines (B-J) correspond to panels (B-J). (Also see Fig. 1b to compare organs in this schema). Dashed circle, heart. **b-e** Coronal CT images. Left of panels, right side of body; upper, dorsal side. **f-j** Horizontal CT images. Left of panels, right side of body; upper, cranial side. L, left side of body; 1, lung; 2, rumen; 3, reticulum; 4, liver; 5, omasum; 6, abomasum; 7, left kidney; 8, spiral loop of colon; 9, jejunum; 10, cecum; 11, right kidney; 12, heart. *Saccus cranialis

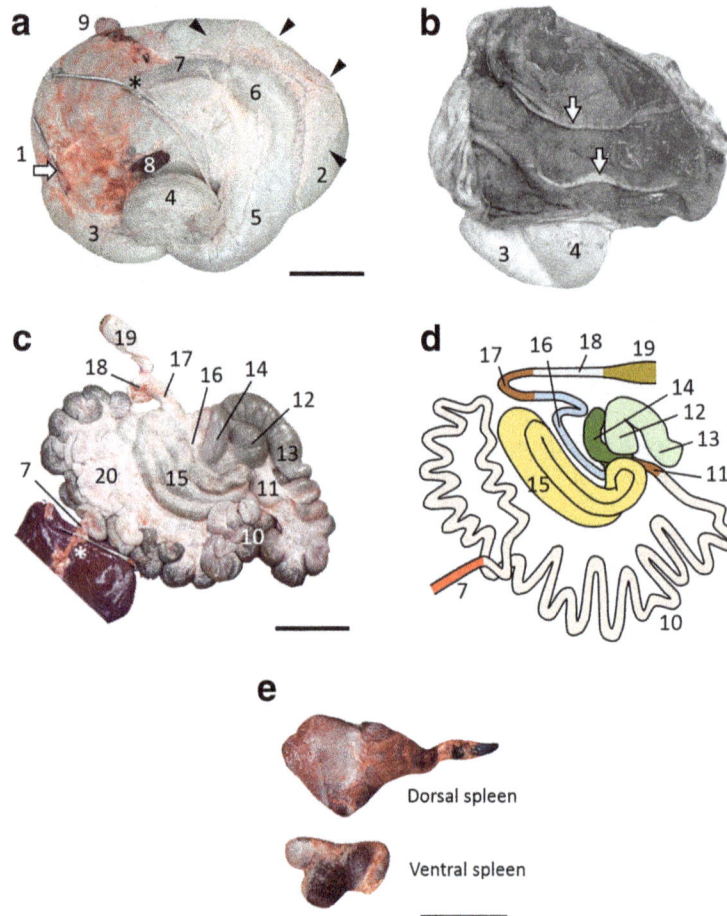

Fig. 3 Morphology of digestive tract and attached spleens. **a** Left lateral view of the isolated digestive tract from esophagus to duodenum and spleens. Arrowheads indicate greater omentum. *Severed part of duodenum conforms to asterisk in panel (**c**). Arrow, severed part of esophagus. Left of panels (**a-d**), cranial; upper, dorsal. **b** Internal structure of rumen with two pillars (arrows). **c** Left lateral view of isolated digestive tract from duodenum to transverse colon and liver. *Severed part of duodenum conforms to asterisk in panel (**a**). **d** Schema of lower digestive tract according to panel (**c**). **e** Morphology of dorsal and ventral spleens. Bar = 20 (**a** and **c**) and 10 (**e**) cm. 1, esophagus; 2, rumen; 3, reticulum; 4, omasum; 5, abomasum, 6, pylorus; 7, duodenum; 8, ventral spleen; 9, dorsal spleen; 10, jejunum; 11, ileum; 12, body of cecum; 13, apex of cecum; 14, proximal loop of colon; 15, spiral loop of colon; 16, distal loop of colon; 17, transverse colon; 18, descending colon; 19, rectum; 20, mesojejunum

that continued anteroventrally from the rumen was recognized the left lateral view of the stomach, and the omasum and abomasum were located at the left ventral side of the rumen (Fig. 3a). The duodenum that continued from the pylorus and ran cranially (Fig. 3a), turned caudally near the porta hepatis (Fig. 3a, c; asterisks) to form the cranial flexure, but not the sigmoid loop, and then joined the jejunum without flexures (Fig. 3c, d, Additional file 1: Figure S3). Therefore, the cranial and descending parts of the duodenum were identified, but the caudal and ascending parts were not. The jejunum was distinguished from the duodenum by the presence of the mesojejunum. The cecum and colon occupied the left dorsal region of the abdominal cavity, with the apex

of the cecum being located more medially than the spiral loop of the colon (Figs. 2e, 3c, d, Additional file 1: Figure S3). Proximal, spiral and distal loops were found in the ascending colon, and a loose distal loop continued to the transverse colon (Fig. 3c, d, Additional file 1: Figure S3). Dorsal and ventral spleens were attached to the cranial part of the rumen (Fig. 3a, e). The ventral spleen was located on the left between the rumen and the omasum, and the dorsal spleen was located at the midline of the rumen (Fig. 3a).

The size of lesser omentum was normal, but the greater omentum was very small (Fig. 3a). The lessor omentum was attached to the omasum and liver as usual, although it was located at left side of the body.

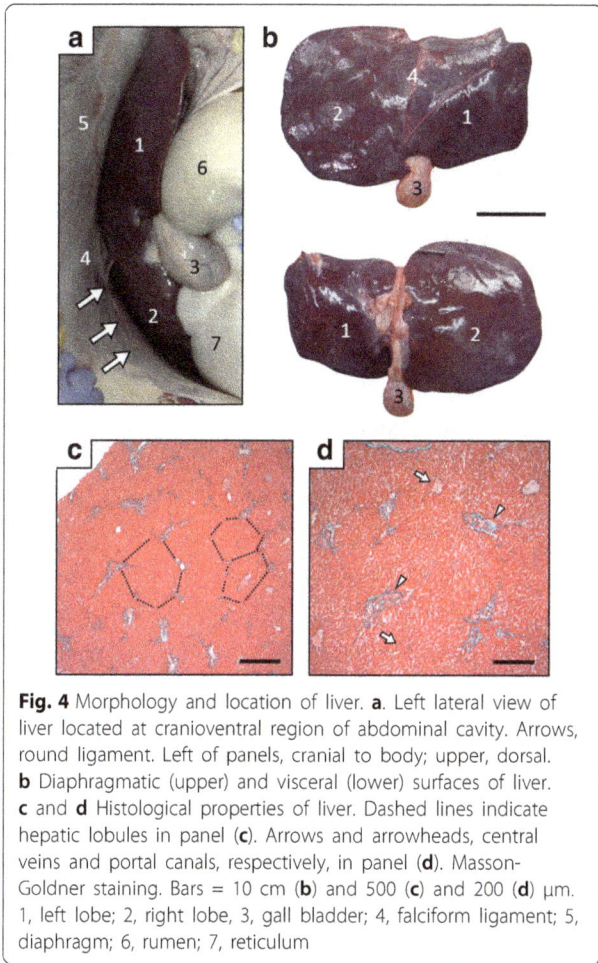

Fig. 4 Morphology and location of liver. **a**. Left lateral view of liver located at cranioventral region of abdominal cavity. Arrows, round ligament. Left of panels, cranial to body; upper, dorsal. **b** Diaphragmatic (upper) and visceral (lower) surfaces of liver. **c** and **d** Histological properties of liver. Dashed lines indicate hepatic lobules in panel (**c**). Arrows and arrowheads, central veins and portal canals, respectively, in panel (**d**). Masson-Goldner staining. Bars = 10 cm (**b**) and 500 (**c**) and 200 (**d**) μm. 1, left lobe; 2, right lobe, 3, gall bladder; 4, falciform ligament; 5, diaphragm; 6, rumen; 7, reticulum

The greater omentum was attached to the greater curvature of abomasum, spread caudally, immediately turned cranially and ended at the left side of rumen.

The liver was located at the left anteroventral area of the abdominal cavity, and the falciform and round ligaments as well as the gall bladder were found in the left lateral view of the cavity (Fig. 4a). The relatively large right lobe and small left lobe were identified by the origin of the falciform ligament, but the caudate lobe was not distinguished (Fig. 4). Because the fissure for round ligament (fissure ligamentum teretis) entered the nearby region of the fossa for gall bladder (fossa vesicae felleae), the quadrate lobe was not recognized (Fig. 4b).

An elongated, pole-like pancreas was abnormally covered with mesojejunum (Fig. 5a–c, Additional file 1: Figure S4) and comprised the body and the right, but not the left lobe (Fig. 5d). The hepatic portal veins passed through the pancreatic notch (Fig. 5a, b, Additional file 1: Figure S4), and the accessory pancreatic duct that joined the descending part of the duodenum protruded from the right lobe (Fig. 5c).

Both the left and right kidneys were retroperitoneal, and the smaller left kidney was secured more cranially than the right (Fig. 6a and b). The shape of the right adrenal gland was irregularly pole-like, whereas the left was comma-shaped as normal (Fig. 6b). The positions and morphological findings of other urogenital organs were normal.

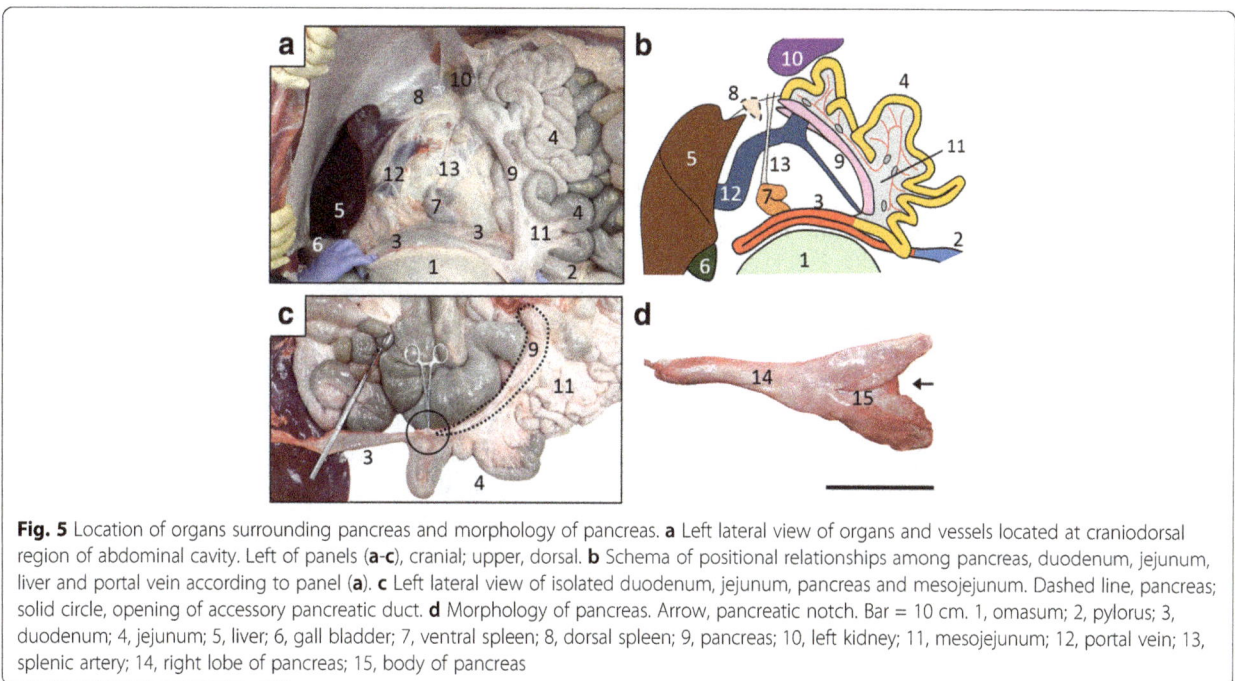

Fig. 5 Location of organs surrounding pancreas and morphology of pancreas. **a** Left lateral view of organs and vessels located at craniodorsal region of abdominal cavity. Left of panels (**a-c**), cranial; upper, dorsal. **b** Schema of positional relationships among pancreas, duodenum, jejunum, liver and portal vein according to panel (**a**). **c** Left lateral view of isolated duodenum, jejunum, pancreas and mesojejunum. Dashed line, pancreas; solid circle, opening of accessory pancreatic duct. **d** Morphology of pancreas. Arrow, pancreatic notch. Bar = 10 cm. 1, omasum; 2, pylorus; 3, duodenum; 4, jejunum; 5, liver; 6, gall bladder; 7, ventral spleen; 8, dorsal spleen; 9, pancreas; 10, left kidney; 11, mesojejunum; 12, portal vein; 13, splenic artery; 14, right lobe of pancreas; 15, body of pancreas

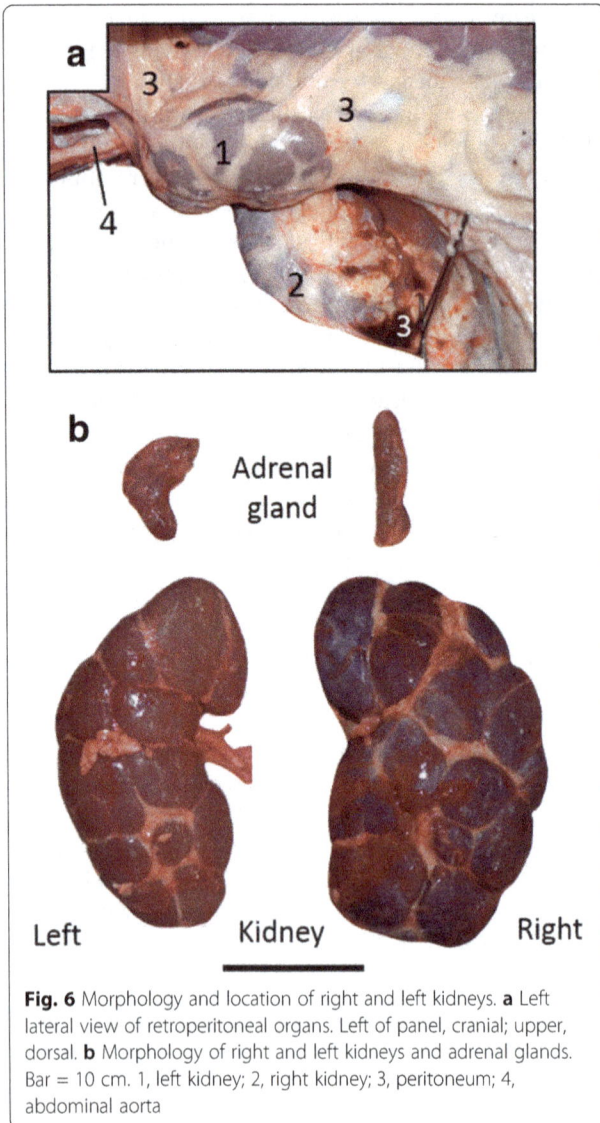

Fig. 6 Morphology and location of right and left kidneys. **a** Left lateral view of retroperitoneal organs. Left of panel, cranial; upper, dorsal. **b** Morphology of right and left kidneys and adrenal glands. Bar = 10 cm. 1, left kidney; 2, right kidney; 3, peritoneum; 4, abdominal aorta

The gastrointestinal tract, liver (Fig. 4c, d), spleen, pancreas, kidney, adrenal gland and uterus were histologically normal.

Diaphragm, vascular system and organs in the thoracic cavity

The diaphragm possessed an aortic hiatus between the left and right crura, an esophageal hiatus through the ventral region of the right crus, and a caval foramen through the central tendon (Fig. 7a). The celiac, cranial mesenteric, renal and caudal mesenteric arteries were normally derived from the abdominal aorta. The caudal vena cava that connected with the renal veins irregularly joined the left azygos vein and passed through the aortic hiatus of the diaphragm (Fig. 7b). Several hepatic veins joined to form the common hepatic vein without a caudal vena cava, and it passed through the caval foramen (Fig. 7c).

The position and morphological findings of the atria and ventricles were normal except for the venous input pathway to the right atrium (Fig. 8a), which received the major systemic cranial vena cava, left azygos and common hepatic veins. The anastomosis of the cranial vena cava and the common hepatic vein formed the sinus venarum cavarum (Fig. 8b). On the other hand, the large azygos vein opened into a distended coronary venous sinus (Fig. 8b). The position and morphological findings of the lungs were normal (Fig. 9).

The heart, lung and trachea were histologically normal. Light and electron microscopy showed that the respiratory epithelium had the cilia (Fig. 10).

Discussion

Atypical arrangements of internal organs are generally categorized as situs inversus totalis and heterotaxy (situs ambiguus), and the latter essentially comprises asplenia and polysplenia syndromes. The present study found situs inversus of the abdominal organs, situs solitus of the thoracic organs and two spleens in a six-month-old Holstein calf, indicating a diagnosis of polysplenia syndrome. Although polysplenia and asplenia syndromes are generally known as left and right isomerism, respectively, in humans, both are often atypical and are accompanied by multiple congenital malformations [1, 2]. The morphological features of the liver, lungs and atria, as well as the position of the apex of the cecum without isomerism in this calf differed from those typical of human polysplenia syndrome. Further studies are needed to compare the characteristics of polysplenia syndrome between cattle and humans.

Three reports have described cattle with two spleens [10, 11, 13], and among six cattle described to date, three had atypically arranged abdominal organs (stomach and liver) indicating heterotaxy, namely, polysplenia syndrome. However, none of these animals had isomerism of the lungs and atria. On the other hand, among three cattle without spleen [11], two had atypically arranged abdominal organs [11] that was recognized as heterotaxy, namely asplenia syndrome. Both of these, as well as a remaining one, had right isomerism of the lungs. These facts suggest that the features of polysplenia syndrome in cattle differ from those in humans with left organ isomerism [8, 9], whereas asplenia syndrome in cattle might be similar to that in humans with typically right isomerism [1, 2]. In addition, cardiovascular malformations accompany not only polysplenia or asplenia syndromes in cattle with situs inversus, but

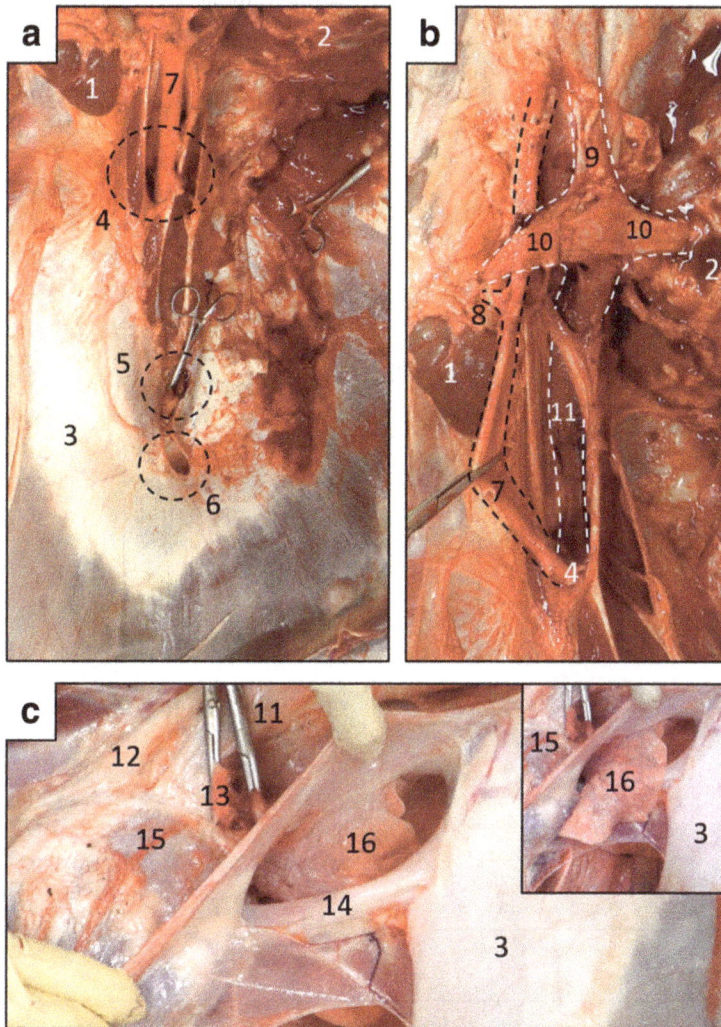

Fig. 7 Morphology of diaphragm and major vascular system. **a** Caudal view of diaphragm. Left of panel, left of body; upper, dorsal. **b** Ventral view of abdominal aorta and caudal vena cava that joins left azygous vein. Black and white dashed lines, arteries and veins, respectively. Left of panel, left of body; upper, caudal. **c** Left lateral view of common hepatic vein that passes through caval foramen into thoracic cavity. Accessory lobe of right lung is evident dorsally. Insert, accessory lobe covering common hepatic vein. Left of panel, cranial to body; and upper, dorsal. 1, left kidney; 2, right kidney; 3, central tendon; 4, aortic hiatus; 5, esophageal hiatus; 6, caval foramen; 7, abdominal aorta; 8, renal artery; 9, caudal vena cava; 10, renal vein; 11, left azygos vein; 12, thoracic aorta; 13, pulmonary trunk; 14, common hepatic vein; 15, left atrium; 16, accessory lobe of right lung

also those with situs solitus of the abdominal organs regardless of having no or two spleens [11]. These facts indicate that association rules are not strict among the number of spleens, atypically arranged abdominal organs and cardiovascular malformations, and that the mechanism through which heterotaxy is generated seems complex and varied, at least in cattle.

Although severe dysplasia of duodenum and pancreas has not been reported as a complicating condition of polysplenia syndrome in cattle, a malformed annular or short pancreas and intestinal malrotation

have been identified in several humans with polysplenia syndrome [14–17]. The present findings indicate that polysplenia syndrome in cattle is also accompanied by multiple organ malformations including duodenal and pancreatic dysplasia.

The calf described herein had abnormalities of the stomach, intestine, liver, kidneys, spleen and cardiovascular system, which, except for severe dysplasia of duodenum and pancreas, seemed broadly similar to those of the three previously described cattle with polysplenia syndrome [10, 11, 13], but there are also some relatively-small differences among these four

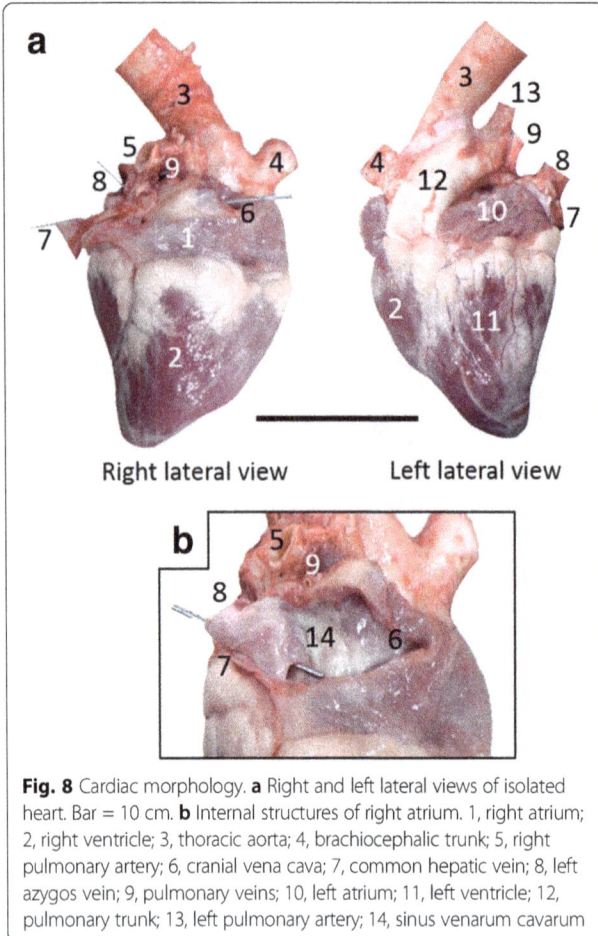

Fig. 8 Cardiac morphology. **a** Right and left lateral views of isolated heart. Bar = 10 cm. **b** Internal structures of right atrium. 1, right atrium; 2, right ventricle; 3, thoracic aorta; 4, brachiocephalic trunk; 5, right pulmonary artery; 6, cranial vena cava; 7, common hepatic vein; 8, left azygos vein; 9, pulmonary veins; 10, left atrium; 11, left ventricle; 12, pulmonary trunk; 13, left pulmonary artery; 14, sinus venarum cavarum

cattle. Caudal vena cava continued to right and left azygos veins in two cattle [11, 13] and the calf reported herein, respectively, while that in the remaining one [10] joined common hepatic vein in the thoracic cavity. Two cattle with polysplenia syndrome [10, 11] possessed four lobes of liver with mirror image, while only three and two lobes with normal image were detected in the cow reported by Boos et al. [13] and the calf in the present case, respectively. In addition, Boos et al. [13] described some histological abnormalities of liver in the cow with polysplenia syndrome. Both two spleens were located at left side of rumen in two cattle [10, 11], while each spleen was attached to the left and right sides in a remaining cow [13], unlike the calf described herein. These differences indicate that various clinical conditions are recognized in cattle with polysplenia syndrome.

Conclusion

The present CT and postmortem findings of a calf with laterality disorders allowed a detailed study of the abdominal organ positions (Fig. 1b) as well as of morphological abnormalities in various abdominal organs and the cardiovascular system. Our findings improve understanding of the malpositions and various types of malformations among the abdominal and thoracic organs of cattle with polysplenia syndrome.

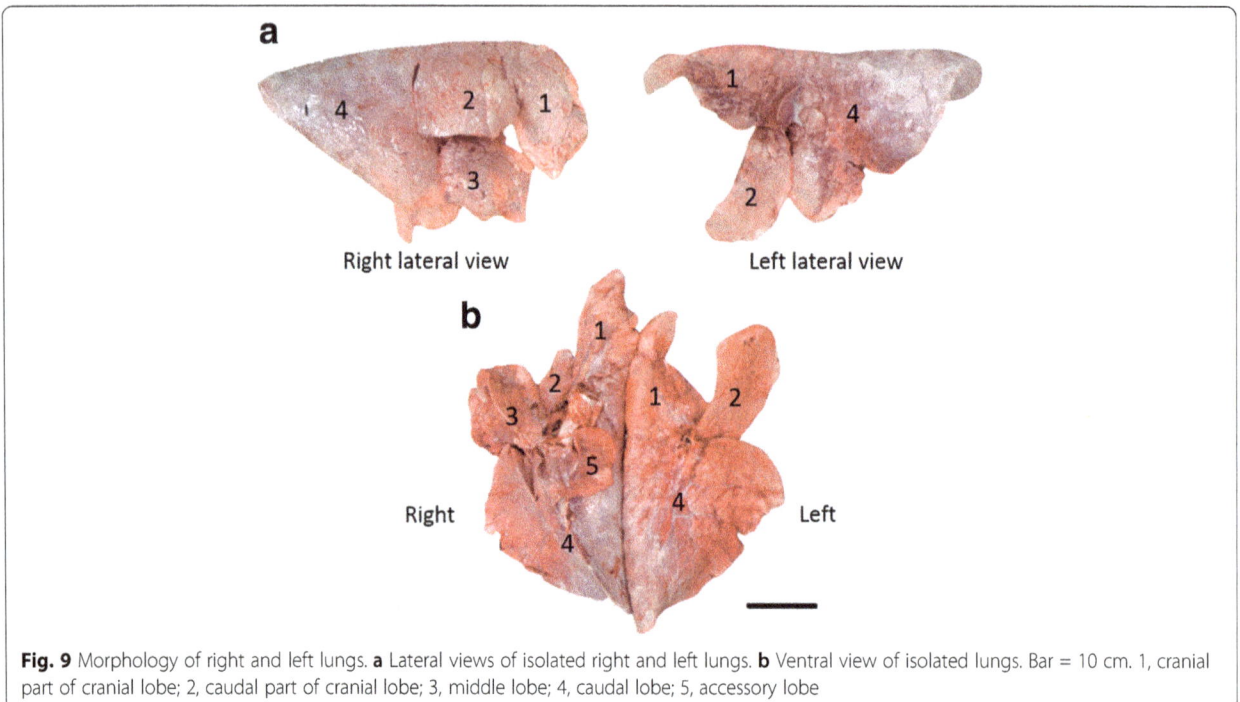

Fig. 9 Morphology of right and left lungs. **a** Lateral views of isolated right and left lungs. **b** Ventral view of isolated lungs. Bar = 10 cm. 1, cranial part of cranial lobe; 2, caudal part of cranial lobe; 3, middle lobe; 4, caudal lobe; 5, accessory lobe

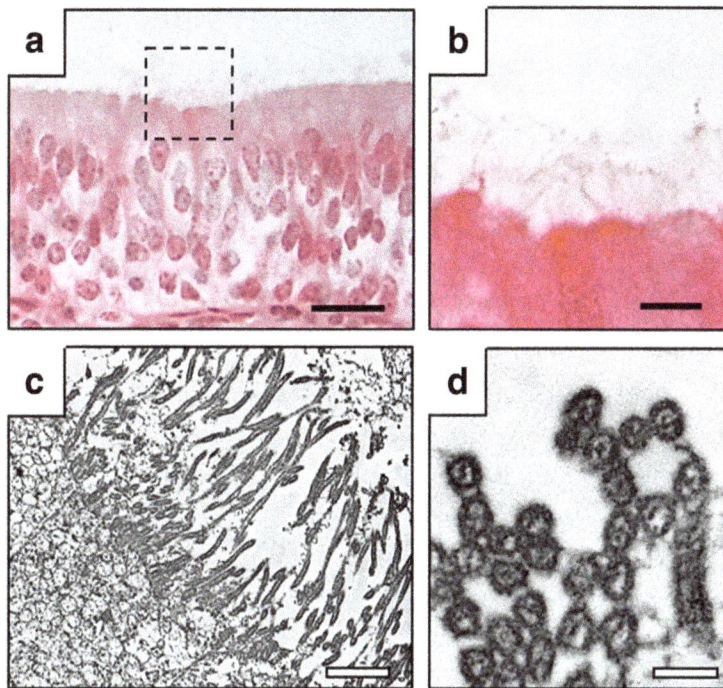

Fig. 10 Histology of respiratory epithelium in trachea of Holstein calf with polysplenia syndrome. **a** Whole image of respiratory epithelium. Dashed box, region shown in panel (**b**) at high magnification. Hematoxylin-eosin staining. **b** cilia of respiratory epithelium. **c** and **d** Back transmission electron microscopic features of cilia, after paraffin embedding. Bars = 20 (**a**), 5 (**b**), 2 (**c**) and 0.3 (**d**) μm

Additional file

Additional file 1: Figure S1. Schematic comparison of abdominal organs between calf presented herein (**A**) and general cattle (**B** and **C**), in association with Fig. 1. 1–22, corresponding to Fig. 1; 23, right kidney; 24, spleen; 25, Grooves of rumen. **Figure S2.** Comparison of computerized tomography (CT) images between calf presented herein (**A** and **C**) and general cattle (**B** and **D**), in association with Fig. 2. 1–6 and asterisks; corresponding to Fig. 2. **Figure S3.** Schematic comparison of digestive tracts between calf presented herein (**A**) and general cattle (**B**), in association with Fig. 3. Numbers are corresponding to Fig. 3. **Figure S4.** Schematic comparison of organs surrounding pancreas between calf presented herein (**A**) and general cattle (**B**), in association with Fig. 5. 1–13, corresponding to Fig. 3 (except 11, coalesced mesojejunum and mesocolon); 14–16, Body, right robe and left lobe of pancreas, respectively.

Abbreviation
CT: Computerized tomographic

Acknowledgements
The authors thank the veterinarians of Tokachi Agricultural Mutual Aid Association (Tokachi NOSAI, Japan) and the staff at the Animal Teaching Hospital, Obihiro University of Agriculture and Veterinary Medicine.

Funding
The Japanese Association of Veterinary Anatomists provided partial financial support for publication fees.

Authors' contributions
DK, TKa and TKi performed the necropsy and described the postmortem findings, and KW performed the histopathological diagnosis. KH and NY performed CT imaging, and DK, TKa, TKi, KH and NY analyzed the images and described the CT procedure. TKa and HI clinically examined the animal and described the clinical findings. DK, KW, MS, NY, HI and NK provided helpful discussion, critically revised the manuscript and added final corrections. All authors approved the final version of the manuscript for submission.

Consent for publication
The calf described herein was owned by a farmer in Tokachi, Japan, and the owner approved the present procedures and publication. The CARE guidelines were adhered to this case report.

Competing interests
The authors declare that they have no competing interests.

Author details
[1]Division of Basic Veterinary Medicine, Obihiro University of Agriculture and Veterinary Medicine, Obihiro, Hokkaido, Japan. [2]Division of Clinical Veterinary Medicine, Obihiro University of Agriculture and Veterinary Medicine, Obihiro, Hokkaido, Japan.

References
1. Zhu L, Belmont JW, Ware SM. Genetics of human heterotaxias. Eur J Hum Genet. 2006;14:17–25.
2. Sutherland MJ, Ware SM. Disorders of left-right asymmetry: heterotaxy and situs inversus. Am J Med Genet C Semin Med Genet. 2009;151C:307–17.
3. Cavrenne R, De Busscher V, Bolen G, Billen F, Clercx C, Snaps F. Primary ciliary dyskinesia and situs inversus in a young dog. Vet Rec. 2008;163:54–5.

4.   Jerram RM, Warman CG, Wu CT. Echocardiographic and radiographic
     diagnosis: complete situs inversus in a cat. Vet Radiol Ultrasound. 2006;47:
     313–5.
5.   Palmers K, van Loon G, Jorissen M, Verdonck F, Chiers K, Picavet MT, Deprez
     P. Situs inversus totalis and primary ciliary dyskinesia (Kartagener's
     syndrome) in a horse. J Vet Intern Med. 2008;22:491–4.
6.   Evans HE. Cyclopia, situs inversus and widely patent ductus arteriosus in a
     new-born pig, *Sus scrofa*. Anat Histol Embryol. 1987;16:221–6.
7.   Kayanuma H, Suganuma T, Shida T, Sato S. A canine case of partial heterotaxia
     detected by radiography and ultrasound. J Vet Med Sci. 2000;62:897–9.
8.   Zwingenberger AL, Spriet M, Hunt GB. Imaging diagnosis-portal vein aplasia
     and interruption of the caudal vena cava in three dogs. Vet Radiol
     Ultrasound. 2011;52:444–7.
9.   Larsen C, Kirk EJ. Abdominal situs inversus in a sheep. N Z Vet J. 1987;35:
     113–4.
10. Fisher KRS, Wilson MS, Partlow GD. Abdominal situs inversus in a Holstein
     calf. Anat Rec. 2002;267:47–51.
11. Okada K, Kuroshima T, Murakami T. Asplenia and polysplenia in cattle. Adv
     Anim Cardiol. 2007;40:39–47.
12. Murakami T, Hagino M, Kaizo S. Situs inversus in a calf. J Jpn Vet Med Assoc.
     2008;61:55–8.
13. Boos A, Geyer H, Müller U, Peter J, Schmid T, Gerspach C, et al. Situs ambiguus
     in a Brown Swiss cow with polysplenia: case report. BMC Vet Res. 2013;9:34.
14. Sriplung H. Polysplenia syndrome: a case with congenital heart block,
     infarction of a splenic mass, and a short pancreas. J Med Assoc Thail. 1991;
     74:355–8.
15. Maier M, Wiesner W, Mengiardi B. Annular pancreas and agenesis of the
     dorsal pancreas in a patient with polysplenia syndrome. AJR Am J
     Roentgenol. 2007;188:W150–3.
16. Kayhan A, Lakadamyali H, Oommen J, Oto A. Polysplenia syndrome
     accompanied with situs inversus totalis and annular pancreas in an elderly
     patient. Clin Imaging. 2010;34:472–5.
17. Ben Ahmed Y, Ghorbel S, Chouikh T, Nouira F, Louati H, Charieg A,
     Chaouachi B. Combination of partial situs inversus, polysplenia and annular
     pancreas with duodenal obstruction and intestinal malrotation. JBR-BTR.
     2012;95:257–60.

11

Evaluation of the limitations and methods to improve rapid phage-based detection of viable *Mycobacterium avium* subsp. *paratuberculosis* in the blood of experimentally infected cattle

Benjamin M. C. Swift[1*], Jonathan N. Huxley[1], Karren M. Plain[3], Douglas J. Begg[3], Kumudika de Silva[3], Auriol C. Purdie[3], Richard J. Whittington[3] and Catherine E. D. Rees[2]

Abstract

Background: Disseminated infection and bacteraemia is an underreported and under-researched aspect of Johne's disease. This is mainly due to the time it takes for *Mycobacterium avium* subsp. *paratuberculosis* (MAP) to grow and lack of sensitivity of culture. Viable MAP cells can be detected in the blood of cattle suffering from Johne's disease within 48 h using peptide-mediated magnetic separation (PMMS) followed by bacteriophage amplification. The aim of this study was to demonstrate the first detection of MAP in the blood of experimentally exposed cattle using the PMMS-bacteriophage assay and to compare these results with the immune response of the animal based on serum ELISA and shedding of MAP by faecal culture.

Results: Using the PMMS-phage assay, seven out of the 19 (37 %) MAP-exposed animals that were tested were positive for viable MAP cells although very low numbers of MAP were detected. Two of these animals were positive by faecal culture and one was positive by serum ELISA. There was no correlation between PMMS-phage assay results and the faecal and serum ELISA results. None of the control animals (10) were positive for MAP using any of the four detection methods. Investigations carried out into the efficiency of the assay; found that the PMMS step was the limiting factor reducing the sensitivity of the phage assay. A modified method using the phage assay directly on isolated peripheral blood mononuclear cells (without PMMS) was found to be superior to the PMMS isolation step.

Conclusions: This proof of concept study has shown that viable MAP cells are present in the blood of MAP-exposed cattle prior to the onset of clinical signs. Although only one time point was tested, the ability to detect viable MAP in the blood of subclinically infected animals by the rapid phage-based method has the potential to increase the understanding of the pathogenesis of Johne's disease progression by warranting further research on the presence of MAP in blood.

Keywords: Johne's disease, Paratuberculosis, Mycobacteria, Bacteriophage, Bacteraemia, Blood test, Detection method

* Correspondence: benjamin.swift@nottingham.ac.uk
[1]School of Veterinary Medicine and Science, Sutton Bonington Campus, Loughborough, Leics LE12 5RD, UK
Full list of author information is available at the end of the article

Background

Mycobacterium avium subsp. *paratuberculosis* (MAP) is a slow-growing bacterium that causes Johne's disease, a wasting disease in ruminants and other animals. A common test for Johne's disease is the serum antibody ELISA test which monitors the humoral immune response of the animal following MAP exposure. However there are well known limitations of this test, for example the sensitivity can be extremely low, especially during subclinical stages of infection [1]. It has been established that there can be a bacteraemic phase in paratuberculosis, which has been demonstrated by both PCR [2] and culture [3, 4]. However the PCR cannot differentiate between live and dead cells and the sensitivity of the PCR assay is limited by inhibitory substances in the blood and the likelihood of there being a low number of MAP cells present [5]. However, studies that have compared the detection of MAP in blood by PCR to ELISAs have found an association but poor correlations between the tests [5, 6]. The long periods of time required for culture of this organism (even using automated systems) means that to date very few studies of blood samples have been completed so that the incidence, intensity and duration of mycobacteraemia is not known in any animal species.

We have previously described the use of a rapid, bacteriophage-based method (phage amplification assay) coupled with PMMS to specifically detect and identify viable MAP cells in the blood of naturally infected animals [7]. The organism was detected in milk and serum ELISA positive animals, but not from a certified Johne's disease free herd that was milk and serum ELISA negative. The PMMS-phage method was employed here as components present in the whole blood inhibited the phage assay to such as extent that the sensitivity of the assay was not useful [7]. Using PMMS it was possible the capture the MAP cells and suspend them in a medium suitable for the phage assay. The stages of natural infection are difficult to ascertain in naturally infected cattle and given the known limitations of the ELISA tests for diagnosis of early infection this was not surprising, but it means that more data is needed to understand the relationship between bacterial load during mycobacteraemia and the blood ELISA results. The aim of this investigation was to apply the PMMS-phage assay to determine whether MAP could be detected in the blood of experimentally exposed cattle, where their stage of infection could be defined. Previously we have demonstrated that the MAP cells are found within the PMBC fraction of blood [7], and we have defined this as bacteraemia even though the MAP cells are intracellular rather than free in the blood stream. It is also possible that the assay would also detect MAP cells free in the bloodstream but to date we have not formally tested this possibility. The presence of viable MAP in cattle blood has previously been demonstrated in serum ELISA positive, inconclusive and negative animals by the PMMS-phage assay in an uncontrolled environment [7], thus the significance of mycobacteraemia is unknown, when compared to tests that are insensitive when used on subclinically infected animals, such as the ELISA test. As it can be assumed that the presence of this organism in the blood of the animal indicates that it has crossed the gut and disseminated, the results would be used to determine if evidence of disseminated infection correlates with the recorded immune response of an animal in a controlled environment.

Results

Use of the blood assay on MAP-exposed, subclinically infected cattle

During our study, none of the animals had weight loss or signs related to Johne's disease. Blood samples were collected at 4 years, 8 months post-exposure from all animals for the PMMS-phage assay. Faecal and blood samples were collected in the same month for faecal culture, PCR and serum antibody ELISA testing. The unexposed cohort remained MAP-negative as determined by faecal culture, serum Ab ELISA and faecal PCR (Table 1) by the end of the experiments. One of the exposed animals (#14) was euthanized due to reasons unrelated to MAP status. Samples from two of the control animals (#3 and #10) produced plaques, however none of these gave a positive result when the IS*900* PCR was performed, indicating that these plaques represented either breakthrough (incomplete inactivation of all extracellular phage particles results in *M. smegmatis* plaques) or detection of a mycobacterium other than MAP [8]. Therefore the phage results agreed with all other tests performed in the negative control herd. Thus none of the control animals were positive for MAP by any of the methods used.

Two of the inoculated animals (#17 and 23) were shedding MAP in their faeces, according to the results of both faecal culture and faecal PCR. Animal #23 was also positive by serum Ab ELISA and animal #17 gave a borderline positive serum Ab ELISA (49 %; cut-off value for a positive test result = 55 %). Using the phage-PCR MAP assay, seven out of the 19 (37 %) MAP-exposed animals that were tested had positive results, indicating that viable MAP cells were detected in their blood samples. However compared to other studies we have performed, only very low numbers of plaques (2–5) were produced in the MAP-positive samples. Interestingly animals #20 and #27, which were positive for the presence of MAP in their blood according to the PMMS-phage assay, had the next two highest (36 % and 41 %, respectively, although classed as negative) serum ELISA test results (Table 1). However there was no overall correlation between the results of phage-PCR and either faecal PCR or serum Ab ELISA.

Table 1 Results from the PMMS-phage assay, faecal culture, faecal PCR and serum ELISA for sub-clinical, experimentally exposed cattle to MAP

			Results			
Sample	Breed	MAP exposure status[a]	PMMS-Phage assay[b]	Faecal culture	Faecal PCR[c]	Serum Ab ELISA (%)[d]
1	Holstein	Control	- (0)	-	-	1.00
2	Holstein	Control	- (0)	-	-	29.34
3	Holstein	Control	- (2)	-	-	3.85
4	Holstein	Control	- (0)	-	-	5.92
5	Holstein	Control	- (0)	-	-	5.42
6	Holstein	Control	- (0)	-	-	3.14
7	Holstein	Control	- (0)	-	-	17.49
8	Holstein	Control	- (0)	-	-	10.49
9	Red/Holstein	Control	- (0)	-	-	3.50
10	Red/Holstein	Control	- (3)	-	-	4.57
11	Holstein	Inoculated	- (0)	-	-	2.57
12	Holstein	Inoculated	+ (2)	-	*	3.28
13	Holstein	Inoculated	- (0)	-	-	5.00
14	Holstein	Inoculated	#[e]	#[e]	#[e]	#[e]
15	Holstein	Inoculated	- (5)	-	-	14.49
16	Holstein	Inoculated	+ (5)	-	-	3.28
17	Holstein	Inoculated	- (0)	+	+	48.68
18	Holstein	Inoculated	+ (2)	-	-	8.57
19	Holstein	Inoculated	- (0)	-	-	6.21
20	Holstein	Inoculated	+ (2)	-	-	35.62
21	Holstein	Inoculated	- (2)	-	-	12.63
22	Holstein	Inoculated	- (0)	-	-	3.50
23	Holstein	Inoculated	- (0)	+	+	118.77
24	Holstein	Inoculated	- (0)	-	-	6.50
25	Holstein	Inoculated	- (0)	-	*	4.64
26	Red/Holstein	Inoculated	+ (4)	-	-	5.21
27	Holstein	Inoculated	+ (5)	-	*	41.11
28	Holstein	Inoculated	- (0)	-	-	2.00
29	Red/Holstein	Inoculated	+ (3)	-	-	13.28
30	Red/Holstein	Inoculated	- (0)	-	*	1.50

[a]Inoculated – animals experimentally exposed to MAP; Control – animals not exposed to MAP
[b]+/- indicates result of combined PMMS-phage –PCR assay. Plaque numbers for each sample given in brackets
[c]* indicates MAP DNA was detected but the quantity detected was below the cut-point for a positive result [16]
[d]Serum Ab ELISA (IDEXX); Positive value > 55 %, suspected value 45–55 %
[e]# Animal 14 was culled due to other illness unrelated to Johne's disease before sample collection

Investigation of PMMS MAP isolation efficiency

As the number of MAP cells detected using the PMMS-phage assay was much lower than recorded in our previous studies [7], the efficiency of the PMMS isolation step was investigated. Cattle strains of MAP were 10-fold serially diluted from 1×10^4 to 1×10^0 viable MAP cells. These cells were inoculated in 7H9 medium and the PMMS method to isolate the MAP cells was carried out. The phage enumeration assay was then used to determine how many MAP cells were captured from the samples. It was found that for this batch of peptide-coated beads the limit of detection of the MAP cells was only 7.3×10^2 pfu.ml^{-1} (Table 2). This indicated that the PMMS MAP isolation method was only isolating roughly 10 % of the MAP cells present in the samples, thus explaining the low levels of plaques recorded in in the blood of MAP-positive samples.

Table 2 Efficiency of phage-PMMS method in isolating and detecting MAP cells compared to MPN

Number of MAP Cells			
MPN	Number of Plaques		
10^4	TNTC	TNTC	TNTC
10^3	88	52	79
10^2	0	5	4
10^1	0	0	0
10^0	0	0	0

MPN most probable number method for determining number of MAP cells [22]
TNTC too numerous to count

Improving the detection of MAP by applying the phage assay to PBMCs isolated from blood

Since the PMMS step requires expensive reagents and optimisation of bead coating, experiments were designed to determine whether MAP cells could be detected from whole blood without using PMMS. It has previously been shown that MAP cells are present in the PBMC's of the blood and not found in the plasma or red blood cell fractions [7]. Initially, simple limit of detection experiments were performed to determine whether the PBMCs inhibited the phage assay. MAP cells were enumerated using the method described by [9] and artificially spiked into modified 7H9 medium and the PBMCs isolated from sheep blood. The phage assay was carried out on the artificially inoculated PBMCs and modified 7H9 medium containing known numbers of MAP cells. The results show that there was no significant difference ($p < 0.05$) in the number of MAP cells detected by the phage assay in the control samples (containing only modified 7H9 medium) and in the PBMCs. Duplicate blood samples tested with the PMMS-phage assay and phage assay without PMMS on PBMCs. The paired samples tested independently with the phage assay, the detection of MAP in PBMCs was more reproducible ($R^2 = 0.92$; Fig. 1b) than isolating MAP cells using the PMMS method on whole blood samples and detecting with the phage assay ($R^2 = 0.54$; Fig. 1a).

Discussion

Cattle infected with MAP may not show clinical signs of disease until years after exposure. Subclinically infected animals may shed MAP into the environment in their faeces and MAP cells can also be present in their milk, but both milk and blood ELISA testing on subclinical animals is notoriously insensitive [10, 11]. It is not known whether all animals that give a positive ELISA test result will go on to develop clinical Johne's disease. The results of this study show the first application of the phage assay on experimentally infected animals and its comparison between faecal shedding and blood ELISA status. The results have shown that animals with variable

blood ELISA status can have detectable levels of viable MAP in their blood. More data are required to better understand the relationship between the host immune response following MAP-exposure and disease progression, however having a clearly defined experimentally controlled study such as this is the first step in determining the usefulness of the phage assay. The ability to be able to detect MAP cells in blood samples from animals and obtain results within days (rather than within months), may aid the understanding of disseminated infection with regards to immune responses during disease progression and development of clinical disease and when this moves beyond a localised gut infection to a systemic phase.

It is also important to consider the type of bacteraemia detected in this study. As demonstrated here and previously, MAP can be detected in the PBMC fraction of blood and not others [7], so it is assumed that the MAP cells detected are internalised within PBMCs. In our experiments to establish the internalisation model, free MAP cells did not co-purify with the PBMCs supporting this model. The possibility exists however that bacteraemic animals have free MAP cells present in their blood that were not detected by our methods.

There were a number of differences in experimental design compared to a previous study in which we found viable MAP cells in whole blood ranging from 3–39 pfu.ml^{-1} [7]. The assay used in the present study was transferred from University of Nottingham laboratory, where the assay was initially established and optimised, to the University of Sydney. The availability of different equipment and the use of different reagents (particularly the peptide-coated beads) meant that some aspects of the method were not fully optimised before the trial was carried out. Furthermore, the animals in this study were castrated males and therefore were not exposed to the same reproductive and lactation stress as the milking dairy cattle used in the previous study [12]. These factors may explain the differences in the plaque number results obtained between the two studies.

The PMMS method requires expensive reagents, such as biotinylated peptides and paramagnetic beads and the use of PMMS introduced more variability into the results gained as tested with paired samples. Using the well-established Ficoll-Paque method to separate PBMCs from whole blood removes the need for PMMS and means that the phage assay can be carried out directly, reducing losses during sample processing. However this does reduce the selectivity of the method and therefore other mycobacteria potentially present in the blood may also be detected by the phage assay. This means that there is a greater reliance on the performance of the end-point PCR identification step to confirm the identity of the cell detected by the phage. Although limited by

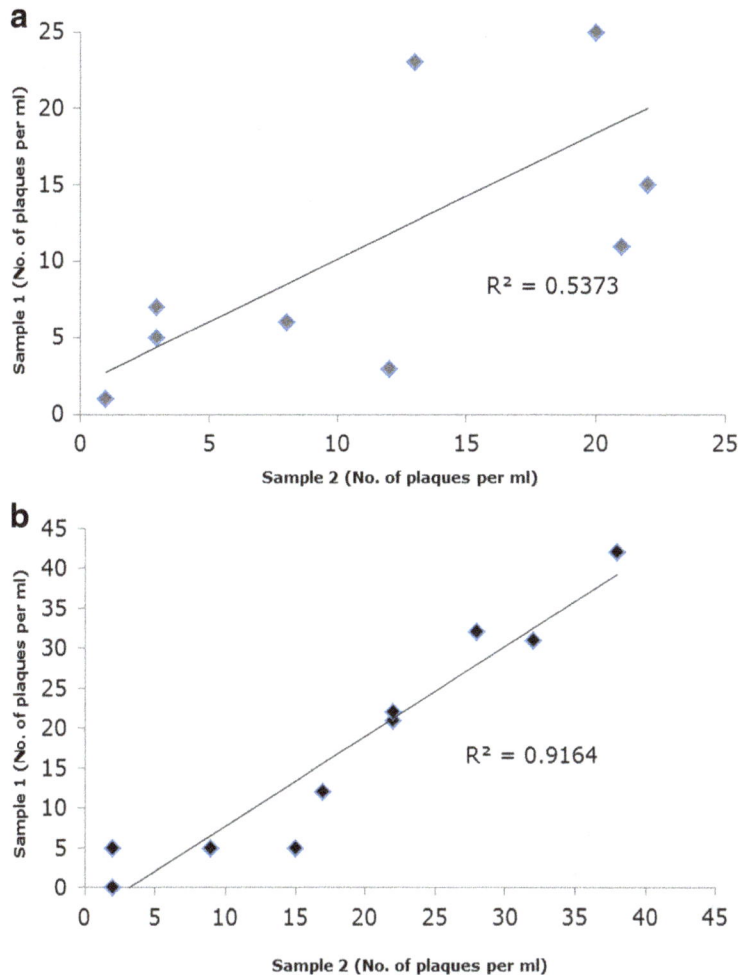

Fig. 1 Reproducibility of detecting MAP from whole blood using PMMS and from PBMCs. Figure show the reproducibility of detecting viable MAP cells using paired samples tested independently with the phage assay using the PMMS to isolate the MAP cells from whole blood (**a**) or detecting the MAP cells with the phage assays directly from PBMCs isolated from sheep blood (**b**)

the inefficiency of the PMMS, the results showed that the assay was specific; despite plaques being detected in two of the unexposed animals, no MAP DNA was detected in these samples, which probably represents some phage break-through in the assay. This has been noted before and emphasises the importance of combining the phage assay with a subsequent PCR-identification step.

In the group of exposed cattle, exposure to MAP occurred over 4.5 years before blood samples were taken for testing and none of the cattle were showing clinical signs of Johne's disease. Approximately one third of MAP-exposed cattle had viable MAP in their blood that was detectable by the PMMS-phage assay. Although two animals were shedding MAP in their faeces, no MAP was detected in the blood of these individuals. This was interesting as it would have been assumed, if the disease had progressed to the stage of detectable level of shedding, that the animals would be more likely to have viable MAP in their

blood. However the results suggested that this was not necessarily the case. In subclinically infected animals especially, faecal shedding is intermittent. This study examined a single time-point and only small volumes of blood were tested. It is also important to note that bacteraemia is a transient event that can happen during any stage of infection, whether that is in the early subclinical stage of infection or later [13]. Thus, these findings may indicate that the presence of viable MAP cells in the blood may also be intermittent relative to when the animals are found to be shedding MAP in their faeces or may be a rare event or may be the different forms MAP are found in the blood may relate to different pathophysiological manifestations that may or may not be related to the lack of correlation between faecal shedding and dissemination into the blood stream. Thus testing additional aliquots from a single time point may assist in determining the true likelihood of detecting viable MAP in the blood.

Conclusions

This is the first study to show that viable MAP may be present in the blood of experimentally MAP-exposed cattle prior to the onset of clinical signs and does not appear to be correlated to other measures, including faecal shedding and serum antibody responses, which are routine diagnostics used for Johne's. The ability to detect viable MAP in the blood of subclinically infected animals may lead to increased sensitivity of diagnosis in the early stages of infection, but this requires validation. Here we have demonstrated the proof of concept in a controlled study where variations such as exposure (time and dose etc.) and environmental factors are limited [14]. Having a tool like the phage assay allows new questions to be asked possibly in a longitudinal study testing the blood, milk and faeces of animals to enable a better understanding of paratuberculosis disease progression, determine when systemic infection occurs in animals, and how this influences the serum and milk ELISAs, and faecal culture or PCR results.

Methods

Bacterial strains, bacteriophage and growth medium

To optimise MAP detection in the peripheral blood mononuclear cells (PBMCs), MAP strains K10 and ATCC 19698 were used. The *Mycobacterium smegmatis* strain was mc^{2}155, which is used routinely in phage assays [15] and the bacteriophage used was D29. All liquid cultures of MAP were prepared in modified 7H9/OADC medium (Becton Dickenson, UK) supplemented with Mycobactin J (2 µg µl-1; Synbiotics Corporation, France) and when performing the phage assay the medium was supplemented with 2 mM CaCl$_2$. MAP strain; CM00/416 was used to both establish the phage assay method in the University of Sydney laboratories and to infect the cattle. MAP strains used for experimental infection trials and for the investigation of peptide mediated magnetic separation (PMMS) isolation efficiency were grown as previously described by Plain [16] and enumerated using the most probable number method (MPN) as described by [17].

Experimentally infected animals and sampling

All animal experiments carried out for this study were approved by The University of Sydney, Australia, animal ethics committee. An existing MAP infection trial being undertaken at the University of Sydney was used to provide samples for analysis (manuscript in preparation). In this trial 30 calves (aged 2–4 months) were age matched then randomly allocated into a group of 20 to be experimentally exposed to MAP (Numbers; 11–30, Table 1) along with a group of 10 age-matched unexposed control animals (Numbers; 1–10, Table 1). With consent, male Holstein and Holstein/Australian Red cross calves

were selected from a property in New South Wales that was unexposed to MAP; there was no previous history of any MAP infection on the farm and both the dams and calves were confirmed to be MAP-negative by faecal culture and faecal PCR. The MAP exposure time points were determined using a previous study [18]. Calves at 3-4 months of age were experimentally inoculated orally with a cattle (C) strain of MAP (CM00/416/C4) using a protocol based on a validated ovine model [19]. The inoculation was carried out in three doses over a period of one month with the total dose of viable MAP administered being 9.46×10^9 MAP cells. The control unexposed group consisted of 10 age-matched calves, and exposed and unexposed cohorts were maintained in different paddocks at the University of Sydney Camden farms to prevent cross-contamination. Control animals were maintained on paddocks where no MAP-infected livestock had been housed in the past. Faecal MAP culture was performed as previously described [20]. All samples tested and presented in this study were obtained from the animals 4 years, 8 months post exposure to allow the potential for MAP infection to establish. Faecal PCR was performed using a validated method as previously described [16] and serum antibody ELISA and results were expressed as signal of the test sample as a proportion of the positive control, corrected for the negative control (S/P %) calculated as described by the manufacturer (IDEXX Laboratories, Maine, USA).

Peptide mediated magnetic MAP isolation and phage detection from blood

MAP cells were isolated from blood and detected using the phage assay using PMMS according to the method described by [7]. Briefly, magnetic beads were coated with peptides aMp3 and aMptD that specifically bind MAP [21]. Coated beads were mixed with 1 ml of whole blood diluted in 9 ml of modified 7H9 medium (10 ml total volume). Samples were mixed for 30 min to allow MAP to bind the beads before they were recovered by centrifugation ($4500 \times g$; 15 min) and washed with 9 ml of modified 7H9 medium before recovered with centrifugation before being resuspended in 1 ml of modified 7H9 medium to carry out the magnetic separation. Beads were then concentrated by magnetic separation and resuspended in 1 ml of modified 7H9 medium prior to testing using the phage amplification assay.

The phage assay and experimental controls were carried out according to [15]. Briefly, 1 ml samples were mixed with D29 bacteriophage (100 µl; 1×10^9 pfu.ml^{-1}) in modified 7H9 Media and incubated for 1 h to allow the phage infection. Any remaining free phage were inactivated using 10 mM (final concentration) ferrous ammonium sulphate (6 min at room temperature whilst being continuously mixed). The ferrous ammonium sulphate was then inactivated with 5 ml of modified 7H9 medium before the samples were mixed with *M. smegmatis* cells (1 ml; 1×10^8 pfu.ml^{-1}) before plating with soft 7H10 agar (0.75 % agar). The number of MAP cells detected was determined by counting the number of plaques formed (data reported as pfu.ml^{-1}) in the lawn of *M. smegmatis* cells. Enumeration of MAP cells in initial inocula was performed by diluting samples until countable numbers of plaques are obtained [9].

PCR identification of MAP cells

After the phage assay, DNA from up to five plaques formed on each plate was extracted using a gel DNA extraction kit (ZymoResearch, UK). Then the nested IS*900* PCR assay was used to confirm that MAP DNA was present as described previously by [7]. The MAP specific IS*900* quantitative PCR was carried out using the method described by [16].

Peripheral blood mononuclear cells isolation

The isolation of the PBMCs from cattle blood was carried out using Ficoll-Paque Plus according to the manufacturer's instructions (GE Healthcare Life Sciences, UK). Briefly 2 ml of whole heparinised blood was mixed with 2 ml of PBS (Dulbecco A). This was carefully layered on 3 ml of Ficoll-Paque Plus in 15 ml falcon tubes. The samples were centrifuged (400 × *g*, for 30 min at 18 °C). After centrifugation the upper layer (plasma) of the sample was removed. The buffy coat layer was carefully removed ensuring the red blood cells were not disturbed. The PBMCs were washed with 6 ml of PBS by centrifugation (100 × *g* for 10 min at 18 °C). The supernatant was removed and the pellet was resuspended in 1 ml of modified 7H9 medium for the phage assay.

Abbreviations
MAP, *Mycobacterium avium* subsp. *paratuberculosis*; MPN, most probable number; PMMS, peptide-mediated magnetic separation

Acknowledgements
This work was supported by Meat and Livestock Australia and by Cattle Council of Australia, Sheepmeat Council of Australia and WoolProducers Australia through Animal Health Australia. BMCS was supported by a SfAM Presidents fund to support travel to Sydney. The authors would like to thank Ms Anna Waldron, Ms Ann-Michele Whittington and Ms Rebecca Maurer for technical support and Mr. Craig Kristo, Mr Nobel Toribio and Mr James Dalton for animal husbandry support.

Authors' contributions
All authors made substantial contributions to the conception and design, acquisition and interpretation of data. Each author has been involved in drafting the manuscript and revising it critically for content and have given final approval of the version to be published and agree to be accountable for all aspects of the work.

Competing interests
The authors declare that they have no competing interests.

Consent for publication
Not applicable.

Author details
[1]School of Veterinary Medicine and Science, Sutton Bonington Campus, Loughborough, Leics LE12 5RD, UK. [2]School of Biosciences, University of Nottingham, Sutton Bonington Campus, Loughborough, Leics LE12 5RD, UK. [3]The University of Sydney, Farm Animal and Veterinary Public Health, Faculty of Veterinary Science, Camden, Australia.

References
1. Collins MT, Gardner IA, Garry FB, Roussel AJ, Wells SJ. Consensus recommendations on diagnostic testing for the detection of paratuberculosis in cattle in the United States. J Am Vet Med Assoc. 2006; 229(12):1912–9.
2. Buergelt CD, Williams JE. Nested PCR on blood and milk for the detection of *Mycobacterium avium* subsp *paratuberculosis* DNA in clinical and subclinical bovine paratuberculosis. Aust Vet J. 2004;82(8):497–503.
3. Bower K, Begg DJ, Whittington RJ. Optimisation of culture of *Mycobacterium avium* subspecies *paratuberculosis* from blood samples. J Microbiol Methods. 2010;80(1):93–9.
4. Bower KL, Begg DJ, Whittington RJ. Culture of *Mycobacterium avium* subspecies *paratuberculosis* (MAP) from blood and extra-intestinal tissues in experimentally infected sheep. Vet Microbiol. 2011;147(1-2):127–32.
5. Pinedo PJ, Rae DO, Williams JE, Donovan GA, Melendez P, Buergelt CD. Association among results of serum ELISA, faecal culture and nested PCR on milk, blood and faeces for the detection of paratuberculosis in dairy cows. Transbound Emerg Dis. 2008;55(2):125–33.
6. Juste RA, Garrido JM, Geijo M, Elguezabal N, Aduriz G, Atxaerandio R, Sevilla I. Comparison of blood polymerase chain reaction and enzyme-linked immunosorbent assay for detection of *Mycobacterium avium* subsp *parataberculosis* infection in cattle and sheep. J Vet Diagn Investig. 2005; 17(4):354–9.
7. Swift BM, Denton EJ, Mahendran SA, Huxley JN, Rees CE. Development of a rapid phage-based method for the detection of viable *Mycobacterium avium* subsp. *paratuberculosis* in blood within 48 h. J Microbiol Methods. 2013;94(3):175–9.
8. Botsaris G, Liapi M, Kakogiannis C, Dodd CER, Rees CED. Detection of *Mycobacterium avium* subsp *paratuberculosis* in bulk tank milk by combined phage-PCR assay: evidence that plaque number is a good predictor of MAP. Int J Food Microbiol. 2013;164(1):76–80.
9. Rees CR, Botsaris G. The use of phage for detection, antibiotic sensitivity testing and enumeration. In: Cardona P-J, editor. Understanding tuberculosis - global experiences and innovative approaches to the diagnosis. Rijeka: InTech; 2012. p. 293–306.
10. Whitlock RH, Wells SJ, Sweeney RW, Van Tiem J. ELISA and fecal culture for paratuberculosis (Johne's disease): sensitivity and specificity of each method. Vet Microbiol. 2000;77(3-4):387–98.
11. Geue L, Kohler H, Klawonn W, Drager K, Hess RG, Conraths FJ. Investigations on suitability of ELISA for the detection of antibodies against *Mycobacterium avium* ssp *paratuberculosis* in bulk milk samples from rhineland-palatinate. Berliner Und Munchener Tierarztliche Wochenschrift. 2007;120(1-2):67–78.
12. Mortensen H, Nielsen SS, Berg P. Genetic variation and heritability of the antibody response to *Mycobacterium avium* subspecies *paratuberculosis* in Danish Holstein cows. J Dairy Sci. 2004;87(7):2108–13.
13. Dennis MM, Antognoli MC, Garry FB, Hirst HL, Lombard JE, Gould DH, Salman MD. Association of severity of enteric granulomatous inflammation with disseminated *Mycobacterium avium* subspecies *paratuberculosis* infection and antemortem test results for paratuberculosis in dairy cows. Vet Microbiol. 2008;131(1-2):154–63.
14. Begg DJ, Whittington RJ. Experimental animal infection models for Johne's disease, an infectious enteropathy caused by *Mycobacterium avium* subsp *paratuberculosis*. Vet J. 2008;176(2):129–45.
15. Stanley EC, Mole RJ, Smith RJ, Glenn SM, Barer MR, McGowan M, Rees CED. Development of a new, combined rapid method using phage and PCR for detection and identification of viable *Mycobacterium paratuberculosis* bacteria within 48 hours. Appl Environ Microbiol. 2007;73(6):1851–7.

16. Plain KM, Marsh IB, Waldron AM, Galea F, Whittington AM, Saunders VF, Begg DJ, de Silva K, Purdie AC, Whittington RJ. High-throughput direct fecal PCR assay for detection of *Mycobacterium avium* subsp. *paratuberculosis* in sheep and cattle. J Clin Microbiol. 2014;52(3):745.

17. Reddacliff LA, Nicholls PJ, Vadali A, Whittington RJ. Use of growth indices from radiometric culture for quantification of sheep strains of *Mycobacterium avium* subsp *paratuberculosis*. Appl Environ Microbiol. 2003; 69(6):3510–6.

18. Purdie AC, Plain KM, Begg DJ, de Silva K, Whittington RJ. Expression of genes associated with the antigen presentation and processing pathway are consistently regulated in early *Mycobacterium avium* subsp. *paratuberculosis* infection. Comp Immunol Microbiol Infect Dis. 2012;35(2):151–62.

19. Begg DJ, de Silva K, Di Fiore L, Taylor DL, Bower K, Zhong L, Kawaji S, Emery D, Whittington RJ. Experimental infection model for Johne's disease using a lyophilised, pure culture, seedstock of *Mycobacterium avium* subspecies *paratuberculosis*. Vet Microbiol. 2010;141(3-4):301–11.

20. Whittington RJ, Marsh I, Turner MJ, McAllister S, Choy E, Eamens GJ, Marshall DJ, Ottaway S. Rapid detection of *Mycobacterium paratuberculosis* in clinical samples from ruminants and in spiked environmental samples by modified BACTEC 12B radiometric culture and direct confirmation by IS900 PCR. J Clin Microbiol. 1998;36(3):701–7.

21. Foddai A, Elliott CT, Grant IR. Maximizing capture efficiency and specificity of magnetic separation for *Mycobacterium avium* subsp *paratuberculosis* cells. Appl Environ Microbiol. 2010;76(22):7550–8.

22. Reddacliff LA, Vadali A, Whittington RJ. The effect of decontamination protocols on the numbers of sheep strain *Mycobacterium avium* subsp *paratuberculosis* isolated from tissues and faeces. Vet Microbiol. 2003;95(4): 271–82.

12

Epidemiological investigation of non-*albicans* *Candida* species recovered from mycotic mastitis of cows in Yinchuan, Ningxia of China

Jun Du[1,2], Xiaoyu Wang[1,2], Huixia Luo[1,2], Yujiong Wang[1,2*], Xiaoming Liu[1,2*] (iD) and Xuezhang Zhou[1,2*]

Abstract

Background: *Candida spp.* is the vital pathogen involved in mycotic mastitis of cows. However the epidemiology and infection of *Candida* species in mycotic mastitis of cow in Ningxia province of China has not been explored. In the present study, the epidemiology, antimicrobial susceptibility and virulence-related genes of non-albicans Candida (NAC) species were investigated.

Methods: A total of 482 milk samples from cows with clinical mastitis in four herds of Yinchuan, Ningxia were collected and used for the isolation and identification of mastic pathogens by phenotypic and molecular characteristics, and matrix-assisted laser desorption ionization-time of flight mass spectrometry. The antimicrobial susceptibility to antifungal agents was also determined by a disk diffusion assay. The presence of virulence-related genes was determined by polymerase chain reaction (PCR).

Results: A total of 60 isolates from nine different *Candida* species were identified from 256 (60/256, 23.44%) milk samples. The most frequently identified species in cows with clinical mastitis groups were *Candida krusei* (*n* = 14) and *Candida parapsilosis* (*n* = 6). Others include *Candida lipolytica*, *Candida lusitaniae*, *Cryptococcus neoformans*. But *no Candida albicans* was identified in this study. Interestingly, All *C. krusei* isolates (14/14) were resistant to fluconazole, fluorocytosine, itraconazole and ketoconazole, 2 out of 14 *C. krusei* were resistant to amphotericin, and 8 out of the 14 were resistant to nystatin. Similarly, all six *C. parapsilosis* isolates were resistant to fluorocytosine, but susceptible to fluconazole, ketoconazole and nystatin; two of the six were resistant amphotericin and itraconazole. Molecularly, all of the *C. parapsilosis* isolates carried eight virulence-related genes, FKS1, FKS2, FKS3, SAP1, SAP2, CDR1, ERG11 and MDR1. All of the *C. krusei* isolates contained three virulence-related genes, ERG11, ABC2 and FKS1.

Conclusion: These data suggested that *Candida* species other than *C. albicans* played a pathogenic role in mycotic mastitis of cows in Yinchuan, Ningxia of China. The high incidence of drug-resistant genes in *C. parapsilosis* and *C. krusei* also highlighted a great concern in public and animal health in this region.

Keywords: Mycotic mastitis, *Candida parapsilosis*, *Candida krusei*, Antimicrobial susceptibility, Virulent gene

* Correspondence: wyj@nxu.edu.cn; lxm1966@nxu.edu.cn; zhouxuezhang@nxu.edu.cn
[1]Key Laboratory of the Ministry of Education for the Conservation and Utilization of Special Biological Resources of Western China, Ningxia University, Yinchuan 750021, Ningxia, China
Full list of author information is available at the end of the article

Background

Cow mastitis is a disease caused by infections of a variety of microorganisms, which causes large economical looses and damages to the breeding industry by decreasing milk productivity and increasing costs of antibiotic treatments and culling [1]. In most cases, fungal infections of the mammary gland are mostly caused by yeast, the main genus of which is Candida [2]. Mastitis infections caused by fungi of the *Candida* genus have long been known in animals, and *Candida*-caused mycotic mastitis was firstly described by Fleischer in 1930 [3]. In this regard, *Candida* species are considered as opportunistic pathogens that colonize the cow udder. In addition, the harness and abuse of antibacterial agents, and treatments of contaminated antibiotic solutions, as well as duct, or other materials brought in contact with the mammary gland also favor yeast colonization in cow udders [4].

Recently, a large number of virulent factors have been found in C. *parapsilosis* and C. *krusei* originated from cow mastitis, including the Fksp subunits of the β-1,3-D-glucan synthase enzyme (FKS1, FKS2, and FKS3), aspartyl proteinases (SAPP1 and SAPP2), the ATP binding cassette (ABC) transporters, *Candida* drug resistance gene 1 (CDR1), major facilitator superfamily (MFS) transporter multiple drug resistance gene 1 (MDR1), 14α-demethylases (ERG11), efflux pump transporters (ABC1 and ABC2). These factors favor the survival and growth of *Candida spp.* in the mammary gland of cows.

The dairy industry is a predominant economics in Ningxia, a province in Western China. The mastitis of cows causes a significant economic loss in this region every year. It has been well documented that mycotic mastitis is an important causation of economic loss to the dairy industry, however the epidemiology of pathogens may vary in different areas. In this respect, there is a paucity of information about antimicrobial resistance and virulent factors of C. *parapsilosis* and C. *krusei* isolates in Ningxia of China, the aim of the present study is therefore to interrogate the epidemiology of fungal infection, and the antimicrobial susceptibility and virulence-related genes of *Candida spp.* in Yinchuan, Ningxia of China.

Methods

Isolation and identification of pathogens

This study was submitted to and approved by the Ethics Committee of Animal Study in Ningxia University. A total of 482 milk samples were collected from the cows with clinical mastitis, which originated from four herds in Yinchuan of Ningxia province, China. Clinical mastitis was defined by swelling, reduced milk flow, and abnormal milk appearance (watery to viscous with clots varying from gray-white to yellowish). Additionally, other signs of infection such as fever, inappetence, ataxia, and depression were also considered. The samples were first plated onto the Sabouraud dextrose broth and incubated at 37 °C for 48 h. The cultures were identified by morphological characteristics (formation of chlamydoconidium, pseudohyphae and germinal tube development), Gram staining, biochemical tests (growth in the presence of 0.1% cyclohexamide (SigmaTM), acidic pH tolerance, urea hydrolysis and carbohydrates assimilation and/or fermentation) and CHROMagar Candida culture [5]. All presumptive *Candida* species isolates were further confirmed by Matrix- assisted laser desorption ionization-time of flight mass spectrometry (MALDI-TOF MS) (VITEK® MS, BioMerieux, France).

Antimicrobial susceptibility tests

The antimicrobial susceptibility was determined by disk diffusion method according to the guideline of the Clinical Laboratory Standards Institute for antifungal susceptibility [6, 7]. A total of six antimicrobial agents were used to evaluate the antimicrobial resistance of the isolates, which included fluorocytosine (1 μg/disk), itraconazole (10 μg/disk), ketoconazole (15 μg/disk), fluconazole (25 μg/disk), nystatin (50 μg/disk), amphotericin (10 μg/disk). Results were recorded as resistant, intermediate and sensitive. The C. *parapsilosis* ATCC 22019 strain and C. *krusei* ATCC 6258 strain were used as references for C. *parapsilosis* strain and C. *krusei*, respectively.

Detection of virulence-related genes

The presence of virulence-related genes in C. *parapsilosis* and C. *krusei* isolates was detected by a PCR assay. The genes of interest, primer sequences, and expected size of fragment of PCR products were given in Tables 1 and 2. The PCR reactions were performed in a final volume of 25 μL of reaction mixture consisted of 50 ng of genomic DNA, 20 pmol of each primer, and 12.5 μL of 2× Taq PCR MasterMix containing 0.1 U of Taq polymerase/μL, 0.5 mM dNTP each, 20 mM Tris-HCl/pH 8.3, 100 mM KCl, 3 mM $MgCl_2$ (Tiangen Biotech, Beijing, China). The cycling conditions were as following: an initial denaturation at 94 °C for 3 min; 30 cycles of denaturation at 94 °C for 30 s, annealing at distinct temperature (Tables 1 and 2) for 30 s, and primer extension at 72 °C for 1 min; and a final extension at 72 °C for 6 min.

Results

Isolation and identification of COW mastitis pathogens

Total of 256 pathogenic yeasts were isolated from 482 milk samples collected from cows with clinical mastitis (256/482, 53.1%) in Yinchuan of Ningxia province. Among them, 60 (60/256, 23.44%) were identified in nine of Candida species, in which the C. *krusei* (23.33%,

Table 1 The PCR primers for amplification of *C. parapsilosis* virulence genes

Gene	Primer sequence[a] (5'–3')[a]	Annealing temperature	Reference
FKS1	F: ATCCAAGATCTTCCGGTGCCTCAA	60°C	[8]
	R: ATCAGCTGACCATGCTGGATATGG		
FKS2	F: AATGGGCAGAGGTTGAGAAGGTAG	60°C	
	R: GGGTTCCAAGCAGGATATGGATCA		
FKS3	F: TCGTAGGTTCGAATCCTGCTGAGA	60°C	
	R: ATGGTGAAGGCGCAACGGTGTAAA		
SAPP1	F: ACTGGACAACAAATTGCAGATG	57°C	[9]
	R: TAAACTGCTTCATTGCTGGTGT		
SAPP2	F: GTCATATGGGGGATTTGCAC	57°C	
	R: CGCTTTGCTGATGTTACCAG		
MDR1	F: TTCGTGATAGTTTTGGTGGTAG	62°C	[10]
	R: TGAACCTGGAGTGAATCTTGT		
CDR1	F: ATTTGCCGACATCCACCGTTAGG	60°C	[11]
	R: ACCATGCTGTTTGCGAGTCCA		
ERG11	F: GTACACCGTCATTACTCTACCCAACA	62°C	
	R: TGCTCCTTTCATTTACAACATCATTT		

[a]F forward, R reverse

14/60) and *C. parapsilosis* 10% (6/60) were two of the most frequent Candida species. In addition, 16.66% isolates (10/60) were classified as *Candida*-like species. Other identified Candida species included *Candida lipolytica* (5/60, 8.33%), *Candida lusitaniae* (5/60, 8.33%), *Candida rugosa* (4/60, 6.67%), *Trichosporon mucoides* (4/60, 6.67%), *Candida sphaerica* (3/60, 5%), *Candida tropicalis* (3/60, 5%) and *Candida utilis* (3/60, 5%).

CHROMagar Candida is a differential culture medium that allows selective isolation of yeasts and simultaneously identifies colonies of *C. albicans*, *C. krusei*, and other *Candida* spp. Results of this study confirm the accuracy of CHROMagar in providing a presumptive identification of *C. krusei* and *C. parapsilosis*, which was consistent with Odds and Bernaerts study [8]. These

Table 2 The PCR primers for amplification of *C. krusei* virulence genes

Gene	Primer sequence[a] (5'–3')[a]	Annealing temperature	Reference
FKS1	F: ACTGCATCGTTTGCTCCTCT	63°C	[12]
	R: GAACATGATCAATTGCCAAC		
ABC1	F: GATAACCATTTCCCACATTTGAGT	60°C	[13]
	R: CATATGTTGCCATGTACACTTCTG		
ABC2	F: CCTTTTGTTCAGTGCCAGATTG	60°C	
	R: GTAACCAGGGACACCAGCAA		
ERG11	F: ATTGCGGCCGATGTCCAGAGGTAT	60°C	
	R: GCGCAGAGTATAAGAAAGGAATGGA		

[a]F forward, R reverse

results of the preliminary identification on the CHROMagar Candida were agree with those by MALDI-TOF MS. *C. krusei* is the only species which grows on Sabouraud's dextrose agar as spreading colonies with a matt or a rough whitish yellow surface, in contrast to the convex colonies of other Candida spp. All 14 isolates of *C. krusei* tested formed colonies after 48 h of incubation at 37 °C on CHROMagar Candida that were typically pale pink, large, rough, flat, and spreading with broad white edges (Fig. 1a). Unlike *C. krusei*, all 6 isolates of *C. parapsilosis* formed white or faint yellow, small, papillae, smooth colony (Fig. 1b). In addition, *C. krusei* and *C. parapsilosis* are both usually found in two basic morphological forms, as yeast and pseudohyphae (Fig. 1a' and b'). Both were frequently present simultaneously in growing cultures and may not be separated easily.

Antimicrobial susceptibility test

The antimicrobial susceptibility test demonstrated that *C. krusei* and *C. parapsilosis* isolates from cow mastitis cases in Yinchuan of Ningxia province had a variable degree of resistance to the antimicrobials (Fig. 2a and b, Tables 3 and 4). *C. krusei* isolates were respectively resistant to fluconazole (14/14, 100%), fluorocytosine (14/14, 100%), itraconazole (14/14, 100%), ketoconazole (14/14, 100%), amphotericin (2/14, 14.3%), and nystatin (8/14, 57.1%) (Table 3). *C. parapsilosis* isolates were respectively resistant to fluorocytosine (6/6, 100%), amphotericin (2/6, 33.3%), itraconazole (2/6, 33.3%), but susceptible to fluconazole (6/6, 100%), ketoconazole (6/6, 100%), nystatin (6/6, 100%) (Table 4).

Detection of Virulence-related Genes

Three virulence-related genes in *C. krusei* and eight virulence-related genes in *C. parapsilosis* isolates were determined by PCR in the present study. The results indicated that all (14/14, 100%) *C. krusei* isolates contained the ERG11 gene, 12 out of 14 (85.7%) *C. krusei* isolates were positive in ABC2 and FKS1 genes, but no ABC1 gene was detected in these *C. krusei* isolates (Fig. 3). All (6/6, 100%) *C. parapsilosis* isolates carried the FKS1, FKS2, FKS3, CDR1, ERG11, SAPP1 and SAPP2 genes; 4 out of the 6 (66.7%) of *C. parapsilosis* isolates contained MDR1 gene (Fig. 4).

Discussion

In this study, Candida species of yeasts were isolated in the 23.44% (60/256) of all samples analyzed. This rate is a marginally higher than the rates reported in Turkey were 12.7% [9], 17.7% [10]; the rates reported in Brazil were 17.3% [4], 12.8% [11]; and the rate reported in Slovenia was 7.5% [12]. *C. krusei* was the predominant species isolated in this study, which was 23.33% (14/60) of the Candida isolates from those with clinical mastitis.

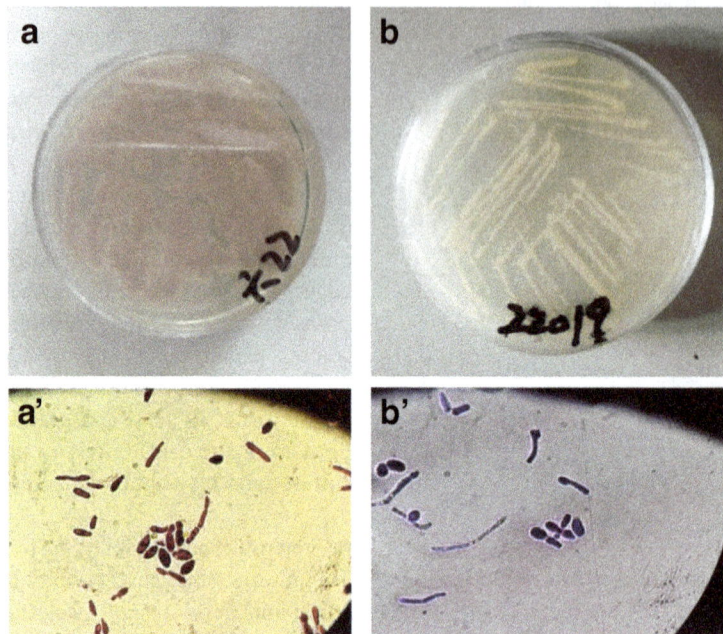

Fig. 1 Representative culture and Morphology of *C. krusei* and *C. parapsilosis* isolates. **a** Representative culture of *C. krusei* on CHROMagar plate; **a'** the morphology of Giemsa staining for yeast and pseudohyphae of a *C. krusei* isolate in **a**; **b** Representative culture of *C. parapsilosis* on CHROMagar plate; **b'** the morphology of Giemsa staining for yeast and pseudohyphae of a *C. parapsilosis* isolate in (**b**)

However, there is some discrepancy among rates that were also reported by others. Pengov et al. reported a 34% [12], De&Marin et al. reported *a* 44.5% [4], and Sartori et al. reported a 34.6% [11] of *C. krusei* isolates from cows with mastitis. Their data reported a higher frequency of *C. krusei* infection than findings in this report. However, the rate of *C. krusei* isolates of our finding was higher than others. Ruz-Peres et al. reported a 18.18% [13], ErbaŞ et al. reported 17.4% [10], Ksouri et al. reported a 15.57% [14], Krukowski et al. reported a 15.5% [15], and Eldesouky et al. reported a 12.2% of *C. krusei* isolates in cow mastitis [16].

C. parapsilosis was the second most frequent *Candida* species in this study. Although often being considered less virulent than *C. albicans*, *C. parapsilosis* is the Candida species with the largest increase in clinical incidence in human beings in recent decades. In animals, *C. parapsilosis* are also of great importance in veterinary medicine because of their ability to infect many animal species, including cows [14, 17, 18]. Indeed, many studies have described the isolation of *C. parapsilosis* from milk samples with a frequency varied from 1.7% to up to 25.4%, depending on the sanitary condition and environmental factors [10, 11, 15, 19].

The discrepancy in determination rates of *Candida* species among different geographic regions might be a consequence effect of several factors, such as the abuse of intramammary antibiotic treatment or the use of dairy

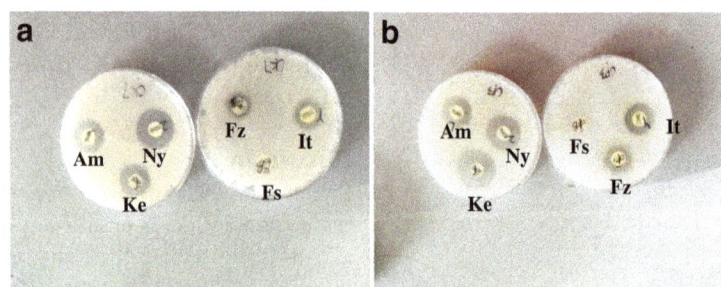

Fig. 2 Representative antimicrobial susceptibility tests by disk diffusion assay for *C. krusei* and *C. parapsilosis* isolates. **a** Representative disk diffusion assay for *C. krusei* isolate cultured on CHROMagar plate with indicated antimicrobial disks; **b** Representative disk diffusion assay for a *C. parapsilosis* isolate cultured on CHROMagar plate with indicated antimicrobial disks. Am, amphotericin; Fs, fluorocytosine; Fz, fluconazole; It, itraconazole; Ke, ketoconazole; Ny, nystatin

Table 3 Results of antimicrobial susceptibility tests of *C. krusei* isolates (*n* = 14)

Antibiotic	Resistant, % (no.)	Intermediate, % (no.)	Susceptible-dose dependent, SDD, % (no.)	Susceptible, % (no.)
Amphotericin	14.3 (*n* = 2)	85.7 (*n* = 12)	0	0
Fluorocytosine	100 (*n* = 14)	0	0	0
Fluconazole	100 (*n* = 14)	0	0	0
Itraconazole	100 (*n* = 14)	0	0	0
Ketoconazole	100 (*n* = 14)	0	0	0
Nystatin	0	42.9 (*n* = 6)	0	57.1 (*n* = 8)

farm made antibiotic infusions for mastitis treatments [15], yeast contaminated food or environment [20], inadequate sanitary practices of milking procedures, and the natural resistance to fluconazole of *C. krusei* [21, 22]. The relative cell-surface hydrophobicity (CSH) of *C. parapsilosis*, adherence to host tissues and plastic surfaces such as milkers or other prosthetic materials [23–25], and the presence of pathogenic strains of *C. krusei* might also contribute the high rate of *C. krusei* and *C. parapsilosis* in milk samples of cow mastitis [26].

Of note, there was no *C. albicans* was identified in this study. This result was in accordance with the report by Erbaş et al. [10]. Although *C. albicans* was considered to be the most frequent cause of fungemia, a number of reports have documented infections caused by *C. parapsilosis, C. krusei* and other NAC species [27], suggesting that the pathogenic role of opportunistic NAC species in this disease. Therefore, it is a necessity to pay more attention in NAC species in order to establish the possible role in mycotic mastitis in the region of Yinchuan, Ningxia of China.

In this study, the antifungal resistance was also investigated. Notably, all of the six *C. parapsilosis* isolates (100%) were resistant to fluorocytosine, and all 14 of *C. krusei* isolates (100%) were resistant to ketoconazole, fluconazole, itraconazole and fluorocytosine. This finding was consistent with a study by Sonmez et al. [28], but was different from reports by others [29–31]. Of interest, these agents are the primary agents against the infection of *Candida* species. It has been reported that the resistance of *C. albicans* to fluorocytosine was up to 10–30%, and the incidence of drug resistant NAC species was increased [10, 28]. For example, the natural or acquired resistance to fluconazole was confirmed in *C. krusei* and

C. glabrata [32]. Our finding in *C. krusei* also supported the results reported in literatures. It was worthy to note that *C. krusei* and *C. parapsilosis* isolates from cow mastitis cases in Ningxia region were more resistant to antifungal agents. It is strongly recommended that fluorocytosine, and other azole antifungal agents should not be used in the treatment with NAC species in dairy cows in Ningxia province of China, owing to the high drug resistance of these *Candida* species. Interestingly, these isolates were from animals that had not been treated with antifungal agents, suggesting that animal isolates were more exposed to anthropic environmental selective pressures, such as the use of azole antifungals in agricultural practices [33]. Therefore, a special attention must be paid to dairy products for monitoring the antifungal susceptibility of Candida spp. and other fungi recovered from animals is extremely important, since they may act as reservoirs of strains causing human disease and may present a risk for immunocompromised patients [34].

Previous reports indicated that a large number of virulent factors have been determined in *C. krusei* and *C. parapsilosis,* including the β-1,3-D-glucan synthase enzyme, aspartyl proteinases, the ATP binding cassette (ABC) transporter, major facilitator superfamily (MFS) transporter, the zinc cluster transcription factor,14α-demethylase, efflux transporters. In the present study, we also observed that most of the *C. krusei* isolates harbored ERG11, ABC2 and FKS1 genes, and most of the *C. parapsilosis* isolates harbored genes of FKS1, FKS2, FKS3, SAPP1, SAPP2, ERG11, CDR1 and MDR1. The Fksp subunits of the β-1,3-D-glucan synthase enzyme are the target of echinocandins [35] and are encoded by the FKS1,FKS2, and FKS3 genes. Many reports indicated that a reduced-susceptibility (RES) to echinocandin was

Table 4 Results of antimicrobial susceptibility tests of *C. parapsilosis* isolates (*n* = 6)

Antibiotic	Resistant, % (no.)	Intermediate, % (no.)	Susceptible-dose dependent, SDD, % (no.)	Susceptible, % (no.)
Amphotericin	33.3 (*n* = 2)	0	33.3 (*n* = 2)	33.3 (*n* = 2)
Fluorocytosine	100 (*n* = 6)	0	0	0
Fluconazole	0	0	0	100 (*n* = 6)
Itraconazole	33.3 (*n* = 2)	0	66.7 (*n* = 4)	0
Ketoconazole	0	0	0	100 (*n* = 6)
Nystatin	0	0	0	100 (*n* = 6)

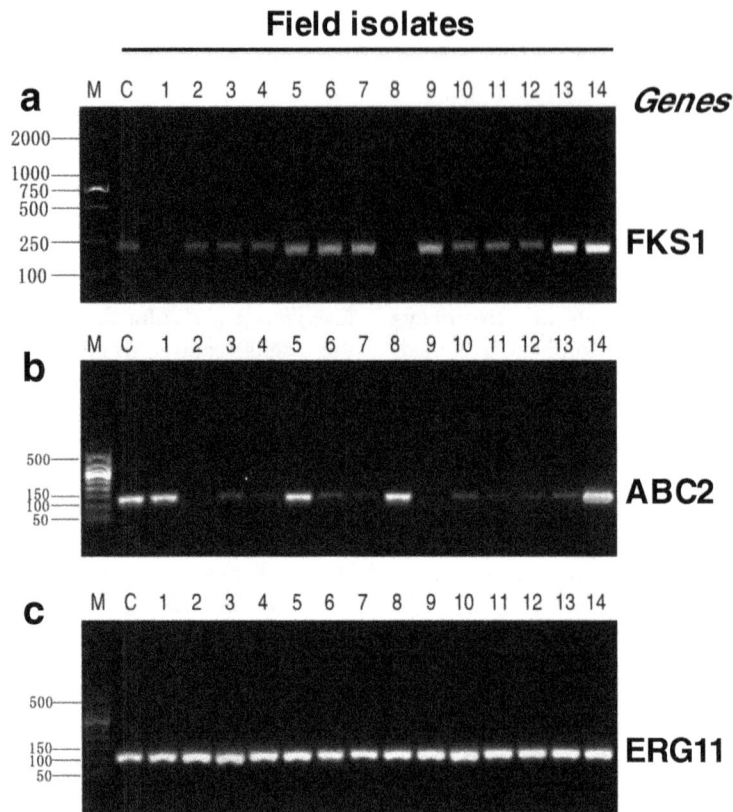

Fig. 3 Virulence-related genes of *C. krusei* isolates determined by PCR assay. The indicated virulence-related genes of (**a**) FSK1, **b** ABC2 and (**c**) EGR11 in 14 field *C. krusei* isolates of this report were detected by PCR assay. C, control *C. krusei* ATCC6258 strain; lanes 1 to 14 represented *C. krusei* isolate 1–14. M, 50 bp DNA ladders

associated with mutations in two conserved regions of fks1 and fks2, and amino acid substitutions in the proteins encoded by these genes were observed within two or three hot spot (HS) regions on each gene [36–38]. In the present study, FKS1, FKS2, and FKS3 genes were detected in all *C. parapsilosis* field isolates. 85.7% (12/14) of *C. krusei* isolates were positive in FKS1 gene. Further studies are needed to confirm the mutation in hot spot (HS) regions on FKS1, FKS2 genes of *C. krusei* and *C. parapsilosis*, in order to elucidate the capability of the M27-A3 guidelines to detect the resistance to echinocandin. The secretion of aspartyl proteinases (Saps) has been considered as important elements of virulence for *C. albicans* [39], and it is also recognized as virulence factors for *C. parapsilosis* [40]. In *C. parapsilosis*, SAPP1, and SAPP2 are two annotated secreted aspartyl proteinase genes [41–43]. These enzymes are involved in providing nutrients for pathogen propagation, tissue colonization and further tissue invasion by rupturing host mucosal membranes [23]. In this study, 100% of *C. parapsilosis* isolates were found to carry SAPP1 and SAPP2 genes. This result was in agreement with previous reports [44].

The common mechanisms of *Candida* resistant to azoles include changes in target enzyme and upregulation of multidrug resistance protein (MDR). The target enzyme of azoles is 14α-lanosterol demethylase (14-DM), which is responsible for the production of an ergosterol precursor and is encoded by the gene *ERG11*. In *C. albicans* and *C. parapsilosis*, the efflux pump genes associated with azole resistance include CDR1, CDR2, and MDR1 [45]. However, the drug-resistant genes in *C. krusei* are involved in ABC1 and ABC2 [46, 47]. The results in this study also showed that 100% of *C. parapsilosis* isolates contained the *CDR1* and *ERG11* genes, and 66.7% of *C. parapsilosis* isolates contained the *MDR1* gene. 100% (14/14) of *C. krusei* isolates contained the *ERG11* gene, and 85.7% (12/14) of *C. krusei* isolates contained the *ABC2* gene, despite the *ABC1* gene was not detect in *C. krusei*. This finding was in line with previous studies in humans by He et al. that ABC2p was suggested to play a more important role in the resistance of *C. krusei* to azoles, instead of ABC1p [46].

Taken together, it was demonstrated that *C. krusei* and *C. parapsilosis* isolates from mycotic mastitis were present in Yinchuan of Ningxia with a potential of multidrug resistance. Further studies are needed to confirm

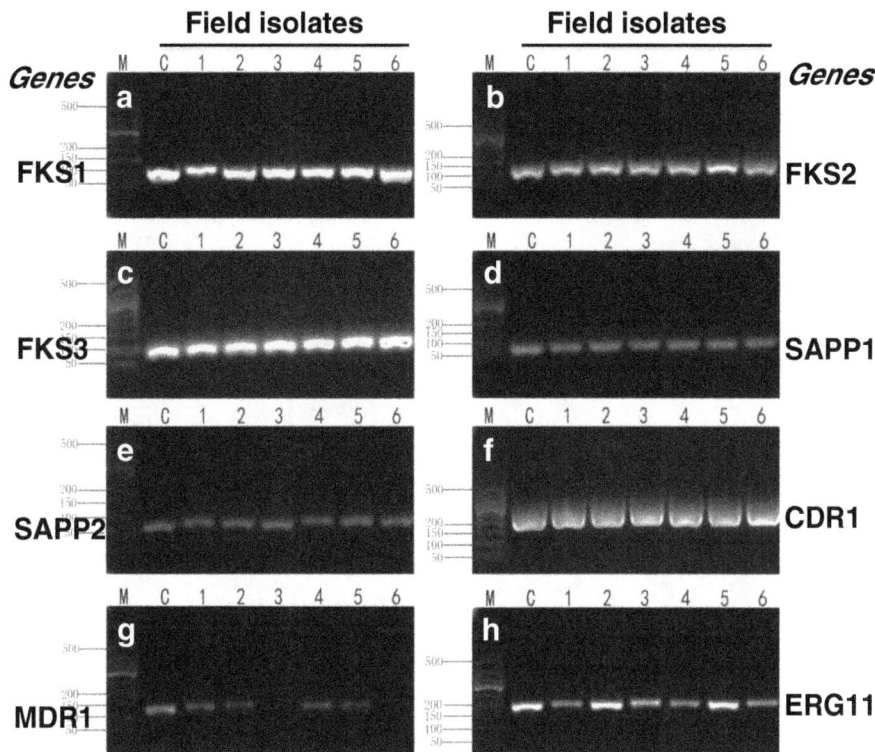

Fig. 4 Virulence-related genes of *C. parapsilosis* isolates determined by PCR assay. The indicated virulence-related genes of (**a**) FSK1, **b** FSK2, **c** FSK3 (**d**) SAPP1, **e** SAPP2, **f** CDR1, **g** MDR1 and (**h**) EGR11 in 6 field *C. parapsilosis* isolates of this report were detected by PCR assay. **c**, *C. parapsilosis* ATCC22019 control strain; lanes 1 to 6 represented *C. parapsilosis* isolate 1–6. M, 2 kb DNA ladders

mutations in ERG11, gain-of-function mutations in transcription factors, such as multidrug resistance regulator 1 (*MRR1*) and transcriptional activator of CDR gene 1 (*TAC1*), and the overexpression of these genes to elucidate the molecular mechanisms of resistance which caused by *C. krusei* and *C. parapsilosis* isolates from mycotic mastitis presented in Yinchuan, Ningxia of China.

Conclusions

A total of 60 isolates obtained from clinical mastitis milk samples in Yinchuan, Ningxia province of China were identified as *Candida* species according to phenotypic characteristics and MALDI-TOF MS. The most frequent *Candida* species found in this study was *C. krusei*, followed by *C. parapsilosis*. Other non-albicans *Candida* (NAC) species were also found in a low frequency. There was no *Candida albicans* was isolated in this study. According to results of this study, NAC may play an important pathogenic role in mycotic mastitis in dairy farms of Yinchuan region in Ningxia province of China. The *C. krusei* and *C. parapsilosis* displayed a strong antifungal resistance with drug-resistant genes. Most of *C. krusei* isolates harbored the ERG11, ABC2 and FKS1 genes, and most of the *C. parapsilosis* isolates harbored the FKS1, FKS2, FKS3, SAPP1, SAPP2, ERG11, CDR1 and MDR1 genes.

Abbreviations

14-DM: 14α-lanosterol demethylase (14-DM); ABC: ATP binding cassette; ATCC: American Type Culture Collection;; C.: Candida; CDR1: Candida drug resistance gene 1; CSH: Cell-surface hydrophobicity; ERG11: 14α-demethylases; FKS: Fksp subunits of the β-1,3-D-glucan synthase enzyme; MALDI-TOF MS: Matrix-assisted laser desorption ionization-time of flight mass spectrometry; MDR1: Multiple drug resistance gene 1; MFS: Major facilitator superfamily; MRR1: Multidrug resistance regulator 1; NAC: Non-albicans Candida; PCR: Polymerase chain reaction; SAP: Secreted aspartyl proteinase; ssp.: Species; TAC1: Transcriptional activator of CDR gene 1

Acknowledgements

The authors warmly thank to all study participants, owners of farms, veterinarians and personnel of the Center of Disease Control and Veterinary Institute in Ningxia Hui Autonomous Region of China, who helped in the realization of this study.

Funding

This study was supported by a grant from the National Natural Science Foundation of China (No.31660728), a grant from the project for the First-class discipline construction (Biology) of Ningxia Universities (NXYLXK2017B05), Major Innovation Projects for Building First-class Universities in Western China (ZKZD2017001) and a grant from Natural Science Foundation of Ningxia (NZ15017). These funding play no role in in the design of the study and collection, analysis, and interpretation of data and in writing the manuscript.

Authors' contributions

JD, YW and XZ conceived and designed the experiments; JD and XW analyzed the data and drafted the manuscript; JD, HL, XL and JW performed experiments and acquired data; YW and XZ interpreted data and critically revised the manuscript. All authors read and approved the final version of the manuscript.

Consent for publication
Not applicable.

Competing interests
The authors declare that they have no competing interests.

Author details
[1]Key Laboratory of the Ministry of Education for the Conservation and Utilization of Special Biological Resources of Western China, Ningxia University, Yinchuan 750021, Ningxia, China. [2]College of Life science, Ningxia University, Yinchuan 750021, Ningxia, China.

References
1. Costa EO, Ribeiro AR, Watanabe ET, Melville PA. Infectious bovine mastitis caused by environmental organisms. Zentralbl Veterinarmed B. 1998;45(2):65.
2. Dworeckakaszak B, Krutkiewicz A, Szopa D, Kleczkowski M, Biegańska M. High Prevalence of Candida Yeast in Milk Samples from Cows Suffering from Mastitis in Poland. Scientific World Journal. 2012;2012(1):196347.
3. Costa EO, Gandra CR, Pires MF, Coutinho SD, Castilho W, Teixeira CM. Survey of bovine mycotic mastitis in dairy herds in the state of Sao Paulo, Brazil. Mycopathologia. 1993;124(1):13–7.
4. De CSR, Marin JM. Isolation of Candida spp. from mastitic bovine milk in Brazil. Mycopathologia. 2005;159(2):251–3.
5. Pfaller MA, Houston, Coffmann. Application of CHROMagar Candida for rapid screening of clinical specimens for Candida albicans, Candida tropicalis, Candida krusei, and Candida (Torulopsis) glabrata. J Clin Microbiol. 1996;34(1):58–61.
6. CLSI. Method for antifungal disk diffusion susceptibility testing of yeasts; approved guideline. Wayne: Clinical and Laboratory Standards Institute; 2004.
7. CLSI. Zone diameter interpretive standards, corresponding minimal inhibitory concentration (MIC) interpretive breakpoints, and quality control limits for antifungal disk diffusion susceptibility testing of yeasts; informational supplement. 2nd ed. Wayne: Clinical and Laboratory Standards Institute; 2008.
8. Odds FC, Bernaerts R. CHROMagar Candida, a new differential isolation medium for presumptive identification of clinically important Candida species. J Clin Microbiol. 1994;32(8):1923.
9. Seker E. Identification of Candida species isolated from bovine mastitic milk and their in vitro hemolytic activity in western Turkey. Mycopathologia. 2010;169(4):303–8.
10. Erbaş G, Parin U, Kirkan Ş, SavaŞAn S, ÖZavci MV, YÜKsel HT. Identification of Candida strains with nested PCR in bovinemastitis and determination of antifungal susceptibilities. Turk J Vet Anim Sci. 2017;41:757–63.
11. Sartori LCA, Santos RC, Marin JM. Identification of Candida species isolated from cows suffering mastitis in four Brazilian states. Arquivo Brasileiro De Medicina Veterinária E Zootecnia. 2014;66(5):1615–7.
12. Pengov A. Prevalence of mycotic mastitis in cows. Acta Veterinaria. 2002; 52(2):133–6.
13. Ruz-Peres M, Benites NR, Yokoya E, Melville PA. Resistência de fungos filamentosos e leveduras isolados de leite CRU bovino à pasteurização e fervura. Vet Zootec. 2010;17(1):62–70.
14. Ksouri S, Djebir S, Hadef Y, Benakhla A. Survey of bovine mycotic mastitis in different mammary gland statuses in two north-eastern regions of Algeria. Mycopathologia. 2015;179(3–4):327–31.
15. Krukowski H, Tietze M, Majewski T, Różański P. Survey of yeast mastitis in dairy herds of small-type farms in the Lublin region, Poland. Mycopathologia. 2001;150(1):5–7.
16. Eldesouky I, Mohamed N, Khalaf D, Salama A, Elsify A, Ombarak R, Elballal S, Effat M, Alshabrawy M. Candida mastitis in dairy cattle with molecular detection of Candida albicans. Kafkas Üniversitesi Veteriner FakÜltesi Dergisi. 2016;22(3):461–4.
17. Costa GMD, Silva ND, Rosa CA, Pereira UDP. Mastite por leveduras em bovinos leiteiros do Sul do Estado de Minas Gerais, Brasil mastitis caused by yeasts in dairy herds in the south of the Minas Gerais state, Brazil. Ciência Rural. 2008;38(7):1938–42.
18. Fadda ME, Pisano MB, Scaccabarozzi L, Mossa V, Deplano M, Moroni P, Liciardi M, Cosentino S. Use of PCR-restriction fragment length polymorphism analysis for identification of yeast species isolated from bovine intramammary infection. J Dairy Sci. 2013;96(12):7692.
19. Zaragoza CS, Olivares RA, Watty AE, Moctezuma Ade L, Tanaca LV. Yeasts isolation from bovine mammary glands under different mastitis status in the Mexican high Plateu. Revista iberoamericana de micologia. 2011;28(2):79–82.
20. Elad D, Shpigel NY, Winkler M, Klinger I, Fuchs V, Saran A, Faingold D. Feed contamination with Candida krusei as a probable source of mycotic mastitis in dairy cows. J Am Vet Med Assoc. 1995;207(5):620.
21. Freydiere AM, Guinet R, Boiron P. Yeast identification in the clinical microbiology laboratory: phenotypical methods. Med Mycol. 2001;39(1):9.
22. Pfaller MA, Andes DR, Diekema DJ, Horn DL, Reboli AC, Rotstein C, Franks B, Azie NE. Epidemiology and outcomes of invasive candidiasis due to non-albicans species of Candida in 2,496 patients: data from the prospective antifungal therapy (PATH) registry 2004-2008. PLoS One. 2014;9(7):e101510.
23. Trofa D, Gacser A, Nosanchuk JD. Candida parapsilosis, an emerging fungal pathogen. Clin Microbiol Rev. 2008;21(4):606–25.
24. Kuhn DM. Comparison of biofilms formed by Candidaalbicans and Candidaparapsilosis on bioprosthetic surfaces. Infect Immun. 2002;70(2): 878–88.
25. Panagoda GJ, Ellepola AN, Samaranayake LP. Adhesion of Candida parapsilosis to epithelial and acrylic surfaces correlates with cell surface hydrophobicity. Mycoses. 2001;44(1–2):29.
26. Spanamberg A, Jr WE, Brayer Pereira DI, Argenta J, Cavallini Sanches EM, Valente P, Ferreiro L. Diversity of yeasts from bovine mastitis in southern Brazil. Revista iberoamericana de micologia. 2008;25(3):154–6.
27. Rodríguez D, Almirante B, Cuenca-Estrella M, Rodríguez-Tudela JL, Mensa J, Ayats J, Sanchez F, Pahissa A. Predictors of candidaemia caused by non-albicans Candida species: results of a population-based surveillance in Barcelona, Spain. Clin Microbiol Infect. 2010;16(11):1676.
28. Sonmez M, Erbas G. Isolation and identification of Candida spp. from mastitis cattle milk and determination of antifungal susceptibilities. Int J Vet Sci. 2017;6(2):104–7.
29. Tavanti A, Hensgens LA, Mogavero S, Majoros L, Senesi S, Campa M. Genotypic and phenotypic properties of *Candida parapsilosis* sensu strictu strains isolated from different geographic regions and body sites. BMC Microbiol. 2010;10(1):1–11.
30. Wawron W, Bochniarz M, Dabrowski R. Antifungal susceptibility of yeasts isolated from secretion of inflamed mammary glands in cows. Pol J Vet Sci. 2010;13(3):487–90.
31. Xiao M, Fan X, Chen SC, Wang H, Sun ZY, Liao K, Chen SL, Yan Y, Kang M, Hu ZD. Antifungal susceptibilities of Candida glabrata species complex, Candida krusei, Candida parapsilosis species complex and Candida tropicalis causing invasive candidiasis in China: 3 year national surveillance. J Antimicrob Chemother. 2015;70(3):802.
32. Gunes I, Kalkanci A, Kustimur S. Comparison of three different commercial kits with conventional methods for the identification of Candida strains to species level. Mikrobiyoloji Bulteni. 2001;35(4):559–64.
33. Rocha MF, Alencar LP, Paiva MA, Melo LM, Bandeira SP, Ponte YB, Sales JA, Guedes GM, Castelo-Branco DS, Bandeira TJ, et al. Cross-resistance to fluconazole induced by exposure to the agricultural azole tetraconazole: an environmental resistance school? Mycoses. 2016;59(5):281–90.
34. Edelmann A, Krüger M, Schmid J. Genetic relationship between human and animal isolates of Candida albicans. J Clin Microbiol. 2005;43(12):6164–6.
35. Douglas CM, D'Ippolito JA, Shei GJ, Meinz M, Onishi J, Marrinan JA, Li W, Abruzzo GK, Flattery A, Bartizal K. Identification of the FKS1 gene of Candida albicans as the essential target of 1,3-beta-D-glucan synthase inhibitors. Antimicrob Agents Chemother. 1997;41(11):2471.
36. Garcia-Effron G, Katiyar SK, Park S, Edlind TD, Perlin DS. A naturally occurring proline-to-alanine amino acid change in Fks1p in Candida parapsilosis, Candida orthopsilosis, and Candida metapsilosis accounts for reduced echinocandin susceptibility. Antimicrob Agents Chemother. 2008;52(7):2305.
37. Martícarrizosa M, Sánchezreus F, March F, Cantón E, Coll P. Implication of Candida parapsilosis FKS1 and FKS2 mutations in reduced Echinocandin susceptibility. Antimicrob Agents Chemother. 2015;59(6):3570–3.
38. Prigitano A, Esposito MC, Cogliati M, Pitzurra L, Santamaria C, Tortorano AM. Acquired echinocandin resistance in a Candida krusei blood isolate confirmed by mutations in the fks1 gene. New Microbiol. 2014;37(2):237–40.

39. Naglik JR, Challacombe SJ, Hube B. Candida albicans secreted Aspartyl proteinases in virulence and pathogenesis. Microbiol Mol Biol Rev. 2003;67(3):400.

40. Horváth P, Nosanchuk JD, Hamari Z, Vágvölgyi C, Gácser A. The identification of gene duplication and the role of secreted aspartyl proteinase 1 in Candida parapsilosis virulence. J Infect Dis. 2012;205(6):923.

41. Dostál J, Dlouhá H, Malon P, Pichová I, Hrusková-Heidingsfeldová O. The precursor of secreted aspartic proteinase Sapp1p from Candida parapsilosis can be activated both autocatalytically and by a membrane-bound processing proteinase. Biol Chem. 2005;386(8):791.

42. Merkerová M, Dostál J, Hradilek M, Pichová I. Cloning and characterization of Sapp2p, the second aspartic proteinase isoenzyme from Candida parapsilosis. FEMS Yeast Res. 2006;6(7):1018–26.

43. Hrusková-Heidingsfeldová O, Dostál J, Majer F, Havlíkova J, Hradilek M, Pichová I. Two aspartic proteinases secreted by the pathogenic yeast Candida parapsilosis differ in expression pattern and catalytic properties. Biol Chem. 2009;390(3):259–68.

44. Dabiri S, Shams-Ghahfarokhi M, Razzaghi-Abyaneh M. SAP(1-3) gene expression in high proteinase producer Candida species strains isolated from Iranian patients with different Candidosis. J Pure Appl Microbiol. 2016;10(3):1891–6.

45. Chen LM, Xu YH, Zhou CL, Zhao J, Li CY, Wang R. Overexpression of CDR1 and CDR2 genes plays an important role in fluconazole resistance in Candida albicans with G487T and T916C mutations. J Int Med Res. 2010;38(2):536.

46. He X, Zhao M, Chen J, Wu R, Zhang J, Cui R, Jiang Y, Chen J, Cao X, Xing Y. Overexpression of both ERG11 and ABC2 genes might be responsible for Itraconazole resistance in clinical isolates of Candida krusei. PLoS One. 2015; 10(8):e0136185.

47. Katiyar SK, Edlind TD. Identification and expression of multidrug resistance-related ABC transporter genes in Candida krusei. Med Mycol. 2001;39(39):109–16.

13

Biosecurity aspects of cattle production in Western Uganda, and associations with seroprevalence of brucellosis, salmonellosis and bovine viral diarrhoea

C. Wolff[1]* (iD), S. Boqvist[1], K. Ståhl[2], C. Masembe[3] and S. Sternberg-Lewerin[1]

Abstract

Background: Many low-income countries have a human population with a high number of cattle owners depending on their livestock for food and income. Infectious diseases threaten the health and production of cattle, affecting both the farmers and their families as well as other actors in often informal value chains. Many infectious diseases can be prevented by good biosecurity. The objectives of this study were to describe herd management and biosecurity routines with potential impact on the prevalence of infectious diseases, and to estimate the burden of infectious diseases in Ugandan cattle herds, using the seroprevalence of three model infections.

Results: Farmer interviews ($n = 144$) showed that biosecurity measures are rarely practised. Visitors' hand-wash was used by 14%, cleaning of boots or feet by 4 and 79% put new cattle directly into the herd. During the 12 months preceding the interviews, 51% of farmers had cattle that died and 31% had noticed abortions among their cows. Interestingly, 72% were satisfied with the health status of their cattle during the same time period. The prevalence (95% CI) of farms with at least one seropositive animal was 16.7% (11.0;23.8), 23.6% (16.9;31.4), and 53.4% (45.0;61.8) for brucella, salmonella and BVD, respectively.

A poisson regression model suggested that having employees looking after the cattle, sharing pasture with other herds, and a higher number of dead cattle were associated with a herd being positive to an increasing number of the diseases. An additive bayesian network model with biosecurity variables and a variable for the number of diseases the herd was positive to resulted in three separate directed acyclic graphs which illustrate how herd characteristics can be grouped together. This model associated the smallest herd size with herds positive to a decreasing number of diseases and having fewer employees.

Conclusion: There is potential for improvement of biosecurity practices in Ugandan cattle production. Salmonella, brucella and BVD were prevalent in cattle herds in the study area and these infections are, to some extent, associated with farm management practices.

Keywords: Serology, Biosecurity, Poisson regression, Additive bayesian network, Endemic infections, Disease control, Uganda, Cattle

* Correspondence: Cecilia.Wolff@slu.se
[1]Department of Biomedical Sciences and Veterinary Public Health, Swedish University of Agricultural Sciences, Uppsala, Sweden
Full list of author information is available at the end of the article

Background

In low-income countries cattle represent a source for nutritious food, income, and may also play a major role in the social context [1]. One threat to livestock production is infectious diseases, endemic or epidemic, which directly affect the farmer but also many formal and informal actors in the often complex value chains of animal products [2]. Outbreaks of infectious, serious and so called Transboundary Animal Diseases (TADs) in cattle may have additional consequences for the farmers due to trade restrictions. Such diseases, and others with high morbidity and clear clinical signs and/or increased mortality are likely to be recognised by farmers, and linked to losses in production [3]. However, endemic infections with less dramatic clinical manifestations, such as reproductive disturbances, diarrhoea, or poor growth, may be part of what farmers perceive as "normal" and not as something possible and/or worthwhile to control, despite negative effects on income and food security.

The prospects for eradication of specific infectious diseases might be beyond what is reasonable in low-income countries, where the livestock production systems differ from systems in high-income countries. Instead, efforts could be made to decrease infection pressure in general, by improved biosecurity, and thus improve productivity.

Biosecurity, i.e. actions to prevent introduction of infectious agents in an area, farm or group of animals, or to limit circulation of those already present, is essential for animal disease control and improved herd health [4]. Many biosecurity measures are not disease-specific, e.g. to keep animals separated from other herds. To successfully suggest changes in biosecurity to cattle farmers one has to know what management routines are currently practiced in the population. In addition, knowledge of actual associations between disease occurrence and management in the local context would give valuable information on current routines with high impact on disease prevalence in cattle.

Uganda is an East African country with great potential for food production and for developing its livestock sector. The bovine disease spectrum is similar to other Sub-Saharan African countries and, in contrast to countries with a more developed dairy production, dominated by infectious diseases. Farmers themselves rank east coast fever (ECF), helminthosis, trypanosomiasis, anaplasmosis, and foot-and-mouth disease (FMD) as major disease problems [5]. As many infectious diseases in cattle are zoonotic and of a public health concern, control of such diseases in cattle is not only an economic issue, but a One Health issue. When disease control actions are carried out, these generally focus on response to outbreaks, e.g. vaccination and movement bans during FMD outbreaks [6]. Research efforts have been targeted mainly at TADs, such as FMD and vector-borne diseases, e.g. trypanosomiasis [7–9]. The knowledge regarding many other endemic and less obvious infectious diseases that are likely to have a high impact on calf mortality, reproduction, growth and milk production, is more limited. Estimating the occurrence of such diseases may provide a better picture of the overall disease burden of infectious diseases in the cattle population. Further, if and to what extent Ugandan cattle farmers practice biosecurity is unknown. This knowledge would be useful for future interventions.

This study is part of a project on biosecurity as a tool for cattle disease prevention. The objectives were: firstly, to describe herd management and biosecurity routines with potential impact on the prevalence of infectious diseases in Ugandan cattle herds and, secondly, to estimate the burden of infectious diseases and any associations with herd management, using three model infections.

Methods

This was a cross-sectional field study with the herd as the unit of interest.

Cattle production in Uganda

Cattle are mainly kept in extensive pasture-based production systems: tethered, in enclosures or with communal grazing. At the last livestock census the country had a population of 11.4 million cattle heads and about 1 in 4 households keep cattle. An estimated 1.5 million cows were milked for household consumption and/or for family income. The average milk production per cow and week was 8.5 l [10]. The low production can partly be attributed to the use of local breeds or local breeds crossed with exotic (mainly European) breeds, poor or sub-optimal feeding, a lack of breeding programmes, general lack of production planning and a high disease burden [1, 11]. People and animals live closely together. Meat and dairy value chains are characterised by short distances to the end consumer, promoting spread of zoonotic infections between cattle and humans via milk, water, soil or direct contact [12, 13].

Extension services for farmers are lacking in capacity in Uganda due to a shortage of qualified veterinarians [14]. Farmers can buy antibiotics, anti-parasitics, tick-treatment or prophylactics over-the-counter in "veterinary shops" where the staff is not required to have any education in veterinary medicine. Moreover, farmers can often not afford veterinary services and wait until they have tried various treatments including medical drugs or traditional remedies before calling a veterinarian [15, 16]. In each district a District Veterinary Officer (DVO) is the governmental representative responsible for animal health and disease control, surveillance and reporting to the Ministry of Agriculture, Animal

Industry and Fisheries. The DVOs can have one or several Veterinary officers and paraveterinarians employed in their unit. Veterinary officers and veterinary assistants/paraveterinarians can also be self-employed but contracted by the DVO for e.g. government vaccination campaigns or externally funded projects.

Selection of study herds

The study area was the neighbouring districts Kabarole, Kamwenge and Kasese in South-Western Uganda, an important area for livestock production (Fig. 1). These districts were chosen because studies on bovine infectious disease are sparse from this specific area and together they include several agro-ecological zones with expected varying herd sizes and management systems (although no detailed data were available). Each district is organised in administrative units of sub-counties, further divided in parishes (or wards in cities) that are subsequently divided in villages (or zones).

At the livestock census in 2008 [10] the estimated number of cattle per district was 67,120, 120,190 and 97,240 in Kabarole, Kamwenge and Kasese, respectively. There were 15,530, 14,100, and 5530 households with at least one bovine animal and the mean (median) herd sizes were 4.3 (3) in Kabarole, 1.8 (8.6) in Kamwenge and 17.6 (11) in Kasese [10]. The DVOs provided estimates of what they perceived as the current distribution of herd sizes in their districts. These herd sizes were larger than in the 2008 census.

To estimate the sample size needed to assess the prevalence of the model diseases at herd level (positive or negative herd) the online tool Epitools [17] was applied. Confidence levels of 95%, 5 to 10% precision, herd sensitivity and specificity in the range of 0.7 to 0.99 and true prevalences from 1 to 50% were evaluated. A total sample size of 180 herds, i.e. 60 per district, was decided upon based on what would be practically possible to manage and allow prevalence estimates of reasonable precision.

A list of all administrative units in Uganda was entered into a Microsoft® Excel (Microsoft Corporation, Redmond, USA) spreadsheet and a simple random selection, using the random number function, of 30 villages was made in each of Kabarole and Kamwenge districts. In Kasese, three quarters of the villages in the initial random sample had, according to the DVO, no cattle

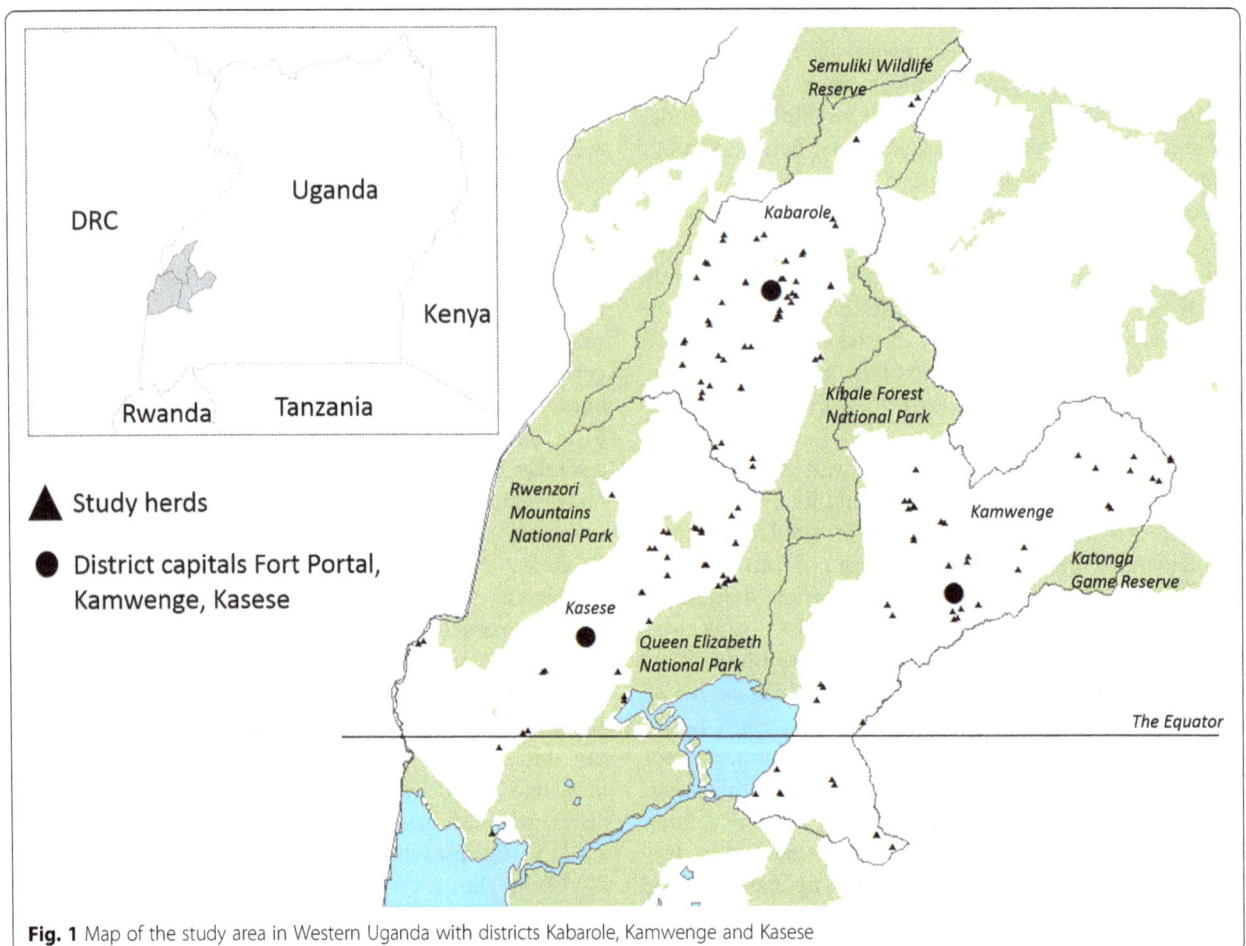

Fig. 1 Map of the study area in Western Uganda with districts Kabarole, Kamwenge and Kasese

because they were located in mountains or in a national park. To reach the desired number of villages, a purposive sample, geographically spread over the district, was made from the list of the remaining villages which the DVO confirmed had at least two cattle herds.

Next, each selected village was visited by a local veterinary officer or veterinary assistant. In Kabarole and Kasese, the village chairperson was asked to list all farmers owning at least two cattle. On a second visit a simple random sample of two herds was made from this list. The selection was made by writing each farmer's name on a piece of paper which was folded and put into a container. The selected farmers were asked if they agreed to participate in the study. If consent was not given, another farmer was selected by the same method. In Kamwenge district, selection of the first farmers was made by driving 5 min east from the village centre and asking the nearest farmer with at least two cattle if (s)he was willing to participate, if not, the next farmer in that direction was asked until two farmers were recruited. This procedure was chosen because of few local team members and poor road conditions.

Data collection
Herds were visited for sampling during January to March 2015 by the first author and a team of 2-5 local veterinary officers and veterinary assistants. The local teams were informed about the purpose of the study before recruiting farmers, and before the data collection visit they were further trained for the interviews, on information to be provided to farmers and biosecurity measures for the team.

In total 144 herds were visited and sampled: 55, 49, and 40 in Kabarole, Kamwenge, and Kasese, respectively. One farm was included because the owner was a prominent member of society and the herd vet, i.e. a local team member, requested that the herd should be part of the project. Six herds were added on the day of sampling; when the originally planned visits could not be carried out the first author asked the local team if they knew any farmers in the village or along the way back to the district headquarters that might be willing to participate. Three of these additional herds had owners that were prominent members of society.

A two-page questionnaire was used for the farmer interviews. The questionnaire was pilot-tested with the DVOs and their assistants before the study, and some questions were revised for improved clarity. The 39 questions were mainly closed (categorical or quantitative) or semi-closed. The topics covered included herd management, reproductive performance, cattle trade, animal mortality, vaccination, and disease prevention measures carried out. The questionnaire (in English) is available as additional file 1.

During the farm visit, a local assistant interviewed the farmer in any of the local languages (Rutoro, Rukhiiga or Rukonzo) with the first author observing and asking for clarification when needed, or the first author asked the questions in English and the local assistant translated. When possible, the answers to the interview questions were verified, e.g. herd size or the presence of other livestock species.

Interview data were coded and entered into a Microsoft® Access database (Microsoft Corporation, Redmond, USA) where possible with restrictions on values valid for entry. After data entry, each interview record was manually checked for data entering errors.

The coordinates of each farm were collected during the farm visit using the GPS tracker app My Tracks on an android mobile device. A map of the study area with the location of all study herds indicated was created in ArcMap 10.3.1 (Esri, Redlands, USA) using shape files available via ArcGis Online and from Free Spatial Data (DIVA-GIS).

Model diseases
Three infections were chosen as model diseases: brucellosis, salmonellosis and bovine viral diarrhoea (BVD). The choice of model diseases was based on expected endemicity in the study area, and availability of commercial serological assays that could be run in Uganda. The diseases have different transmission routes, together representing those of many other infections. They also differ in contagiousness but can be prevented by good biosecurity. The clinical manifestations include reproductive disorders such as abortions, infertility and prolonged calving intervals (brucella, salmonella, BVD), calf morbidity and mortality due to respiratory illness and diarrhoea (salmonella, BVD), loss in milk yield (brucella, salmonella, BVD), lameness and swelling of joints (brucella). Infections can be clinically manifest or subclinical. Salmonellosis and brucellosis are also important zoonoses [18, 19]. When introduced in a herd, infection can persist in individual animals and/or remain in the herd if new susceptible animals are added. Animals that have been infected with either of these agents remain seropositive for years (brucella and BVD) or months (salmonella).

The seroprevalence of brucellosis in Ugandan cattle has been estimated to range from 6.5% on herd level and 5.0% on animal level [20] to 55.6% at herd level and 15.8% at animal level [21], and varying from 0 to 78% at herd level depending on year and area [22]. No reports from the current study area could be found but in two neighbouring districts the animal prevalence was estimated to 14% [23]. However, Kabi et al. [24] reported that the seroprevalence in Ugandan cattle varies between agro-ecological zones. The within-herd prevalence of bovine brucellosis has been estimated to 26% in peri-urban and urban Kampala [20] and from 1 to 90% in Mbarara district [21].

For salmonella and BVDV, peer reviewed reports of seroprevalence in Ugandan or African cattle appear sparse. However, an MSc study suggested a seroprevalence at herd level of 57 and of 24% on individual level for salmonella, and 39% at herd level and 23% on individual level for BVD, in urban and peri-urban Kampala [25]. The seroprevalence of BVD has been reported as 92% at herd level in Cameroon [26], and 12% at the individual level in Tanzania [27].

Sample collection

Epitools was used to calculate the number of animals to be tested in a herd to detect at least one positive animal [28]. Based on the herd size information from the DVOs, herd sizes from 10 to 100 animals were evaluated with a confidence level of 95%, desired herd sensitivity of 95%, and test sensitivity of 98%. The final sample size was as follows: in herds with up to 20 cattle, all animals were sampled; in herds with 21 to 50 cattle, 20 animals were sampled; and in herds with more than 50 cattle, 30 animals were sampled.

Where possible, and based on intra-herd epidemiology of the respective diseases and expected seroconversion, animals between approximately 5 months and 2 years of age were prioritised for blood sampling, to get the recent infection history of the herd. Such animals are hereafter referred to as "young" individuals and older animals as "adult". The age group was determined based on the owners' information and the local teams' judgement, but age could not be ascertained in more detail.

Four farms had only one animal on the day of the herd visit because one animal had been sold or had died since recruitment. For 17 herds, fewer cattle than planned were bled for reasons including: cattle escaped, were too aggressive to handle, only calves were available, or because the farmer was unable to assist the team.

Blood was collected with Vacutainer® from the jugular or coccygeal vein into sterile tubes with no additive. Samples were stored in a cool box during transport to the field laboratory where they were centrifuged at 3000 rpm for 15 min and sera transferred to Greiner cryo vials. Within 2 days the refrigerated sera were transferred to a deep freezer (exact temperature varied but always below the freezing point) until transported to the laboratory at Makerere University in Kampala. At the laboratory, samples were stored in −20° C or −80° C until analysed.

All handling of animals including sampling was carried out under the direct supervision of the District Veterinary office staff in accordance with their national mandate.

Serological assays

Serum samples were analysed at Makerere University in Kampala, using commercial ELISAs. The assays were run according to the manufacturers' instructions. The person performing the analyses was blinded to information about the samples except the sampling date and district of origin.

For brucella, the SVANOVIR® Brucella-Ab I-ELISA (Svanova, Uppsala, Sweden) which detects antibodies to *B. abortus* and *B. melitensis* was used. According to the manufacturer, the test had a sensitivity (Se) of 100% in brucella-positive individuals and a specificity (Sp) of 100% in a brucella-negative population.

For salmonella, the PrioCHECK® Salmonella Ab bovine ELISA was used (Thermo Fischer Scientific, Waltham, Massachusetts, USA). It detects antibodies directed against *Salmonella* serotypes belonging to the B and D group, specifically the O-antigens: 1, 4, 5, 12 and 1, 9 and 12. The manufacturer has not provided the Se and Sp of this assay.

For BVD the SVANOVIR® BVDV-Ab (Svanova, Uppsala, Sweden) which detects both genotypes of BVD virus was used. According to the manufacturer the Se is 100% and the Sp 98.2% for serum samples.

For BVD it has been shown that it is sufficient to test a small number of individuals when the objective is to detect an ongoing or recent infection in the herd [29] and up to seven samples per herd were analysed, fewer for herds with <7 cattle. To save costs, serological analyses for brucella and salmonella were initially also performed on up to seven samples per herd. From herds where no positive samples were detected in this first round, more samples were analysed so as not to miss positive herds. For Brucella, all remaining samples from negative herds were analysed in a second round. For salmonella, up to nine more samples were analysed from the negative herds in a second round, while the remaining samples from herds that were still salmonella-negative were analysed in a third round. The number of samples included in each round of analysis was based on how many samples could be analysed with each ELISA kit. When possible, samples analysed first were from individuals of <2 years of age for salmonella and BVD to get the most recent infection history of the herd. For Brucella, individuals >2 years of age were prioritised because reproductively active (i.e. sexually mature) animals are at higher risk of clinical infection.

The manufacturers' cut-off (percent positivity, PP) for positive test result was used as cut-off for the individual animals' test result (PP 60 for brucella, 35 for salmonella and 10 for BVD). To check consistency and quality of the ELISA analyses, the distribution of PP values below the cut-off for a positive test, i.e. all negative individual samples, were compared between runs/plates (of the same ELISA) by visual inspection of plots of the distribution of values, by summary measures and by testing for difference with the Kruskal-Wallis test to ensure that negative samples were clearly below the cut-off and no individual plate/run deviated towards the cut-off-value.

Data management and analysis

Data management and statistical analyses were performed in the statistical package R [30].

Descriptive statistics including tabulations, summary measures and plots of distributions of variables were produced at herd level and, for the serological data, also at the individual animal level.

Of the eight categories for herd size, the three categories for the largest herd sizes were merged to one (> 50 cattle) because of few observations. The five remaining categories reflected, roughly, different management systems. The dichotomous variables for use of own bull, communal bull, artificial insemination (AI) and "other" were combined to one categorical variable with the categories own bull, communal bull, and "other". For seven variables there was very little variation, i.e. all responses were similar and therefore not suspected to explain any variation in herd serological status. These variables were not evaluated further. They included, for example, the use of tick spray for disease prevention, if cattle were kept on pasture, the number of cattle sold due to disease and the number of cattle returned from market unsold.

The remaining 29 variables were evaluated for associations between herd management and infectious disease burden. For each disease, a herd was defined as positive if at least one individual animal had a positive test result. A variable for the number of diseases a herd was positive to (0-3) was created. Apparent prevalence of positive herds for each disease with exact binomial confidence intervals were calculated in total, including only young or only adult individuals, and per district. Differences between districts were examined with Fisher's exact test as were differences between districts for the number of diseases (0, 1, 2, or 3) a herd was positive to.

Differences between categories of categorical interview variables was assessed by Chi2 test or Fisher's exact test and for numerical variables by the Kruskal Wallis test. All 29 variables were also evaluated in univariable poisson regression models for the outcome number of diseases a herd was positive to. If there was a difference with a p-value <0.2 the variable was evaluated further in the multivariable poisson regression model. Multicollinearity was evaluated by a Spearman rank correlation matrix including all the continuous variables, by a Spearman rank correlation matrix including all the binary variables, and for the categorical variables pairwise combinations were evaluated by the Chi-square test or Fisher exact test [31]. Where two variables were judged

as correlated (correlation coefficient > 0.9 [31] or $p < 0.05$ in the Chi-square or Fisher exact test) the one which was highest in the hierarchy was selected for further evaluation, e.g. number of dead cattle above the number of dead calves and district above herd size. Herd size was correlated to several other variables but did not fulfil the criteria for a confounder as it was not on any causal pathway to the outcome. The initial multivariable poisson model included the binary (yes/no) variables for employees caring for the cattle, if cattle were tethered for pasture, if pasture was shared with other cattle herds, if cattle were watered from an open water source (e.g. a river), if there were pigs on the farm, and if the farmer was satisfied with the health status of the cattle in the previous 12 months. The initial model also included the categorical variables for bull and district, and the continuous variables for number of calves born, cattle that died, cattle that were sold or given away, and cattle bought or received, during the previous 12 months.

The model was reduced by removing the least significant variable one at a time and comparing it to the previous by a likelihood ratio test. The parameter estimates were checked for changes of >25% which would suggest the removed variable was a confounder and to be kept in the model. All removed variables were added to the final model one at a time to evaluate confounding. First order interactions between all main effects in the final model were tested. The final model was evaluated for overdispersion and residuals were visually examined. The number of observations included in the final model was 144.

Because of the many statistical tests carried out, a conservative significance level of $p < 0.01$ was applied when interpreting test results.

As the focus of our interest was biosecurity, to further explore the direct and indirect associations between herd routines that influence biosecurity and the serological status of herds, an Additive Bayesian Network (ABN) was fitted to the interview and serological data. An ABN is a multivariate model that describes the complex relationships of the direct and indirect connections between random variables. The ABN methodology is an extension of usual GLM models to include multiple response variables [32]. An ABN has two parts; the structure (qualitative) and the model parameters (quantitative). Each node comprises a generalized linear model where the arcs from parents of that node represent the covariates in the regression model. An ABN illustrates both direct and indirect associations between variables and each outcome variable [33]. A directed acyclic graph (DAG) graphically represents the joint probability distribution of all random variables in the data. The model parameters are represented by local probability distributions for all the variables in the network. A general introduction to ABN in veterinary epidemiology is given by Lewis et al. [33] and Lewis and Ward [34]. Other applications in veterinary epidemiology include associations between on-farm biosecurity practice and equine influenza [35], cattle herd risk factors and salmonella [36], and identification of *E. coli* antimicrobial resistance patterns [32].

Interview responses representing biosecurity measures, a variable classifying the herd as negative or positive to at least one of the diseases, herd size and district were included - in total 14 variables. The number of variables was reduced to adapt to the recommendation not to exceed 10% of the number of observations (personal communication M Pittavino, Institute of Mathematics, Applied Statistics Group, University of Zurich, Switzerland). As required for the analyses, multinomial variables were split into several binomial variables, and observations with missing data removed ($n = 4$). A hill climber search [37] with 1000 iterations was applied to identify a globally optimal consensus DAG for the data. The network score (the log marginal likelihood) was estimated using Laplace approximations at each node and used as the goodness of fit measure. The means and variances hyper parameters for the Gaussian priors were fixed at zero and 1000, respectively. The number of allowed parents per node, i.e. the number of covariates in the regression model of a node, was increased for each new search, starting at one, until the network score did not increase further. Arcs between binomial variables derived from the same multinomial variable, and arcs with the herd serology status as parent node were banned. The arcs in the consensus DAG were retained for the next search. The analysis was performed with the R library abn [38].

Results

Herd sizes, production and management data from the farmer interviews are presented in Tables 1 and 2. As seen in the livestock census 2008 [14], small herds of less than five dominated in Kabarole district while Kasese district had the largest herds. The median number of animals per herd corresponded to those estimated in the census, except for Kasese where the median size in this study was 21–50 animals and the median figure reported in 2008 was 11 animals [10]. The most commonly stated main purpose of the cattle was dairy and beef (59.7%, $n = 86$), while dairy was the second most common (34.7%, $n = 50$). However, dairy was less common in Kasese (17.5%, $n = 7$) and the two farmers that stated beef production as the main purpose were in this district. Other purposes included e.g. draught power or not stated ($n = 6$). The most frequent breed was mixed breed (local and exotic) which was the main breed in 76 (53%) herds. The second most common breed was local breed,

which was the main breed in 57 (40%) herds. Exotic breed was the most common breed in 11 (8%) herds. Most cattle were managed by the owner and his/her family but about a third of the cattle herds in Kabarole, half of those in Kamwenge and two thirds of those in Kasese also had employees taking care of the cattle. In Kabarole, most farmers (76.4%, $n = 42$) used a communal bull but ten farmers (18.2%) had their own bull and seven (12.7%) also stated that they use AI. Only one farmer in each of the districts Kamwenge and Kasese used AI and in Kamwenge about half of the farmers had their own bull and half used a communal bull. In Kasese, two thirds of the farms had their own bull and one third used a communal bull.

The majority of farmers had sold or given away cattle the last year, mainly for slaughter. According to the local field teams there is a weekly livestock market in Kamwenge but no such market in Kabarole nor in Kasese, except for one on the border to the Democratic Republic of Congo where cattle transported from Tanzania are also traded. However, it was noted that the weekly markets often had a section for livestock, poultry and meat. The farmer interviews gave the same picture where 18 famers in Kamwenge had sold cattle at a market but only two in Kabarole and one in Kasese.

In addition, almost half of the farmers had bought or received new cattle, mainly from other farmers and less frequently from a livestock market. The majority (79.2%, $n = 114$) said that they always put new animals directly together with the rest of the herd. Of the 11 farmers that used some type of quarantine for new cattle, a separate pasture or paddock or at another home was stated as the quarantine location. Seven farmers (one in Kabarole, two in Kamwenge and four in Kasese) said they kept their cattle separated from other cattle herds (Table 1). Biosecurity measures like visitors' hand-wash or cleaning of boots or feet were rarely practiced.

Overall, almost three quarters of farmers were satisfied with the health status of their cattle the preceding 12 months, however in Kasese this figure was only 50%. In addition, 123 (85%) stated that they had given antibiotics or other drugs as prevention to healthy cattle and 142 (99%) had used tick spray during the same time period (Table 1). Of the farmers that stated they had had their cattle vaccinated ($n = 22$), the most frequent diseases stated as vaccinated against were FMD ($n = 10$), blackleg ($n = 5$), ECF ($n = 2$) and one each for brucella, anaplasma and lumpy skin disease. The time of vaccination was often not clear and could be up to "several years ago".

The types of wildlife some farmers (41.7%, $n = 60$) stated as present in their pastureland were mainly monkeys and baboons ($n = 26$) and a wide range of animals mentioned by one or a few farmers including wildcats,

Table 1 Description of some of the non-numerical parameters regarding cattle herd management and performance in Western Uganda

	Kabarole n (%)	Kamwenge n (%)	Kasese n (%)	Total study area n (%)
Total	55	49	40	144
Herd size*				
2-5	35 (64)	13 (26.5)	2 (5)	50 (34.7)
6-10	11 (20)	14 (28.6)	7 (17.5)	32 (22.2)
11-20	4 (7.3)	11 (22.4)	6 (15)	21 (14.6)
21-50	5 (9.1)	6 (12.2)	14 (35)	25 (17.4)
51-100	0	3 (6.1)	9 (22.5)	12 (8.3)
101-150	0	0	0	0
151-200	0	2 (4.1)	2 (5)	4 (2.8)
Purpose of cattle *				
Beef	0	0	2 (5)	2 (1.4)
Dairy	24 (43.6)	19 (38.8)	7 (17.5)	50 (34.7)
Dairy and beef	29 (52.7)	30 (61.2)	27 (67.5)	86 (59.7)
Other	2 (3.6)	0	4 (10)	6 (4.2)
Cattle care by				
Husband/wife*	54 (98.2)	35 (71.4)	21 (52.5)	110 (76.4)
Other family*	43 (78.2)	26 (53.1)	18 (45.0)	87 (60.4)
Employees*	18 (32.7)	27 (55.1)	26 (65.0)	71 (49.3)
Decision regarding cattle made by				
Husband/wife	54 (98.2)	47 (95.9)	39 (97.5)	140 (97.2)
Other family*	22 (40.0)	12 (24.4)	3 (7.5)	37 (25.7)
Employees	0	1 (2.0)	0	1 (0.7)
You, family, employees have contact with other's cattle*	26 (47.2)	18 (36.7)	29 (72.5)	73 (50.7)
Visitors to cattle				
Wash hands before	6 (10.9)	5 (10.2)	9 (22.5)	20 (13.9)
Clean boots or feet	0	3 (6.1)	2 (5.0)	5 (3.5)
Share equipment with other farmers	20 (36.4)	11 (22.4)	8 (20.0)	39 (27.1)
Farmers in area help each other with cattle	21 (38.2)	21 (42.9)	22 (55.0)	64 (44.4)
New cattle put directly with the herd				
Always	47 (85.5)	37 (75.5)	30 (75.0)	114 (79.2)
Never	2 (3.6)	5 (10.2)	4 (10.0)	11 (7.6)
Sometimes	1 (1.8)	0	1 (2.5)	2 (1.4)
Never bought	3 (5.5)	5 (10.2)	4 (12.5)	13 (9.0)
Missing value	2 (3.6)	2 (4.1)	0	4 (2.8)
Preventive measures for healthy cattle last 12 months				
Tick spray/ dip	54 (98.2)	48 (98.0)	40 (100)	142 (98.6)
Traditional medicine	1 (1.8)	2 (4.1)	0	3 (2.1)
Cattle kept separate from other cattle herds	1 (1.8)	2 (4.1)	4 (10)	7 (4.9)
Antibiotics or other drugs*	39 (70.9)	45 (91.8)	39 (97.5)	123 (85.4)
Deworming	53 (96.4)	46 (93.9)	39 (97.5)	138 (95.8)
Vaccination*	8 (14.5)	2 (4.1)	12 (30.0)	22 (15.2)
Has own bull*	10 (18.2)	24 (49.0)	26 (65.0)	60 (41.7)
Use communal bull*	42 (76.4)	24 (49.0)	14 (35.0)	80 (55.6)

Table 1 Description of some of the non-numerical parameters regarding cattle herd management and performance in Western Uganda *(Continued)*

	Kabarole n (%)	Kamwenge n (%)	Kasese n (%)	Total study area n (%)
Use artificial insemination	7 (12.7)	1 (2.0)	1 (2.5)	9 (6.3)
Cattle on pasture	51 (92.7)	49 (100)	40 (100)	140 (97.2)
Cattle on pasture tethered*	16 (29.1)	11 (22.4)	1 (2.5)	28 (19.4)
Fenced pastures*				
Yes	24 (43.6)	16 (32.7)	13 (32.5)	53 (36.8)
Yes, partly	19 (34.5)	22 (44.9)	5 (12.5)	46 (31.9)
No	8 (14.5)	11 (22.4)	22 (55)	41 (28.5)
Shared pasture	18 (32.7)	19 (38.8)	24 (60.0)	61 (42.4)
Use pasture in national park*	1 (1.8)	0	17 (42.5)	18 (12.5)
Wildlife on the pasture	22 (40.0)	15 (30.6)	23 (57.5)	60 (41.7)
Water to cattle from open water*	23 (41.8)	9 (18.4)	35 (87.5)	67 (46.5)
Water to cattle from a well*	27 (49.1)	42 (85.7)	2 (5.0)	71 (49.3)
Water to cattle from tap	9 (16.4)	1 (2.0)	4 (10.0)	14 (9.7)
Other livestock on the farm or in the household	52 (94.5)	47 (95.9)	35 (87.5)	134 (93.1)
goat	45 (81.8)	40 (81.6)	28 (70.0)	113 (78.5)
chicken	40 (72.7)	39 (79.6)	28 (70.0)	107 (74.3)
pig*	18 (32.7)	11 (22.4)	2 (5.0)	31 (21.5)
dog*	24 (43.6)	5 (10.2)	8 (20.0)	37 (25.7)
Farmer was satisfied with the health of the cattle over the last 12 months*	43 (78.2)	41 (83.7)	20 (50.0)	104 (72.2)

* Statistically significant ($p < 0.01$) difference between districts at Fisher exact test
Data from an interview with the farm owner or manager of 144 cattle herds in Western Uganda. All variables refer to the last 12 months prior to the sampling occasion. The study was performed in 2015

foxes, elephants, buffalos, wild pigs, mongoose, antelopes, hyenas, rabbits, lions, hippos, and crocodiles or "animals from Queen Elizabeth National Park".

Water sources differed somewhat between the three districts. In Kabarole 16.4% ($n = 9$) of the farmers stated that they gave the animals water from a tap, while 49.1% ($n = 27$) said they took water from a well and 41.8% ($n = 23$) that the cattle drank from an open water source. In Kamwenge, the corresponding figures were 2.0% ($n = 1$) tap water, 85.7% ($n = 42$) from a well and 18.4% ($n = 9$) from an open source while in Kasese they were 10.0% (n = 4) tap water, 5.0% ($n = 2$) from a well and 87.5% ($n = 35$) from an open water source.

Serological results

Serum was collected from in total 1567 individual cattle; 901 young and 666 adults. The median number (1st; 3rd quartile) of individuals per herd was 8 (4;20). The median number (1st; 3rd quartile) of analysed samples per herd was 7 (4;12), 7 (4;16) and 7 (4;7) for brucella, salmonella and BVD, respectively.

The apparent herd prevalences of the three diseases are presented in Table 3. The difference between districts was statistically significant ($p < 0.001$) for brucella but not for salmonella and BVD ($p > 0.05$). Kasese had

the highest apparent herd prevalence of brucella and Kabarole the lowest. Approximately half of the herds in all three districts were positive for BVD. The apparent herd prevalence when test results from young or adult individuals were assessed separately are shown in Table 4. There was a trend that herds in Kasese tested positive to a higher number of diseases than herds in Kamwenge. Herds in Kabarole were to a larger extent negative or only positive to one disease (Table 5). For BVD, 48 of 138 herds had at least one positive young individual and of those, 24 had two or more positive young individuals, suggesting an ongoing infection in the herd.

Associations between herd management and biosecurity and infectious disease burden

The poisson model suggested that having employees to care for the cattle was associated with a herd being positive to a higher number of diseases with an RR estimate (95% CI) of 1.9 (1.3;2.8). In addition, an increasing number of dead cattle the previous 12 months and sharing pasture both had RR estimates (95% CI) of 1.1 (1.0;1.1).

The globally best fitting DAG from the multivariate model is presented in Fig. 2. It included ten nodes and seven arcs in three separate networks. Variables for quarantine for new cattle, having visitors in contact with

Table 2 Description of some of the numeric parameters regarding cattle herd management and performance in Western Uganda

	Kabarole		Kamwenge		Kasese		Total	
	n^a	Median[b] (Q1;Q3)	n^a	median (Q1;Q3)	n^a	median (Q1;Q3)	n^a	median (Q1;Q3)
Calves born*	48	1 (1;3)	46	4 (1.25;6)	39	6 (3;14)	133	3 (1;7)
Abortions*	14	1 (1;1)	12	2 (1;3.25)	18	2 (1.25;3)	44	1.5 (1;3)
Dead cattle*	23	1 (1;1)	26	2 (1;2)	25	3 (1;5)	74	(2 (1;3)
Dead calves	7	1 (1;1.5)	14	2 (1;5.25)	18	2 (1;3.75)	39	2 (1;3)
Sold or given away cattle*	38	1 (1;2)	44	3 (2;4.25)	33	4 (2;6)	115	3 (1.5;4)
Cattle sold for slaughter*	26	1 (1;2)	30	3 (2;4)	30	3 (2;5)	86	2 (1;4)
Cattle sold at market	2	3 (2;4)	18	2.5 (2;3.75)	1	1 (1;1)	21	2 (2;4)
Cattle returned unsold from market	0		2	1.5 (1.25;1.75)	0		2	
Sold or given to another farmer	18	1.5 (1;2)	12	3.5 (1;8.5)	5	3 (3;3)	35	2 (1;3.5)
Cattle sold due to poor performance	5	1 (1;2)	9	2 (1;3)	11	2 (1.5;3.5)	25	2 (1;3)
Cattle sold due to disease	3	1 (1;1)	4	1 (1;1.25)	5	4 (1;5)	12	1 (1;2.5)
Cattle bought or received	15	2 (1;2)	32	2 (1;4)	18	2 (1;3)	65	2 (1;3)
Cattle bought from market	2	1 (1;1)	12	1 (1;2)	4	1.5 (1;2.25)	18	1 (1;2)
Cattle bought/received from other farmer	13	2 (1;2)	22	2.5 (1;4.74)	14	2 (1;2.75)	49	2 (1;4)
Cattle bought with known disease	0	–	6	1 (1;1)	1	1 (1;1)	7	1 (1;1)

*Statistically significant ($p < 0.01$) difference between districts at Kruskal Wallis test
[a]Number of responses with value >0, included in the median value
[b]Median value based on responses >0
Data from an interview with the farm owner or manager. All variables refer to the last 12 months prior to the sampling occasion. The study included 144 farms and was performed in Western Uganda in 2015

the cattle wash hands or boots, and the categories of herd sizes >5 were not connected to any other nodes and are not shown. The global fit of the model did not increase beyond 1 parent (network score = –1258). The first network illustrates how small farms (2–5 cattle) were less likely to have employees care for their cattle, or to be positive to at least one disease, as well as more likely to be from district 1 (Kabarole). The second network of variables illustrates how farmers who help each other were more likely to also share equipment, the cattle to have contact with other cattle and to share pasture. The third network illustrates how farms in district 2 (Kamwenge) were more likely to have bought cattle. The second and third networks illustrate how herd parameters are associated with each other, although not to herd serological status.

Table 3 Frequency and apparent prevalence (%) of seropositive cattle herds in Western Uganda

	Kabarole n (%)	Kamwenge n (%)	Kasese n (%)	Total n (%)[a]
Total	55	49	40	144
Brucella[b]	3 (5.5)	5 (10.2)	16 (40)	24 (16.7 (11.0;23.8))
Salmonella	8 (14.5)	16 (32.7)	10 (25.0)	34 (23.6 (16.9;31.4))
BVD	28 (50.9)	27 (55.1)	22 (55.0)	77 (53.4 (45.0;61.8))

[a]With exact binomial confidence intervals of %
[b]Significant difference ($p < 0.001$) between districts with two-sided Fisher's exact test

Discussion
Potential for improved herd biosecurity
A shared characteristic of the three model diseases is that they can be controlled and prevented by biosecurity measures. Zoonotic infections such as salmonellosis and

Table 4 Frequency and prevalence (%) of seropositive cattle herds based on tested young or adult individuals

	Herds with young[a] individuals		Herds with adult[a] individuals	
	n positive herds (n tested)	prevalence positive herds (%)[b]	n positive herds (n tested)	prevalence positive herds (%)[b]
Brucella	4 (123)	3.3 (0.9;8.1)	21 (137)	15.3 (9.7;22.5)
Salmonella	15 (138)	10.9 (6.2;17.3)	20 (129)	15.5 (9.7;22.9)
BVD	48 (138)	34.8 (26.9;43.4)	40 (89)	44.9 (34.4;55.9)

[a]young <2 years, adult >2 years
[b]With exact binomial confidence intervals
Prevalence of herds that were sero-positive to brucella, salmonella, and BVDV. A herd was classified as positive if at least one sampled individual animal tested positive in the respective ELISA, using the manufacturers' cut-offs. The number of sampled animals per herd was; all for herds with up to 20 cattle, 20 for herds with 21 to 50 cattle, and 30 if there were more than 50 cattle. When feasible, individuals up to 2 years of age were sampled. First, samples from seven individuals were tested serologically prioritising young individuals for salmonella and BVD, and older for brucella. If all were negative, for brucella the remaining samples from the herd were analysed. For salmonella another nine samples were analysed, if the herd was still negative any remaining samples were analysed. The study was performed in Western Uganda in 2015

Table 5 Number of diseases cattle herds were positive to

Number of ELISA tests for which the herd had a positive test result *	Kabarole n (%)	Kamwenge n (%)	Kasese n (%)	Total n (%)
0	22 (40.0)	13 (26.5)	11 (27.5)	46 (31.9)
1	28 (50.9)	26 (53.1)	12 (30.0)	66 (45.8)
2	4 (7.3)	8 (16.3)	15 (37.5)	27 (18.8)
3	1 (1.8)	2 (4.1)	2 (5.0)	5 (3.5)

* Difference between districts tested with two-sided Fisher's exact test; $p < 0.007$
Frequency and prevalence of herds that were sero-positive (at least one individual with a positive test result) to none, one or several of brucella, salmonella or BVD in each study district. The study was performed in cattle herds in Western Uganda in 2015

brucellosis are particularly important to control in countries such as Uganda with a large rural population with close contact with livestock, and dairy and meat value chains that may lack the capacity to control these pathogens [39, 40]. Studies on herd biosecurity routines and basic production aspects in Ugandan cattle farms are sparse. The results from the farmer interviews indicate that there is a huge potential for improved biosecurity. For example, the proportion of farmers that had introduced new cattle in the last 12 months was high while very few stated that they used a quarantine for new cattle. Given that most farmers had received only a few animals, it should be possible to keep them isolated for a

limited time period. Most cattle that were sold were destined for slaughter while most cattle that were brought in came from another farmer, and only a few from the cattle market. Cattle markets are relatively rare in the study area but the finding is, nevertheless, positive from a disease control point of view. The proportion of farmers who had traded diseased cattle was low, which is also positive from a disease control perspective. This figure may however be an under-estimation due to reluctance to divulge such habits if it was perceived to be regarded as poor practice by the researchers.

In Kabarole district, land ownership allows more farmers to have paddocked pasture for their cattle. This should reduce the risk of disease transmission because the herd is separated from other herds. For farmers that use communal grazing, a disease prevention measure could be for the community to treat their cattle as one epidemiological unit and to keep their common herd separate from the herds of other communities. A well or open water were the most common sources of water, and watering points are sites where herds mix. Agreements could be made between communities to visit the watering point at different times, to separate herds in space and time and thus reduce the risk of direct disease transmission.

Biosecurity routines, such as handwashing and cleaning boots for visitors to the cattle, were rarely used.

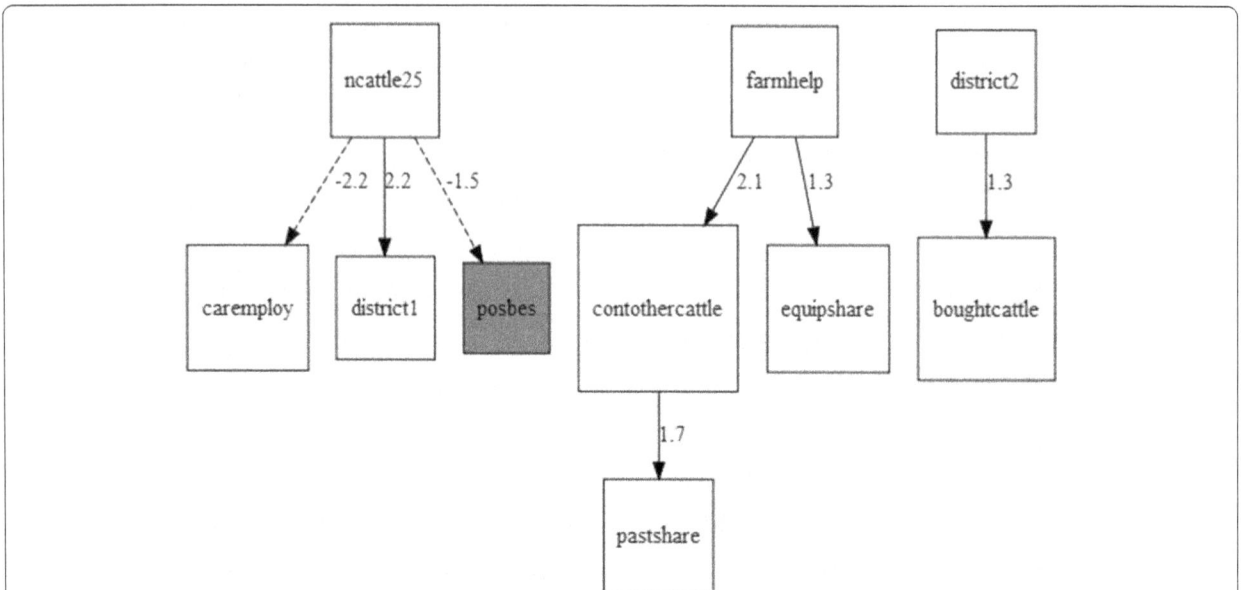

Fig. 2 Final global Directed Acyclic Graph (DAG) from additive Bayesian network modelling. The DAG illustrates direct and indirect associations between herd status (positive/negative serology) to at least one of brucella, salmonella or BVD, and farmer interview answers relating to herd biosecurity. A heuristic search with 1000 iterations was run with up to 2 parents allowed. Arcs that were included in at least 50% of local models are shown. Variables with no arcs are not shown. The numbers on the arcs represent the posterior marginal densities; a dashed line indicate negative association. All variables were binomial (no/yes). Variable coding: posbes = herd sero-positive to at least one of the diseases, district1 = Kabarole, district2 = Kamwenge, ncattle25 = herd size 2-5 cattle, careemploy = employees care for cattle, pastshare = share pasture with other cattle herds, boughtcattle = bought or received cattle the last 12 months, equipshare = share equipment with other farms, contothecattle = people from the farm has contact with cattle form other farms, farmhelp = farmers in the area help each other with their cattle

Many basic measures such as an extra pair of boots for visitors, asking visitors to wash their hands before handling cattle, a small paddock to keep new or diseased cattle isolated, and to only buy healthy looking cattle are all relatively low-cost investments which, if they were practiced, would reduce the risk of introduction of infectious diseases to the herd. Advice on preventive actions is difficult to convey, since what is sought is a non-event (absence of disease or production loss) and hence the incentives are not tangible. This is a particular challenge in settings where farmers might not feel empowered or in charge of their own destiny and where uncontrollable disasters and political unrest are recurrent. In addition, biosecurity measures have to be practically feasible for farmers to adopt them, and knowledge about what the local population of farmers can actually do, or think they can't do, is needed. This will be the focus of some of our future studies.

Farmer knowledge of infectious diseases and biosecurity was not directly assessed. Surveys from other low-income countries with small-scale farming have suggested a general lack of knowledge and access to information on infectious diseases and control measures. Studies in Uganda have found that small-holders with pigs had knowledge about African swine fever but did, however, not implement biosecurity [14, 41].

The risk factor analysis suggested an association between the number of diseases a herd was serologically positive for and the following factors: employees caring for the cattle, shared pasture and a higher number of dead cattle. While this association does not necessarily reflect a causation, the first two could be explained as related to indirect and direct contacts with other herds, while the third is most likely an indication of the presence of disease in the herd and hence a "risk factor" that is in fact a symptom. However, risk factors are often correlated and individual effects may be difficult to discern. The ABN is one approach to tackle this challenge. The results from the ABN suggested that a herd size of two to five cattle was associated with a herd being negative to all three diseases, and not having employees caring for the cattle. The DAG from the ABN illustrates connections between variables that the univariate analysis did not capture. This method could potentially be used to create "farmer profiles" in the study population and suggest how, within a small geographical area, farmers vary in their management. For biosecurity interventions it would be very helpful to know which practices are likely "to come in a package" and thus need to be addressed as such. Moreover, it is important to know if farmers should be addressed as a community rather than individual farmers because their cattle are kept as one epidemiological unit. This was supported by the ABN results where a cluster with shared pasture, shared equipment and farms helping each other emerged. A larger sample size might had given enough power to detect more risk factors in the poisson regression model and to generate more complex DAGs.

During the field work it was noted how heterogeneous the farmer population is in the study area. Not only due to parameters that were assessed in the interviews, but to factors related to cultural group, socio-economic status, and experience as cattle owners. This could complicate the assessment of differences between farms with healthy and less heathy cattle, i.e. make identification of single risk factors for disease harder to identify. The three districts differed in some aspects, where a few could be seen in the DAGs (Fig. 2), e.g. Kabarole having smaller farms managed by the farmer and his/her family. There were some characteristics that were different for the farms in Kasese when compared to those in Kabarole and Kamwenge. The herds were larger and used more open water sources and pasture in natural parks or game reserves. They appeared to be more likely to be managed by the husband/wife of the family with less involvement of other relatives, and employees taking care of the cattle. They sold more cattle and the farmers and their employees were more frequently in contact with other cattle. Still, the overall heterogeneity of the farms and farmers made subgroups less distinguishable by the current study design.

Serological results

The results indicate that the three model diseases are endemic at a moderate (brucella and salmonella) to high (BVD) prevalence in the Ugandan cattle population. The diseases were chosen because they have different epidemiology and transmission routes and affect herd productivity.

The prevalence of salmonella in animal production is high in many parts of the world. Production systems that cause continuous exposure to salmonella are usually intensive [42, 43] but in more extensive systems the animals may be exposed to environmental bacteria, for example where they are gathered at night or in the milking kraal. The prevalence of salmonella isolated from mesenteric lymph nodes from healthy cattle slaughtered in an abattoir in north-east Nigeria was 61% [44], although the authors remark that this was probably an over-estimation because of the stress animals had been subjected to before being culled. Nevertheless, salmonella can be expected to circulate in smallholder farming systems in Africa, which the current results support, and measures to prevent exposure of the animals to salmonella from other herds and/or the environment would be expected to reduce the prevalence.

BVD has not been extensively investigated in smallholder systems in Africa. A study among small-scale

family farms in northern Turkey revealed a seroprevalence of 32% [45]. In Kerala province in India, a study
among smallholder dairy cattle revealed a prevalence of
24.7%, but higher among animals with infertility problems [46]. There are some reports of BVD being endemic in African cattle, from South Africa [47] and
Cameroon [26]. A Swedish MSc thesis detected serological reactions among cattle in peri-urban farming systems in Kampala [25]. In addition, serological reactions
to BVDV have been detected in African wildlife such as
kudu and eland [48]. BVD has an immunosuppressive
effect and in populations with a high infection pressure,
e.g. sub-Saharan Africa, this impact on health and performance is likely to be even more important. Reducing
the high seroprevalence found in this study might therefore have a positive effect in cattle herds in Uganda.

There are various reports on the presence of brucellosis in African cattle. In Uganda, serosurveys show different herd prevalence levels ranging between 5 and
100% in different regions, agro-ecological zones, and different animal management systems [21, 22, 24, 49] and
it is not always clear to what extent vaccination was
practised in the study populations. No reports on prevalence from the particular area of this study have been
found but it was clear that brucellosis is endemic.

There was a difference in prevalence between districts
for Brucella as well as for the number of diseases herds
were positive to, both had the highest numbers in Kasese
district. Sharing pasture remained in the final poisson
model of the outcome number of diseases a herd was
positive to. Kasese had many farms with communal
grazing, however district did not remain in the model.
The ABN suggested that herds that were negative to all
three diseases, were linked to fewer employees, small
herd size, and being located in Kabarole. Brucella has
been shown to be more prevalent in pastoralist communities with communal grazing, both in livestock and in
humans [18], agreeing with the current results.

Interestingly, a clear majority of all farmers (72%) were
satisfied with the health status of their cattle the last
year, although this figure was lower for the farmers in
Kasese. This supports that a high burden of clinical and
subclinical infections with negative effects on productivity might be considered as normal in the study population. Similarly, abortions could be part of what is
experienced as "normal", although abortions were more
frequent in Kasese. Abortions were likely under-reported
by farmers because pregnancy checks are hardly ever
performed and abortions in early stages of pregnancy
would be noted as a prolonged calving interval. This, in
turn, might not be noted either, because there were generally no farm records. In a previous study, Kenyan
farmers have been suggested to report too low numbers
of abortions in their cattle [50].

Study limitations and practical challenges with impact on methodology and results

There are challenges to carrying out epidemiological
field studies in low-income countries. Failure to acknowledge these challenges may lead to invalid results being
presented as solid data.

Firstly, in the absence of formal or informal animal
registers, any sampling frame will be incomplete and the
level of representativity is often difficult to assess. Even
when a lot of effort is put into applying a probability
sampling method, practical obstacles such as inaccess-
ibility due to poor roads and failure to communicate regarding visiting time often lead to replacement of
selected farms in a manner that sometimes impact on
the sampling method. In the present study, the aim was
a random sample of villages, which for practical reasons
was not fully achieved. However, the team members that
recruited farmers did at that point not know which factors that would be evaluated, and the first author had no
knowledge about the villages in the area. This would
have avoided any subconscious selection bias. Two herds
per village were sampled which, in theory, would call for
village to be included as a random variable in models.
However, village is not an epidemiological unit in
Uganda but an administrative and geographical delimita-
tion of the rural landscape where farms are continuously
scattered and usually not aggregated near a village
centre. Hence, no actual clustering on village level is to
be expected.

Secondly, cultural clashes and language barriers may
lead to misunderstandings that affect the results of interviews and questionnaires. The extent of this is difficult
to assess. Including control questions to detect bias or
misunderstandings may lead to further complications
and distrust. One factor not to be ignored in this type of
studies is the power imbalance and skewed relation
when European, well-educated researchers perform field
studies in farms in low-income countries. It is well-
known that interview subjects might say what they think
the interviewer (or researcher present in the interview
situation) wants to hear, in favour of the most truthful
answer. One should have in mind that highly unequal
relationships such as in this study could increase the response bias. Efforts were made at each interview to inform the farmer that their name would not be published,
who the team members were and what organisation they
represented for the study. Still, it cannot be known
whether the farmers felt confident enough to supply
truthful answers to all questions.

The aim of 60 sampled herds per district was not
reached, mainly in Kamwenge and Kasese. This was because of the unexpectedly large amount of time needed
for transportation between farms, time spent at each
farm visit, and farmers that had forgotten about the

agreed visit and/or had already let their cattle out on pasture when the team arrived. Because of budget and time limitations, the field work could not be extended until all 180 farms had been visited. Efforts were made to cover as many geographical areas as possible during sampling. Only one farmer declined to participate in the study at the first recruitment visit. The authors believe the results to fairly represent the situation of bovine infectious diseases and cattle management in the study area, as well as other regions in Uganda with similar conditions.

Serology does not indicate the current infection status but rather the infection history of the herd. By prioritising individuals of less than 2 years' age for salmonella and BVD a more recent history could be assessed. Nevertheless, a cross-sectional study is a snapshot of serological status and a longitudinal study design might have given more information about infection dynamics in the population. It is not known if there are seasonal variations in three model diseases in the study area, a fact that would clearly affect the results. However, cattle production itself does not appear to change by season in the study area so there is no obvious reason for seasonality in the disease prevalence.

The herd sensitivity is lower for farms with few tested individuals and hence the risk of false negatives increases. However, a cut-off of one positive individual per herd reduces herd specificity, especially with the sample analysis approach used in this study. Increasing the cut-off for a herd to be classified as positive to two seropositive individuals (of any age) changed the number of positive herds from 24 to 12 for brucella, from 34 to 6 for salmonella, and from 77 to 37 for BVD. This reduction of >50%, highlights the importance of case definition for prevalence estimates. In addition, changing the cut-off could influence a risk factor analysis. For BVD, at least three positive samples from individuals has been used as a cut-off for the herd to be likely to have persistently infected (PI) animals, i.e. an ongoing infection, as herds without PI animals tend to self-clear from the infection [51]. However, in the current study the objective was not to identify herds with PI animals but herds with animals that were seropositive, i.e. had undergone the infection.

The ELISAs used for the serological analyses were not evaluated in the study population. It is known that tests validated in non-exposed populations perform differently in endemic situations and this could have affected the results in the current study [52]. However, there are no available commercial tests for the diseases in this study that are properly evaluated in African settings and perhaps no market incentive to perform such evaluations of the current tests. The salmonella ELISA does not detect all serotypes and hence the real seroprevalence of salmonella is likely to be higher.

Conclusions

There is room for improvement of biosecurity practices in Ugandan cattle production. However, to take proper action to improve biosecurity and cattle health in any region, knowledge of the current status is necessary, i.e. disease prevalence and a description of management systems. Design and evaluation of interventions for such improvement would benefit from the creation of farmer profiles, using data on disease prevalence and herd management practices. As in many studies on biosecurity, few significant associations were seen in the risk factor analysis but this does not justify lack of efforts to improve biosecurity.

Despite the challenges and limits of this study, it is clear that salmonella, brucella and BVD are present in cattle in the study area. Moreover, the prevalence of these infections is, to some extent, associated with farm management practices. Brucellosis and salmonellosis are important zoonotic infections and reducing the prevalence within and between herds would benefit the human population directly as well as allow for healthier livestock.

Abbreviations

AI: Artificial insemination; BVD: Bovine viral diarrhoea; DAG: Directed acyclic graph; DVO: District veterinary officer; ECF: East Coast fever; FMD: Foot-and-mouth disease; TAD: Transboundary animal disease

Acknowledgements

The authors would like to thank the District Veterinary Officers Drs Salvatory Abigaba, Alfred Kamanyire and Kalule Godfrey and their respective teams for dedicated assistance with the field work, Mr. Johnson Mayega for skilful laboratory work, and the participating farmers for generously sharing their experiences and time.

Funding

The study was financially supported by the non-commercial, not-for-profit, governmental Swedish Research Council (Vetenskapsrådet), Grant No. 348-2013-6608. The funding body had no role in study design, data collection, analysis, or interpretation, or writing of the manuscript.

Authors' contributions

CW and SSL designed the study with support from KS and SB. CW planned and carried out the field work, statistical analysis and drafted the manuscript. CM was responsible for the laboratory work. All authors revised the manuscript and approved the final version of the manuscript.

Ethics approval and consent to participate

Informed consent was given (orally) by all participating farmers after they were informed about the purpose of the study and that the identities of individual farmers were not to be included in any reports. Moreover, farmers were informed that if they wished, they would receive the serological results from their own herd. The local team members later visited all study herds to share the results and to give general biosecurity advice orally, and in an information leaflet that was left with the farmer. No other incentives were given to the farmers.

All handling of animals including sampling was carried out under the direct supervision of the District Veterinary office staff in accordance with their national mandate. The district veterinary office, under the Ministry of Agriculture Animal Industry and Fisheries (MAAIF) has the official mandate to design effective and efficient animal disease control strategies and carry out investigations related to animal diseases in the country (The Ugandan Animal Disease Act, Ch 38, part III, point 9). Therefore the need for approval by a local research ethics committee was deemed unnecessary.

Informed oral consent was obtained from the veterinary officers who were trained and interviewed the participating farmers.

Consent for publication
Not applicable.

Competing interests
The authors declare that they have no competing interests.

Author details
[1]Department of Biomedical Sciences and Veterinary Public Health, Swedish University of Agricultural Sciences, Uppsala, Sweden. [2]Department of Disease Control and Epidemiology, National Veterinary Institute, Uppsala, Sweden. [3]College of Natural Sciences, Makerere University, Kampala, Uganda.

References
1. Perry B, Grace D. The impacts of livestock diseases and their control on growth and development processes that are pro-poor. Philos Trans R Soc B. 2009;364(1530):2643–55.
2. Rich KM, Perry BD. The economic and poverty impacts of animal diseases in developing countries: new roles, new demands for economics and epidemiology. Prev Vet Med. 2011;101(3-4):133–47.
3. Baluka SA. Economic effects of foot and mouth disease outbreaks along the cattle marketing chain in Uganda. Veterinary World. 2016;9(6):544–53.
4. Mee JF, Geraghty T, O'Neill R, More SJ. Bioexclusion of diseases from dairy and beef farms: risks of introducing infectious agents and risk reduction strategies. Vet J. 2012;194(2):143–50.
5. Ocaido M, Otim CP, Okuna NM, Erume J, Ssekitto C, Wafula RZO, Kakaire D, Walubengo J, Monrad J. Socio-economic and livestock disease survey of agropastoral communities in Serere County, Soroti District, Uganda. Livest Res Rural Dev. 2005;17(8). http://www.lrrd.org/lrrd17/8/ocai17093.htm.
6. Mwiine FN, Ayebazibwe C, Olaho-Mukani W, Alexandersen S, Balinda SN, Masembe C, Ademun Okurut AR, Christensen LS, SØrensen KJ, TjØrnehØj K. Serotype specificity of antibodies against foot-and-mouth disease virus in cattle in selected districts in Uganda. Transbound Emerg Dis. 2010;57(5):365–74.
7. Selby R, Bardosh K, Picozzi K, Waiswa C, Welburn SC. Cattle movements and trypanosomes: Restocking efforts and the spread of Trypanosoma brucei rhodesiense sleeping sickness in post-conflict Uganda. Parasit Vectors. 2013;6(1):281.
8. Angubua Baluka S, Hisali E, Wasswa F, Ocaido M, Mugisha A. Socio-economic risk factors associated with foot and mouth disease, and contagious bovine pleuropneumonia outbreaks in Uganda. Livest Res Rural Dev. 2013;25(12). http://www.lrrd.org/lrrd25/12/balu25214.htm.
9. Ayebazibwe C, Tjørnehøj K, Mwiine FN, Muwanika VB, Ademun Okurut AR, Siegismund HR, Alexandersen S. Patterns, risk factors and characteristics of reported and perceived foot-and-mouth disease (FMD) in Uganda. Trop Anim Health Prod. 2010;42(7):1547–59.
10. Ministry of Agriculture AIF. The National Livestock Census Report 2008. Kampala: Statistics UBo; 2010.
11. Ekou J. Dairy production and marketing in Uganda: current status, constraints and way forward. Afr J Agric Res. 2014;9(10):881–8.
12. Butaye P, Michael GB, Schwarz S, Barrett TJ, Brisabois A, White DG. The clonal spread of multidrug-resistant non-typhi salmonella serotypes. Microbes Infect. 2006;8(7):1891–7.
13. Laxminarayan R, Duse A, Wattal C, Zaidi AKM, Wertheim HFL, Sumpradit N, Vlieghe E, Hara GL, Gould IM, Goossens H, et al. Antibiotic resistance-the need for global solutions. Lancet Infect Dis. 2013;13(12):1057–98.
14. Nantima N, Davies J, Dione M, Ocaido M, Okoth E, Mugisha A, Bishop R. Enhancing knowledge and awareness of biosecurity practices for control of African swine fever among smallholder pig farmers in four districts along the Kenya–Uganda border. Trop Anim Health Prod. 2016;48(4):727–34.
15. Nabukenya I, Rubaire-Akiiki C, Olila D, Ikwap K, Höglund J. Ethnopharmacological practices by livestock farmers in Uganda: Survey experiences from Mpigi and Gulu districts. J Ethnobiol Ethnomed. 2014;10(1). doi:10.1186/1746-4269-10-9.
16. Kateete DP, Kabugo U, Baluku H, Nyakarahuka L, Kyobe S, Okee M, Najjuka CF, Joloba ML. Prevalence and Antimicrobial Susceptibility Patterns of Bacteria from Milkmen and Cows with Clinical Mastitis in and around Kampala, Uganda. PLoS One. 2013;8(5). http://journals.plos.org/plosone/article?id=10.1371/journal.pone.0063413.
17. Sample size to estimate a true prevalence with an imperfect test. http://epitools.ausvet.com.au/content.php?page=PrevalenceSS. Accessed 2 Oct 2014.
18. Osoro EM, Munyua P, Omulo S, Ogola E, Ade F, Mbatha P, Mbabu M, Ng'ang'a Z, Kairu S, Maritim M, et al. Strong association between human and animal brucella seropositivity in a linked study in Kenya, 2012-2013. Am J Trop Med Hyg. 2015;93(2):224–31.
19. Afema JA, Byarugaba DK, Shah DH, Atukwase E, Nambi M, Sischo WM. Potential sources and transmission of salmonella and antimicrobial resistance in Kampala, Uganda. PLoS One. 2016;11(3):e0152130.
20. Makita K, Fèvre EM, Waiswa C, Eisler MC, Thrusfield M, Welburn SC. Herd prevalence of bovine brucellosis and analysis of risk factors in cattle in urban and peri-urban areas of the Kampala economic zone, Uganda. BMC Vet Res. 2011;7:60.
21. Bernard F, Vincent C, Matthieu L, David R, James D. Tuberculosis and brucellosis prevalence survey on dairy cattle in Mbarara milk basin (Uganda). Prev Vet Med. 2005;67(4):267–81.
22. Kashiwazaki Y, Ecewu E, Imaligat JO, Mawejje R, Kirunda M, Kato M, Musoke GM, Ademun RAO. Epidemiology of bovine brucellosis by a combination of rose Bengal test and indirect ELISA in the five districts of Uganda. J Vet Med Sci. 2012;74(11):1417–22.
23. Miller R, Nakavuma JL, Ssajjakambwe P, Vudriko P, Musisi N, Kaneene JB. The Prevalence of Brucellosis in Cattle, Goats and Humans in Rural Uganda: A Comparative Study. Transbound Emerg Dis. 2015;63(6):e197-e210.
24. Kabi F, Muwanika V, Masembe C. Spatial distribution of Brucella antibodies with reference to indigenous cattle populations among contrasting agro-ecological zones of Uganda. Prev Vet Med. 2015;121(1-2):56–63.
25. Jönsson E: Seroprevalence and risk factors for bovine brucellosis, salmonellosis and bovine viral diarrhea in urban and peri-urban areas of Kampala, Uganda. 2013.
26. Handel IG, Willoughby K, Land F, Koterwas B, Morgan KL, Tanya VN, BMd B. Seroepidemiology of bovine viral Diarrhoea virus (BVDV) in the Adamawa region of Cameroon and use of the SPOT test to identify herds with PI calves. PLoS One. 2011;6(7):e21620.
27. Msolla P, Sinclair JA, Nettleton P. Prevalence of antibodies to bovine virus diarrhoea-mucosal disease virus in Tanzanian cattle. Trop Anim Health Prod. 1988;20(2):114–6.
28. Design prevalence required to achieve target herd or system (population) sensitivity. http://epitools.ausvet.com.au/content.php?page=HerdSens5. Accessed 5 Nov 2014.
29. Houe H, Baker JC, Maes RK, Ruegg PL, Lloyd JW. Application of antibody titers against bovine viral diarrhea virus (BVDV) as a measure to detect herds with cattle persistently infected with BVDV. J Vet Diagn Investig. 1995;7(3):327–32.
30. R Core Team. R: a language and environment for statistical computing, version 3.2.2. Vienna: R Foundation for Statistical Computing; 2015.
31. Dohoo IR, Martin W, Stryhn H. Veterinary epidemiologic research. Ch 15.4.2. Charlottetown: AVC Inc.; 2009.
32. Ludwig A, Berthiaume P, Boerlin P, Gow S, Léger D, Lewis FI. Identifying associations in Escherichia Coli antimicrobial resistance patterns using additive Bayesian networks. Prev Vet Med. 2013;110(1):64–75.
33. Lewis FI, Brülisauer F, Gunn GJ. Structure discovery in Bayesian networks: an analytical tool for analysing complex animal health data. Prev Vet Med. 2011;100(2):109–15.

34. Lewis FI, Ward MP. Improving epidemiologic data analyses through multivariate regression modelling. Emerg Themes Epidemiol. 2013;10(1):1–10.

35. Firestone SM, Lewis FI, Schemann K, Ward MP, Toribio J-ALML, Dhand NK. Understanding the associations between on-farm biosecurity practice and equine influenza infection during the 2007 outbreak in Australia. Prev Vet Med. 2013;110(1):28–36.

36. Ågren ECC, Frössling J, Wahlström H, Emanuelson U, Sternberg Lewerin S. A questionnaire study of associations between potential risk factors and salmonella status in Swedish dairy herds. Prev Vet Med. 2017;143:21–9.

37. Lewis FI, McCormick BJJ. Revealing the complexity of health determinants in resource-poor settings. Am J Epidemiol. 2012;176(11):1051–9.

38. Lewis FI, Pittavino M, Furrer R. abn: Modelling Multivariate Data with Additive Bayesian Networks. 2016. http://www.r-bayesian-networks.org/.

39. Mugizi DR, Boqvist S, Nasinyama GW, Waiswa C, Ikwap K, Rock K, Lindahl E, Magnusson U, Erume J. Prevalence of and factors associated with Brucella sero-positivity in cattle in urban and peri-urban Gulu and Soroti towns of Uganda. J Vet Med Sci. 2015;77(5):557–64.

40. Rock KT, Mugizi DR, Ståhl K, Magnusson U, Boqvist S. The milk delivery chain and presence of Brucella spp. antibodies in bulk milk in Uganda. Trop Anim Health Prod. 2016;48(5):985–94.

41. Chenais E, Boqvist S, Sternberg-Lewerin S, Emanuelson U, Ouma E, Dione M, Aliro T, Crafoord F, Masembe C, Ståhl K. Knowledge, Attitudes and Practices Related to African Swine Fever Within Smallholder Pig Production in Northern Uganda. TransboundEmerg Dis. 2015;64(1):101-15.

42. Nielsen LR, Baggesen DL, Aabo S, Moos MK, Rattenborg E. Prevalence and risk factors for salmonella in veal calves at Danish cattle abattoirs. Epidemiol Infect. 2011;139(7):1075–80.

43. Cummings KJ, Warnick LD, Elton M, Gröhn YT, McDonough PL, Siler JD. The effect of clinical outbreaks of salmonellosis on the prevalence of fecal salmonella shedding among dairy cattle in New York. Foodborne Pathog Dis. 2010;7(7):815–23.

44. Jajere SM, Adamu NB, Atsanda NN, Onyilokwu SA, Gashua MM, Hambali IU, Mustapha FB. Prevalence and antimicrobial resistance profiles of salmonella isolates in apparently healthy slaughtered food animals at Maiduguri central abattoir, Nigeria. Asian Pac J Trop Dis. 2015;5(12):996–1000.

45. Tutuncu H, Yazici Z. Screening for persistently infected cattle with bovine viral diarrhea virus in small-holder cattle farms located in Samsun Province, northern Turkey. J Anim Plant Sci. 2016;26(1):291–3.

46. Kulangara V, Joseph A, Thrithamarassery N, Sivasailam A, Kalappurackal L, Mattappillil S, Syam R, Mapranath S. Epidemiology of bovine viral diarrhoea among tropical small holder dairy units in Kerala, India. Trop Anim Health Prod. 2015;47(3):575–9.

47. Njiro SM, Kidanemariam AG, Tsotetsi AM, Katsande TC, Mnisi M, Lubisi BA, Potts AD, Baloyi F, Moyo G, Mpofu J, et al. A study of some infectious causes of reproductive disorders in cattle owned by resource-poor farmers in Gauteng province, South Africa. J S Afr Vet Assoc. 2011;82(4):213–8.

48. Scott TP, Stylianides E, Markotter W, Nel L. Serological survey of bovine viral diarrhoea virus in Namibian and South African kudu (Tragelaphus strepsiceros) and eland (Taurotragus oryx). J S Afr Vet Assoc. 2013;84(1). doi: 10.4102/jsava.v84i1.937.

49. Magona JW, Walubengo J, Galiwango T, Etoori A. Seroprevalence and potential risk of bovine brucellosis in zerograzing and pastoral dairy systems in Uganda. Trop Anim Health Prod. 2009;41(8):1765–71.

50. Walker JG, Ogola E, Knobel D. Piloting mobile phone-based syndromic surveillance of livestock diseases in Kenya. Épidémiologie et Santé Animale 2011(No.59/60):19-21.

51. Stahl K, Lindberg A, Rivera H, Ortiz C, Moreno-Lopez J. Self-clearance from BVDV infections - a frequent finding in dairy herds in an endemically infected region in Peru. Prev Vet Med. 2008;83(3-4):285–96.

52. Greiner M, Kumar S, Kyeswa C. Evaluation and comparison of antibody ELISAs for serodiagnosis of bovine trypanosomosis. Vet Parasitol. 1997;73(3-4):197–205.

14

Metronidazole for the treatment of *Tritrichomonas foetus* in bulls

David Love[1], Virginia R. Fajt[2], Thomas Hairgrove[3], Meredyth Jones[1] and James A. Thompson[1*]

Abstract

Background: *Tritrichomonas foetus* is a sexually transmitted protozoon that causes reproductive failure, among cattle, so disruptive that many western US states have initiated control programs. Current control programs are based on the testing and exclusion of individual bulls. Unfortunately, these programs are utilizing screening tests that are lacking in sensitivity. Blanket treatment of all the exposed bulls and adequate sexual rest for the exposed cows could provide a more viable disease control option. The objectives of this study were twofold. The first objective was to demonstrate effectiveness for metronidazole treatment of a bull under ideal conditions and with an optimized treatment regime. This type of study with a single subject is often referred to as an n-of-1 or single subject clinical trial. The second objective of the current study was to review the scientific basis for the banning of metronidazole for use in Food Animals by the Animal Medicinal Drug Use Clarification Act of 1994 (AMDUCA).

Results: Results from an antimicrobial assay indicated that metronidazole at a concentration of 0.5 μg/mL successfully eliminated in vitro protozoal growth of bovine *Tritrichomonas foetus*. The estimated effective intravenous dose was two treatments with 60 mg/kg metronidazole, 24 h apart. A bull that had tested positive for *Tritrichomonas foetus* culture at weekly intervals for 5 weeks prior to treatment was negative for *Tritrichomonas foetus* culture at weekly intervals for five consecutive weeks following this treatment regimen. An objective evaluation of the published evidence on the potential public health significance of using metronidazole to treat *Tritrichomonas foetus* in bulls provides encouragement for veterinarians and regulators to consider approaches that might lead to permitting the legal use of metronidazole in bulls.

Conclusion: The study demonstrated successful inhibition of *Tritrichomonas foetus* both in vitro and in vivo with metronidazole. The current status of metronidazole is that the Animal Medicinal Drug Use Clarification Act of 1994 prohibits its extra-label use in food-producing animals. Veterinarians and regulators should consider approaches that might lead to permitting the legal use of metronidazole in bulls.

Keywords: *Tritrichomonas feotus*, metronidazole, cattle

Background

Tritrichomonas foetus TF) is a sexually transmitted protozoon that causes reproductive failure so disruptive among cattle that virtually all US states west of the Mississippi River have initiated control programs [1]. Each of these control programs is based on the testing and exclusion of individual bulls. The resultant loss of bulls as a result of the various test and removal programs has been substantive. For example, over 1000 bulls were culled from Texas ranches in 2010 [2]. In addition to the cost of replacing bulls, test and removal programs will often fail to control TF in cattle populations because the sensitivity of the testing is low [1]. The pursuit of a more sensitive test has been vigorous but results, thus far, show that newer tests can be very sensitive in experimental conditions, but show limited success in field conditions [3, 4]. The current standard for regulatory testing in Texas is a single negative polymerase chain reaction (PCR) test which, based on the available literature, has a test sensitivity very likely in the 70–80% range. While the identification of infected individuals is currently lacking sensitivity, the identification of infected populations can be very accurate. The accuracy of a test for identifying infected

---

* Correspondence: jthompson@cvm.tamu.edu
[1]Department of Large Animal Clinical Sciences, College of Veterinary Medicine and Biomedical Sciences, Texas A&M University, College Station, TX 77843, USA
Full list of author information is available at the end of the article

populations depends on test specificity, the disease infectivity and the population size [5]. For detecting TF-infected herds, every bull could be cultured and positive cultures confirmed by a TF-specific PCR thus resulting in a nearly perfect specificity for individual bulls [6]. Thus, multiple sire breeding groups could be evaluated very accurately because the disease is highly contagious and multiple bulls would be affected. Blanket treatment of all the exposed bulls and adequate sexual rest for the exposed cows could provide a more viable disease control option than test and removal. In the early 1980's, both dimetridazole (50 mg/kg PO for 5 days) and ipronidazole (60 g IM) were shown to be 100% effective against TF in bulls [7, 8]. At that time, blanket treatments were recommended. Since then, the nitroimidazole family, including dimetridazole and ipronidazole, has been prohibited from use in animals intended for consumption as food. In humans, another nitroimidazole, metronidazole, is considered nearly 100% effective for *Trichomonas vaginalis* (VG) by the Centers for Disease Control and Prevention (CDC) and is the sole recommended treatment. Very few other drugs have even been evaluated [9]. The CDC also recognizes the low sensitivity of testing, and recommends that all sexual partners of a patient with TV be treated with metronidazole [10]. Despite the obvious potential, no treatment trials for TF have been found in the available veterinary literature since 1985, leaving this treatment potential untapped.

In 1954, Rhone-Poulenc discovered that azomycin (2-nitroimidazole), isolated from a *Streptomyces*species, had weak in vitro activity against TV. The company investigated over 200 related chemicals and discovered that metronidazole (1-(2-hydroxyethyl)-2-methyl-5-nitroimidazole) was the most promising trichomonacide [11]. Over six decades later metronidazole is still widely used. It has a limited spectrum that encompasses various protozoans (like *Tritrichomonas* spp.) as well as most Gram-negative and Gram-positive anaerobic bacteria. It is both cost effective and readily available. It is currently produced and marketed in various forms for human use, including as a solution for intravenous injection. Under the provisions of AMDUCA and 21 Code of Federal Regulations (CFR) part 530, the FDA can prohibit extralabel use of an entire class of drugs in selected animal species. According to this mandate, nitroimidazoles (including metronidazole) are not allowed extralabel in any food-producing animal. Because no products containing metronidazole are currently approved in the U.S., this ban on extralabel use is effectively a ban on all use in cattle.

The first objective of this study was to perform in vitro susceptibility testing for metronidazole on bovine-origin TF and to use these data along with published pharmacokinetic data in cattle to estimate an effective dose and to pilot test the dose in the infected bull. Showing that metronidazole could potentially be efficacious in vivo for treatment of TF infections would complement the second objective which was to review the scientific information relevant to the banning of metronidazole in Food Animals by the Animal Medicinal Drug Use Clarification Act of 1994 (AMDUCA).

Methods

Procedure 1: in vitro susceptibility assay

Commercial injectable 5 mg/mL metronidazole solution was diluted with isotonic saline into five stock solutions: 0.01 mg/mL, 0.005 mg/mL, 0.0025 mg/mL, 0.00125 mg/mL, and 0.000625 mg/mL. From each stock solution, 0.2 mL was added to five separate commercially available culturing pouches, [1] each containing 3.8 mL of culture media, for a total of five culture media pouches per stock concentration. Five culture media pouches were used as a control group, and were inoculated with 0.2 ml of isotonic saline. Final concentrations of metronidazole in the pouches were: 0 µg/ml, 0.0313 µg/mL, 0.0625 µg/mL, 0.125 µg/mL, 0.25 µg/mL and 0.5 µg/mL.

A second researcher, who was masked to drug concentrations, inoculated each pouch with 20,000 viable organisms from a single standard culture taken from the bull that had been designated for a treatment trial. Organism numbers were estimated using a Neubauer Chamber.[2] Pouches were contained in an upright position and incubated at 35 °C. Pouches were evaluated for TF viability at 24-h intervals for 7 days by the masked researcher. An absence of all typical TF motility was used as the criterion for growth inhibition. The evaluation at 48 h was chosen as the critical cut-point.

Procedure 2: treatment of infected Bull

An 891 kg, Charolais-cross bull identified as positive for TF by the Texas Bovine Trichomoniasis Control Program was selected. The bull had previously been cultured positive multiple times over 6 months. The bull was restrained in a squeeze chute and an intravenous catheter was placed in the left jugular vein. A volume of 10.8 l of 5 mg/mL (approximately 60 mg/kg; 54 g total dose) commercially available metronidazole solution was administered intravenously over a period of one hour. The treatment was repeated 24 h later. Pre-treatment cultures were taken at weekly intervals for 5 weeks including just prior to the first intravenous injection. Post-treatment cultures were obtained on the 6th day following the second intravenous injection and then at weekly intervals for 4 weeks. The clinical cultures, including both pre- and post-treatment were performed as recommended in the Texas Trichomonas Bull Test Program. Briefly, an aggressive scraping was collected from the fornix of the prepuce using an infusion pipette and a 12-ml syringe providing negative pressure during the scraping. The collection of material from the scraping was immediately inoculated into an InPouch TF[1] culture system and cultured in an upright position for 7 days at 37 °C. Every 24 h, the culture media was examined under a light microscope and cultures were considered positive when the presence of any typical TF motility was observed. Potential bias was controlled by having the clinician most experience with sample collection and culture evaluations (JAT) perform all the testing and evaluation. The study also avoided bias by using culture rather than PCR. Test results by PCR can be falsely positive as a result of non-viable organisms.

Statistics

Proportions of viable cultures were compared using Bayesian methods. For $i = 2$ proportions, the counts of positive results (r_i) were modeled as binomial with rate parameter μ_i and number of trials (N_i) such that:

$$r_i \sim binomial(\mu_i, , N_i)$$

r and N were provided by the data and μ was assigned a prior that was vague normal with a mean of zero and precision of 0.0001, on the logit scale. To compare the two proportions, the odds ratio was estimated, and the odds ratio was considered statistically significant at $P < 0.05$; if the exceedance probability was greater than 95% that the odds ratio drawn from the full posterior distribution was greater than 1. In the use of Bayesian P-values this is called the posterior predictive P-value [12].

Results

In vitro susceptibility assay

Culture viability at 48 h was 0% for 0.5 µg/mL metronidazole and 100% for 0 µg/mL metronidazole. The highest two concentrations were each more likely to result in a non-viable culture than each of the lower concentrations ($P < 0.05$; Table 1).

Treatment of infected Bull

All five pre-treatment cultures were positive, and all five post-treatment cultures were negative. The proportions

Table 1 Culture viability by metronidazole concentration

Metronidazole concentration	Viable cultures at 48 h
0.0 µg/ml	5/5[a]
0.0313 µg/mL	4/5[a]
0.0625 µg/mL	4/5[a]
0.125 µg/mL	5/5[a]
0.25 µg/mL	1/5[b]
0.5 µg/mL	0/5[b]

Proportions with different superscripts were significantly different ($P < 0.05$)

of positive culture before and after treatment were significantly different ($P < 0.05$).

Discussion

N-of-1 or single subject clinical trials consider an individual patient as the sole unit of observation in a study investigating the efficacy or side-effect profiles of interventions. The ultimate goal of an n-of-1 trial is to determine the optimal or best intervention for an individual patient using objective data-driven criteria. The first goal of this study was to demonstrate efficacy of metronidazole in an individual bull. The evaluation used a 48-h susceptibility assay for TF isolated from the patient and subsequent treatment of the patient with the dose of metronidazole that was estimated to be effective. In the opinion of the authors, a clinical trial with more animals was not warranted prior to a full discussion among the veterinary profession regarding the potential for disease control and the current legal implications for such treatment. This discussion should include a critical review of the published literature relevant to the legal status of metronidazole use in cattle.

Metronidazole at a concentration of 0.5 µg/mL successfully eliminated viable protozoal growth of bovine TF in 5/5 cultures and a concentration of 0.25 µg/mL eliminated viable protozoal growth in 4/5 cultures after 48-h. Based on published pharmacodynamics parameters, a dose of 60 mg/kg would provide a blood concentration of greater than 0.25 µg/mL for nearly a 24-h period [13]. A second dose at 24 h was given and should have maintained a blood concentration of greater than 0.25 µg/mL for most of a 48-h period. This dose regimen resulted in repeated negative cultures in a previously culture-positive bull. Studies are needed to verify the effectiveness with this regimen in a population of bulls, but this n-of-1 study demonstrates feasibility. More than 40 years ago, a small number of studies showed promising potential for treating TF in bulls with metronidazole [14–17]. While treatments appeared to be effective, problems resulted from the poor solubility and the acidity of metronidazole in solution. Complications following intravenous treatment included tachypnea and tachycardia and muscle trembling [14, 17]. The product evaluated in the current study, while dilute and requiring a large volume for injection, was isosmotic and pH buffered. The bull being treated maintained steady heart and respiratory rates during the treatment.

Metronidazole is currently available as an approved human drug, in a formulation that was well-tolerated by the patient in the current study. However, the FDA currently prohibits extralabel use of the entire family of nitroimidazoles in food-producing animals. The FDA prohibits drugs based on either of two statutes: either "an acceptable analytical method (for evaluating tissue residues) needs to be established and such a method has not or cannot be established" or "the extra-label use of the drug or drug class presents a public health risk" [18]. It is thought that nitroimidazoles are banned due to the latter statute, namely their potential to cause cancer in humans.

The most desirable solution to improve reproductive health and TF control would be the development of a new animal drug with label indications for treatment of TF in bulls. Alternatively, an expedient solution would be to remove metronidazole from the banned list or, more specifically, metronidazole treatment of TF in bulls could be exempted. The AMDUCA, Section 530.21 states that "a prohibition may be a general ban on the extralabel use of the drug or class of drugs or may be limited to a specific species, indication, dosage form, route of administration, or combination of factors." Logically, the inverse would also be true and that a limited exemption could be permitted. An exemption for this dosage form, species, and indication would not constitute a risk to public health because humans could be protected from exposure from meat. First, the dosage form requires large volume of intravenous injection which is very likely to remain under control of licensed veterinarians who have become adequately sensitized to the issues of tissue residues. Second, the veterinarian can recommend prolonged drug withdrawal times and these should be acceptable to informed owners because these bulls will usually remain in service for at least several months of a breeding season. Third, the number of bulls that will be treated will be relatively few. For bulls that do not respond to treatment, a satisfactory withholding time for slaughter should be established.

While the veterinary profession could prevent human exposure to meat residues of metronidazole or its metabolites, others may choose to debate the entire body of evidence that metronidazole is a human carcinogen. There is a large body of literature to consider before engaging in this debate. As early as 1977, the International Agency for Research in Cancer (IARC) [19] had decided that metronidazole was carcinogenic in rodents, citing Rustia and Shubik [20]. The countering viewpoint was presented immediately and vigorously in 1977 [21], 1979 [22] and 1981 [23]. However, the IARC confirmed their decision in 1982 [24], citing a second paper by Rustia and Shubik [25] and their earlier decision [19]. In 1986, the FDA provided considerable discussion to support the decision to remove a nitroimidazole, dimetridazole, as an approved animal drug [26]. The cited work included five papers said to show that metronidazole was carcinogenic in rodents [20, 25, 27–29]. By 1987, the IARC repeated their earlier assessment and cited a list [27, 28, 30] virtually identical to the references provided by the FDA [31]. In disputing the cancer risk in rodents, Roe provides very specific criticisms, claiming that virtually all long-term exposures of rodents

to metronidazole show an increase in both life span and weight gain and these two factors were the more likely causes of any variation in cancer incidence [32]. In one study, the median survival time increased from 83 weeks for the control rats (did not get metronidazole) to 122 weeks for rats receiving metronidazole [25]. Thus, the treated animals lived nearly 50% longer. This potential bias is more than just increased time-at-risk for the treated animals because the rate of cancer development (i.e., incident cases per unit time) increases with age. Weight gain has also had a strong association with cancer in rodents with the rate of cancer elevated 6–8 times when mice are fed *ad libitum* versus restricted feeding [33]. The debate as to the carcinogenicity of metronidazole in rodents will not to be satisfactorily resolved, based on the conflicting and often biased literature that currently exists. In more recent years, the debate has centered on the broader issues such as the usefulness of rodent studies in evaluating human health risks, even if the risks estimated for rodents are accurate for the reported doses. There is an overwhelming number of drugs currently used with evidence of carcinogenicity in rodents. For example, one review of 535 marketed pharmaceuticals showed that more than half of them had been judged carcinogenic in rodents [34]. Recently, several papers have reviewed the history and problems using largely unstandardized testing and analyses of results from cancer studies in rodents [35–37]. Emerging from this evaluation has been a general criticism of inferring human cancer risk from rodent studies as well as recommendations to standardize the use of rodent studies in cancer risk assessment [38]. There have been epidemiologic studies evaluating the human cancer risk and metronidazole is not considered a risk factor for human cancer [39]. This entire body of literature should be re-evaluated and we encourage the veterinary profession to be aware of these issues and join the debate. We also urge the FDA to grant the veterinary profession an opportunity to treat TF in bulls with metronidazole.

Conclusion

The current study showed potential for treating TF in bulls with metronidazole.

Endnotes

[1] InPouch TF™ Biomed Diagnostics, PO Box 2366, 1388 Antelope Road, White City, OR 97503

[2] Azer Scientific, Inc., 701 Hemlock Road, Morgantown, PA 19543

Abbreviations

AMDUCA: Animal Medicinal Drug Use Clarification Act of 1994; FDA: United States Food and Drug Administration; IARC: International Agency for Research on Cancer; TF: *Tritrichomonas foetus*; TV: *Trichomonas vaginalis*

Acknowledgements
The authors thank Dr. Elena Gart for providing Russian to English translations.

Funding
This research was supported in part by funding provided by the Department of Large Animal Clinical Sciences.

Authors' contributions
DL participated in study design, study conduct and manuscript preparation and provided leadership in the in vitro sensitivity testing. VF participated in study design, study conduct and manuscript preparation and provided leadership in the determination of metronidazole dose. TH participated in study design, study conduct and manuscript preparation and provided leadership in the acquisition of the patient and gaining permission to treat the patient. MJ participated in study design, study conduct and manuscript preparation and provided leadership in the treatment of the patient. JT participated in all aspects of the study and provided leadership in the acquisition of funding to conduct the study. All authors approved the manuscript.

Competing interests
The authors declare that they have no competing interests.

Consent for publication
Not applicable.

Author details
[1]Department of Large Animal Clinical Sciences, College of Veterinary Medicine and Biomedical Sciences, Texas A&M University, College Station, TX 77843, USA. [2]Department of Veterinary Physiology and Pharmacology, College of Veterinary Medicine and Biomedical Sciences, Texas A&M University, College Station, TX 77843, USA. [3]Texas A&M AgriLife Extension Service, Department of Animal Science, College of Agriculture and Life Sciences, Texas A&M University, College Station, TX 77843, USA.

References
1. Yao CQ. Diagnosis of Tritrichomonas foetus-infected bulls, an ultimate approach to eradicate bovine trichomoniasis in US cattle? J Med Microbiol. 2013;62:1–9.
2. Szonyi B, Srinath I, Schwartz A, Clavijo A, Ivanek R. Spatio-temporal epidemiology of Tritrichomonas foetus infection in Texas bulls based on state-wide diagnostic laboratory data. Vet Parasitol. 2012;186(3–4):450–5.
3. Cobo ER, Favetto PH, Lane VM, Friend A, VanHooser K, Mitchell J, BonDurant RH. Sensitivity and specificity of culture and PCR of smegma samples of bulls experimentally infected with Tritrichomonas foetus. Theriogenology. 2007;68(6):853–60.
4. Perez A, Cobo E, Martinez A, Campero C, Spath E. Bayesian estimation of Tritrichomonas foetus diagnostic test sensitivity and specificity in range beef bulls. Vet Parasitol. 2006;142(1–2):159–62.
5. Martin SW, Shoukri M, Thorburn MA. Evaluating the health-status of herds based on tests applied to individuals. Prev Vet Med. 1992;14(1–2):33–43.
6. Campero CM, Dubra CR, Bolondi A, Cacciato C, Cobo E, Perez S, Odeon A, Cipolla A, BonDurant RH. Two-step (culture and PCR) diagnostic approach for differentiation of non-T. foetus trichomonads from genitalia of virgin beef bulls in Argentina. Vet Parasitol. 2003;112(3):167–75.
7. Kimsey PB, Darien BJ, Kendrick JW, Franti CE. Bovine trichomoniasis — diagnosis and treatment. J Am Vet Med Assoc. 1980;177(7):616–9.
8. Skirrow S, Bondurant R, Farley J, Correa J. Efficacy of ipronidazole against trichomoniasis in beef bulls. J Am Vet Med Assoc. 1985;187(4):405–7.
9. Stover KR, Riche DM, Gandy CL, Henderson H. What would we do without metronidazole? Am J Med Sci. 2012;343(4):316–9.

10. Workowski KA, Berman SM. Centers for disease control and prevention sexually transmitted disease treatment guidelines. Clin Infect Dis. 2011;53:S59–63.
11. Roe FJC. Metronidazole — review of uses and toxicity. J Antimicrob Chemoth. 1977;3(3):205–12.
12. Zhang JL. Comparative investigation of three Bayesian p values. Comput Stat Data An. 2014;79:277–91.
13. Bhavsar SK, Malik JK. Pharmacokinetics of metronidazole in calves. Brit Vet J. 1994;150(4):389–93.
14. Gasparini G, Vagni M, Tardani A. Treatment of bovine Trichomoniasis with metronidazole. Vet Rec. 1963;75(37):940–3.
15. Reshetnyak BZ, Bartenev VS, Bibikov FA, Pahomova HG. Metronidazole - an effective drug for trichomoniasis in cattle (Russian). Veterinariia. 1969;46(3):86–90.
16. Chodorkovsky A, Zabolotzky V, Timchenko K, Olchovik V, Yamkovoy A, Boyko R, Burdusha I. Study of the efficacy of metronidazole in trichomoniasis (Russian). Veterinariia. 1970;10:104–5.
17. Golikov AV, Bayuta NV. Use of trichopol in tichomoniasis of sire bulls (Russian). Veterinariia. 1972;11:37–40.
18. CFR. Extralable drug use in animals. Fed Regist. 1996;61:57732–46.
19. IARC. Metronidazole. IARC Monographs on the Evaluation of the Carcinogenic Risk of Chemicals to Man. 1977;13:113–22.
20. Rustia M, Shubik P. Induction of lung tumors and malignant lymphomas in mice by metronidazole. J Natl Cancer I. 1972;48(3):721–9.
21. Roe FJC. Metronidazole: tumorigenicity studies in mice, rats and hamsters. In: Metronidazole: Proceedings, Montreal, May 26–28 1976 Excerpta Medica, Amsterdam, Netherlands; 1977. p. 132–137.
22. Roe FJC. A critical appraisal of the toxicology of metronidazole. In: Metronidazole: Royal Society of Medicine International Congress and Symposium series no 18. London: Academic Press Inc and Royal Society of Medicine; 1979. p. 215–22.
23. Roe JFC. Safety evaluation of metronidazole from the viewpoint of general toxicity and carcinogenicity 1981 appraisal. In: European Metronidazole. Symposium. 1981. p. 41–47.
24. IARC. Metronidazole. IARC Monogr Eval Carcinog Risks Hum. 1982. (Supplement 4):160–162.
25. Rustia M, Shubik P. Experimental induction of hepatomas, mammary-tumors, and other tumors with metronidazole in noninbred Sas-Mrc(Wi)Br rats. J Natl Cancer I. 1979;63(3):863–8.
26. CFR. Dimetridazole; opportunity for hearing. Fed Regist. 1986;52:45244–62.
27. Cavaliere A, Bacci M, Amorosi A, Delgaudio M, Vitali R. Induction of lung-tumors and lymphomas in Balb/C mice by metronidazole. Tumori. 1983; 69(5):379–82.
28. Cavaliere A, Bacci M, Vitali R. Induction of mammary-tumors with metronidazole in female sprague-dawley rats. Tumori. 1984;70(4):307–11.
29. Rust J. An Assessment of Metronidazole Tumorigenicity: Studies in the Mouse and Rat. In: Finegold SM, McFadzean JA, Roe FJC, editors. Metronidazole Proceedings of hte International Metronidazole Confreence Montreal. Quebec, Canada Amsterdam: Excerpta Medica; 1977. p. 138–44.
30. Akareem AM, Fleiszer DM, Richards GK, Senterman MK, Brown RA. Effect of long-term Metronidazole (Mtz) therapy on experimental colon cancer in rats. J Surg Res. 1984;36(6):547–52.
31. IARC. Metronidazole. IARC Monogr Eval Carcinog Risks Hum. 1987. (Supplement 7):250–252.
32. Roe F. The Leon-Golberg-Memorial-Lecture-recent advances in toxicology relevant to carcinogenesis - 7 cameos. Food Chem Toxicol. 1993;31(11):909–24.
33. Roe FJC. Are nutritionists worried about the epidemic of tumors in laboratory-animals. Proc Nutr Soc. 1981;40(1):57–65.
34. Brambilla G, Mattioli F, Robbiano L, Martelli A. Update of carcinogenicity studies in animals and humans of 535 marketed pharmaceuticals. Mutat Res-Rev Mutat. 2012;750(1):1–51.
35. Marone PA, Hall WC, Hayes AW. Reassessing the two-year rodent carcinogenicity bioassay: A review of the applicability in human risk and current perspectives. Regul Toxicol Pharmacol. 2014;68(1):108–18.
36. Jacobs AC, Hatfield KP. History of chronic toxicity and animal carcinogenicity studies for pharmaceuticals. Vet Pathol. 2013;50(2):324–33.
37. Alden CL, Lynn A, Bourdeau A, Morton D, Sistare FD, Kadambi VJ, Silverman L. A critical review of the effectiveness of rodent pharmaceutical carcinogenesis testing in predicting for human risk. Vet Pathol. 2011;48(3):772–84.
38. Morton D, Sistare FD, Nambiar PR, Turner OC, Radi Z, Bower N. Regulatory forum commentary: alternative mouse models for future cancer risk assessment. Toxicol Pathol. 2014;42(5):799–806.

15

Dynamic network measures reveal the impact of cattle markets and alpine summering on the risk of epidemic outbreaks in the Swiss cattle population

Beatriz Vidondo*[iD] and Bernhard Voelkl

Abstract

Background: Livestock herds are interconnected with each other via an intricate network of transports of animals which represents a potential substrate for the spread of epidemic diseases. We analysed four years (2012–2015) of daily bovine transports to assess the risk of disease transmission and identify times and locations where monitoring would be most effective. Specifically, we investigated how the seasonal dynamics of transport networks, driven by the alpine summering and traditional cattle markets, affect the risk of epidemic outbreaks.

Results: We found strong and consistent seasonal variation in several structural network measures as well as in measures for outbreak risk. Analysis of the consequences of excluding markets, dealers and alpine pastures from the network shows that markets contribute much more to the overall outbreak risk than alpine summering. Static descriptors of monthly transport networks were poor predictors of outbreak risk emanating from individual holdings; a dynamic measure, which takes the temporal structure of the network into account, gave better risk estimates. A stochastic simulation suggests that targeted surveillance based on this dynamic network allows a higher detection rate and smaller outbreak size at detection than compared to other sampling schemes.

Conclusions: Dynamic measures based on time-stamped data—the outgoing contact chain—can give better risk estimates and could help to improve surveillance schemes. Using this measure we find evidence that even in a country with intense summering practice, markets continue being the prime risk factor for the spread of contagious diseases.

Keywords: Cattle networks, Alpine pastures, Markets, Surveillance, Contact chain

Background

Many livestock diseases can spread through direct contact between animals. Movement of animals from one herd to another can, therefore, lead to the spreading of highly infectious diseases [1–5]. The transport network of individuals from one holding to another is an important factor determining the spread of infectious diseases like, bovine tuberculosis, bovine viral diarrhoea, foot-and-mouth disease, and bovine coronavirus, amongst others [3, 6, 7]. Consequently, national registers for livestock transports have been implemented in many countries and the incorporation of network information in epidemic models and surveillance schemes has become an active area of research [8–13]. In these models the network is characterized as a graph, where holdings are represented by nodes and transports between holdings are edges. This allows for the calculation of quantitative descriptors of the connectivity (see Table 1 for definitions) and sub-structuring of the network as well as the positions of individual holdings in that network [14]. Dynamic approaches take into account the time sequence of the animal movements and allow more realistic models of transmission processes [15–18]. For example, Nörenmark et al. [19] and Frössling et al. [20] investigated to what extent a dynamic network measure—the ingoing contact chain—could be a useful measure when setting up strategies for disease control and for risk based surveillance.

* Correspondence: Beatriz.vidondo@gmx.ch
Veterinary Public Health Institute, University of Bern, Schwarzenburgstrasse 155, CH-3097 Liebefeld, Switzerland

Table 1 Definition and sources of the network analysis terms used in this study

Type / Name	Definition	Source
Holding centrality metrics		
In-degree	Number of individual sources providing animals to a specific holding	[14]
Out-degree	Number of individual recipients obtaining animals from a specific holding	[14]
Betweenness	The frequency a livestock operation is in the shortest path between pairs of holdings in the network	[47]
Closeness	The inverse of the sum of the shortest geodesic distances from a source holding to all other reachable holdings in the network	[48]
Static network measures of connectivity		
Geodesic	The shortest path length between two holdings	[49]
Path length	A path between farm A and farm C is the number of steps required to travel from A to C. In a path, holdings and animal transports cannot be repeated between a source and a destination. Related terms are geodesic and average path length	[49]
Average path length (APL)	The shortest path, or geodesic, among two holdings averaged over all pairs of holdings in the network	[50]
Density	Sum of the number of all links divided by the number of possible links in the network	[14]
Clustering coefficient (CC)	If a neighbour is defined as a the holding in direct contact with the holding of interest, the clustering coefficient represents the proportion of one's neighbours who are also neighbours of one another	[50]
Components	Maximally connected subregions of a network in which all pairs of holdings are directly or indirectly linked	[46]
Giant strongly connected component (GSCC)	The largest component in the network in which all pairs of holdings are linked via directed paths	[43, 46]
Giant weakly connected component (GWCC)	The largest component in the network in which all pairs of holdings are linked regardless of the direction of the link	[43, 46]
Assortativity	Correlation between the degrees of linked premises	[51]

In cattle networks, markets have been identified as major contributors to network connectivity and thus, potential disease spread [21, 22]. Their exclusion has been shown to substantially reduce potential outbreak size [22, 23]. Likewise, Webb and colleagues [15] found that agricultural shows increase the potential for epidemic outbreaks in British sheep. Both markets and shows can be described as gathering events, where animals from many holdings are brought to the same location for a short period of time and are again distributed to many holdings, afterwards. Gathering events of both human and animals can drastically increase the risk of acquiring and transmitting disease [24, 25]. In network terms gathering events can be described as 'hubs'—nodes with high in- and out-degree, connecting a high number of holdings. The existence of hubs in a network can lead to a highly skewed or heavy tailed distribution, which has important implication for the spread of diseases [26, 27]. Some cattle networks have, indeed, degree distributions which are almost scale-free [23, 28, 29]. Switzerland's cattle industry is characterized by high market activity. Even though relatively few in numbers, cattle markets occur at different periodicities and have a wide range of catchment areas. They also vary in terms of veterinary regulations. The markets events range from weekly or biweekly regional trade markets that last a few hours, with a high proportion of animals being taken to slaughter, to yearly national or international events that last a whole weekend.

In addition to cattle markets, the practise of summering herds on high alpine pastures (alps) represents another type of gathering event—much smaller in extent than the markets but also much more numerous. The annual practice of summering to access forage at high pastures dates from ancient times and about half of all Swiss cattle farms participate in it [30]. The animals are walked up on foot in April/May and down in September/October. Usually, the animals from different farms form a single herd that roams during the day and overnights in a single stable with shared milking facilities. Thus, alps can be considered as another type of gathering event where disease could be transferred [31, 32]. In contrast to that of the markets, the periodicity of the alps is yearly, and their duration is around three months. As small gathering events, alps can be expected to contribute to network connectivity, yet due to differences in size and duration their contribution might differ from that of markets. While the impact of markets on network connectivity has been studied previously [21, 23], the role of alpine summering has not yet been evaluated.

While alpine summering has a prominent role in cattle management in Switzerland, Austria, Liechtenstein and the alpine regions of France, Italy and Germany (Bavaria), the practice of bringing herds to higher summer pastures—often leading to a mixing of herds—exist in the Molise, Apulia and Abruzzo regions in Italy, the Pyrenees, the Cantabrian mountains in Spain, the Caucasus and Pontic mountains, the Zagros mountain range in Iran, along the Himalaya and Hindu-Kush mountain ranges, in Norway and Sweden, and—to a lesser extent— also in Wales, Scotland and Ireland. In the case of a severe disease outbreak, or an emergent disease, it would be crucial to know how both types of traditional practices—markets and summering—contribute to disease spread. Here we aim to quantify the relative importance of alpine summering in relation to market related transports for the risk of epidemic outbreaks. We use both a static and a dynamic network analysis approach. Additionally, we investigate seasonal and long-term variation in the network structure of the cattle transport network, its implication for the risk of epidemic outbreaks and its implication for disease surveillance.

Methods

(a) Data acquisition

Daily animal transports from January 1st 2012 to December 31st 2015 were extracted from the nationwide Schweizer Tierverkehr Datenbank (Swiss registry for animal movements) using Microsoft SQL Server 2014. For the present analysis, every animal's history was queried since the start of data recording in 1999 to make sure it was complete. Double entries were excluded. As the slaughterhouses constitute sinks and we are interested in potential disease spread between living animals, the end transports to slaughter were excluded. As we focus on within country transmission, import and export from abroad were excluded.

SQL routines were used to prepare the so-called edge lists containing origin, destination and date of transport for the four years of data. Using information from the Swiss holdings registry [33] holdings were categorized as markets (including trade markets, auctions and exhibitions), dealers, alps and farms. Preliminary data analysis helped to identify one holding originally categorized as farm that showed markedly outlying (10 times higher) values of transports. A closer investigation showed that the respective holder owned three different locations, two with stables and permanent presence of animals and one large hall regularly used for markets but also for other purposes. We, therefore, re-classified the holding at this last location as a market.

(b) Static network description

We created a static network for each month, where each active holding was represented by a vertex (node) with a directed edge between two nodes representing a movement of one or more animals from the premises of origin to the destination within the respective month. We chose one month to make our results comparable to those from other countries where authors have also analysed monthly networks. It has furthermore been argued elsewhere [23] that the duration of a month will create networks large and dense enough (but not too large and dense) to make a network approach feasible and –at the same time– it seems unlikely that, given the current surveillance schemes in place, any severe disease could spread undetected for much longer than a month. A holding is considered as 'active', when it occurred at least once in the edge list –either as a holding of origin or destination. The constructed network is directed, as origin and destination of the transport are specified, and simple, as multiple movements from one holding to another within a month are not reflected by multiple edges between nodes nor in weights attributed to the edges. The number of animals moved is also not considered, so the links are unweighted, for reasons described in the next section. For each monthly network of the four-year period we calculated the following network measures: density, giant strongly connected component (GSCC), giant weakly connected component (GWCC), clustering coefficient (CC), average path length (APL) and degree assortativity (Assort, see Table 1 for definitions). Furthermore, we calculated for each holding for each month the following nodal network measures: in-degree, out-degree, degree product (in-degree × out-degree), betweenness, in-degree closeness centrality, and out-degree closeness centrality (see Table 1 for definitions).

In order to evaluate the relative importance of the different holding categories, we constructed three further 'reduced' networks by consecutively removing markets (and all edges leading from or to markets), dealers (and all edges leading from and to dealers), and alps (and all edges leading from or to alps) from the network. Hence, apart from the full network we got one network without markets, one network without markets and dealers and one network without markets, dealers and alps (farms only).

(c) Dynamic network measures

Static networks, where edges represent the accumulated movement activity over an extended time do not acknowledge the temporal order of movement events. This can be problematic, specifically when it comes to modelling transmission processes on the network [15–17, 19]. We, therefore, calculated for each node the so-called accessible world [15, 34] or outgoing contact chain [19, 35, 36] over a 30-day period –taking the daily time structure stamps of the events into account–, starting with the first day of each month. The outgoing contact chain is constituted by all destination holdings that can be reached from a certain holding, with the following assumptions: 1) an outbreak would spread undetected and uncontrolled for 30 days, 2) a holding can reach another holding if an animal is transported from that holding to the other holding, 3) at the arrival of an infected animal, all animals at a particular holding get infected immediately and are able to further transmit the infection when leaving the holding, 4) transports are recorded on a daily basis and a holding can only transmit the disease to other holdings, when it was itself infected before animals were moved from it to other holdings, 5) when there were transports from and to a holding on the same day, we assumed that animals were leaving the holding after new animals arrived; i.e. animals arriving transmit the disease to animals leaving on the same day. In this respect, this calculation represents a simple Susceptible-Infectious (SI) model [37, 38] for a worst-case scenario of an epidemic disease that is transmitted via animal movements, only. With these stringent assumptions, the process does not precisely model any specific disease, but it is rather to be seen as a generic model. Importantly, as this process does not incorporate any stochastic element, an exact value can be calculated for each node, given a specific network.

While the outgoing contact chain is a rather generic infection process model, its assumptions are reasonably compatible with a fast spreading disease such as Bluetongue, Foot and Mouth Disease or Lumpy Skin Disease – at least for an initial stage, before effective measures are taken. The length of the outgoing contact chain after 30 days, starting at a specific holding A, gives the number of holdings that would get infected under such a transmission process. We refer to this number as the 'outbreak size' of holding A. The choice of 30 days is an arbitrary one, though it is intended to make the results comparable to those built on static monthly measures as they are frequently encountered in the literature. This holding's property is not constant along time but varies with season and year. Preliminary analysis showed that the distribution for this nodal metric is highly skewed, so neither mean nor median values give a sensible network-wide summary statistic for the overall risk of the outbreak of a large-scale epidemic. Thus, to derive a network-wide measure for outbreak risk, we calculated the proportion of the total number of active nodes with an outbreak size of 100 or larger (which we call hubs thereafter). This measure is conceptually similar to the reachability ratio proposed in [16, 39] and presented in [40], however it differs insofar as these authors present averages and maximum reachability ratios calculated over all nodes. Associations between different measures were assessed by computing Spearman correlation coefficients and generalized linear regression models.

(d) Stochastic outbreak simulations

To gauge the difference between predictions based on static and dynamic network descriptors, we added a stochastic simulation study that compared detection rate and outbreak size at detection of targeted surveillance

based on static and dynamic network measures. As a baseline or reference, we also considered random surveillance. For this simulation disease was seeded on one randomly chosen holding out of all active holdings of the respective year and disease spread was modelled as a stochastic process with the disease spreading from one holding A to another holding B on a given day with probability $p = 0.8$ if a movement of animals from A to B occurred on that day. The number of animals transported from one holding to another was not considered here. Incubation time was assumed to be zero, meaning that on the very day when one or more contagious animals were moved to a holding, animals on that holding can get infected and can also spread the disease further if animals were moved to another holding on the same day or later. Additionally, each day the disease spreads to a randomly chosen holding with probability $q = 0.05$, acknowledging that the disease can also be conveyed by other means than movements of contagious animals. Note that this is a very simple model, not taking into account specific infection dynamics of existing diseases. The parameter values for p and q were arbitrarily chosen, p being a high value and q being a low value. The maximum time for disease spread was set to 30 days with daily updates of infected holdings. Inspired by the work of [41], for each simulation a small number ($n = 100$) of holdings was selected as surveillance targets. Surveillance targets were either chosen randomly from all active holdings (random surveillance), randomly chosen from holdings with high scores for one of the six nodal network metrics (top 5% percentile, static targeted surveillance), or randomly chosen from all holdings with an outbreak size of 100 or more (dynamic targeted surveillance). We use outgoing measures rather than ingoing for convenience. A similar analysis could be carried out using the equivalent ingoing metrics however, as our aim was only to compare the performance of static versus dynamic metrics, this should be of no further concern, here. The simulation was repeated 1000 times for each month of the study period. As measures of the effectiveness of the surveillance scheme we evaluated (i) how often the epidemic was detected at one of the surveillance targets within 30 days, and (ii) the number of infected holdings at the time of detection.

(e) Software packages and code

Monthly static network measures were calculated using the software R (version 3.1.2) -package igraph (version 3.2.2, https://r-project.org). The plot for the principal component analysis was made using packages ggplot2 and ggfortify. Calculation of the outgoing contact chain and stochastic simulations were done with the software *Mathematica* version 10.2 from Wolfram Research Inc.

Results

The total number of active holdings per year decreased slightly from 47,212 holdings in 2012 to 44,688 holdings in 2015. Out of those holdings 84.1% were farms, 15.1% alpine pastures, 0.24% dealers and 0.58% markets. The decrease in the number of holdings over the years is mainly due to a loss of active farms at a rate of about 800 farms per year, reflecting the general consolidation trends in the European livestock economy. In total, the Tierverkehr Datenbank recorded between 619,273 transports of animals from one holding to another in 2012 and 598,659 transports in 2015 (Table 2). During the four-year study period 52.4% of Swiss cattle farms participated in the summering practice. Overall 20.2% of the alps hosted animals from a single farm, while the remaining 79.8% of alps hosted animals from several (median: 4 farms, inter quartile range: 2–8) farms. Likewise, 44.2% of cattle farms participated in markets by either selling or purchasing animals at markets. At a typical market, one can find animals from 120 (median; IQR: 44–262) different farms. The number of active holdings (Fig. 1) shows a clear seasonality with one peak in early summer and a second one in autumn, coinciding with the summering pattern. The seasonal pattern mainly disappears after excluding alps except for a trough in summer that is still visible for the farms-only data.

(a) Monthly static networks

For each month of the four-year study period we calculated static network measures for the directed transport networks. The monthly degree distributions are highly skewed towards minimum values, (Fig. 2). Deviations from a linear trend in these double-logarithmic plots can be observed. Specifically, the in-degree distributions declined faster than a power-law distribution, similar to the networks described in [42]. With respect to the monthly static network measures (Fig. 3, panels a and c), both GSCC and GWCC vary strongly with season showing a trough during the summer months of July and August when the summering practice keeps animals high up in the pastures. The exclusion of the markets and dealers almost halves the GSCC from 8.64 to 4.54% and reduces the GWCC only slightly from 90.2 to about 85.9%. For both GWCC and GSCC the seasonal pattern largely disappears with the additional exclusion of the alps.

Table 2 Average number of animal transports between holding categories per year

	Farm	Alp	Dealer	Market
Farm	415,170	45,690	6066	67,028
Alp	64,808	4342	111	483
Dealer	2730	26	129	100
Market	7070	61	50	22

With very low values, the density increases in winter and in the farms-only network in summer (Fig. 3, panel b). The clustering coefficient increases in spring and autumn, especially after having excluded the markets and dealers. The seasonality also disappears with the additional exclusion of the alps (Fig. 3, panel d). Accordingly, the APL increases from a mean of 8.77 for the full network to 9.95 for the farm subnetwork (Fig. 3, panel E). The negative values of the disassortativity index indicate a tendency of high degree nodes to be linked to low degree nodes. The full network (Fig. 3, panel f, blue line) shows a clear seasonal pattern with maximal disassortativity each September. These spikes of highest disassortativity accentuate further when excluding markets and dealers (Fig. 3, panel f, green line), but disappear after exclusion of the alps, suggesting that the markets might render the network overall less disassortative. In any case, the values become closer to zero for the network composed of farms only. In conclusion, the alps are responsible for the marked seasonality in network properties and this trend becomes stronger when excluding markets and dealers. The farms only subnetwork is devoid of a clear seasonal fluctuation except in summer when the density shows maximum values.

The seasonality in the network characteristics becomes also apparent in a principal component analysis (PCA) of network-wide measures and numbers of active holdings per category over the 48 months of the study. The first two components explained together 78.9% of the variation (PC1: 50.8%, PC2: 28.1%) and the bi-plot shows that these two components together clearly discern the different seasons of the yearly cycle, with the winter months (November–April) loading low and Spring (May–June) loading high on PC1, while Summer (July – August) loading high and Autumn (September–October) loading low on PC2 (Fig. 4).

(b) Dynamic network measures

For each holding we evaluated the outgoing contact chain over 30 days, starting at the first day of each month. The length of this chain can be interpreted as a measure for the maximal outbreak size after 30 days considering a highly contagious disease requiring close proximity between individuals for transmission. The outbreak sizes per holding within a single month varied considerably from one, indicating that there was no transmission from this holding within 30 days, to 3754. Outbreak sizes for holdings also varied in the course of the year: 90.7% of all those holdings which had an outbreak size of 100 or larger during one month had an outbreak size of only one during another month of the same year. The distribution of outbreak sizes is highly skewed, with on average 54.5% of the holdings having an outbreak size of one (i.e. there was no transport to any

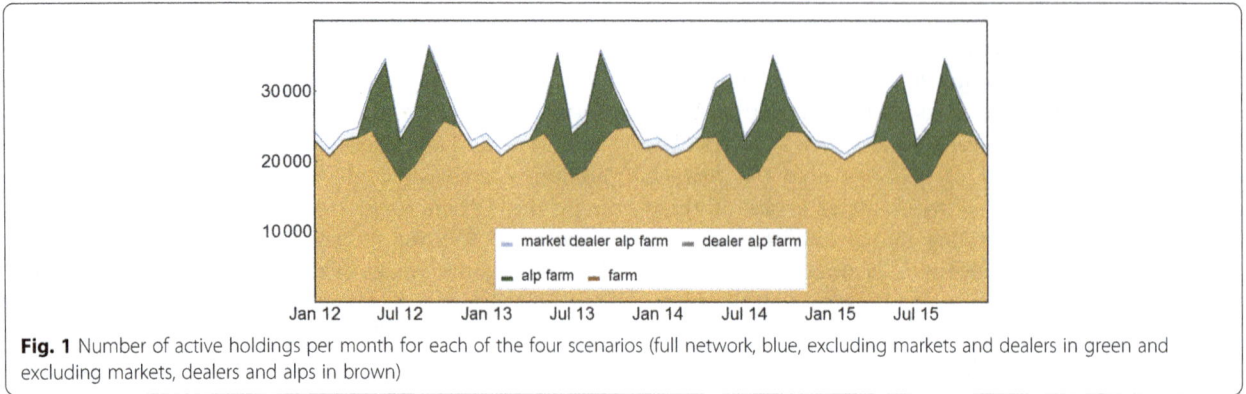

Fig. 1 Number of active holdings per month for each of the four scenarios (full network, blue, excluding markets and dealers in green and excluding markets, dealers and alps in brown)

other holding in that respective month) and 85.1% had an outbreak size of less than ten, meaning that even within 30 days of undetected transmission, a highly contagious disease would not have spread to more than ten holdings. We refer to holdings with an outbreak size of 100 or larger as 'hubs', as such holdings have the potential to convey a disease to many holdings at any time of an epidemic outbreak. Note, that the categorization of a node of being a hub or not can change from month to month depending on the actual transports and, consequently, the contact chain of the respective month. The percentage of holdings qualifying as hubs is given in Fig. 5. Depending on the month, only between 2.1 and 12.0% of the holdings qualify as hubs, suggesting that in most cases an outbreak starting at a single holding would spread only to a limited number of other holdings within the focal period. Figure 5 shows, further, a strong seasonality, with peaks in spring and autumn and low numbers in summer and around Christmas and New-year.

In our attempt to quantify the contribution of the different holding categories to the overall connectivity of the network and the risk of epidemic outbreaks we consecutively removed all markets and all transports from- and to markets from the network, then all dealers and finally all alps and all transports from and to alps. The number of hubs for these reduced networks is also shown in Fig. 5, and can be interpreted as an estimate for outbreak risk if, for disease-control or other reasons, a certain practise, like the conduct of markets or the summering on alpine pastures, would have been banned or given up. As indicated by the light blue area, removing markets from the network has a substantial impact on the number of hubs, despite making up only a tiny proportion of all holdings. In contrast, dealers, who make up a similar small proportion of the population of holdings, have a much smaller impact (light grey area). Simulated removal of alps from the transport network further reduced the number of hubs during late spring and autumn –at the times when animals were brought to the summer pastures and when they returned to the farms. On the other hand, it had no effect during winter, when the alps were closed and during the summer months, when summering practice seems to lead to an overall reduction of movements and, hence, outbreak risk.

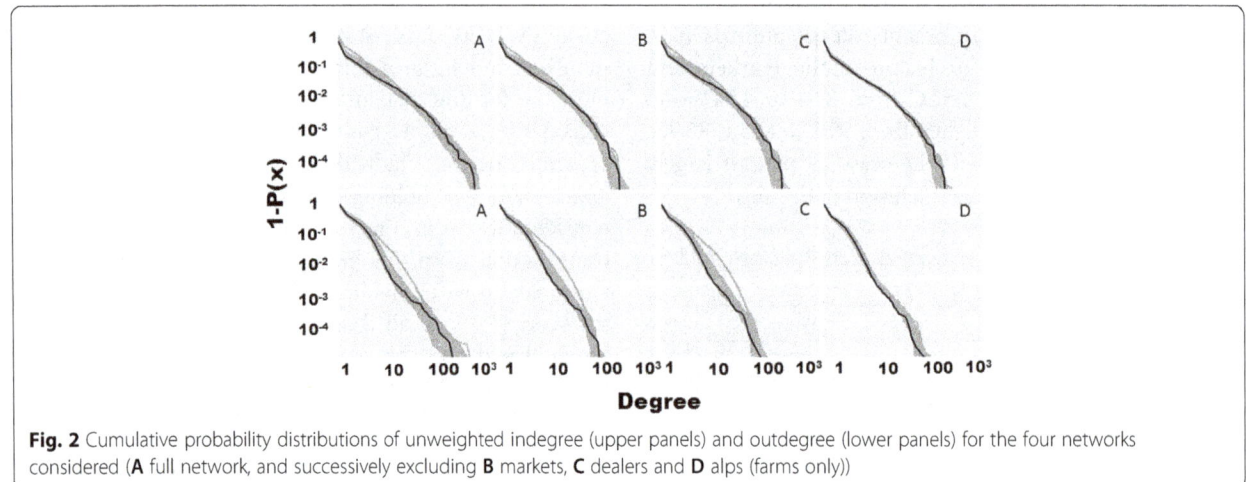

Fig. 2 Cumulative probability distributions of unweighted indegree (upper panels) and outdegree (lower panels) for the four networks considered (**A** full network, and successively excluding **B** markets, **C** dealers and **D** alps (farms only))

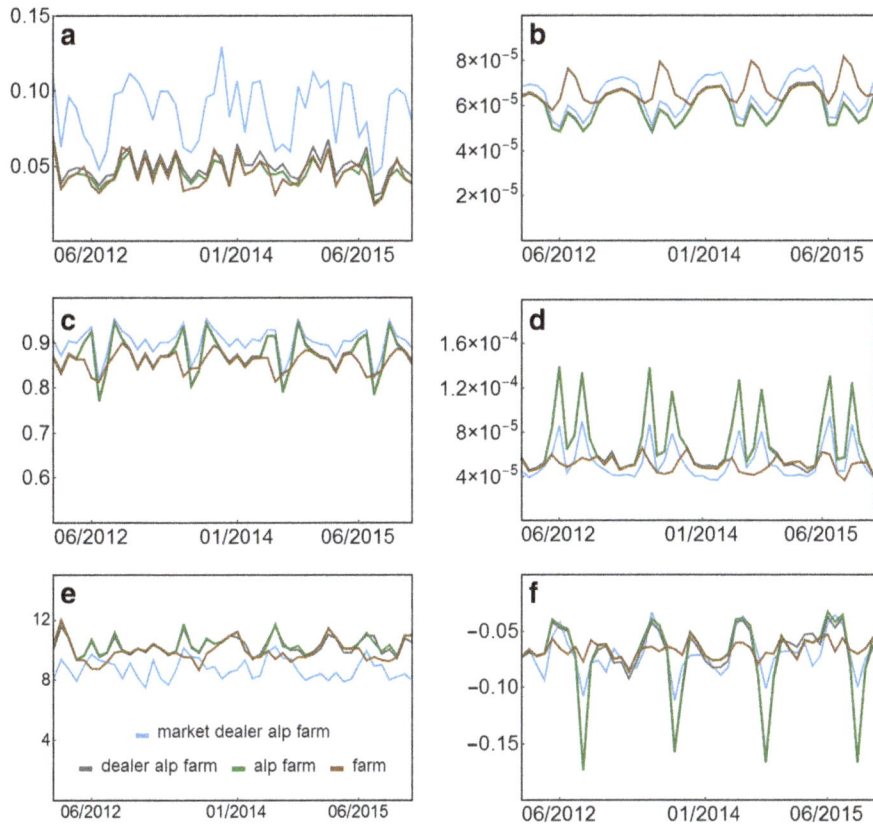

Fig. 3 Monthly static network measures, (**a**) giant strongly connected component GSSC, (**b**) density, (**c**) giant weakly connected component GWCC, (**d**) clustering coefficient, (**e**) average path length, (**f**) assortativity, starting from Jan 1st 2012 to December 31st 2015. Complete network in blue, network without markets in grey, network without markets and dealers (alps and farms network) in green and finally without alps (farms only network) in brown. Please note that the grey line is hardly visible as it is under the green line

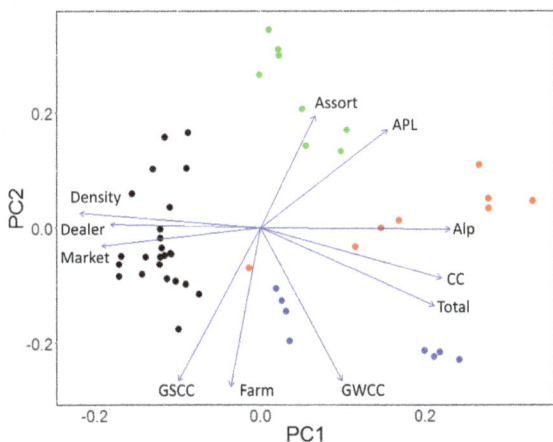

Fig. 4 Principal component analysis of network measures and number of holdings for the 48 monthly networks. Black: Winter (November–April), red: Spring (May–June), green: Summer (July – August), blue: Autumn (September–October), Assort: degree assortativity, APL: average path length, CC: clustering coefficient, Dens: density, Tot: total number of active holdings per month; Farm, Alp, Dealer, Market: number of active holdings per month in each category

(c) Associations between different metrics

(i) Network-wide measures

A linear regression model with the number of hubs (i.e. holdings with an outbreak size of ≥100) as dependent variable and the monthly network-wide measures GSCC, GWCC, clustering coefficient, average path length and assortativity and their interactions as independent variables could explain over 80% of the total variance (adjusted R^2 = 0.805, n = 48). The best single predictor for the number of hubs was the GSCC with R^2 = 0.551. The absolute number of movements between holdings and the number of active holdings per month were less reliable predictors for the number of hubs with R^2 = 0.208, and R^2 = 0.576 respectively. The number of active farms was the single best predictor: R^2 = 0.437. Most of the static monthly network measures were correlated with the number of active holdings per month and the number of hubs per month (Table 3).

(ii) nodal network measures

Given the heterogeneity of the transport network and the variation in outbreak size between individual

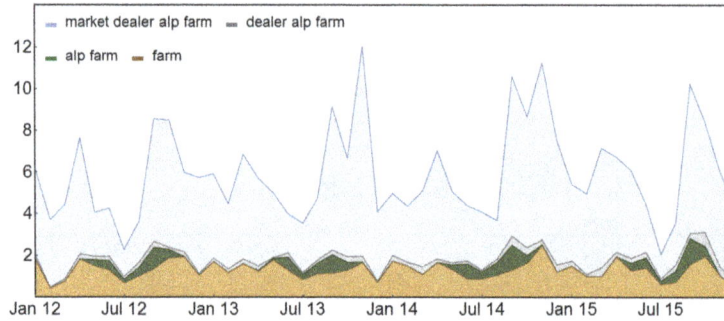

Fig. 5 Percentage of total active holdings that render outbreaks larger than 100 other holdings after 30 days of disease spread, calculated at the start of every month from January 1st 2012 to December 31st 2015. Different colors depict the different scenarios, all holdings (upper line), markets contribution (blue area), dealers contribution (grey area), alps contribution (green area), and finally, farms only in brown color

holdings, we asked to what extent nodal network measures (i.e. measures attributed to individual holdings) can be predictive of the outbreak size for a holding in a given month. A generalized linear model with outbreak size as the dependent variable, a log-link function (family Poisson), month and all six static monthly nodal measures (see below) and their interactions as independent variables resulted in an adjusted $R^2 = 0.199$, while post-hoc models with a single network measure, each, delivered even lower proportions of variance explained: in-degree $R^2 = 0.003$, out-degree $R^2 = 0.021$, degree product $R^2 = 0.003$, betweenness centrality $R^2 = 0.008$, in-degree closeness $R^2 = 0.038$, out-degree closeness: $R^2 = 0.108$. All these associations between static nodal measures and outbreak size are far too weak for making them reliable predictors for the outbreak size to be expected for a disease spreading from a specific holding. In conclusion, simple network measures as the monthly number of active holdings as well as monthly static network measures

Table 3 Correlation coefficients between the monthly ($n = 48$) number of active holdings (total, number of active farms, alps, dealers and markets) and static monthly network measures (GSCC: Giant strongly connected component, GWCC: giant weakly connected component, CC: clustering coefficient, APL: average path length, Assort: degree assortativity)

	Total	Farm	Alp	Dealer	Market
Total		0.25	0.88	−0.64	−0.67
Farm			−0.23	0.18	0.25
Alp				−0.73	−0.81
Dealer					0.52
GSCC	−0.03	0.76	−0.40	0.29	0.40
GWCC	0.72	0.69	0.39	−0.27	−0.23
Density	−0.90	−0.02	−0.89	0.68	0.66
CC	0.94	0.06	0.91	−0.69	−0.73
APL	0.35	−0.47	0.58	−0.41	−0.44
Assort	−0.01	−0.34	0.16	−0.06	−0.28

are informative about the overall outbreak risk in a given month. However, nodal measures of monthly networks do not inform about the danger emanating from a specific holding.

(c) Surveillance examples

In order to gauge whether holding characterization based on dynamic network measures can be of practical importance, we ran a simulation where sentinel holdings were either selected based on the outgoing contact chain, static network measures of monthly networks, or randomly selected from all active holdings. Surveillance based on dynamic measures (randomly selecting 100 holdings from all holdings with outbreak size of 100 or more) detected an epidemic on average in 83.5% of the simulations, while the detection rate was markedly lower for both static targeted surveillance, and random surveillance. Furthermore, the median outbreak size at the time point of detection was lower for the dynamic targeted surveillance than for any other scheme (Table 4). The higher efficiency of surveillance based on the outgoing contact chain is according to our expectations: as outbreak size reflects the accessible world, taking into account the temporal structure of the movements, this measure allows to identify those holdings from which a disease would spread rapidly. Surprisingly, surveillance based on monthly network measures was even less effective than surveillance based on randomly selected sentinel holding. At the moment we do not have an explanation for this finding. This simulation does not constitute a systematic investigation into surveillance strategies, as we employed only a rather simple transmission model and because we confined our investigation to a single set of parameter values instead of exploring the parameter space more broadly. However, these results can at least deliver a proof of concept.

Table 4 Comparison of the effectiveness of different surveillance schemes. IQR: interquartile range

Sentinel selection	Detection rate % (median)	IQR	Outbreak size (median)	IQR
outgoing contact chain	83.5	72.4–92.9	50	35–89
in-degree	54.7	32.2–67.2	104	56–173
out-degree	52.6	40.7–66.4	132	49–214
degree prod	56.0	42.4–77.6	114	54–158
betweenness	54.6	31.7–68.7	91	34–159
in-closeness	64.5	38.7–81.2	85	47–151
out-closeness	52.1	35.9–70.3	118	41–208
random	70.3	57.7–85.6	112	57–143

Epidemic outbreaks were simulated using stochastic simulations with 1000 repetitions for each months and each scenario. Surveillance was based on 100 holdings which were either randomly selected from all holdings with an outgoing contact chain larger than 100, from all holdings in the top 5% percentile for one of the static network measures in the respective month, or from all active holdings

Discussion

We found strong seasonality for several network measures –both static measures for monthly networks and dynamic measures like the proportion of hubs. For example, in autumn the risk of larger outbreaks is five times larger than in summer. Fluctuations in monthly static network measures such as the GSCC are in line with this picture.

The alpine summering practice has a strong impact on the seasonal changes in the transportation network. Simulated removal of alps from the network reduces seasonal variations in the connectivity (GWCC), clustering and assortativity substantially. Yet, when it comes to the risk of epidemic outbreaks, markets play a much more prominent role than the alps. This has an important implication with respect to outbreak control interventions, such as transport bans.

By using unweighted data, our focus is on fast spreading and highly contagious diseases, where the transport of a single contagious animal would be sufficient to infect all other animals in the holding of destination. Disease specific models that take the transmission dynamics and etiopathology of the specific disease into account would be required to evaluate the contribution of holding types in the case of slow spreading diseases [18, 43]. Furthermore, we would like to mention that our approach (simulated removal of nodes, see also [12, 29, 44, 45] for further use of this approach) does not account for any potential increased activity of the holdings that remain in the network when excluding a particular holding category. This is, of course, a simplification and a study by Robinson and colleagues [46] reported that after the introduction of mandatory standstill periods in Great Britain (a response to the foot and mouth disease epidemic in 2001), movement patterns have shifted constantly resulting even in an increase of the GSCC and hence the potential size of the next epidemic outbreak.

The prominence of markets as potential hubs for disease transmission, as it is suggested by their contribution to the overall outbreak sizes, is well in line with results found in other studies [22, 23]. It is important to keep in mind that large market events are subject to strict regulations to prevent disease spread, including medical surveillance and spatial structuring to minimize contact between animals from different holdings. Thus our findings should be interpreted as potential risk, if no control measures and regulations were in place, but not as a measure for the actual risk given current best practice. In this respect, our results clearly stress the importance of the precautionary measures taken at large markets and auctions. Network-wide metrics of monthly transport networks were clearly correlated with the proportion of hubs (i.e. outgoing contact chains with a length of 100 or more), which we took as a risk measure. However, the number of active holdings, specifically markets and alps, also showed a strong seasonal pattern that was correlated with outbreak risk. From a mechanistic perspective, it is clear that changes in the risk of epidemic outbreaks are a consequence of changes in connectivity of the network, which is a consequence of the seasonal cycle. Thus, while several network measures allow predictions about times of increased outbreak risk, the same feat could be achieved by simply counting holdings per category, or –even more simply– consulting the calendar. Nodal static network measures of the monthly transport networks give only poor predictions for the risk emanating from a specific holding. Of the six static nodal measures considered, only out-degree closeness had a noteworthy predictive value for outbreak risk.

The reason for this discrepancy can be found in the temporal dynamics of the network. If a transport from holding B to holding C is recorded for time point t, and another transport from holding A to holding B for the time point $t + 1$, then holding A is connected to both holdings B and C in the monthly static network. However, due to the temporal order of the connections a disease could not spread from A to C. The outgoing contact chain takes this temporal dynamic into account and, as such, it gives a more sensible network metric reflecting its potential contribution to an epidemic outbreak. We simulated outbreaks with a simple model, which contained two stochastic components: one for the transmission of the disease due to animal transports from one holding to the other, and a second one for transmission due to environmental contamination or other unknown reasons. Yet, targeted surveillance based on the dynamic outgoing contact chain could detect epidemics more often and more effectively than both targeted surveillance based on holdings selected based on static network properties and random surveillance. Our results are in line with recent propositions for selection criteria of surveillance targets based on cluster analysis of the transport network [18, 41]. Even though the methodologies differ in details, all these studies accumulate evidence of how promising dynamic or temporal approaches are for risk evaluation and surveillance.

Conclusions

We provided a detailed description of the Swiss cattle transport network. Static descriptors of monthly transport networks give only poor predictors for the outbreak risk emanating from individual holdings; yet a dynamic measure based on time-stamped data—the outgoing contact chain—can give better risk estimates and could help to improve surveillance schemes. Using this measure we find evidence that even in a country with intense summering practice, markets continue being the prime risk factor for the spread of contagious diseases.

Abbreviations

APL: Average path length; Assort: Degree assortativity; CC: Clustering coefficient; GSCC: Giant strongly connected component; GWCC: Giant weakly connected component; IQR: Interquantile range

Acknowledgements

We thank the company Identitas AG for access to data, in particular Christian Beglinger and Sean Redmond for support and discussion. Discussions with Gertraud Schüpbach from the Veterinary Public Health Institute of the University of Bern, Rahel Struchen and Sara Schärrer from the Swiss Federal Office for Food Safety and Veterinary Affairs provided helpful insights for the interpretation of the results. We thank Colin Garroway for comments on the manuscript.

Availability of data, code and materials

Information about the Swiss animal movement database (in German, Tier Verkehr Datenbank (TVD) or French, Banque de données sur le traffic des animaux (BDTA)) can be found at the internet portal Agate from the Swiss Confederation: https://www.agate.ch/portal/web/agate/home. Raw data were accessed after establishing a confidentiality agreement with the company Identitas AG and the data owners: https://www.identitas.ch. Email: info@identitas.ch. Data cannot be made publicly available for ethical and legal reasons. Interested and qualified researchers would be able to access these data in the same manner as the authors. We confirm that we did not have any special access privileges that others would not have.

Funding

No special funding was needed for the current study.

Authors' contributions

BV acquired and extracted the data and carried out data analysis using R; conceived and designed of the study and drafted the manuscript; BV conceived and designed the study; carried out data analysis using Mathematica and drafted the manuscript. All authors gave final approval for publication.

Consent for publication

Not applicable.

Competing interests

The authors declare that they have no competing interests. We have no competing interests.

References

1. Green DM, Kiss IZ, Mitchell AP, Kao RR. Estimates for local and movement-based transmission of bovine tuberculosis in British cattle. Proc R Soc B Biol Sci. 2008;275(1638):1001–5. Available from: htp://dx.doi.org/10.1098/rspb.2007.1601
2. Cowie CE, Marreos N, Gortázar C, Jaroso R, White PCL, Balseiro A. Shared risk factors for multiple livestock diseases: a case study of bovine tuberculosis and brucellosis. Res Vet Sci. 2014;97(3):491–7. Available from: http://dx.doi.org/10.1016/j.rvsc.2014.09.002
3. Gilbert M, Mitchell A, Bourn D, Mawdsley J, Clifton-Hadley R, Wint W. Cattle movements and bovine tuberculosis in Great Britain. Nature. 2005;435(May):491–6. Available from: http://dx.doi.org/10.1038/nature03548
4. Craft ME. Infectious disease transmission and contact networks in wildlife and livestock. Philos Trans R Soc Lond B Biol Sci. 2015;370(1669):1–12. Available from: http//dx.doi.org/10.1098/rstb.2014.0107
5. Price S, Garner T, Cunningham A, Langton T, Nichols R. Reconstructing the emergence of a lethal infectious disease of wildlife supports a key role for spread through translocations by humans. Proc R Soc B Biol Sci. 2016;283:20160952.
6. Ortiz-Pelaez A, Pfeiffer DU, Soares-Magalhães RJ, Guitian FJ. Use of social network analysis to characterize the pattern of animal movements in the initial phases of the 2001 foot and mouth disease (FMD) epidemic in the UK. Prev Vet Med. 2006;76(1–2):40–55. Available from: http://dx.doi.org/10.1016/j.prevetmed.2006.04.007
7. House T, Keeling MJ. Insights from unifying modern approximations to infections on networks. J R Soc Interface. 2011;8(54):67–73. Available from: http://dx.doi.org/10.1098/rsif.2010.0179
8. Dubé C, Ribble C, Kelton D, McNab B. A review of network analysis terminology and its application to foot-and-mouth disease modelling and policy development. Transbound Emerg Dis. 2009;56(3):73–85. Available from: http://dx.doi.org/10.1111/j.1865-1682.2008.01064.x
9. Firestone SM, Ward MP, Christley RM, Dhand NK. The importance of location in contact networks: describing early epidemic spread using spatial social network analysis. Prev Vet Med. 2011;102(3):185–95. Available from: http://dx.doi.org/10.1016/j.prevetmed.2011.07.006
10. Dawson PM, Werkman M, Ellen B-P, Tildesley MJ. Epidemic predictions in an imperfect world: modelling disease spread with partial data. Proc R Soc B Biol Sci. 2015;282(1808):20150205. Available from: http://dx.doi.org/10.1098/rspb.2015.0205
11. Kao RR, Danon L, Green DM, Kiss IZ. Demographic structure and pathogen dynamics on the network of livestock movements in Great Britain. Proc R Soc B Biol Sci. 2006;273(1597):1999–2007. Available from: http://www.pubmedcentral.nih.gov/articlerender.fcgi?artid=1635475&tool=pmcentrez&rendertype=abstract
12. Natale F, Giovannini A, Savini L, Palma D, Possenti L, Fiore G, et al. Network analysis of Italian cattle trade patterns and evaluation of risks for potential disease spread. Prev Vet Med. 2009;92:341–50.
13. Lentz HHK, Selhorst T, Sokolov IM. Spread of infectious diseases in directed and modular metapopulation networks. Phys Rev E - Stat Nonlinear, Soft Matter Phys. 2012;85(6):1–9. Available from: http://dx.doi.org/10.1103/PhysRevE.85.066111

14. Wasserman S, Faust K. Social network analysis: Methods and applications. Cambridge Univ Press [Internet]. 1994;1:116. Available from: http://dx.doi.org/10.1525/ae.1997.24.1.219

15. Webb CR. Investigating the potential spread of infectious diseases of sheep via agricultural shows in Great Britain. Epidemiol Infect. 2006;134(1):31–40. Available from: http://dx.doi.org/10.1017/S095026880500467X

16. Holme P, Saramäki J. Temporal networks. Phys Rep. 2012;519(3):97–125. Available from: http://dx.doi.org/10.1016/j.physrep.2012.03.001

17. Bajardi P, Barrat A, Natale F, Savini L, Colizza V. Dynamical patterns of cattle trade movements. PLoS One. 2011;6(5) Available from: http://dx.doi.org/10.1371/journal.pone.0019869

18. Bajardi P, Barrat A, Savini L, Colizza V. Optimizing surveillance for livestock disease spreading through animal movements. J R Soc Interface. 2012;9(76): 2814–25. Available from: http://dx.doi.org/10.1098/rsif.2012.0289

19. Nöremark M, Håkansson N, Sternberg S, Lindberg A, Jonsson A. Network analysis of cattle and pig movements in Sweden : Measures relevant for disease control and risk based surveillance. Prev Vet Med. 2011;99:78–90. Available from: http://dx.dooi.org/10.1016/j.prevetmed.2010.12.009

20. Frössling J, Ohlson A, Björkman C, Håkansson N, Nöremark M. Application of network analysis parameters in risk-based surveillance – examples based on cattle trade data and bovine infections in Sweden. Prev Vet Med. 2012;105(3): 202–8. Available from: http://dx.doi.org/10.1016/j.prevetmed.2011.12.011

21. Christley RM, Robinson SE, Lysons R, French NP. Network analysis of cattle movement in Great Britain. Proc Soc Vet Epidemiol Prev Med. 2005:234–44.

22. Robinson SE, Christley RM. Exploring the role of auction markets in cattle movements within Great Britain. Prev Vet Med. 2007;81(1–3):21–37. Available from: http://dx.doi.org/10.1016/j.prevetmed.2007.04.011

23. Mweu MM, Fournié G, Halasa T, Toft N, Nielsen SS. Temporal characterisation of the network of Danish cattle movements and its implication for disease control : 2000–2009. Prev Vet Med. 2013;110(3–4): 379–87. Available from: http://dx.doi.org/10.1016/j.prevetmed.2013.02.015

24. Anderson R, May R. Infectious diseases of humans. Dynamics and control. New York: Oxford University Press; 1991.

25. Lloyd-Smith JO, Schreiber SJ, Kopp PE, Getz WM. Superspreading and the effect of individual variation on disease emergence. Nature. 2005;438(7066): 355–9. Available from: http://dx.doi.org/10.1038/nature04153

26. May RM, Lloyd AL. Infection dynamics on scale-free networks. Phys Rev E. 2001;64:1–4. Available from: http://dx.doi.org/10.1103/PhysRevE.64.066112

27. Barabási A-L, Albert R. Emergence of Scaling in Random Networks. Science (80-). 1999;286(October):509–12. Available from: http://dx.doi.org/10.1126/science.286.5439.509

28. Natale F, Savini L, Giovannini A, Calistri P, Candeloro L, Fiore G. Evaluation of risk and vulnerability using a disease flow centrality measure in dynamic cattle trade networks. Prev Vet Med. 2011;98(2–3):111–8. Available from: http://dx.doi.org/10.1016/j.prevetmed.2010.11.013

29. Dutta BL, Ezanno P, Vergu E. Characteristics of the spatio-temporal network of cattle movements in France over a 5-year period. Prev Vet Med. 2014; Available from: http://dx.doi.org/10.1016/j.prevetmed.2014.09.005

30. Schärrer S, Widgren S, Schwermer H, Lindberg A, Vidondo B, Zinsstag J, et al. Evaluation of farm-level parameters derived from animal movements for use in risk-based surveillance programmes of cattle in Switzerland. BMC Vet Res. 2015;11:149. Available from: http://dx.doi.org/10.1186/s12917-015-0468-8

31. Braun U, Schönmann M, Ehrensperger F, Hilbe M, Brunner D, Stärk KD, et al. Epidemiology of bovine virus diarrhoea in cattle on communal alpine pastures in Switzerland. Zentralbl Veterinarmed A. 1998;45(8):445–52.

32. Siegwart N, Hilbe M, Hässig M, Braun U. Increased risk of BVDV infection of calves from pregnant dams on communal alpine pastures in Switzerland. Vet J. 2006;172(2):386–8. Available from: http://dx.doi.org/10.1016/j.tvjl.2005.07.018

33. Switzerland, Federal Office for Food Safety and Veterinary Affairs [Website]. (2016, May 17). Agrarpolitisches Informationssystem (AGIS). Retrieved March 9, 2018, from https://www.blw.admin.ch/blw/de/home/politik/datenmanagement/agis.html.

34. Webb CR. Farm animal networks : unraveling the contact structure of the British sheep population. Prev Vet Med. 2005;68:3–17. Available from: http://dx.doi.org/10.1016/j.prevetmed.2005.01.003

35. Büttner K, Krieter J, Traulsen A, Traulsen I. Static network analysis of a pork supply chain in northern Germany — characterisation of the potential spread of infectious diseases via animal movements. Prev Vet Med. 2013;110(3–4):418–28. Available from: http://dx.doi.org/10.1016/j.prevetmed.2013.01.008

36. Büttner K, Salau J, Krieter J. Quality assessment of static aggregation compared to the temporal approach based on a pig trade network in Northern Germany. Prev Vet Med. 2016;129:1–8. Available from: http://dx.doi.org/10.1016/j.prevetmed.2016.05.005

37. Hamer H. Epidemic disease in England: the evidence of variability and of persistency of type, ser. Milroy lectures. Lancet. 1906;1(3):733–9.

38. Hethcote H. The mathematics of infectious diseases. SIAM Rev. 2000;42:599–653.

39. Holme P. Network reachability of real-world contact sequences. Phys Rev E - Stat Nonlinear, Soft Matter Phys. 2005;71(4):1–8. Available from: http://dx.doi.org/10.1103/PhysRevE.71.046119

40. Dutta BL, Ezanno P, Vergu E. Characteristics of the spatio-temporal network of cattle movements in France over a 5-year period. Prev Vet Med. 2014;117(1):79–94. Available from: http://dx.doi.org/10.1016/j.prevetmed.2014.09.005

41. Schirdewahn F, Colizza V, Lentz HHK, Koher A, Belik V, Hövel P. Surveillance for outbreak detection in livestock-trade networks. In: Masuda N, Holme P, editors Temporal network epidemiology. Singapore: Springer; 2017.

42. Amaral LAN, Scala A, Barthe M. Classes of small-world networks. Proc Nat Acad Sci USA. 2000;97(21) Available from: http://dx.doi.org/10.1073?pnas.200327197

43. Kao RR, Green DM, Johnson J, Kiss IZ. Disease dynamics over very different time-scales: foot-and-mouth disease and scrapie on the network of livestock movements in the UK. J R Soc Interface. 2007;4(16):907–16. Available from: http://dx.doi.org/10.1098/rsif.2007.1129

44. Büttner K, Krieter J, Traulsen A, Traulsen I. Efficient interruption of infection chains by targeted removal of central Holdings in an Animal Trade Network. PLoS One. 2013;8(9) Available from: http://dx.doi.org/10.1371/journal.pone.0074292

45. Iyer S, Killingback T, Sundaram B, Wang Z. Attack Robustness and Centrality of complex networks. PLoS One. 2013;8(4) Available from: http://dx.doi.org/10.1371/journal.pone.0059613

46. Robinson SE, Everett MG, Christley RM. Recent network evolution increases the potential for large epidemics in the British cattle population. J R Soc Interface. 2007;4(15):669–74. Available from: http://dx.doi.org/10.1098/rsif.2007.0214

47. Freeman LC. Centrality in social networks conceptual clarification. Soc Networks. 1978;1(3):215–39.

48. Christley RM, Pinchbeck GL, Bowers RG, Clancy D, French NP, Bennett R, et al. Infection in social networks: using network analysis to identify high-risk individuals. Am J Epidemiol. 2005;162(10):1024–31. Available from: http://dx.doi.org/10.1093/aje/kwi308

49. De Nooy W, Mrvar A, Batagelj V. Exploratory social network analysis with Pajek. New York: Cambridge University Press; 2005.

50. Watts DJ, Strogatz SH. Collective dynamics of "small world" networks. Nature. 1998;393(6684):440–2. Available from: http://dx.doi.org/http://dx.doi.org/10.1038/30918

51. Newman MEJ. Assortative mixing in networks. Phys Rev Lett. 2002;89(20): 208701. Available from: http://dx.doi.org/10.1103/PhysRevLett.89.208701

16

Potential for rapid antibody detection to identify tuberculous cattle with non-reactive tuberculin skin test results

W. Ray Waters[1] [iD], H. Martin Vordermeier[2], Shelley Rhodes[2], Bhagwati Khatri[2], Mitchell V. Palmer[1], Mayara F. Maggioli[1], Tyler C. Thacker[1], Jeffrey T. Nelson[3], Bruce V. Thomsen[3], Suelee Robbe-Austerman[3], Doris M. Bravo Garcia[3], Mark A. Schoenbaum[4], Mark S. Camacho[5], Jean S. Ray[6], Javan Esfandiari[7], Paul Lambotte[7], Rena Greenwald[7], Adrian Grandison[7], Alina Sikar-Gang[7] and Konstantin P. Lyashchenko[7*]

Abstract

Background: Bovine tuberculosis (TB) control programs generally rely on the tuberculin skin test (TST) for ante-mortem detection of *Mycobacterium bovis*-infected cattle.

Results: Present findings demonstrate that a rapid antibody test based on Dual-Path Platform (DPP®) technology, when applied 1-3 weeks after TST, detected 9 of 11 and 34 of 52 TST non-reactive yet *M. bovis*-infected cattle from the US and GB, respectively. The specificity of the assay ranged from 98.9% ($n = 92$, US) to 96.0% ($n = 50$, GB) with samples from TB-free herds. Multi-antigen print immunoassay (MAPIA) revealed the presence of antibodies to multiple antigens of *M. bovis* in sera from TST non-reactors diagnosed with TB.

Conclusions: Thus, use of serologic assays in series with TST can identify a significant number of TST non-reactive tuberculous cattle for more efficient removal from TB-affected herds.

Keywords: Antibody, Bovine tuberculosis, Dual path platform, Multi-antigen print immunoassay, Tuberculin skin test, *Mycobacterium bovis*

Background

Tuberculosis (TB) in humans and animals may result from exposure to bacilli within the *Mycobacterium tuberculosis* complex such as *M. tb*, *M. bovis*, *M. africanum*, *M. microti*, *M. caprae*, *M. orygis*, *M. suricattae*, *M. mungi*, or *M. canetti* [1, 2]. *M. bovis* is the species most often isolated from tuberculous cattle. Despite intensive and costly control efforts over many decades, bovine TB persists in most countries adversely affecting animal health, welfare, and trade as well as the livelihoods of producers. Persistence of bovine TB in livestock populations also demands the maintenance of costly federal and regional regulatory networks. Control strategies rely largely on ante-mortem testing and slaughter inspection to identify livestock herds at

risk. With cattle, the principal ante-mortem tests for presumptive diagnosis of bovine TB are immunoassays that detect cell-mediated responses, including both in vivo [i.e., tuberculin skin test (TST)] and in vitro [i.e., interferon gamma release assay (IGRA)] methods [3–5]. In many countries, TST is applied as the primary test and IGRA may be used as an ancillary test in cattle to maximize the number of infected animals identified or as a confirmatory test [6]. The most common applications for ante-mortem testing include routine surveillance to identify *M. bovis*-infected herds, test and removal schemes, movement tests, epidemiologic trace-back testing, and in TB-affected herds to delineate animals going to a slaughter plant versus being condemned for rendering. While used extensively for over 100 years in cattle, the TST does have a number of severe shortcomings. The sensitivity of TST ranges broadly from 55 to 97% depending on the type and technical variations of test applied, quality of purified protein derivative (PPD), environmental exposure/burden

* Correspondence: KLyashchenko@chembio.com
[7]Chembio Diagnostic Systems, Inc., Medford, NY, USA
Full list of author information is available at the end of the article

to atypical mycobacteria, and many other factors [4, 5]. Thus, improved ante-mortem tests and/or testing strategies for bovine TB are greatly needed.

Over the past decade, a new generation of serologic tests designed to detect antibodies to multiple *M. bovis* specific antigens have emerged for application in cattle [7–13]. Of these, an ELISA using MPB83 and MPB70 antigens (*M. bovis* Ab Test, IDEXX Laboratories, Westbrook, Maine, US [10]) is approved by the *Office International des Epizooties* and US Department of Agriculture for discretionary use in cattle; however, application of this test has been limited primarily to confirmation of infection. Injection of PPDs for TST significantly boosts antibody responses in *M. bovis* infected cattle, including animals without prior detectable antibody responses [10, 11, 14–16]. The enhanced IgG responses are elicited by *M. bovis* specific antigens (e.g., MPB83 and MPB70) and characterized by accelerated antibody affinity maturation [17]. The boosted antibody responses wane beginning ~1-2 months after PPD administration; however, they can be further increased upon PPD re-injection [17]. Despite these advances, existing antibody assays generally lack diagnostic sensitivity, especially in early infection, and thus require further improvements [4–6].

In the present study, sera from *M. bovis*-infected cattle identified as TST non-reactors in naturally-exposed herds within the US (*n* = 11) and GB (*n* = 52) were evaluated for antibody responses to *M. bovis* specific proteins using a next generation immunochromatographic test based on Dual-Path Platform (DPP°) technology developed by Chembio Diagnostic Systems, Inc. (Medford, New York, US) [9]. Findings demonstrate the potential for use of antibody assays to detect *M. bovis* infection in TST non-reactive cattle within TB-affected herds.

Methods
Naturally-infected herds
Great Britain
Sera (*n* = 127) from GB were obtained from *M. bovis*-infected cattle detected during routine surveillance, including multiple herds and animals of diverse age, gender, breed, and management systems. All animals received a single intradermal comparative cervical test (SICCT) and of these, 52 animals were SICCT negative, IGRA positive (blood collected for IGRA and serum ~60 days post-SICCT) with tuberculous lesions detected postmortem. The other 75 animals were SICCT positive and not tested by IGRA (serum collection circa 3 weeks post-SICCT at the abattoir) with tuberculous lesions upon postmortem and *M. bovis* isolated from lesions. Serum samples were also collected from 50 SICCT negative, IGRA negative cattle located in a TB-free region of GB.

Texas, US
A cow with tuberculous lesions (later confirmed as *M. bovis* upon mycobacterial culture) was detected upon routine inspection at an abattoir in 2014. The source herd of this cow was determined to consist of approximately 11,000 Holstein dairy cattle. On postmortem examination of initial caudal-fold tuberculin test (CFT) reactors, a tuberculous lesion rate (confirmed by histopathology) of 1.5% was found. Based on epidemiologic risk factors, it was determined by regulatory officials and dairy management that destocking the dairy would be the best approach to rid it of *M. bovis*. A CFT and collection of blood for serological testing was completed 10-15 days prior to postmortem examination at destocking. A 2% overall prevalence of TB (visible tuberculous lesions confirmed by histopathology) was noted in cattle among CFT positive (35%) and CFT negative cattle (0.4%). From the CFT false-negative cattle, 7 serum samples obtained within 3 weeks after PPD injection were available for serologic analysis.

Michigan, US
Two cattle herds, one beef and the other dairy (mostly Jersey), within the TB-endemic region of Michigan (Northeast corner of the lower peninsula) were identified as TB-affected via routine surveillance in 2015. Animals within these two herds had received CFTs yearly for TB surveillance prior to 2015. Upon identification of tuberculous animals in 2015, CFTs were applied more frequently (~4-6 month interval) in both herds. Based on the presence of gross lesions upon depopulation, prevalence of TB was estimated to be ~21% (17/81) and 9% (53/561) for the beef and dairy herds, respectively. Four serum samples collected from *M. bovis*-infected CFT non-reactors within 3 weeks after the last CFT administration were available for serologic testing.

Multi-antigen print immunoassay (MAPIA)
MAPIA was performed as described previously [18]. Briefly, a panel of ten *M. tuberculosis*-complex antigens immobilized on nitrocellulose membrane included ESAT-6 (Rv3875), CFP10 (Rv3874), MPB64 (Rv1980c), MPB70 (Rv2875), MPB83 (Rv2873), CFP10/ ESAT-6 fusion protein, MPB70/MPB83 fusion protein, MPB70/ CFP10/Rv0934 fusion protein, bovine PPD (bPPD), and *M. bovis* culture filtrate (MBCF). Strips were cut and blocked with 1% nonfat milk in PBS with 0.05% Tween 20 for 1 h prior to incubation with serum samples diluted 1:40 in blocking solution for 1 h. After washing, strips were incubated with peroxidase-conjugated Protein G (Sigma, St. Louis, MO) diluted 1:1000 for 1 h, washed, and developed with 3,3′,5,5′-tetramethyl benzidine (TMB) (Kirkegaard & Perry Laboratories, Inc., Gaithersburg, MD).

Dual-path platform (DPP®) assay

Bovine IgG and IgM antibodies to CFP10/ESAT-6 and MPB70/MPB83 protein fusions were detected as described previously [17] using goat anti-bovine IgG and anti-bovine IgM antibodies (Kirkegaard & Perry Laboratories Inc.) conjugated to colloidal gold nanoparticles by Chembio standard procedure. Sera were diluted 1:20 in sample buffer for testing by DPP assay and results were recorded at 15 min using an optical reader to measure test band reflectance in relative light units (RLU), as previously described [17]. Using pre-established cut-off values of 20 RLU for CFP10/ESAT-6 antigen and 40 RLU for MPB70/MPB83 antigen (same for both IgM and IgG antibody detection), DPP assay readouts were expressed as signal-to-cutoff ratios, with any values ≥ 1.00 being interpreted as a reactive result and any value <1.00 being considered as a non-reactive result.

Data analysis

Diagnostic performance of the serologic assays was assessed against the gold standard of *M. bovis* culture and/or IS-6110 PCR by calculating test sensitivity and specificity using available software [19] and presented with the 95% confidence intervals (CI). Fisher's exact test was used for analysis of antigen recognition by bovine antibodies in the present study.

Results

Antibody detection in TST non-reactors versus TST reactors: Samples from Great Britain

With sera from *M. bovis* infected cattle in GB, the overall IgG reactivity rates in the DPP assay were similar for SICCT reactors (60%, *n* = 75) and non-reactors (65.4%, *n* = 52). The DPP assay specificity assessed with sera from 50 SICCT negative, IGRA negative cattle from TB non-endemic regions of GB was 96% (Table 1). Of note, the response rate by *M. bovis*-infected cattle to each fusion antigen differed significantly (*p* < 0.01, Fisher's exact test) based on TST reactivity. The ratio of MPB70/MPB83 to CFP10/ESAT-6 reactivity was approximately 3.6:1 for TST reactors versus 1:1 for TST non-reactors (Table 1). In line with the above, only 1 of 75 TST

reactors but 13 of 52 TST non-reactors produced antibody solely to CFP10/ESAT-6 yet not to MPB70/MPB83. Thus, the immunodominance of serologically related MPB70 and/or MPB83 proteins typically detected in *M. bovis* infected cattle after PPD administration for SICCT was less evident with TST non-reactors, as reactivity rates to CFP10/ESAT-6 and MPB70/MPB83 were essentially equivalent, and there were considerably more CFP10/ESAT-6 antibody responders within the TST non-reactor subset. This observation demonstrates the potential value for use of additional antigens to maximize the sensitivity of serologic tests, particularly with TST non-reactors.

Antibody detection in TST non-reactors: Opportunistic samples from the United States

During 2015 – 2016, two herds from disparate regions of the US were under-going whole herd depopulation due to *M. bovis* infection within the herds. Antemortem testing was used to delineate animals going to a slaughter plant (test negative) versus being condemned for rendering (test positive). Serum samples were available for serologic analysis from 11 CFT non-reactive adult cows with gross tuberculous lesions from the two herds [Texas (*n* = 7) and Michigan (*n* = 4)]. Infection was confirmed in these animals by histopathology, mycobacterial culture, and/or IS-6110 PCR (Table 2). Each of the animals had received a CFT ~6 months prior to the non-reactive CFT; of which 4/7 (Texas) and 0/4 (Michigan) were reactive at that earlier time point, thus indicating TST reversion in four of the animals. Serum samples collected ~1-3 weeks after the last CFT were tested with the DPP assay for the presence of IgM and IgG antibodies to CFP10/ESAT-6 and MPB70/MPB83. Only 4/11 animals had IgM to MPB70/MPB83, whereas none produced IgM to CFP10/ESAT-6. In contrast, relatively potent IgG responses were elicited by MPB70/MPB83 and CFP10/ESAT-6 in 8/11 and 3/11 animals, respectively (Table 3). IgM readouts were generally lower than those obtained for IgG, suggesting little added value of IgM antibody detection from a serodiagnostic sensitivity perspective. The one exception was with animal #364 which had a high IgM and a borderline

Table 1 IgG reactivity rates in tuberculin skin test reactors and non-reactors found among *M. bovis*-infected cattle in GB

Animal group[a]	No. of animals	Individual antigen reactivity rates in DPP assay[b]		DPP assay reactivity rate[b] (%, 95% CI)
		CFP10/ESAT-6	MPB70/MPB83	
M. bovis infected, TST Reactors	75	12 (16%)	43 (57.3%)	45 (60%; 95% CI: 48,71.2)
M. bovis infected, TST Non-Reactors	52	21 (40.4%)	21 (40.4%)	34 (65.4%, 95% CI: 50.9, 78)
Total *M. bovis* infected	127	33 (26%)	64 (50.4%)	79 (62.2%, 95% CI: 53.2, 70.6)
TB-free, TST Non-Reactors	50	2 (4%)	2 (4%)	2 (4%, 95% CI: 0.5, 13.7)

[a]Sera obtained from animals in multiple herds and of diverse age, gender, breed and management systems
[b]Data are presented as number (percent) positive per group for each antigen or assay

Table 2 Diagnostic characterization of TST false-negative cattle identified in US herds infected with *M. bovis*

State	Animal ID #	CFT[a] results		Postmortem examination results			
		Jul-2015	Jan-2016	Gross lesions	Histopathology	PCR[b]	Culture
TX	755	Pos	Neg	Present	TB compatible[c], IS6110 Pos[d]	Neg	Neg
	272	Pos	Neg	Present	TB compatible	Pos	ND[e]
	976	Neg	Neg	Present	TB compatible	Pos	ND
	857	Pos	Neg	Present	TB compatible	Pos	ND
	676	Pos	Neg	Present	TB compatible	Pos	ND
	889	Pos	Neg	Present	TB compatible	Pos	ND
	352	Neg	Neg	Present	TB compatible	Pos	ND
MI[f]		Sep-2015	Jan-2016				
	855	Neg	Neg	Present	TB compatible	ND	*M. bovis*
	370	Neg	Neg	Present	TB compatible	ND	*M. bovis*
	364	Neg	Neg	Present	Not compatible	ND	*M. bovis*
		May-2015	Oct-2015				
	809	Neg	Neg	Present	TB compatible	ND	*M. bovis*

[a]CFT, caudal fold test (i.e., intradermal *M. bovis* PPD injected into the caudal skin fold and response determined by palpation 72 h after injection) was applied <3 wks prior to collection of serum
[b]PCR, polymerase chain reaction for either *M. tb* complex IS6110 or 1081 DNA on fresh tissue
[c]Microscopic granulomatous lesions consistent with bovine TB and containing acid-fast bacilli
[d]IS6110 DNA by polymerase chain reaction on formalin-fixed tissue
[e]ND, not done
[f]Animals # 855, 370, and 364 were from a beef herd while #809 was from a dairy herd

IgG signal (Table 3). Development of transient IgM responses early in the course of experimental *M. bovis* infection, as well as shortly after TST administration, has been demonstrated in previous studies using different strains of *M. bovis* [11, 17, 20]. Thus, the present findings in the US demonstrated antibody responses to *M. bovis* antigens in ~82% of TST non-reactive cattle with confirmed TB. The specificity of

the DPP assay evaluated on 92 samples from TB-free states within the US was 98.9% (95% CI: 94.1, 99.9) or somewhat higher than that found in GB (96.0%; 95% CI: 86.3, 99.5), although not statistically significant (p = 0.3).

Antigen recognition by IgG antibodies in sera from TST non-responders

Serum samples from the US cattle were also analyzed by MAPIA to determine antigen recognition patterns. Sera from 10 of 11 cattle reacted with multiple recombinant antigens of *M. bovis* (Fig. 1). The one animal (#855) which had a negative result in the DPP assay, exhibited IgG binding only to MBCF in MAPIA. Animal #857, which was the second DPP non-reactor in this group (Table 3), displayed in MAPIA weak reactivity with the two fusion antigens containing MPB70. Based on line intensity, the magnitude of antibody responses and antigen recognition profiles varied among the animals. The most reactive antigens included MPB70 and MPB83 proteins, MPB70/MPB83 and MPB70/CFP10/Rv0934 hybrids, as well as bPPD and MBCF. For *M. bovis*-infected TST non-reactors, the role of CFP10/ESAT-6 in eliciting antibody responses was not significant in the US (Table 3, Fig. 1) as compared to the GB set of specimens (Table 1). Overall, the antigen recognition results supported the immunodominance of MPB70 and MPB83 proteins in *M. bovis* infection of cattle [8, 11, 17, 20].

Table 3 Quantitative measure of IgM and IgG responses produced by TST false-negative cattle in the US

Animal ID #	Reactivity in the DPP assay[a]			
	CFP10/ESAT-6		MPB70/MPB83	
	IgM	IgG	IgM	IgG
755	0	0.35	0	**6.15**
272	0	0	**1.15**	**1.08**
976	0	**1.50**	0.75	**7.45**
857	0	0	0.15	0
676	0	0	**1.48**	**4.85**
889	0	**2.05**	0	**19.15**
352	0	0.30	0	**5.00**
855	0	0	0	0
370	0	0	**3.10**	**14.95**
364	0	0.50	**6.78**	**1.58**
809	0	**14.45**	0.55	**6.23**

[a]Data are presented as signal-to-cutoff ratios with ≥1.00 considered as antibody reactive results (shown in bold)

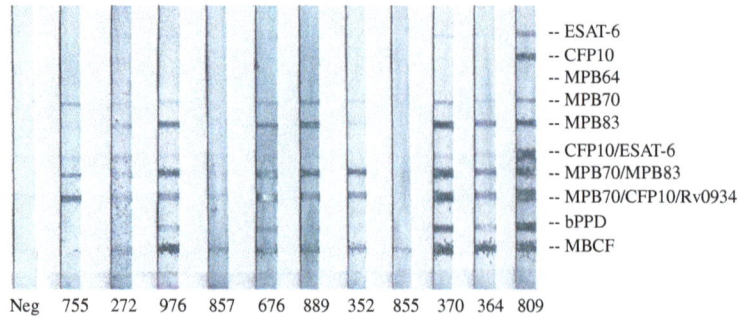

Fig. 1 MAPIA testing of TST non-reactive cattle diagnosed with *M. bovis* infection in the US. Assay was performed as described in Methods. Antigens printed onto nitrocellulose membrane are shown on the right. Results are presented for sera from a negative control (on the left), 7 animals from TX, and 4 animals from MI. Animal ID numbers are shown on the bottom (see Table 2 for diagnostic characterization). Visible bands on the strips indicate the presence of IgG to corresponding antigen(s). Intensity of the bands generally correlates with the antibody level

Discussion

A major impediment to the control of bovine TB is the relatively poor accuracy of current ante-mortem tests compounded by difficulties in reliably detecting tuberculous lesions and/or the agent in all infected animals upon slaughter surveillance. TST non-reactive cattle are particularly problematic when applying test and remove strategies in TB-affected herds. According to Lepper et al. [21], "Anergy to tuberculin is defined as the failure of an animal with visible evidence of tuberculosis to show a palpable cutaneous delayed hypersensitivity response to a tuberculin, at the time when the test is read." Such TST non-reactive animals, however, may still be responsive on other cell-mediated (e.g., IGRAs [22, 23]) or antibody detection immunoassays [24, 25]. In general, TST non-responsiveness is more common in animals from herds with high within-herd prevalence and in animals at advanced stages of disease [21, 22]. Desensitization as a result of repeated short interval application of TSTs may also lead to reduced TST responses associated with increased interleukin-10 (IL-10) and decreased IL-1β production to TB antigens [26]. Our findings further demonstrate that certain *M. bovis*-infected cattle can escape detection by TST by reverting from reactors to non-reactors within several months; however, a significant proportion of these animals can be identified by ancillary tests, such as the antibody tests described herein or IGRA [27].

In the present study, *M. bovis* specific antibody was detected by the DPP assay in ~82% and ~65% of *M. bovis*-infected, TST non-reactive cattle from the US and GB, respectively. As reported previously, an in-house MPB83-based ELISA detected 9 of 20 (45%) SICCT non-reactors diagnosed with bovine TB in GB [28]. Similarly, the MPB70/83 antibody reactivity rate in the group of TST non-reactive cattle in GB was ~40% as demonstrated by DPP assay (for MPB70/MPB83 antigen test line only, present study) and also shown independently

by a commercial ELISA (*M. bovis* Ab Test, IDEXX Laboratories, Westbrook, Maine) using a cocktail of MPB70 and MPB83 proteins (unpublished data). Importantly, integrating CFP10/ESAT-6 to supplement MPB70/MPB83 antigen in the DPP assay enhanced the overall test sensitivity to ~65% in TST non-reactive cattle in GB, thereby highlighting the added benefit of combining multiple antigens in serologic assays.

Conclusions

The present findings demonstrate the potential for use of antibody tests in TB-affected herds to rapidly identify *M. bovis*-infected, but TST non-reactive cattle. The DPP assay may also be considered for use in series with TST as a movement test, particularly as the assay may be applied pen-side without the need for laboratory equipment and results are available within 20 min. For example, Mexican cattle are required to have a negative CFT within 60 days of entry into the US. Thus, the DPP assay could be applied after CFT in Mexico or at the US/Mexico border as a further safeguard against entry of *M. bovis*-infected cattle into the US. For this application, given a very low disease prevalence in the large number of cattle crossing the US border with recent CFT negative results, a viable serologic assay would need to have an extremely high specificity (>99.9%) to provide an acceptable positive predictive value at a low pre-test probability. In the case of the DPP assay, the target specificity can be established and validated by having a cut-off value adjusted to meet this key requirement without a significant loss of diagnostic sensitivity. Given the limited number of samples available for the study, present findings should be considered preliminary and more extensive studies particularly in other bovine TB endemic countries are warranted to further verify the utility of this approach. These studies should also include sera from cattle infected with non-tuberculous mycobacteria (e.g., *M. avium* subsp. *paratuberculosis*, *M.*

kansasii, etc.) to determine the possible interaction of these mycobacteria on this approach, particularly as injection of PPDs for SICCT in cattle vaccinated with heat inactivated Johne's disease vaccine may induce false positive responses to MPB83/70-based antibody assays [29]. With that said, present findings clearly demonstrate that use of serologic assays in series with TST can identify a significant number of TST non-reactive tuberculous cattle for more efficient removal from TB-affected herds.

Abbreviations

CFP10: Culture filtrate protein 10 kDa protein; CFT: Caudal fold test; CI: Confidence interval; DPP: Dual path platform; ESAT-6: Early secretory antigenic target 6 kDa protein; GB: Great Britain; IGRA: Interferon-γ release assay; MAPIA: Multi-antigen print immunoassay; MPB70: Mobility protein of bovis, 70; MPB83: Mobility protein of bovis, 83; PPD: Purified protein derivative; RLU: Relative light units; SICCT: Single intradermal comparative cervical test; TB: Tuberculosis; TST: Tuberculin skin test; US: United States

Acknowledgements

We thank the Texas Animal Health Commission and Michigan Department of Agriculture for procurement of samples and providing supportive data. We thank Allison Lasley from USDA-ARS and Kent Munden from USDA-APHIS-VS for excellent technical support.

Funding

This work was partially supported by the USDA, National Institute of Food and Agriculture, Small Business Innovative Research [Award #2015-33610-23505, #2016-33610-25688], and the Department for Environment, Food and Rural Affairs (Defra) UK including funding from the devolved authorities. HMV is a Jenner Investigator.

Authors' contributions

WRW and KPL conceived the studies and wrote the initial draft of the manuscript. All authors reviewed and provided input into the crafting of the manuscript. JE, PL, RG, AG, ASG, and KPL performed all DPP and MAPIA assays. WRW, TC, MM, MP, MS, MC, JN, and JR organized collection and distribution of serum samples and supportive data from the US. SR, BK and HMV organized collection of serum samples and supportive data from the UK. BT performed microscopic analysis of tissues from samples collected in the US. DMG and SRA provided mycobacterial culture and PCR data. All authors read and approved the final manuscript.

Competing interests

USDA is an equal opportunity provider and employer. Mention of trade names or commercial products in this publication is solely for the purpose of providing specific information and does not imply recommendation or endorsement by the U.S. Department of Agriculture. JE, PL, RG, AG, ASG, and KPL are employees of Chembio Diagnostic Systems, Inc.

Consent for publication

All authors have read and approved the manuscript for submission and publication. Administrators from each agency (i.e., USDA-ARS, USDA-APHIS, APHA, DEFRA) and Chembio have approved the manuscript for submission and publication.

Author details

[1]National Animal Disease Center, Agricultural Research Service, United States Department of Agriculture (USDA), Ames, IA, USA. [2]Tuberculosis Research Group, Animal and Plant Health Agency, Addlestone, UK. [3]National Veterinary Services Laboratories, Animal and Plant Health Inspection Service (APHIS), USDA, Ames, IA, USA. [4]Veterinary Services (VS), APHIS, USDA, Fort Collins, CO, USA. [5]VS, APHIS, USDA, Raleigh, NC, USA. [6]VS, APHIS, USDA, East Lansing, MI, USA. [7]Chembio Diagnostic Systems, Inc., Medford, NY, USA.

References

1. Alexander KA, Sanderson CE, Larsen MH, Robbe-Austerman S, Williams MC, Palmer MV. Emerging tuberculosis pathogen hijacks social communication behavior in the group-living banded mongoose (*Mungos mungo*). MBio. 2016;7:e00281–16.
2. Clark C, van Helden P, Miller M, Parsons S. Animal-adapted members of the *Mycobacterium tuberculosis* complex endemic to southern African subregion. J South African Vet Assn. 2016;87:a11322.
3. Buddle BM, de Lisle GW, Waters WR, Vordermeier HM. Diagnosis of *Mycobacterium bovis* infection in cattle, vol. 9. In: Mukundan H, Chambers MA, Waters WR, Larsen MH, editors. Many Hosts of Mycobacteria: Tuberculosis, Leprosy, and other Mycobacterial Diseases of Man and Animals. Oxfordshire: CABI Publishing; 2015. p. 168–84.
4. de la Rua-Domenech R, Goodchild AT, Vordermeier HM, Hewinson RG, Christiansen KH, Clifton-Hadley RS. Ante mortem diagnosis of tuberculosis in cattle: a review of the tuberculin tests, gamma-interferon assay and other ancillary diagnostic techniques. Res Vet Sci. 2006;81:190–210.
5. Schiller I, Oesch B, Vordermeier HM, Palmer MV, Harris BN, Orloski KA, et al. Bovine tuberculosis: a review of current and emerging diagnostic techniques in view of their relevance for disease control and eradication. Transbound Emerg Dis. 2010;57:205–20.
6. Bezos J, Casal C, Romero B, Schroeder B, Hardegger R, Raeber AJ, et al. Current ante-mortem techniques for diagnosis of bovine tuberculosis. Res Vet Sci. 2014;97:S44–52.
7. Green LR, Jones CC, Sherwood AL, Garkavi IV, Cangelosi GA, Thacker TC, et al. Single-antigen serological testing for bovine tuberculosis. Clin Vaccine Immunol. 2009;16:1309–13.
8. Lyashchenko K, Whelan AO, Greenwald R, Pollock JM, Andersen P, Hewinson RG, et al. Association of tuberculin-boosted antibody responses with pathology and cell-mediated immunity in cattle vaccinated with *Mycobacterium bovis* BCG and infected with *M. bovis*. Infect Immun. 2004;72:2462–7.
9. Lyashchenko KP, Greenwald R, Esfandiari J, O'Brien DJ, Schmitt SM, Palmer MV, et al. Rapid detection of serum antibody by dual-path platform VetTB assay in white-tailed deer infected with *Mycobacterium bovis*. Clin Vaccine Immunol. 2013;20:907–11.
10. Waters WR, Buddle BM, Vordermeier HM, Gormley E, Palmer MV, Thacker TC, et al. Development and evaluation of an enzyme-linked immunosorbent assay for use in the detection of bovine tuberculosis in cattle. Clin Vaccine Immunol. 2011;18:1882–8.
11. Waters WR, Palmer MV, Thacker TC, Bannantine JP, Vordermeier HM, Hewinson RG, et al. Early antibody responses to experimental *Mycobacterium bovis* infection of cattle. Clin Vaccine Immunol. 2006;13:648–54.
12. Whelan C, Shuralev E, O'Keeffe G, Hyland P, Kwok HF, Snoddy P, et al. Multiplex immunoassay for serological diagnosis of *Mycobacterium bovis* infection in cattle. Clin Vaccine Immunol. 2008;15:1834–8.
13. Whelan C, Whelan AO, Shuralev E, Kwok HF, Hewinson G, Clarke J, et al. Performance of the Enferplex TB assay with cattle in great Britain and assessment of its suitability as a test to distinguish infected and vaccinated animals. Clin Vaccine Immunol. 2010;17:813–7.
14. Casal C, Díez-Guerrier A, Álvarez J, Rodriguez-Campos S, Mateos A, Linscott R, et al. Strategic use of serology for the diagnosis of bovine tuberculosis after intradermal skin testing. Vet Microbiol. 2014;170:342–51.
15. Lightbody KA, Skuce RA, Neill SD, Pollock JM. Mycobacterial antigen-specific antibody responses in bovine tuberculosis: an ELISA with potential to confirm disease status. Vet Rec. 1998;142:295–300.
16. Lightbody KA, McNair J, Neill SD, Pollock JM. IgG isotype antibody

responses to epitopes of the *Mycobacterium bovis* protein MPB70 in immunised and in tuberculin skin test-reactor cattle. Vet Microbiol. 2000;75:177–88.

17. Waters WR, Palmer MV, Stafne MR, Bass KE, Maggioli MF, Thacker TC, et al. Effects of serial skin testing with purified protein derivative on the level and quality of antibodies to complex and defined antigens in *Mycobacterium bovis*-infected cattle. Clin Vaccine Immunol. 2015;22:641–9.

18. Lyashchenko KP, Singh M, Colangeli R, Gennaro ML. A multi-antigen print immunoassay for the development of serological diagnosis of infectious diseases. J Immunol Methods. 2000;242:91–100.

19. Lowry R. VassarStats: Website for Statistical Computation. Clinical Research Calculators. http://vassarstats.net/. Accessed 5 June 2017.

20. Waters WR, Thacker TC, Nelson JT, DiCarlo DM, Maggioli MF, Greenwald R, et al. Virulence of two strains of *Mycobacterium bovis* in cattle following aerosol infection. J Comp Pathol. 2014;151:410–9.

21. Lepper AW, Pearson CW, Corner LA. Anergy to tuberculin in beef cattle. Aust Vet J. 1977;53:214–6.

22. Houlihan MG, Dixon FW, Page NA. Outbreak of bovine tuberculosis featuring anergy to the skin test, udder lesions and milkborne disease in young calves. Vet Rec. 2008;163:357–61.

23. Sinclair JA, Dawson KL, Buddle BM. The effectiveness of parallel gamma-interferon testing in New Zealand's bovine tuberculosis eradication programme. Prev Vet Med. 2016;127:94–9.

24. Lilenbaum W, Fonseca LS. The use of Elisa as a complementary tool for bovine tuberculosis control in Brazil. Braz J Vet Res Anim Sci. 2006;43:256–61.

25. Plackett P, Ripper J, Corner LA, Small K, de Witte K, Melville L, et al. An ELISA for the detection of anergic tuberculous cattle. Aust Vet J. 1989;66:15–9.

26. Coad M, Clifford D, Rhodes SG, Hewinson RG, Vordermeier HM, Whelan AO. Repeat tuberculin skin testing leads to desensitisation in naturally infected tuberculous cattle which is associated with elevated interleukin-10 and decreased interleukin-1 beta responses. Vet Res. 2010;41:14.

27. Wood PR, Jones SL. BOVIGAM: an in vitro cellular diagnostic test for bovine tuberculosis. Tuberculosis (Edinb). 2001;81:147–55.

28. Coad M, Downs SH, Durr PA, Clifton-Hadley RS, Hewinson RG, Vordermeier HM, et al. Blood-based assays to detect *Mycobacterium bovis*-infected cattle missed by tuberculin skin testing. Vet Rec. 2008;162:382–4.

29. Coad M, Clifford DJ, Vordermeier HM, Whelan AO. The consequences of vaccination with the Johne's disease vaccine, Gudair, on diagnosis of bovine tuberculosis. Vet Rec. 2013;172(10):266.

17

Assessing changing weather and the El Niño Southern Oscillation impacts on cattle rabies outbreaks and mortality in Costa Rica (1985–2016)

Sabine E. Hutter[1], Annemarie Käsbohrer[1]*[iD], Silvia Lucia Fallas González[2], Bernal León[3], Katharina Brugger[1], Mario Baldi[4], L. Mario Romero[5], Yan Gao[6] and Luis Fernando Chaves[7,8]

Abstract

Background: Rabies is a major zoonotic disease affecting humans, domestic and wildlife mammals. Cattle are the most important domestic animals impacted by rabies virus in the New World, leading to thousands of cattle deaths per year and eliciting large economic losses. In the New World, virus transmission in cattle is primarily associated with *Desmodus rotundus*, the common vampire bat. This study analyses the association of weather fluctuations and the El Niño Southern Oscillation (ENSO), with the occurrence and magnitude, in terms of associated mortality, of cattle rabies outbreaks. Data from the 100 cattle rabies outbreaks recorded between 1985 and 2016 in Costa Rica were analyzed. Periodograms for time series of rabies outbreaks and the El Niño 4 index were estimated. Seasonality was studied using a seasonal boxplot. The association between epidemiological and climatic time series was studied via cross wavelet coherence analysis. Retrospective space-time scan cluster analyses were also performed. Finally, seasonal autoregressive time series models were fitted to study linear associations between monthly number of outbreaks, monthly mortality rates and the El Niño 4 index, temperature, and rainfall.

Results: Large rabies mortality occurred towards the Atlantic basin of the country. Outbreak occurrence and size were not directly associated with ENSO, but were sensitive to weather variables impacted by ENSO. Both, ENSO phases and rabies outbreaks, showed a similar 5 year period in their oscillations. Cattle rabies mortality and outbreak occurrence increased with temperature, whereas outbreak occurrence decreased with rainfall. These results suggest that special weather conditions might favor the occurrence of cattle rabies outbreaks.

Conclusions: Further efforts are necessary to articulate the mechanisms underpinning the association between weather changes and cattle rabies outbreaks. One hypothesis is that exacerbation of cattle rabies outbreaks might be mediated by impacts of weather conditions on common vampire bat movement and access to food resources on its natural habitats. Further eco-epidemiological field studies could help to understand rabies virus transmission ecology, and to propose sound interventions to control this major veterinary public health problem.

Keywords: *Bos taurus*, Costa Rica, Rhabdovirus, Climate change, Wavelets

* Correspondence: annemarie.kaesbohrer@vetmeduni.ac.at
[1]Institute of Veterinary Public Health, Department for Farm Animals and
Veterinary Public Health, University of Veterinary Medicine, Veterinärplatz 1,
1210 Wien, Austria
Full list of author information is available at the end of the article

Background

Rabies is a major zoonotic disease worldwide, with a high human and domestic animal death toll. Rabies virus is a negative-sense single stranded ribonucleic acid (RNA) virus belonging to the *Lyssavirus* genus in the Rhabdoviridae family [1]. Recently, it was estimated that globally canine rabies causes approximately 59,000 human deaths [2]. Furthermore, over 99% of all virus transmission to humans comes from dogs, the rest mainly coming from bats [3].

The problem of humans being at risk of rabies virus transmission has existed in the New World for centuries [4, 5]. Nowadays, due to an elimination programme for dog transmitted human rabies initiated in 1983 [6, 7], the main mode of zoonotic rabies transmission in the New World is via bites from infected common vampire bats to other mammal species, including humans and livestock [3]. This was also the case in Costa Rica, where dog rabies has not been a problem for decades [8].

The common vampire bat, *Desmodus (D.) rotundus*, is the major reservoir and vector of rabies virus in the New World [9–14]. *D. rotundus* bloodfeeds on mammals and transmits rabies virus, with livestock animals being greatly affected by this lethal disease [15]. In fact, common vampire bat bites and rabies transmission are limiting factors for livestock, mainly cattle production, as shown in an economic evaluation for Mexico [6, 11]. In 1968 it was estimated that over 500,000 cattle died from bat-transmitted rabies in Latin America [16]. Following the establishment of bat control methods and cattle rabies vaccination campaigns, rabies virus associated cattle mortality has plummeted [6, 16]. By 1983, rabies was responsible for 9904 cattle deaths, for 1831 in 1993 and for 1580 in 2006 in Latin America [6, 16]. Nevertheless, in Latin America, the decline has not been always monotonic, with surges in rabies virus associated cattle mortality, like in 2000, when 6088 cattle heads died by rabies, and in 2002, when cattle deaths totaled 3327 animals [13, 16]. The economic loss associated with cattle deaths by rabies has also been significant all over Latin America, adding to an estimate of 100 million US dollars in 1966 [17] and between 44 and 50 million US dollars by 1986 [18].

The epidemiology of numerous cattle rabies outbreaks has been studied across Latin America [4, 7, 9], in places so far apart like northeastern Mexico [19] and southern Brazil [20]. In Costa Rica, between 1985 and 2014, over 70 cattle rabies outbreaks were reported, including 723 cattle deaths, placing a considerable burden on the local livestock industry [8]. All of these outbreaks have been linked by the unequivocal presence of common vampire bats, and this might be enhanced by a couple of factors. The first factor is that common vampire bat dispersal is directly linked with a preference for bovine meals, probably because fenced-in cattle are a predictable resource when compared with free-ranging natural hosts [21].

This situation generates perfect conditions for transmission in domestic animals not observed in wildlife species or in free range grazing livestock herds [22]. The second factor is that common vampire bats are organisms sensitive to weather changes, despite having an endothermic nature. For example, it is known that temperature influences the geographic distribution of the common vampire bat [23]. Common vampire bats tend to be limited to low elevations in the tropics, mainly because of food availability, but also because of limitations on the maximum meal weight they can carry in flight, and the low energetic cost for keeping common vampire bat body temperature constant [23]. Common vampire bats move between caves depending on the presence of roosts and climatic conditions [24]. For example, high humidity seems essential for common vampire bat roosting, in dry areas common vampire bats roost in water wells and leave the wells as they become dry, concentrating in wells with water during the dry season, and dispersing to other wells as they get filled with water during the rainy season [25]. Moreover, foraging time [26] and roost size [24] of common vampire bats seem to follow seasonal patterns. For example, common vampire bats spend more time attached to their prey during the dry season than they do during the wet season, likely ingesting more blood during the dry season [26]. Although controversial [27, 28], the birth of common vampire bats seems to be seasonal, despite the fixed 205 day gestation period recorded for vampires. However, when vampire births occur during the rainy season, rabies transmission could be enhanced via increased bat movement [25]. Nevertheless, little attention has been given to the potential impacts that anomalous weather conditions could have on the likelihood of cattle rabies outbreaks, despite abundant observations have shown that common vampire bat biology, and rabies transmission, could be sensitive to environmental changes.

ENSO can lead to extreme climatic conditions, such as severe droughts and floods, and more generally can alter weather patterns in a manner that creates ideal ecological conditions for disease transmission [29–31]. In Central America, several studies have found a differential impact of ENSO on the incidence of malaria [32, 33] and leishmaniasis [34], making this region ideal to test hypotheses about differential ENSO impacts on disease transmission. To the best of our knowledge, only one study has analysed rabies cases in relation to ENSO, where 416 human rabies cases mainly associated to dogs were recorded in Venezuela during 2002–2004 [35]. This study showed an increased human rabies incidence during the cold ENSO phase (La Niña in 2004) when compared to the hot ENSO phase (El Niño in 2002) [35], a suggestive observation despite data limitations for statistical inferences. Thus, provided that global climate change, here defined as the increase in frequency and intensity of extreme weather events driven by phenomena like ENSO, is known to affect livestock infectious diseases [36], such as Rift Valley Fever, and many other infectious diseases [37, 38] and the existence of detailed records about cattle rabies outbreaks in Costa Rica [8] creates an ideal scenario to test the hypothesis that ENSO phases might have differential impacts on the occurrence and size of cattle rabies outbreaks. Here, it is studied if the frequency of rabies outbreaks and cattle mortality per outbreak are associated with ENSO driven changes in weather patterns in Costa Rica between 1985 and 2016.

Methods

Study area

Costa Rica is located in Central America, bordering Nicaragua to the North, Panama to the Southeast, the Pacific Ocean to the West, and the Caribbean Sea (Atlantic Ocean) to the East. This small country of 51,100 km^2 has a tropical climate. According to the last agricultural census, in 2014, there were around 37,000 cattle farms with 1,300,000 cattle heads in Costa Rica [39]. The number of cattle heads has decreased 37.5% since the last census in 1984. Several factors have driven this herd size decline, including a cattle crisis in the 1980s, mainly driven by a price reduction for meat [40]. In addition, the creation of national parks and biological reserves since the 1970s, in accordance with Costa Rica's land use development strategy, promoted land use change from cattle ranching to biodiversity conservation areas [41, 42].

Epidemiological data

Data on cattle rabies cases from 1985 to 2016 from Servicio Nacional de Salud Animal (SENASA), the Costa Rican Veterinary Authority were used. Their database includes the data of all registered outbreaks, the number and species of dead animals attributed to rabies, and the georeferenced location of the outbreak. Meanwhile, the geolocation of outbreaks was variable, as in some instances it was based on the centroid of the county where the outbreak happened, while in other instances it was based on the coordinates of the cattle ranch itself, measured with a Global Positioning System (GPS) [8].

Detailed information on rabies diagnostic testing was available from the laboratory database [8]. During the study period diagnostic tools became more robust by the inclusion of more sensitive and specific methods. Starting in 1983 with the detection of Negri bodies in brain samples processed with Sellers staining, the diagnostic method changed to mice inoculation and immune fluorescence testing in 1993. The latter method was enhanced by the addition of several deoxyribonucleic acid (DNA) based molecular techniques in 2013, described in more detail elsewhere [8].

In this study, an outbreak was defined as the occurrence of at least one reported confirmed rabies virus cattle death at one farm. For the analysis, three time series were constructed using information from the 100 rabies outbreaks that were reported in cattle herds of Costa Rica between January 1985 and December 2016.

The first time series was a monthly record of cattle deaths due to rabies virus infection. The second time series recorded the monthly number of outbreaks, while the third series had the monthly mortality by rabies virus infection. In this last time series, mortality was defined as the percent of deaths, due to rabies virus infection, in the farm where the outbreak was detected, i.e., for each farm was estimated as follows: (number of recorded deaths due to rabies * 100) / number of cattle heads present in the farm. When two or more outbreaks occurred in the same month, mortality was estimated by adding deaths and total cattle heads across all affected farms.

For 33 of the outbreaks, all before 2000, the number of cattle heads in farms where deaths by rabies were recorded needed to be imputed. This was done because spatially explicit data about cattle heads in Costa Rica are only available after 2013, the time when the agricultural holdings database (Sistema de Reconocimentos Ambientales, SIREA) was established [8, 43], and also to consider that cattle density has been reduced in the country. Therefore, under the valid assumption that neighbouring farms have similar herd sizes [39], a fully cross-validated regression tree was employed to estimate the missing herd sizes as function of farm coordinates and altitude. Briefly, the regression tree is a quantitative method where a set of rules are derived by looking at patterns of association between the independent variables and a response variable, having the ability to capture non-linear relationships between variables when compared with linear regression [44]. Full cross-validation means that parameters are estimated by fitting a set of models where, one at a time, each observation is left out of the model fitting and then that observation is predicted with parameters estimated from all other observations so that parameters are chosen by minimizing the difference between all predicted values and observations [45].

Meteorological data

For the analysis, the definition of cold and hot ENSO phases from the National Oceanic and Atmospheric Administration (NOAA) was used [46]. This definition is based on the Oceanic Niño Index, the 3 month running average of temperature anomalies in the Niño 3.4 region (5°N-5°S, 120°-170°W), and where anomalies are based on 30 year base periods, which are updated every 7 years. The analyses employed monthly anomalies from the El Niño 4 index [47] which are based on measurements from the following area: 5°N-5°S, 160°E-150°W. This selection was based on previous research [32, 34, 49, 50] that has shown this index has the highest association with weather anomalies in Central America. For temperature and rainfall, gridded data, available at the Royal Netherlands Meteorological Institute (KNMI) climate explorer [51], were employed. Data were downloaded for the land surface contained in the area enclosed by the following coordinates: 11.00°N-8.55°N, 86.00°W-83.00°W, which roughly corresponds to Costa Rica and contains all the recorded outbreaks between 1985 and 2016. More specifically, data from the following two databases at KNMI were used for: (i) temperature: NOAA data from the Global Historical Climatology Network version 2 and the Climate Anomaly Monitoring System (GHCN_CAMS 2 m model), with a spatial resolution of 0.5° [52]; (ii) rainfall: data from the NOAA Climate Data Record (CDR) of Satellite-Gauge Precipitation from the Global Precipitation Climatology Project (GPCP), V2.3, with a spatial resolution of 2.5° [53].

Statistical analysis

Cattle rabies outbreak and mortality cycles were studied with frequency domain time series methods, which are tools to study the cyclic behaviour of time series [54]. Periodograms, which show the distribution of power (i.e., variance) among different frequencies, i.e., the inverse of the period for a given cycle, were estimated by taking the Fourier transform of the time series. In a periodogram, a peak indicates a dominant frequency in the cyclic behaviour of a time series [54]. Periodograms were estimated for the annual time series rabies outbreaks and the El Niño 4 index, assuming that these two time series were stationary, i.e., with a constant mean and variance [54]. The seasonality of the epidemiological and meteorological data was assessed using a seasonal boxplot, i.e., a graph where boxplots for the 12 months in the year are sequentially drawn [54]. Meanwhile, the association between epidemiological and climatic time series was studied using cross wavelet coherence analysis. This is a time frequency time series analysis technique that can assess the non-stationary, i.e., changing through time, association between two time series. In this analysis, the association between cycles of different periods between two time series is studied through time, showing periods of time when the cycles are significantly coupled or not [48–50, 54–56].

Retrospective space-time cluster analyses were performed with the scan statistic [57]. Discrete poisson models [58] were implemented for the study period 1985–2016. For the analyses elliptic scanning windows were used. The elliptic window is the preferred choice when looking for clusters over a space that is asymmetric, i.e., like the map of Costa Rica where land occurs along a southeast to northwest axis, and the elliptic window has the advantage of converging to a circle when there is a cluster around a focal point, or an ellipse when clusters follow a pattern similar to a line [59].

The scan statistic was employed to search for high rates. Following the suggestion by Kulldorf et al. [57], the scan statistic was estimated with a scanning window whose maximum radius for the long axis of the ellipse generated an area covering half of the points where cattle rabies outbreaks were located. The number of reported rabies deaths and the population at risk for each outbreak were used in order to detect whether outbreaks were purely at random, or if clusters of high mortality existed. Four models were estimated: two models were set up with an annual aggregation of cases, and two with a monthly aggregation of cases. For the models of annual and monthly data aggregation, one model considered observations including the imputed values for the estimated populations with the regression trees, and one included only information for outbreaks where the population at risk had also been registered.

Finally, seasonal autoregressive (SAR) time series models were developed to study the linear association between the monthly number of outbreaks, the monthly mortality rates and the El Niño 4 index, temperature, and rainfall. The protocol to develop SAR models started by fitting "null models" that accounted for periodicities of the focal time series (outbreaks or mortality) by appropriately considering the correlation structure of the time series as inferred from the inspection of the auto-correlation function (ACF) and partial auto-correlation function (PACF) of the time series. The output of ACF is a plot that depicts the correlation of a time series with itself at different time lags, while the PACF output depicts a similar correlation but only considering consecutive time lags [54]. The resulting "null models" were then used to pre-whiten, also known as filtering, the time series of the climatic covariates. This process removes any common auto-correlation structure from the climatic time series; thus, when a cross-correlation function (CCF) is estimated from the residuals of the null model and the pre-whiten time series, no spurious association — emerging from the time series having a similar auto-correlation structure — is found. The CCF is a graphical representation of the correlation between two time series at different lags [54]. Following the inspection of the cross-correlation function, we built "full models" that considered all covariates at the significant lags ($p < 0.05$) with the highest correlations. These full models were then simplified by a process of backward elimination [60], where the model with the lowest Akaike Information Criterion (AIC) from a set of models with the same number of parameter was further simplified until a minimum AIC was reached [50]. The AIC is a trade-off function that considers the number of parameters and the likelihood of a model, allowing the selection of models based on AIC minimization [50]. To ease model interpretation, the mean of climatic time series were removed when fitting the SAR time series models [49], something that equals intercepts to the average value of the focal time series studied [54].

Software

All geographic information systems procedures, mapping, regression trees, cross-wavelet analyses and the time series models were done with the statistical software R version 3.4.0. The scan cluster analyses were performed with SaTScan [57].

Results

During the study period (1985–2016), there were 9 El Niño events, the two biggest events, i.e., those accounting for the most extreme ENSO fluctuations, occurring in 1998 and 2015 (Fig. 1a). During this time period, corresponding to the peak power value of approximately 0.2 cycles per year, ENSO had an approximate oscillation period of 5 years, as suggested by a periodogram of the monthly anomalies from the El Niño 4 index (Fig. 1b). During the same time period, rabies outbreaks in cattle occurred all over Costa Rica (Fig. 2a). As shown in Fig. 2a, most of the outbreaks had between 1 and 5 rabies deaths, but there were two large outbreaks of 139 and 194 deaths in 1985 and 2003, respectively (see Additional file 1: Figure S1A), which occurred in the Atlantic basin of the country. Figure 2a shows that outbreaks were very common between 1986 and 1991, 2001 and 2005 as well as between 2011 and 2015. Figure 2a also shows that most outbreaks occurred in the lowlands of Costa Rica. Additional file 2: Figure S2 is a video showing the month, year and size of each individual outbreak and ENSO phase. Further inspection of cattle rabies death records suggests that most deaths have occurred in April and July (Additional file 1: Figure S1B), which have been associated with the large outbreaks of 1985 and 2003 (Additional file 1: Figure S1A and C). Nevertheless, there is no seasonality in the number of rabies deaths recorded, based on the

median values (zero) of the boxplots for cattle mortality in the whole territory of Costa Rica (Additional file 1: Figure S1D). Indeed, it seems that rabies deaths randomly happen throughout the year.

Figure 2b shows results from the annual scan spatio-temporal cluster analysis about rabies mortality in cattle. Two datasets were considered, a dataset for which denominators, i.e., cattle herd size were all known, and a dataset for which some denominators were imputed. In the two analyses, clusters of cattle rabies deaths were identified in northeastern Costa Rica, in the Atlantic basin. There was an important spatial overlap between the only cluster identified with the dataset without imputed denominators (which only spanned 2003) and Cluster 1 from the analysis with imputed denominators (which spanned from 1985 to 1999). Cluster 2 coincided temporally (from 2003 to 2015) and to a lesser extent spatially with the cluster from the analysis without imputed data. Similar results were observed when the cluster analysis was performed for monthly data of cattle rabies deaths (Additional file 3: Figure S3), where the 2003 outbreak of 194 dead cows was identified as an important cluster independently of considering the imputed denominators, and clusters were also mainly located in the Atlantic basin of Costa Rica. The calculated relative risks, where relative risk is an adimensional proportion defined as the ratio between the probability of cattle death by rabies happening inside the spatial cluster when compared with the probability of such event outside the cluster area, reflect the increased risks for cattle deaths by rabies virus infections in these identified areas when compared with the surrounding landscape.

The incidence of outbreaks, independently of the number of rabies deaths, was variable across the study period, with one to nine outbreaks per year

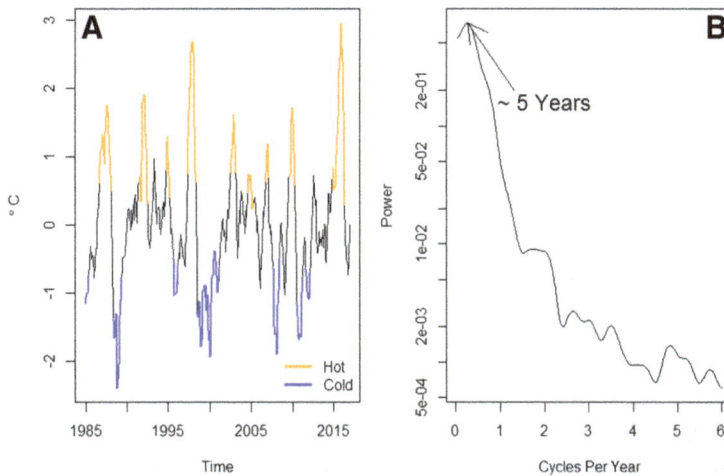

Fig. 1 The El Niño Southern Oscillation (ENSO) dynamics (1985–2016). **a** Monthly time series of the El Niño 4 index from 1985 to 2016. ENSO events are highlighted; hot phases are shown in orange, cold phases in blue. **b** Periodogram for the monthly anomalies from the El Niño 4 index. The peak power value corresponds at 0.2 cycles per year, indicating that ENSO had an approximate oscillation period of 5 years over the study period

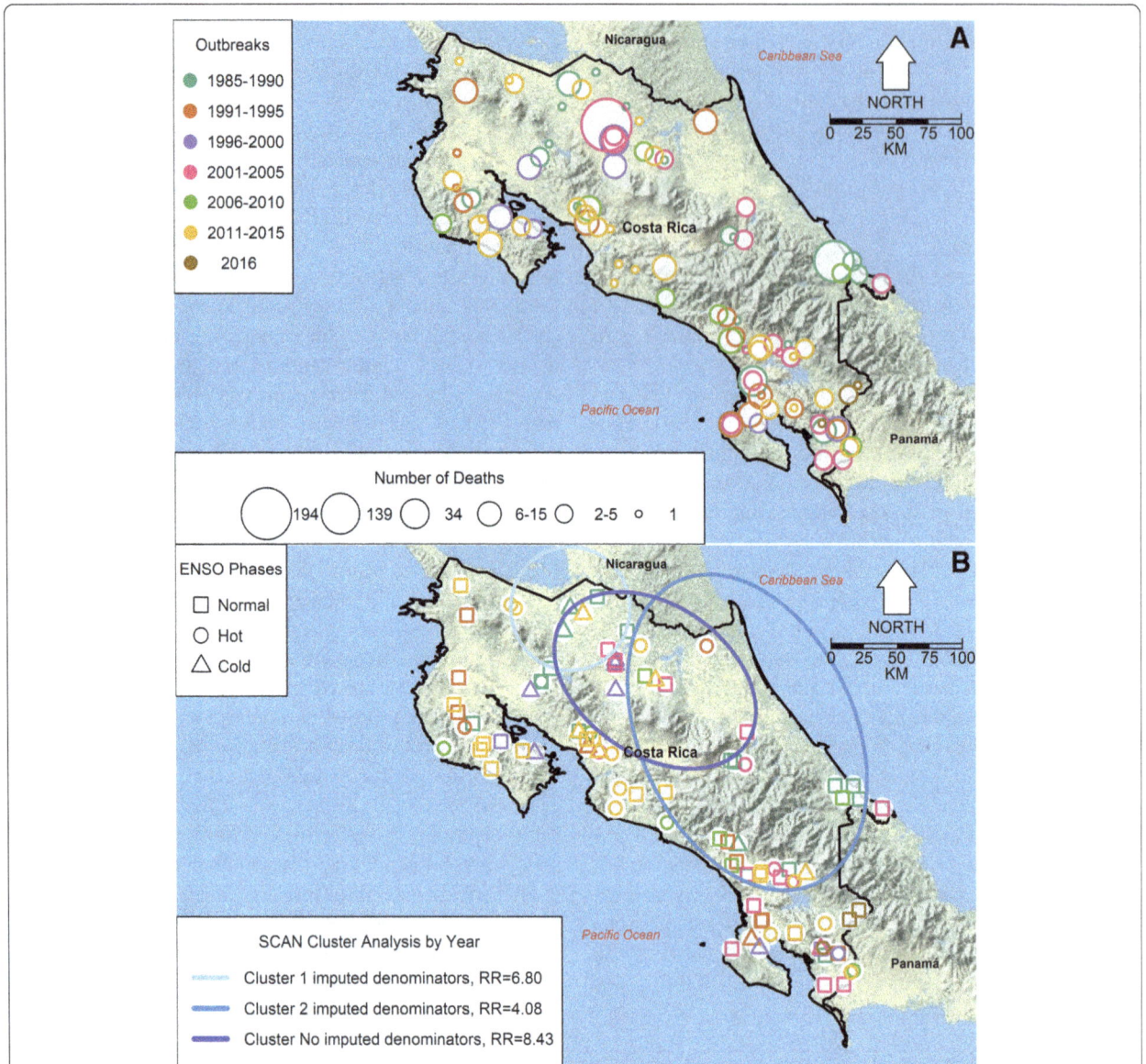

Fig. 2 Geolocation of cattle rabies outbreaks in Costa Rica (1985–2016). **a** Map of Costa Rica with the period and size, i.e., number of rabies deaths, of cattle rabies outbreaks. Most outbreaks occurred between 2001 and 2005 and 2011–2015. Numbers of rabies deaths were grouped in 6 categories and most of the outbreaks had between 1 to 5 deaths. **b** Map of Costa Rica with the El Niño Southern Oscillation (ENSO) phase in which cattle rabies outbreaks occurred (square-normal, circle-hot and triangle-cold). Disease clusters from the annual SaTScan spatio-temporal analysis are indicated by circles. Clusters were estimated with and without imputed denominators. These maps were made using as base a public domain map from the US National Park Service [86]

occurring during the studied years, and with cyclical fadeouts every few years up to 2002 (Fig. 3a).

There is a periodicity of approximately 5 years in the number of annual outbreaks as suggested by the annual outbreak number time series periodogram (Fig. 3b), the same oscillation period observed for ENSO. Outbreaks occurred more frequently in the months of August, September and May (Fig. 3c), and this seems to be an incipient seasonal pattern, yet at any given month no outbreaks regularly happened (Fig. 3d).

The lack of seasonality in cattle rabies deaths, and outbreaks is in sharp contrast with the seasonal weather of Costa Rica. Temperature (Fig. 4a) and rainfall (Fig. 4b) show a clear unimodal pattern, with two seasons. Temperature reaches a maximum in April and is lowest in December (Fig. 4a). For rainfall, there is a dry season between December and April, followed by a wet season the rest of the year (Fig. 4b). Nevertheless, mortality associated with cattle rabies outbreaks (Fig. 4c), although proportionally higher in May, August and September in specific years, is not regularly higher in any given month of the year.

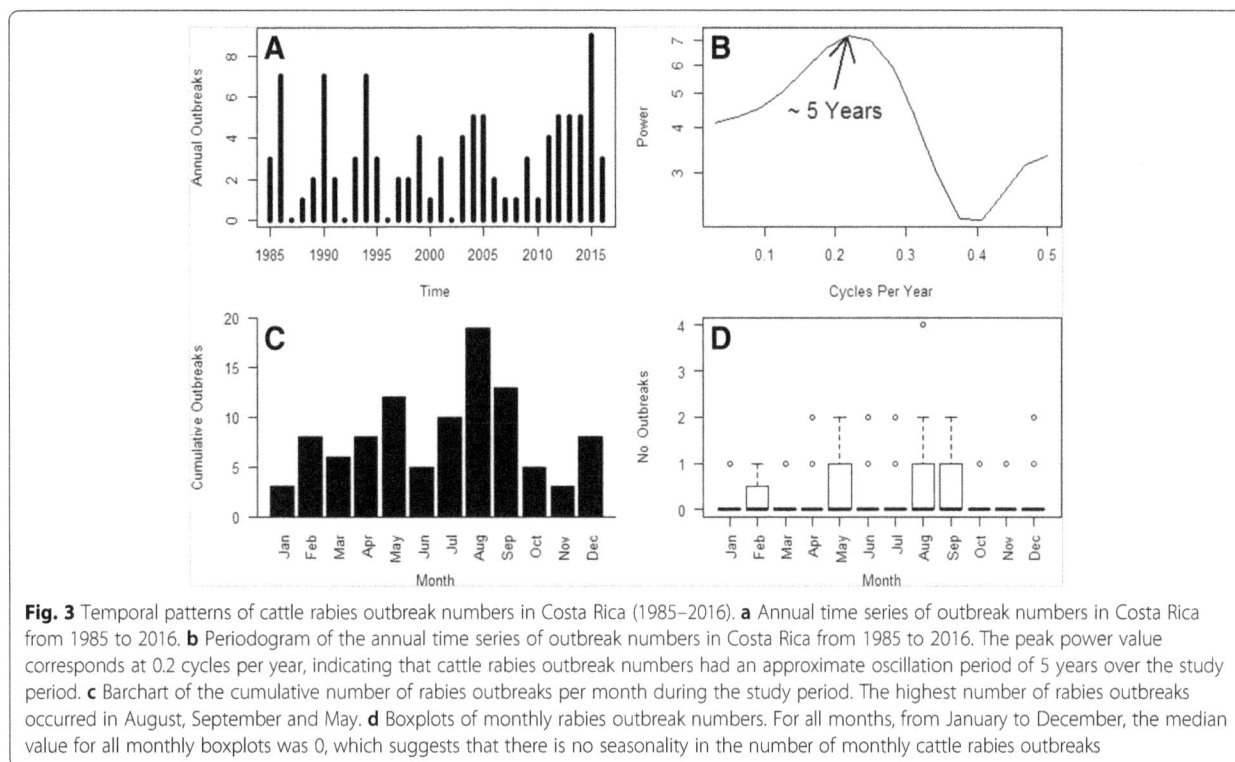

Fig. 3 Temporal patterns of cattle rabies outbreak numbers in Costa Rica (1985–2016). **a** Annual time series of outbreak numbers in Costa Rica from 1985 to 2016. **b** Periodogram of the annual time series of outbreak numbers in Costa Rica from 1985 to 2016. The peak power value corresponds at 0.2 cycles per year, indicating that cattle rabies outbreak numbers had an approximate oscillation period of 5 years over the study period. **c** Barchart of the cumulative number of rabies outbreaks per month during the study period. The highest number of rabies outbreaks occurred in August, September and May. **d** Boxplots of monthly rabies outbreak numbers. For all months, from January to December, the median value for all monthly boxplots was 0, which suggests that there is no seasonality in the number of monthly cattle rabies outbreaks

The signature of ENSO in temperature (Fig. 5a) suggests that the hot ENSO phase is associated with higher than average temperatures in Costa Rica. Meanwhile, the cold phase of ENSO is associated with increased rainfall in Costa Rica (Fig. 5b). After 1998, outbreaks of cattle rabies seemed to be more prone to occur during the hot and cold phases of ENSO (Fig. 5c), and some cattle rabies outbreaks with high mortality rates occurred during the cold phase of ENSO (Fig. 5d).

The cross wavelet coherence analysis suggests that interannual cycles of temperature and ENSO are highly coherent at periods of 2–5 years over the studied period (Fig. 6a). For rainfall (Fig. 6b), cycles of 3–6 years of ENSO and rainfall were highly coherent around the 1998 ENSO event, and have become increasingly coherent for the same periods (2–4 years) since 2006. Cattle rabies outbreaks have been coherent at larger time scales, 8–10 years, with ENSO (Fig. 6c),

and cycles have become increasingly associated at scales of 2–4 years after 2010. Monthly cattle rabies mortality shows no major significant association with ENSO during the studied period (Fig. 6d). The cross wavelet coherence analysis suggests that number of outbreaks (Fig. 6e) and cattle rabies mortality (Fig. 6f) between 1998 and 2005, for interannual cycles of periods of 4–6 years, were highly associated with temperature. Moreover, for outbreak numbers, the association was also significant after 2010 for cycles of periods of 2–4 years (Fig. 6e). Rainfall was associated with number of outbreaks (Fig. 6g) for cycles of periods of 4–6 years from 2006 to 2008. Meanwhile, there was no major significant association between cycles in rainfall and mortality (Fig. 6h) during the study period.

The number of outbreaks time series was not strongly autocorrelated, i.e., there was no clear pattern of significantly ($p < 0.05$) decreasing auto-correlation with time

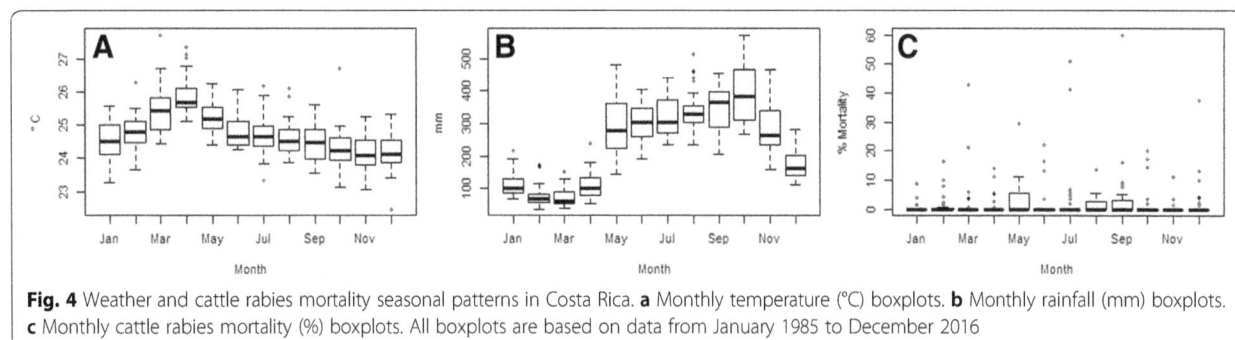

Fig. 4 Weather and cattle rabies mortality seasonal patterns in Costa Rica. **a** Monthly temperature (°C) boxplots. **b** Monthly rainfall (mm) boxplots. **c** Monthly cattle rabies mortality (%) boxplots. All boxplots are based on data from January 1985 to December 2016

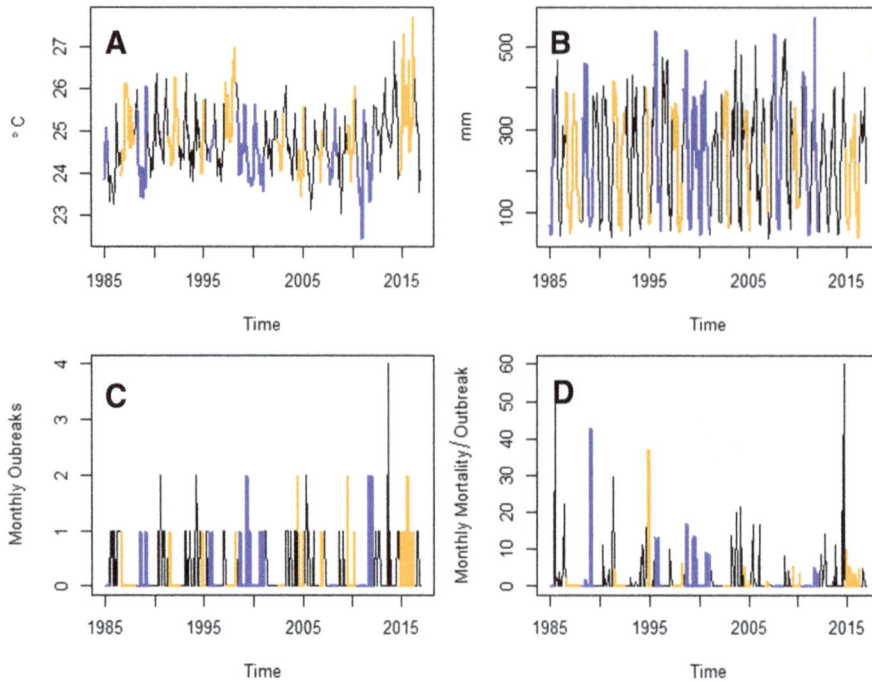

Fig. 5 The El Niño Southern Oscillation (ENSO) events and monthly weather and cattle rabies time series (1985–2016). **a** Temperature (°C). **b** Rainfall (mm). **c** Cattle rabies outbreak numbers. **d** Cattle rabies mortality (%). ENSO events are highlighted, hot phases are shown in orange, cold phases in blue, and normal months in black

lag (Additional file 4: Figure S4A), and showed periodicities every 2 months (Additional file 4: Figure S4B), which was considered when fitting a null SAR time series model (Additional file 5: Table S1). Moreover this time series had no significant linear association with ENSO 4 over the studied period and up to 22 months of lag (Additional file 4: Figure S4C). A similar pattern of low auto-correlation was observed for the outbreak mortality time series (Additional file 4: Figure S4D), which showed periodicities of 2 months (Additional file 4: Figure S4E), also considered when fitting a null SAR time series model (Additional file 6: Table S2). The mortality time series also had no significant association with ENSO 4 up to 22 months of lag (Additional file 4: Figure S4F). When considering covariates to develop a full model for the number of outbreaks time series (Additional file 5: Table S1), lags 4 and 17 of temperature were considered (Fig. 7a) and lags 6, 11 and 16 of rainfall (Fig. 7b). Meanwhile, the full model for the mortality rate only considered lag 17 of temperature as covariate (Fig. 7c), since no significant ($p < 0.05$) association was found with rainfall (Fig. 7d).

The best time series model for the number of outbreaks (Table 1) had temperature with 17 months of lag and rainfall with 16 months of lag as covariate, having a positive coefficient for temperature and a negative one for rainfall. For cattle rabies mortality, the best model considered temperature with 17 months of lag as covariate (Table 2)

with a positive coefficient. These results indicate that outbreaks and mortality are more likely to increase following hot-spells. Outbreaks are also less prone if the hot spell is followed by rain in an amount that offsets the positive impact of hot temperatures from the previous time lag (Table 1).

Discussion

The similar period, around 5 years, of interannual oscillations in ENSO and cattle rabies outbreaks is suggestive of an association between changes in the occurrence of rabies outbreaks and ENSO phases. This type of association has been widely found for several diseases, e.g., malaria [32], dengue [61], cutaneous leishmaniasis [34, 49] in Central America, but also elsewhere, e.g. Asia [62], Africa [29, 63, 64] and Oceania [65]. Although this formal analysis did not find signatures of ENSO on outbreak occurrence and cattle mortality dynamics, these two time series were associated with temperature and rainfall (outbreak occurrence only). As further shown by the analysis, the dynamics of temperature and rainfall in the studied area had interannual cycles associated with ENSO, thus implying the observed dynamics might be sensitive to ENSO. In the studied data, no seasonality in the number of monthly rabies deaths and recorded outbreaks was observed.

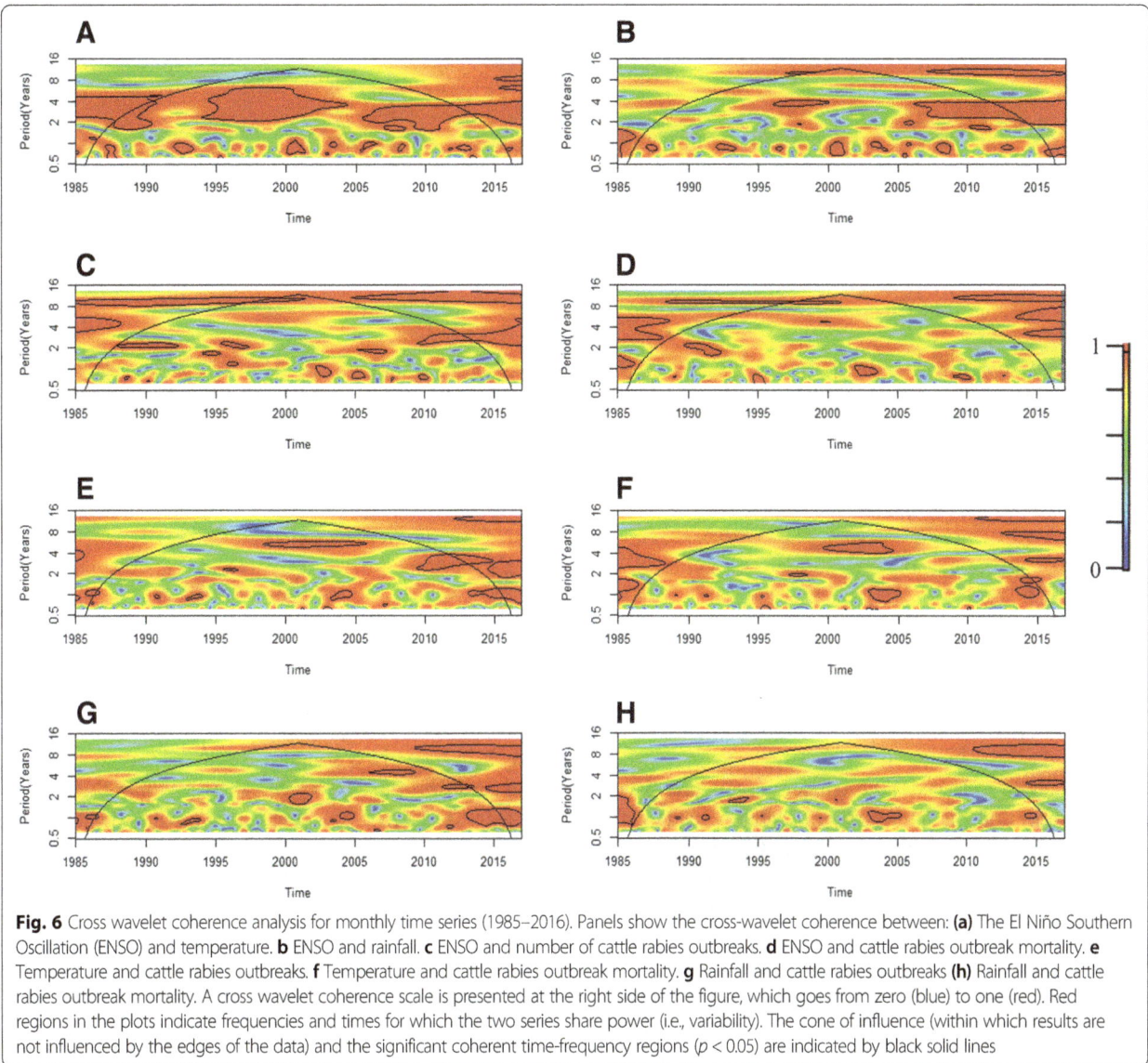

Fig. 6 Cross wavelet coherence analysis for monthly time series (1985–2016). Panels show the cross-wavelet coherence between: (**a**) The El Niño Southern Oscillation (ENSO) and temperature. **b** ENSO and rainfall. **c** ENSO and number of cattle rabies outbreaks. **d** ENSO and cattle rabies outbreak mortality. **e** Temperature and cattle rabies outbreaks. **f** Temperature and cattle rabies outbreak mortality. **g** Rainfall and cattle rabies outbreaks (**h**) Rainfall and cattle rabies outbreak mortality. A cross wavelet coherence scale is presented at the right side of the figure, which goes from zero (blue) to one (red). Red regions in the plots indicate frequencies and times for which the two series share power (i.e., variability). The cone of influence (within which results are not influenced by the edges of the data) and the significant coherent time-frequency regions (p < 0.05) are indicated by black solid lines

Clusters identified with the scan statistic were mainly located in the Atlantic basin of Costa Rica, something suggested in a preceding description of most of the studied dataset [8]. The largest outbreaks and high mortality clusters occurred in the Atlantic basin of Costa Rica, a fact that might be related with common vampire bat ease of movement or dispersal, in an area slightly more humid [25] than the Pacific basin of Costa Rica [66]. Similarly, cattle rabies geolocation patterns might reflect common vampire bat abundance patterns, which are known to be associated with specific environments, defined by the niche of bats [13, 67, 68], and where vampire abundance could increase rabies virus transmission [14, 69]. An additional reason could be that surveillance and prevention might be less effective in the Atlantic basin of Costa Rica. For example, underreporting could delay appropriate action and promote a reduced

coverage of cattle rabies vaccination, which in turn might lead to larger outbreaks, given an increased susceptibility in host populations without rabies vaccination, as predicted by mathematical models dealing with risk perception and infectious disease transmission prevention [70]. Actually, the large 1985 outbreak occurred when no control programme was in operation. In 2003, high cattle mortality was likely fuelled by late reports to the veterinary authority, as well as, a difficult access to the study area, a situation that was exacerbated by heavy rainfall which delayed, by at least a month, cattle vaccination. We suspect common vampire bats were exceptionally abundant, and likely linked with a high rabies virus prevalence. The fact that the only rabies virus isolation from a common vampire bat in Costa Rica was made where the 2003 outbreak occurred might support this further, but pure coincidence can not be ruled out

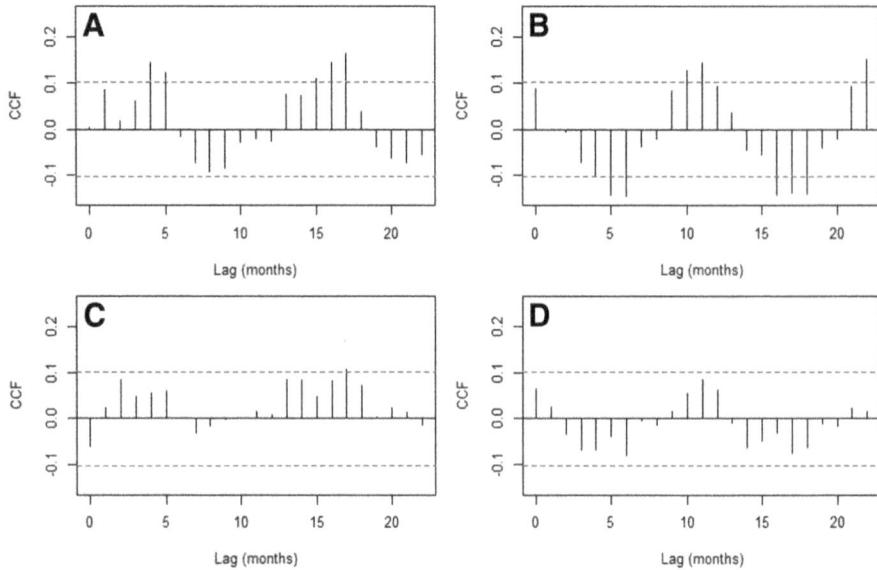

Fig. 7 Cross correlation functions between weather variables and cattle rabies. **a** Temperature and cattle rabies outbreaks (most significant correlations at lags 4 and 17 months). **b** Rainfall and cattle rabies outbreaks (most significant correlations at lags 6, 11 and 16 months). **c** Temperature and cattle rabies outbreak mortality (significant at lag 17 months). **d** Rainfall and cattle rabies outbreak mortality (no significant lag). Blue dashed lines indicate the 95% confidence limits for correlations that can be expected by random

as frequency of bat testing for rabies virus is not available.

This analysis indicated that temperature might have a positive impact on the transmission of rabies virus to cattle, as suggested by the positive association of temperature with outbreak occurrence and cattle rabies mortality. The lag of this association roughly corresponds to two common vampire bat generations (16 to 17 months), and is in accordance with observations about increases in survival and recruitment of common vampire bats following hot temperatures [25]. Specifically, common vampire bat pre-weaning time is around 3 months, a period when vampire pups mainly feed on milk [71]. Suckling vampire pups are more likely to survive at higher temperatures given the increased blood foraging in hotter/dryer conditions by adult common vampire bats [26]. Here, it is important to note that food deprivation for two to 3 days (e.g. through adverse weather

conditions such as strong rains) might become fatal for young bats [72], which also have a high failure rate to obtain blood meals on their own [73]. Therefore, any change that increases the survival of young vampire pups might be critical to increase rabies virus transmission by increasing the size of common vampire bat populations. Pre-weaning vampire pups also start to receive regurgitated blood during their first 3 months of age, a key event for rabies virus transmission within common vampire bats [71, 74]. Indeed, food deprivation in common vampire bats can be compensated through reciprocal blood sharing by roost mates [73]. Thus, a likely increase in vampire pup survival during the pre-weaning period, followed by the six to 7 months common vampire bats need to reach adulthood, when bats are between 9 and 10 months old, and also become sexually mature, with a diet exclusively based on fresh blood, could lead to a larger cohort of reproducing common vampire bats [14, 75]. This larger cohort then takes seven additional months (thus adding up to around 16 to 17 months since

Table 1 Parameter estimates for the best time series model of monthly cattle rabies outbreak numbers (1985–2016). The first column shows the estimated parameters and time lag (in months). The second column depicts the estimates. Covariates for the best model included temperature (with 17 months of lag), and rainfall (with 16 months of lag)

Parameter (Lag)	Estimate ± S.E.
Intercept	−0.069 ± 0.097
Φ SAR (3)	0.1403 ± 0.0506
Temp (17)	0.1301 ± 0.0365
Rain (16)	−0.0006 ± 0.0002

SAR Seasonal autoregressive, *Temp* Temperature, *S.E.* Standard Error

Table 2 Parameter estimates for the best time series model of monthly cattle rabies mortality (1985–2016). The first column shows the estimated parameters and time lag (in months). The second column depicts the estimates. Covariates for the best model included temperature (with 17 months of lag)

Parameter (Lag)	Estimate ± S.E.
Intercept	−0.0608 ± 1.2243
Φ SAR (2)	0.1186 ± 0.0506
Temp (17)	1.0421 ± 0.4621

SAR Seasonal autoregressive, *Temp* Temperature, *S.E.* Standard Error

birth) for the gestation of a new *D. rotondus* vampire cohort [76] which will require additional blood as they try to raise their pups. Most likely this will increase blood foraging pressure, thus, rendering plausible an increase in rabies virus transmission 16–17 months after hotter temperatures might have promoted breeding and increased the survival of pre-weaning vampire pups.

Meanwhile, a mechanism where the delays are due to changes in the rabies virus incubation period in cattle or common vampire bats is unlikely, provided experimental studies have shown incubation periods in the order of one to 2 months in cattle [77], and two to 4 weeks in bats [78]. Nevertheless, rabies virus incubation period is known to be highly variable depending upon inoculation site [78] and viral loads [79]. The variability in rabies virus incubation period could be one of the reasons behind the lack of a seasonality in cattle rabies outbreaks, although the occurrence of large outbreaks where the onset of clinical signs is more or less synchronous suggests that, at least in outbreak foci, cattle have a homogenous incubation period [80]. The occurrence time for the two largest cattle rabies outbreaks, which happened during April and July, and the frequent occurrence of outbreaks during the months of August, September and May, further suggest that common vampire bat ecology plays a major role explaining the lagged effects of temperature on cattle rabies outbreaks. For example, Turner [28] found that most bats were pregnant or lactating in the months of April–May and July–October, an observation coinciding with records from Costa Rica [26], suggesting increased transmission might be related to maternal care of pre-weaning common vampire bats, a possibility supported by the increased rabies virus exposure of juvenile and subadult common vampire bats, which have shown higher rabies virus seroprevalence than adults [14].

In contrast to the positive impacts of temperature on outbreak occurrence and size, this analysis showed that rainfall had a negative impact on cattle rabies outbreak frequency in the SAR time series models. This negative association might emerge from rainfall impacts on bat foraging. For example, a study found that rainfall increased flight metabolism, while reducing or ceasing foraging activities [81]. Reductions in vampire foraging time could then diminish the exposure of cattle to rabies virus infected common vampire bats. Similarly, the location of cattle by common vampire bats could become more difficult following rainy periods, when a reduction in cattle surface body temperature could interfere with common vampire bat thermoreceptors used to locate their prey [82]. Furthermore, abundant rain after a hot spell might diminish vampire fecundity, via a reduction in common vampire bat movement and mating between roosts [24]. For example, a study in Costa Rica showed that common vampire bats

stayed away from their roost more time during the dry than the wet season [26].

Limitations of this study include the following: (i) The mortality analysis considered estimated denominators in a fraction of the data. Nevertheless, as shown by the scan cluster analysis, the results were robust when those "estimated" denominators were ignored. (ii) Data on cattle mortality were less robust compared to outbreak data which may hide the true number of rabies cases. (iii) The impact of rabies vaccination on cattle mortality patterns was not assessed due to lacking systematic information about vaccination rates in affected herds. (iv) It was assumed that rabies virus was only transmitted by common vampire bats [12, 20]. However, it is possible that other wildlife species might have been involved in rabies virus transmission, as have been documented for raccoons, skunks and foxes [5, 76, 83–85]. In Costa Rica, a cattle rabies case was associated with the rabies virus strain of *Tadarida brasiliensis*, an insectivorous bat species [8]. In that sense, sequencing rabies viruses isolated in cattle rabies outbreaks might help to identify other species potentially involved in rabies virus transmission. Phylogenetic analyses could also help to better understand geographical and temporal structures of outbreaks, and whether common vampire bats and other wildlife host species are involved in local enzootic cycles or epizootic waves [73, 83]. Further efforts are also necessary to understand rabies virus spillovers across species and the role that bat immunity has on the persistence of rabies virus among common vampire bats under changing environments, to fully untangle the impacts that changing rainfall and temperature patterns have on common vampire bat population structure and dynamics, as well as, rabies virus transmission.

Finally, the results of this study clearly indicate that the occurrence, and mortality of cattle rabies outbreaks are associated with weather fluctuations in Costa Rica. Further eco-epidemiological field studies are necessary to articulate the mechanisms behind the associations observed in this study, especially regarding the hypothesis that outbreaks, and their magnitude, might be driven by changes in the population dynamics of vampire bats triggered by ENSO phases. The resulting information from those studies will be useful not only to understand rabies virus transmission ecology, but also to propose appropriate intervention measures to control this major veterinary public health problem.

Conclusions

Further efforts are necessary to articulate the mechanisms underpinning the association between weather changes and cattle rabies outbreaks. One hypothesis is that exacerbation of cattle rabies outbreaks might be mediated by impacts of weather conditions on common vampire bat movement and access to food resources on its natural habitats. Further eco-epidemiological field studies could help to understand rabies virus transmission ecology, and to propose sound interventions to control this major veterinary public health problem.

Additional files

Additional file 1: Figure S1. Temporal patterns of cattle rabies outbreak mortality in Costa Rica (1985–2016). (A) Annual time series of cattle deaths from 1985 to 2016. Peak values occurred in 1985 and 2003, with 149 and 193 cattle deaths, respectively. (B) Bar chart of monthly cumulative cattle deaths from 1985-2016. The highest number of deaths occurred in April and July. (C) Monthly time series of cattle deaths from 1985 to 2016. It reflects both outbreaks during a year and cattle rabies deaths per month. Peak values occurred in 1985 and 2003. Note that in year 1995, although there is no peak value of cattle rabies death, there were many small outbreaks. (D) Boxplots of monthly cattle rabies deaths from 1985 to 2016. For all months from January to December, the median value of the boxplots is 0, which suggests that there is no seasonality in monthly cattle rabies deaths.

Additional file 2: Figure S2. Video showing the geolocation and size of cattle rabies outbreaks (1985-2016). In the video the current outbreak is shown in red while past outbreaks are in blue. Circles are proportional to the number of cattle rabies deaths (for guidance see the inset legend at the bottom of the map) and for every outbreak, the month and year, the exact number of cattle rabies deaths, and the El Niño Southern Oscillation (ENSO) phase are indicated in the upper left corner of the map.

Additional file 3: Figure S3. Monthly scan cluster analysis. Cluster 1 (circle with dash line in green color) and cluster 2 (circle with dash line in red color) were calculated with data for which denominators were imputed; cluster 3 (circle with dash line in purple color) and cluster 4 (circle with dash line in pink color) were calculated with no imputed denominators. Relative risks (RR) reflect the magnitude of the difference of the deaths in the cluster to regular deaths, which is the expected death assuming they occurred as driven by a homogenous Poisson process. This figure was made using a public domain map from the US National Park Service (https://www.nps.gov/hfc/carto/data-sources.cfm) as background.

Additional file 4: Figure S4. Correlation functions. (A) Auto-correlation function (ACF) of the monthly cattle rabies outbreak number time series; it shows the data were not strongly auto-correlated. (B) Partial auto-correlation function (PACF) of the monthly cattle rabies outbreaks time series data shows that there is a periodicity of 3 months. (C) Cross-correlation function (CCF) of the monthly cattle rabies outbreak number time series data with the ENSO4 index data. This CCF shows no significant linear association over the studied period, when considering up to 22 months of lag. (D) ACF of the monthly outbreak mortality time series. This time series shows low autocorrelation. (E) PACF of the monthly outbreak mortality time series. This time series shows a periodicity of 2 months. (F) CCF of the monthly outbreak mortality time series data with the ENSO4 index data. This CCF shows no significant linear association over the studied period, when considering up to 22 months of lag. In all panels blue dashed lines indicate the 95% confidence limits for correlations that can be expected by random.

Additional file 5: Table S1. Selection of the best monthly cattle rabies outbreaks time series model. Columns indicate the type of model (models): full or the backward elimination round. The Akaike Information Criterion (AIC) is a model selection criterion which is minimized by the best model. The AIC for the best models of each selection round are bolded. o and x indicate, respectively, the presence or absence of a variable in a model. Temp is an abbreviation for temperature. Time lags are in months.

Additional file 6: Table S2. Selection of the best monthly cattle mortality time series model. Columns indicate the type of model: Null, or

full model. The Akaike Information Criterion (AIC) is a model selection criterion which is minimized by the best model. The best model has its AIC bolded. o and x indicate, respectively, the presence or absence of a variable in a model. Temp is an abbreviation for temperature. Time lags are in months.

Abbreviations

ACF: Auto-correlation function; AIC: Akaike Information Criterion; AR: Autoregressive; CCF: Cross-correlation function; CDR: Climate Data Record; *D.: Desmodus*; DNA: Deoxyribonucleic acid; ENSO: El Niño Southern Oscillation; GHCN_CAMS 2 m model: Global Historical Climatology Network version 2 and the Climate Anomaly Monitoring System; GPCP: Global Precipitation Climatology Project; GPS: Global Positioning System; KNMI: Royal Netherlands Meteorological Institute; PACF: Partial auto-correlation function; RNA: Ribonucleic acid; RR: Relative risks; S.E.: Standard error; SAR: Seasonal autoregressive; SaTScan: Software for spatial, temporal, or space-time scan statistics; SENASA: Servicio Nacional de Salud Animal; SIREA: Sistema de Reconocimentos Ambientales; Temp: Temperature; US: United States

Acknowledgements

The authors wish to acknowledge the Ministry of Agriculture and Livestock of Costa Rica, as well as the SENASA officials which contributed to the rabies surveillance programme and made this study possible.

Authors' contributions

LFC conception and design of the study, conducting the analysis, interpretation of the results, writing the manuscript, revising critically the manuscript; SEH conception and design of the study, conducting the analysis, interpretation of the results, writing the manuscript, revising critically the manuscript; AK conception and design of the study, interpretation of the results, revising critically the manuscript; SFG, BL, KB, MB, MR, YG interpretation of the results, contribution to writing the manuscript, revising critically the manuscript. All authors read and approved the manuscript.

Consent for publication

Not applicable. Manuscript does not contain information on individuals.

Competing interests

The authors declare that they have no competing interests.

Author details

[1]Institute of Veterinary Public Health, Department for Farm Animals and Veterinary Public Health, University of Veterinary Medicine, Veterinärplatz 1, 1210 Wien, Austria. [2]Laboratorio de Pruebas de Paternidad, Caja Costarricense del Seguro Social (CCSS), San José, Costa Rica. [3]Servicio Nacional de Salud Animal (SENASA), Heredia, Costa Rica. [4]Research Institute of Wildlife Ecology, University of Veterinary Medicine, Vienna, Austria. [5]Centro de Investigación en Enfermedades Tropicales (CIET), Universidad de Costa Rica, San Pedro de Montes de Oca, Costa Rica. [6]Centro de Investigaciones en Geografía Ambiental, Universidad Nacional Autónoma de México, 58190 Morelia, Michoacán, Mexico. [7]Instituto Costarricense de Investigación y Enseñanza en Nutrición y Salud, Apartado Postal 4-2250, Tres Ríos, Cartago, Costa Rica. [8]Programa de Investigación en Enfermedades Tropicales (PIET), Escuela de Medicina Veterinaria, Universidad Nacional, Heredia, Costa Rica.

References

1. Schnell MJ, McGettigan JP, Wirblich C, Papaneri A. The cell biology of rabies virus: using stealth to reach the brain. Nat Rev Micro. 2010;8(1):51–61.

2. Hampson K, Coudeville L, Lembo T, Sambo M, Kieffer A, Attlan M, Barrat J, Blanton JD, Briggs DJ, Cleaveland S, et al. Estimating the global burden of endemic canine rabies. PLoS Negl Trop Dis. 2015;9(4):e0003709.

3. Jackson AC. Human rabies: a 2016 update. Curr Infect Dis Rep. 2016;18(11):38.

4. Schneider MC, Romijn PC, Uieda W, Tamayo H, DFd S, Belotto A, JBd S, Leanes LF. Rabies transmitted by vampire bats to humans: an emerging zoonotic disease in Latin America? Rev Panam Salud Publica. 2009;25(3):260.

5. Krebs JW, Wilson ML, Childs JE. Rabies- epidemiology, prevention, and future research. J Mammal. 1995;76(3):681–94.

6. Belotto A, Leanes L, Schneider M, Tamayo H, Correa E. Overview of rabies in the Americas. Virus Res. 2005;111(1):5–12.

7. Vigilato MAN, Cosivi O, Knöbl T, Clavijo A, Silva HMT. Rabies update for Latin America and the Caribbean. Emerg Infect Dis. 2013;19(4):678–9.

8. Hutter SE, Brugger K, Sancho Vargas VH, González R, Aguilar O, León B, Tichy A, Firth CL, Rubel F. Rabies in Costa Rica: documentation of the surveillance program and the endemic situation from 1985 to 2014. Vector-Borne Zoonotic Dis. 2016;16(5):334–41.

9. Ellison JA, Gilbert AT, Recuenco S, Moran D, Alvarez DA, Kuzmina N, Garcia DL, Peruski LF, Mendonça MT, Lindblade KA, et al. Bat rabies in Guatemala. PLoS Negl Trop Dis. 2014;8(7):e3070.

10. Pawan JL. The transmission of paralytic rabies in Trinidad by the vampire bat (Desmodus rotundus Murinus Wagner, 1840). Ann Trop Med Parasitol. 1936;30(1):101–30.

11. Anderson A, Shwiff S, Gebhardt K, Ramírez AJ, Shwiff S, Kohler D, Lecuona L. Economic evaluation of vampire bat (Desmodus rotundus) rabies prevention in Mexico. Transbound Emerg Dis. 2014;61(2):140–6.

12. Johnson N, Aréchiga-Ceballos N, Aguilar-Setien A. Vampire bat rabies: ecology, epidemiology and control. Viruses. 2014;6(5):1911–28.

13. Lee DN, Papeş M, Van Den Bussche RA. Present and potential future distribution of common vampire bats in the Americas and the associated risk to cattle. PLoS One. 2012;7(8):e42466.

14. Streicker DG, Recuenco S, Valderrama W, Gomez Benavides J, Vargas I, Pacheco V, Condori Condori RE, Montgomery J, Rupprecht CE, Rohani P, et al. Ecological and anthropogenic drivers of rabies exposure in vampire bats: implications for transmission and control. Proc R Soc B Biol Sci. 2012;279(1742):3384–92.

15. Mayen F. Haematophagous bats in Brazil, their role in rabies transmission, impact on public health, livestock industry and alternatives to an indiscriminate reduction of bat population. J Vet Med B Infect Dis Vet Public Health. 2003;50(10):469–72.

16. Arellano-Sota C. Vampire bat-transmitted rabies in cattle. Rev Infect Dis. 1988;10:S707.

17. Steele J. International aspects of veterinary medicine and its relation to health nutrition and human welfare. Mil Med. 1966;131(9 P 1):765–78.

18. Acha PN, Szyfres B. Zoonosis y enfermedades transmisibles comunes al hombre ya los animales. Pub Científica. 2003;284–92.

19. Martínez-Burnes J, López A, Medellín J, Haines D, Loza E, Martínez M. An outbreak of vampire bat-transmitted rabies in cattle in northeastern Mexico. Can Vet J. 1997;38(3):175–7.

20. Kobayashi Y, Ogawa A, Sato G, Sato T, Itou T, Samara SI, Carvalho AAB, Nociti DP, Ito FH, Sakai T. Geographical distribution of vampire bat-related cattle rabies in Brazil. J Vet Med Sci. 2006;68(10):1097–100.

21. Voigt CC, Kelm DH. Host preference of the common vampire bat (Desmodus rotundus; Chiroptera) assessed by stable isotopes. J Mammal. 2006;87(1):1–6.

22. Wallace R. Big farms make big flu: dispatches on influenza, agribusiness, and the nature of science: NYU Press; 2016.

23. McNab BK. Energetics and the distribution of vampires. J Mammal. 1973;54(1):131–44.

24. Trajano E. Movements of cave bats in Southeastern Brazil, with emphasis on the population ecology of the common vampire bat, Desmodus rotundus (Chiroptera). Biotropica. 1996;28(1):121.

25. Lord RD. Seasonal reproduction of vampire bats and its relation to seasonality of bovine rabies. J Wildl Dis. 1992;28(2):292–4.

26. Young AM. Foraging of vampire bats (Desmodus rotundus) in Atlantic wet lowland Costa Rica. Revista de Biologia Tropical. 1971;18:73–88.

27. Wainwright M. The Nat Hist of Costa Rican mammals. Florida, US: Zona Tropical; 2002.

28. Turner DC: The vampire bat; a field study in behavior and Ecology 1975.

29. Anyamba A, Linthicum KJ, Small JL, Collins KM, Tucker CJ, Pak EW, Britch SC, Eastman JR, Pinzon JE, Russell KL. Climate teleconnections and recent patterns of human and animal disease outbreaks. PLoS Negl Trop Dis. 2012;6(1):e1465.

30. Patz JA, Campbell-Lendrum D, Holloway T, Foley JA. Impact of regional climate change on human health. Nature. 2005;438(7066):310–7.

31. Chretien J-P, Anyamba A, Small J, Britch S, Sanchez JL, Halbach AC, Tucker C, Linthicum KJ: Global climate anomalies and potential infectious disease risks: 2014-2015. PLoS Currents 2015. doi: https://doi.org/10.1371/currents. outbreaks.95fbc4a8fb4695e049baabfc2fc8289f.

32. Hurtado LA, Cáceres L, Chaves LF, Calzada JE. When climate change couples social neglect: malaria dynamics in Panamá. Emerging Microbes & Infections. 2014;3(4):e27.

33. Hurtado LA, Calzada JE, Rigg CA, Castillo M, Chaves LF. Climatic fluctuations and malaria transmission dynamics, prior to elimination, in Guna Yala República de Panamá. *Malaria Journal.* 2018;17(1):85.

34. Yamada K, Valderrama A, Gottdenker N, Cerezo L, Minakawa N, Saldaña A, Calzada JE, Chaves LF. Macroecological patterns of American cutaneous leishmaniasis transmission across the health areas of Panamá (1980–2012). Parasite Epidemiology and Control. 2016;1(2):42–55.

35. Rifakis PM, Benitez JA, Rodriguez-Morales AJ, Dickson SM, De-La-Paz-Pineda J. Ecoepidemiological and social factors related to rabies incidence in Venezuela during 2002-2004. International Journal of Biomedical Science : IJBS. 2006;2(1):1–6.

36. Bett B, Kiunga P, Gachohi J, Sindato C, Mbotha D, Robinson T, Lindahl J, Grace D. Effects of climate change on the occurrence and distribution of livestock diseases. Preventive veterinary medicine. 2017;137:119–29.

37. Patz JA, Olson SH. Climate change and health: global to local influences on disease risk. Ann Trop Med Parasitol. 2006;100:535–49.

38. Liang L, Gong P. Climate change and human infectious diseases: a synthesis of research findings from global and spatio-temporal perspectives. Environ Int. 2017;103:99–108.

39. VI Censo Nacional Agropecuario. [http://www.inec.go.cr]. Accessed 21 Dec 2017.

40. León Sáenz J. Historia económica de Costa Rica en el Siglo XX. Vol.2. La economía rural. San Jose: Universidad de Costa Rica; 2012.

41. Rosero-Bixby L, Palloni A. Population and deforestation in Costa Rica. Popul Environ. 1998;20(2):149–85.

42. Arturo Sánchez-Azofeifa G, Daily GC, Pfaff ASP, Busch C. Integrity and isolation of Costa Rica's national parks and biological reserves: examining the dynamics of land-cover change. Biol Conserv. 2003;109(1):123–35.

43. SIREA (Sistema Integrado de Registro de Establecimientos Agropecuarios). [Information provided by Bernal Leon].

44. Olden Julian D, Lawler Joshua J, Poff NL. Machine learning methods without tears: a primer for ecologists. Q Rev Biol. 2008;83(2):171–93.

45. Kuhn M, Johnson K. Applied predictive modeling. New York: Springer; 2013.

46. Cold & Warm Episodes by Season. [www.cpc.ncep.noaa.gov/products/analysis_monitoring/ensostuff/ensoyears.shtml]. Accessed 21 Dec 2017.

47. NOAA Climate Prediction Center. [http://www.cpc.ncep.noaa.gov/data/indices/ersst3b.nino.mth.81-10.ascii]. Accessed 21 Dec 2017.

48. Maraun D, Kurths J. Cross wavelet analysis: significance testing and pitfalls. Nonlinear Proc Geophys. 2004;11:505–14.

49. Chaves LF, Calzada JE, Valderrama A, Saldaña A. Cutaneous leishmaniasis and sand fly fluctuations are associated with El Niño in Panamá. PLoS Negl Trop Dis. 2014;8(10):e3210.

50. Chaves LF, Pascual M. Climate cycles and forecasts of cutaneous leishmaniasis, a nonstationary vector-borne disease. PLoS Med. 2006;3(8):e295.

51. KMNI Climate Explorer. [http://climexp.knmi.nl/start/cgi]. Accessed 21 Dec 2017.

52. GHCN_CAMS Gridded 2m Temperature (Land). [https://www.esrl.noaa.gov/psd/data/gridded/data.ghcncams.html]. Accessed 21 Dec 2017.

53. GPCP Version 2.3 Combined Precipitation Data Set. [https://www.esrl.noaa.gov/psd/data/gridded/data.gpcp.html]. Accessed 21 Dec 2017.

54. Shumway RH, Stoffer DS. Time series analysis and its applications. In: Springer Texts in Statistics. New York: Springer New York; 2011.

55. Torrence C, Compo G. A practical guide to wavelet analysis. Bull Am Meteor Soc. 1998;79:61–78.

56. Cazelles B, Chavez M, Berteaux D, Ménard F, Vik J, Jenouvrier S, Stenseth N. Wavelet analysis of ecological time series. Oecologia. 2008;156(2):287–304.

57. Kulldorff M, Rand K, Gherman G, Williams G, DeFrancesco D. SaTScan-software for the spatial and space time scan statistics, version 2.1. Bethesda, Madison, National Cancer Institute; 1998.

58. Kulldorff M, Heffernan R, Hartman J, Assunção R, Mostashari F. A space–time permutation scan statistic for disease outbreak detection. PLoS Med. 2005;2(3):e59.

59. Kulldorff M, Huang L, Pickle L, Duczmal L. An elliptic spatial scan statistic. Stat Med. 2006;25(22):3929–43.

60. Faraway JJ: Linear models with R: CRC press; 2014.

61. Fuller D, Troyo A, Beier J. El Nino southern oscillation and vegetation dynamics as predictors of dengue fever cases in Costa Rica. Environ Res Lett. 2009;4(1):014011.

62. Chuang T-W, Chaves LF, Chen P-J. Effects of local and regional climatic fluctuations on dengue outbreaks in southern Taiwan. PLoS One. 2017;12(6):e0178698.

63. Chaves LF, Satake A, Hashizume M, Minakawa N. Indian ocean dipole and rainfall drive a Moran effect in East Africa malaria transmission. J Infect Dis. 2012;205(12):1885–91.

64. Chaves LF, Hashizume M, Satake A, Minakawa N. Regime shifts and heterogeneous trends in malaria time series from western Kenya highlands. Parasitology. 2012;139:14–25.

65. Chaves LF, Kaneko A, Taleo G, Pascual M, Wilson ML. Malaria transmission pattern resilience to climatic variability is mediated by insecticide-treated nets. Malar J. 2008;7(1):100.

66. Vargas G: Geografía de Costa Rica: EUNED; 2006.

67. Ramoni-Perazzi P, Muñoz-Romo M, Chaves LF, Kunz TH. Range prediction for the giant fruit-eating bat, Artibeus amplus (Phyllostomidae: Stenodermatinae) in South America. Stud Neotropical Fauna Environ. 2012;47(2):87–103.

68. Escobar LE, Peterson AT, Papeş M, Favi M, Yung V, Restif O, Qiao H, Medina-Vogel G. Ecological approaches in veterinary epidemiology: mapping the risk of bat-borne rabies using vegetation indices and night-time light satellite imagery. Vet Res. 2015;46(1):92.

69. Delpietro HA, Marchevsky N, Simonetti E. Relative population densities and predation of the common vampire bat (Desmodus rotundus) in natural and cattle-raising areas in north-East Argentina. Preventive Veterinary Medicine. 1992;14(1–2):13–20.

70. Predescu M, Levins R, Awerbuch-Friedlander T. Analysis of a nonlinear system for community intervention in mosquito control. Discrete and Continuous Dynamical Systems-Series B. 2006;6(3):605–22.

71. Schmidt U, Manske U. Die Jugendentwicklung der Vampirfledermaeuse. Z Saeugetierk. 1973;38:14733.

72. Freitas MB, Welker AF, Millan SF, Pinheiro EC. Metabolic responses induced by fasting in the common vampire bat Desmodus rotundus. J. Comp. Physiol. B. 2003;173(8):703–7.

73. Wilkinson GS. Reciprocal food sharing in the vampire bat. Nature. 1984;308(5955):181–4.

74. Reid F. A field guide to the mammals of Central America and Southeast Mexico. Oxford: Oxford University Press; 1997.

75. Greenhall AM, Schmidt U. Natural history of vampire bats. Florida: CRC Press; 1988.

76. Schmidt U. Die Tragzeit der Vampirfledermäuse (Desmodus rotundus). Z Saugetierk. 1974;3:9129–32.

77. Martell M, Di Batalla C, Baer G, Acuna J. Experimental bovine paralytic rabies—"derriengue". Vet Rec. 1974;95(23):527–30.

78. Moreno JA, Baer GM. Experimental rabies in the vampire bat. The American Journal of Tropical Medicine and Hygiene. 1980;29(2):254–9.

79. Almeida MF, Martorelli LFA, Aires CC, Sallum PC, Durigon EL, Massad E. Experimental rabies infection in haematophagous bats Desmodus rotundus. Epidemiol Infect. 2005;133(3):523–7.

80. Hudson LC, Weinstock D, Jordan T, Bold-Fletcher NO. Clinical features of experimentally induced rabies in cattle and sheep. J Veterinary Med Ser B. 1996;43(1–10):85–95.

81. Voigt CC, Schneeberger K, Voigt-Heucke SL, Lewanzik D: Rain increases the energy cost of bat flight. Biol Lett 2011:rsbl20110313.

82. Kürten L, Schmidt U. Thermoperception in the common vampire bat (Desmodus rotundus). J. Comp. Physiol. A Neuroethol. Sens. Neural Behav. Physiol. 1982;146(2):223–8.

83. Smith DL, Lucey B, Waller LA, Childs JE, Real LA. Predicting the spatial dynamics of rabies epidemics on heterogeneous landscapes. Proc Natl Acad Sci U S A. 2002;99(6):3668–72.

84. Badilla X, Pérez-Herra V, Quirós L, Morice A, Jiménez E, Sáenz E, Salazar F, Fernández R, Orciari L, Yager P. Human rabies: a reemerging disease in Costa Rica? Emerg Infect Dis. 2003;9(6):721.

85. Escobar L, Peterson AT, Favi M, Yung V, Medina-Vogel G. Bat-borne rabies in Latin America. Rev Inst Med Trop Sao Paulo. 2015;57:63–72.

86. Data Sources & Accuracy for National Park Service Maps. [https://www.nps.gov/hfc/carto/data-sources.cfm]. Accessed 21 Dec 2017.

18

Treatment of corn with lactic acid or hydrochloric acid modulates the rumen and plasma metabolic profiles as well as inflammatory responses in beef steers

You Yang, Guozhong Dong[*] ⓘ, Zhi Wang, Junhui Liu, Jingbo Chen and Zhu Zhang

Abstract

Background: High-grain diets that meet the energy requirements of high-producing ruminants are associated with a high risk of rumen disorders. Mild acid treatment with lactic acid (LA) has been used to modify the degradable characteristics of grains to improve the negative effects of high-grain diets. However, the related studies mainly focused on dairy cows and explored the effects on rumen fermentation, production performance, ruminal pH and so forth. And up to date, no studies have reported the hydrochloric acid (HA) treatment of grains for ruminant animals. Therefore, based on metabolomics analysis, the aim of this study was to evaluate the effects of treatment of corn by steeping in 1% LA or 1% HA for 48 h on the rumen and plasma metabolic profiles in beef steers fed a high corn (48.76%) diet with a 60:40 ratio of concentrate to roughage. The inflammatory responses of beef cattle fed LA- and HA-treated corn were also investigated.

Results: Based on ultra-high-performance liquid tandem chromatography-quadrupole time-of-flight mass spectrometry (UHPLC-QTOF/MS) metabolomics and multivariate analyses, this study showed that steeping corn in 1% LA or 1% HA modulated the metabolic profiles of the rumen. Feeding beef steers corn steeped in 1% LA or 1% HA was associated with lower relative abundance of carbohydrate metabolites, amino acid metabolites, xanthine, uracil and DL-lactate in the rumen; with higher ruminal pH; with lower concentrations of acetate, iso-butyrate and iso-valerate; and with a tendency for lower ruminal lipopolysaccharide (LPS) concentrations. Moreover, the data showed lower concentrations of plasma C-reactive protein, serum amyloid A, haptoglobin, interleukin (IL)-1β and IL-8 in beef steers fed 1% LA- or HA-treated corn. The 1% LA treatment decreased the concentrations of plasma LPS, LPS-binding protein and tumour necrosis factor-alpha and the relative abundance of L-phenylalanine, DL-3-phenyllactic acid and tyramine in plasma. The 1% HA treatment decreased the relative abundance of urea in plasma and increased the relative abundance of all amino acids in the plasma.

Conclusions: These findings indicated that LA or HA treatment of corn modulated the degradation characteristics of starch, which contributed to improving the rumen and plasma metabolic profiles and to decreasing inflammatory responses in beef steers fed a high-concentrate diet.

Keywords: Metabolomics, Beef steers, Lactic acid, Hydrochloric acid, Corn

* Correspondence: gzdong@swu.edu.cn
College of Animal Science and Technology, Southwest University,
Chongqing 400716, People's Republic of China

Background

For ruminants, the efficiency of dietary energy utilization is greater for starch derived from cereals than for cellulose derived from forages [1]. Today's intensive feedlot management systems encourage large amounts of cereal grains in the diets of steers to enhance fattening. As a result, excessive amounts of carbohydrates cause volatile fatty acid (VFA) accumulation, ruminal pH depression, and ruminal microbiota dysbiosis, which could increase the possibility of developing subacute ruminal acidosis (SARA) [2, 3]. When SARA occurs, a large amount of bacterial lipopolysaccharide (LPS) is released in the rumen and the large intestine [4, 5]. When rumen LPS is translocated into the blood circulation, an inflammatory response is activated [5, 6]. Entry of LPS into blood circulation can also result in metabolic disorders [7]. The negative effects of high-grain diets will decrease the long-term production of beef cattle.

Thus, strategies to prevent high-grain diets from causing SARA are highly desirable. Over the years, substantial research efforts have been made to modulate the rumen degradability of grain by physical and chemical processing, aiming to improve grain feed efficiency and its health benefits [8–10]. Reynolds considered that the majority of the supplemental energy arising from post-ruminal starch digestion is used with high efficiency to support body adipose and protein retention [1]. Deckardt reviewed the most important achievements in the chemical processing to date and considered that mild acids are a promising method [11]. Among these organic acids, lactic acid (LA) has received more attention [12–14], and the results demonstrated that LA processing modulated rumen fermentation patterns, increased rumen pH and productivity, and improved cow health. However, this method might require more research to confirm its mechanism of lowering the amount of rumen fermentable starch from the metabolomics point of view. In addition, no previous investigations on the effects of steeping grain in LA on beef cattle have been conducted.

Hydrochloric acid (HA) is an inorganic acid and the component of the gastric acid secreted by the abomasum. Moreover, HA has a higher acidity than LA and the pK_a value of HA and LA is − 7.0 and 3.9, respectively [15]. Therefore, we hypothesized that HA treatment should be as effective as LA treatment for slowing the starch degradation rate of grain in the rumen.

Metabolomics is an emerging field of "omics" science that uses high-throughput approaches coupled with multivariate analysis to extract comprehensive metabolic information and measured metabolic phenotypes in mammals, plants, and microbes [16–18]. Therefore, based on the fact that corn is the main grain feedstuff in many countries, the aim of this study was to evaluate the effects of feeding corn steeped in LA or HA on the metabolic profiles of the rumen and plasma in beef steers by Ultra-high-performance liquid tandem chromatography-quadrupole time-of-flight mass spectrometry (UHPLC-Q-TOF-MS). In addition, the inflammatory responses of beef cattle fed LA- and HA-treated corn was investigated.

Methods

Animals, diet and experimental design

Eighteen Charolais × Luxi hybrid steers of the Southwest University (Chongqing, China) experimental station were selected for this experiment. Steers born in the same season of calving and weaned and grown under the same conditions of management and feeding were used in the experiment. At the beginning of the experiment, the animals were 13 months of age and their initial body weight was 368 ± 52 kg with a similar body condition score. These steers were randomly assigned to 1 of 3 diets according to the completely randomized design. The 3 diets were identical, with the only difference being the treatment of corn: 48.76% (dry matter basis) corn grain in the diet was steeped for 48 h in an equal quantity (i.e., in a 1:1 ratio, wt/vol) of tap water (CON), 1% lactic acid (LA) or 1% hydrochloric acid (HA) before being mixed with other ingredients of the total mixed ration (TMR). The LA (L(+)-lactic acid, 85%, wt/wt) used in this study was purchased from Chengdu Cologne Chemical Co., Ltd. (Chengdu, China), and HA (hydrochloric acid, 37%, wt/wt) was purchased from Chongqing Chuandong Chemical Co., Ltd. (Chongqing, China). The steers were offered a TMR with a concentrate to roughage ratio of 60:40 (Table 1).

The experimental period was 32 d, with the first 7 d used for an adaptation period. The steers were housed in individual tie stalls and had free access to feed and water. Diets were fed two times daily at 0630 and 1730 h to allow 5–10% orts. During the trial, the amount of feed offered and orts was recorded each day, and dry matter intake (DMI) for each cattle was calculated based on the dry matter (DM) contents of the diets. The DMI data were then used to calculate average daily dry matter intake (ADMI). After the experiment, all beef cattle were kept by stockmen of the university experimental station.

Sample collection

Blood samples (10 mL per steer) were obtained from the jugular vein on day 20 before the morning feeding and were collected into sterile and pyrogen-free tubes containing sodium heparin. Blood samples were stored on ice and then centrifuged at 4 °C and 3000×g for 15 min to separate plasma. The plasma was transferred to 2 mL pyrogen-free tubes and immediately stored at − 80 °C until further analysis.

Four beef steers in each treatment were selected randomly for collection of the rumen fluid at the end of the

Table 1 Ingredients and nutrient composition of the experimental diets

Items	Diet[a]		
	LA	HA	CON
Ingredients, % of DM			
Chinese wildrye	11.50	11.50	11.50
Sorghum distiller's grains	28.50	28.50	28.50
Corn grain (water-treated)	–	–	48.76
Corn grain (LA-treated)	48.76	–	–
Corn grain (HA-treated)	–	48.76	–
Rapeseed meal	8.61	8.61	8.61
Salt	0.30	0.30	0.30
Limestone	2.05	2.05	2.05
Premix[b]	0.28	0.28	0.28
Nutrients, % of DM			
Crude protein	13.5	13.8	13.6
Ether extract	4.2	4.2	4.2
Neutral detergent fibre	34.5	34.6	34.4
Acid detergent fibre	14.3	13.7	14.4
Crude ash	5.1	4.8	4.8
Calcium	0.35	0.36	0.38
Total phosphorus	0.31	0.29	0.31
Nonfiber carbohydrate[c]	42.7	42.6	43.0
Starch in corn grain	63.4	66.0	71.8

[a]LA is the treatment diet based on corn grain steeped for 48 h in an equal quantity of tap water containing 1% lactic acid (wt/vol), HA is the treatment diet based on corn grain steeped for 48 h in an equal quantity of tap water containing 1% hydrochloric acid (wt/vol), and CON is the control diet containing corn grain steeped for 48 h in an equal quantity of tap water
[b]Formulated to provide the following (per kg diet): Fe (as ferrous sulfate), 50 mg; Cu (as copper sulfate), 10 mg; Mn (as manganese sulfate), 20 mg; Co (as cobaltous sulfate), 0.1 mg; Zn (as zinc sulfate), 30 mg; I (as potassium iodate), 0.5 mg; Se (as sodium selenite), 0.1 mg; vitamin A, 2240 IU; vitamin D_3, 500 IU; vitamin E, 40 IU; riboflavin, 6.2 mg; nicotinic acid, 22 mg; D-pantothenic acid, 22 mg; vitamin B_{12}, 0.02 mg; biotin, 0.15 mg; and choline, 0.92 mg
[c]Nonfiber carbohydrate (%) = 100 - (% neutral detergent fibre + % crude protein + % ether extract + % crude ash)

experimental period. Three hours after the morning feeding, the rumen fluid was collected by an oral stomach tube equipped with a strainer and a 150 mL pyrogen-free syringe as described by Shen et al. [19]. When the length of the tube inserted into the rumen was approximately 200 cm, the rumen fluid was obtained with a 150 mL pyrogen-free syringe. The initial 50 mL rumen fluid was discarded to avoid saliva contamination. Subsequently, approximately 150 mL of rumen fluid was strained through four layers of sterile cheesecloth and filtrate was divided into two parts: one part was immediately used to determine the pH value by the portable pH meter (Rex PHS-3E, Shanghai INESA Scientific Instrument Co., Ltd., Shanghai, China), while the other part was transferred into sterile and pyrogen-free centrifuge tubes (approximately 50 mL per cattle) and

centrifuged at 4 °C and 10,000×g for 30 min, and the supernatant was stored in 2 mL pyrogen-free tubes at – 80 °C until further analysis.

Analyses of volatile fatty acids
VFA concentrations in the rumen fluid were determined using gas chromatograph (GC-2010, Shimadzu, Japan) with SH-RTX-WAX capillary columns (30 m × 0.25 μm × 0.25 mm, Shimadzu, Japan). One mL ruminal fluid supernatant was added to 0.2 mL metaphosphoric acid solution (25%), and vortexed for 30 s. After kept at 5 °C for 5 h, the samples were centrifuged at 14500×g for 15 min at 4 °C. The supernatant was transferred into a 1-mL glass vial for gas chromatograph analysis. One μL aliquot of the analyte was injected in split mode (50:1). The samples were run at a programmed temperature gradient (100 °C initial temperature for 1 min, with a 5 ° C rise per min until 190 °C, and 190 °C for 15 min). N_2 gas was used as the carrier gas, and the column flow rate was $1 \, mL \, min^{-1}$. The temperature of the injector and detector was 230 °C and 240 °C, respectively.

Analyses of lipopolysaccharide, cytokines, and acute phase proteins
The LPS concentrations in the rumen fluid and plasma; the concentrations of acute phase proteins in plasma, including C-reactive protein (CRP), serum amyloid A (SAA), haptoglobin (Hp), and LPS-binding protein (LBP); and plasma cytokines concentrations, including interleukin (IL)-1β, IL-6, IL-8 and tumour necrosis factor-alpha (TNF-α), were determined using a commercially available bovine ELISA kit. A bovine LPS ELISA kit was purchased from Shanghai Preferred Biotechnology Co., Ltd. (Shanghai, China). Bovine SAA, CRP, Hp and LBP ELISA kits were purchased from Nanjing Jianchen Saihao Biotechnology Co., Ltd. (Nanjing, China). Bovine IL-1β, IL-6, IL-8 and TNF-α ELISA kits were purchased from Shanghai Jinma Biotechnology Co., Ltd. (Shanghai, China). According to the manufacturer's instructions, all samples were tested in duplicate, and the optical density values were read at 450 nm on an automatic microplate reader (EIX808IU; Biotek, Winooski, VT).

Metabolomics analysis
All samples of rumen fluid and plasma were thawed at 4 °C on ice. Then, 100 μL of sample was taken, placed in an Eppendorf tube, and then extracted with 300 μL of methanol and 20 μL of internal standard substances for 30 s vortexes. Then, the extracts were ultrasound treated for 10 min (incubated in ice water) and incubated for 1 h at – 20 °C to precipitate proteins. Then, the samples were centrifuged at 12000×g for 15 min at 4 °C. In total, 200 μL of supernatant was transferred to LC-MS vials

and stored at $-80\,°C$ until the UHPLC-QTOF/MS analysis [20].

LC-MS/MS analyses were carried out on a UHPLC system (1290, Agilent Technologies) with a UPLC BEH amide column (2.1 mm × 100 mm × 1.7 μm, Waters) coupled to a Triple TOF 6600 (Q-TOF, AB Sciex). Elution was performed with a mobile phase of A (25 mM ammonium acetate and 25 mM ammonium hydroxide in water, pH = 9.75) and B (acetonitrile) under the following gradient program: 0 min, 5% A and 95% B; 7 min, 35% A and 65% B; 9 min, 60% A and 40% B; 9.1 min, 5% A and 95% B; and 12 min, 5% A and 95% B. The flow rate was 0.5 mL/min, and the injection volume was 2 μL for positive ion mode (POS) and 3 μL for negative ion mode (NEG).

A Triple TOF mass spectrometer was used to acquire MS/MS spectra on an information-dependent basis (IDA) during an LC/MS experiment. In this mode, the acquisition software (Analyst TF 1.7, AB Sciex) continuously evaluated the full scan survey MS data as it collected and triggered the acquisition of MS/MS spectra depending on preselected criteria. In each cycle, 12 precursor ions whose intensity was greater than 100 were chosen for fragmentation at collision energy (CE) of 30 V (15 MS/MS events with a product ion accumulation time of 50 msec each). The electron spray ionization (ESI) source conditions were set as follows: ion source gas 1 as 60 Psi, ion source gas 2 as 60 Psi, curtain gas as 35 Psi, source temperature as 650 °C, ion spray voltage floating (ISVF) 5000 V or -4000 V in positive or negative mode, respectively [21].

Data processing and statistical analyses

Data of ADMI; the rumen pH values and the concentrations of VFA; the LPS concentrations in the plasma and rumen fluid; plasma concentrations of SAA, CRP, Hp and LBP; and plasma concentrations of IL-1β, IL-6, IL-8 and TNF-α were analysed using ANOVA of SAS (SAS Institute, 2000). Data are presented as the means ± standard deviation. Means among treatments were compared using Duncan's multiple range test. Statistical significance was defined as $P \leq 0.05$.

MS raw data (.d) files were converted to the mzXML format by ProteoWizard (http://proteowizard.source-forge.net/downloads.shtml) and processed by R package XCMS (https://xcmsonline.scripps.edu/landing_page.php?pgcontent=mainPage). The preprocessing results generated a data matrix that consisted of the retention time (RT), mass to charge ratio (m/z) values, and peak intensity [22]. The R package CAMERA was used for peak annotation after XCMS data processing. An in-house MS2 database was applied for metabolite identification [23].

The SIMCA software package (V14.1, Umea, Sweden) was used for pattern recognition multivariate analysis, including principal component analysis (PCA) and orthogonal partial least-squares discriminant analysis (OPLS-DA). Based on the OPLS-DA results, metabolites were plotted according to their importance in separating the dietary groups, and each metabolite received a value called variable importance for the projection (VIP). If VIP exceeded 1.0, the metabolite was first selected as the changed variable. These variables were then assessed using one-way ANOVA analysis. If $P \leq 0.05$, the variables were defined as the significantly differential metabolites among the 3 groups [24].

Results

Feed intake

The ADMI of the CON, LA and HA groups was 6.82 ± 0.71, 6.33 ± 0.85 and 6.50 ± 0.69 kg, respectively. There was no significant difference in ADMI among the three diets ($P = 0.53$).

The ruminal pH and the concentrations of volatile fatty acids in the rumen

Data for the ruminal pH and VFA contents are shown in Table 2. Compared with the ruminal pH value of the CON group, those of the LA and HA groups were higher ($P < 0.01$). Both LA-treated and HA-treated groups had lower acetate, iso-butyrate and iso-valerate ($P < 0.05$), while propionate, butyrate and valerate were not affected ($P > 0.05$).

The lipopolysaccharide concentrations in the rumen fluid and plasma

The effects of LA and HA on LPS concentrations in the rumen fluid and plasma are shown in Table 3. The lowest LPS concentrations in plasma were observed in beef cattle fed the LA-treated corn ($P < 0.05$).

Plasma acute phase proteins concentrations

As shown in Table 4, compared with the CON group, the LA and HA groups had significantly decreased CRP, SAA and Hp concentrations in plasma ($P < 0.05$). Moreover, the LA treatment had decreased LBP concentrations in plasma relative to the other groups ($P < 0.05$).

Plasma cytokines concentrations

The plasma cytokines concentrations of the different diets are reported in Table 5. No differences in plasma IL-6 concentrations among the three diets were observed ($P > 0.05$), but the concentrations of plasma IL-1β and IL-8 in the LA and HA groups were lower than those in the CON group ($P < 0.05$). The plasma TNF-α concentration in the LA group was the lowest among the experimental diets ($P < 0.05$). Although the plasma

Table 2 The ruminal pH and volatile fatty acids concentrations in the rumen of beef steers fed different diets

Items	Diet[1]			P-value
	LA	HA	CON	
Ruminal pH	6.74 ± 0.16^A	6.73 ± 0.17^A	6.22 ± 0.07^B	0.0008
Acetate (mmol/L)	35.68 ± 2.79^A	37.58 ± 5.15^A	52.89 ± 9.03^B	0.0066
Propionate (mmol/L)	26.08 ± 2.83	27.92 ± 9.41	28.46 ± 6.79	0.8787
Butyrate (mmol/L)	7.38 ± 0.68	7.11 ± 1.94	9.27 ± 1.70	0.1532
Iso-butyrate (mmol/L)	0.64 ± 0.10^a	0.59 ± 0.03^a	0.85 ± 0.20^b	0.0408
Iso-valerate (mmol/L)	1.40 ± 0.17^A	1.74 ± 0.51^A	2.75 ± 0.47^B	0.0033
Valerate (mmol/L)	0.61 ± 0.14	0.66 ± 0.20	0.91 ± 0.28	0.1627

A-B Means within a row differ ($P < 0.01$)
a-b Means within a row differ ($P < 0.05$)
[1]LA is the treatment diet based on corn grain steeped for 48 h in an equal quantity of tap water containing 1% lactic acid (wt/vol), HA is the treatment diet based ong corn grain steeped for 48 h in an equal quantity of tap water containing 1% hydrochloric acid (wt/vol), and CON is the control diet containing corn grain steeped for 48 h in an equal quantity of tap water

TNF-α concentration in the HA group was numerically lower than that in the CON group, the difference did not attain a significant level ($P > 0.05$).

Multivariate analysis of rumen fluid and plasma metabolites

The ionization source of LC-QTOFMS is ESI, including positive and negative ion modes (POS; NEG). Multivariate statistical analyses were used to explore metabolic differences among the CON, LA and HA groups. PCA, which is an unsupervised pattern recognition method, was performed to examine the intrinsic variation in metabolic patterns among the three groups. Under the supervised OPLS-DA, variable information can be extracted for the classification of large amounts of samples and for reducing unwanted systematic noise. PCA and OPLS-DA analyses of UHPLC-QTOF/MS metabolic profiles of the rumen fluid samples are illustrated in Fig. 1a, b, c and d. For ruminal samples, the distributions of the LA and HA groups overlapped and could not be well separated from each other, but the responses of beef cattle fed the CON diets were further apart from those corresponding to the LA and HA diets (Fig. 1a, b, c and

Table 3 The lipopolysaccharide (LPS) concentrations in rumen fluid and plasma of beef steers fed different diets

Items	Diet[1]			P-value
	LA	HA	CON	
LPS (rumen, EU/mL)[2]	21,284.45± 1253.38	23,326.95± 2945.41	25,392.15± 1553.99	0.0572
LPS (plasma, EU/mL)	0.39 ± 0.06^a	0.48 ± 0.08^b	0.49 ± 0.06^b	0.0380

A-B Means within a row differ ($P < 0.01$)
a-b Means within a row differ ($P < 0.05$)
[1]LA is the treatment diet based on corn grain steeped for 48 h in an equal quantity of tap water containing 1% lactic acid (wt/vol), HA is the treatment diet based on corn grain steeped for 48 h in an equal quantity of tap water containing 1% hydrochloric acid (wt/vol), and CON is the control diet containing corn grain steeped for 48 h in an equal quantity of tap water
[2]EU endotoxin unit

d). The PCA score plot of plasma samples in the three groups could not clearly differentiate (Fig. 1e and g); however, they could be well discriminated using the OPLS-DA model (Fig. 1f and h), especially between the LA group and the other two groups.

Significantly different metabolites among the three diets

As shown in Fig. 2a and b, a total of 85 rumen metabolites (POS: 43 and NEG: 42) with statistical significance ($P < 0.05$) and VIP > 1 were identified as major differentiating substances among the 3 groups. Most rumen metabolites were markedly increased by the CON diet relative to the LA or HA diet. First, all carbohydrate metabolites were found at the highest relative abundance in the beef steers fed the CON diet; these carbohydrate metabolites included 3-alpha-mannobiose, D-fructose, D-lyxose, D-maltose, D-mannose, D-tagatose, D-threitol, isomaltose, L-sorbose, sucrose, trehalose, pyruvate and so forth. The relative abundance of these carbohydrate metabolites was the lowest in the LA group. Moreover, amino acid metabolites of beef cattle fed the CON diet were also found at the highest relative abundance; these metabolites included Ala-Gly, Ala-Lys, Arg-Ala, Arg-Ile, D-proline, Gly-Lys, L-asparagine, L-leucine, Phe-Glu and Tyr-Thr, except for His-Met and Phe-Ala. With regard to nucleotide metabolites, the relative abundances (%) of thymine, uracil and xanthine in the CON group were more than 11.84-fold and 8.56-fold, 2.57-fold and 1.89-fold, 3.40-fold and 2.48-fold higher than those in the LA and HA groups, respectively (Additional file 1: Table S1). Finally, the relative abundances of DL-lactate, beta-lactic acid and glutaric acid were the highest in the CON group, followed by the HA group, and the lowest in the LA group.

Figure 3 lists 25 plasma different metabolites from the three groups. Several compounds in the plasma have attracted attention as significantly different metabolites. For example, the relative abundance of tyramine and DL-3-phenyllactic acid in the LA group was the lowest

Table 4 The acute phase proteins concentrations in plasma of beef steers fed different diets

Items	Diet[1]			P-value
	LA	HA	CON	
CRP[2] (µg/mL)	22.95 ± 4.57^A	15.04 ± 5.30^A	36.03 ± 9.35^B	0.0003
SAA (µg/mL)	80.01 ± 13.71^a	76.94 ± 12.11^a	105.38 ± 27.25^b	0.0386
LBP (µg/mL)	13.06 ± 4.96^a	22.04 ± 5.33^b	21.37 ± 5.35^b	0.0160
Hp (µg/mL)	373.86 ± 125.63^a	489.60 ± 209.89^a	701.49 ± 133.22^b	0.0098

$^{A-B}$Means within a row differ ($P < 0.01$)
$^{a-b}$Means within a row differ ($P < 0.05$)
[1]LA is the treatment diet based on corn grain steeped for 48 h in an equal quantity of tap water containing 1% lactic acid (wt/vol), HA is the treatment diet based on corn grain steeped for 48 h in an equal quantity of tap water containing 1% hydrochloric acid (wt/vol), and CON is the control diet containing corn grain steeped for 48 h in an equal quantity of tap water
[2]CRP C-reactive protein, SAA serum amyloid A, LBP lipopolysaccharide-binding protein, Hp haptoglobin

compared with those in the other groups (Additional file 2: Table S2). L-citrulline, L-leucine, L-methionine, and L-phenylalanine, which are involved in amino acid metabolism, were also the lowest in the LA group, but the related histidine metabolites, such as His-Ser and L-Histidine, were the lowest in the CON group, followed by the LA group, and the highest in the HA group (Additional file 2: Table S2). The relative abundance of above amino acids in the plasma was the highest in the HA group, and the relative abundance of urea in plasma was the highest in the LA group and the lowest in HA group (Additional file 2: Table S2).

Discussion

Although corn steeped in 1% LA or 1% HA had higher acidity (Additional file 3: Table S3), the results showed that there was no significant difference in ADMI among the three treatments. Mickdam et al. [25] also reported that diets including barley treated with 0.5% or 1% LA did not exert a negative effect on DMI in early-lactating cows.

Feeding a high-grain diet can result in the accumulation of organic acids in the rumen and the weakening of rumen buffering, which can lead to a decrease in the rumen pH [26, 27]. Our results demonstrated that steeping corn grain in 1% LA or 1% HA significantly increased the rumen pH. The greater rumen pH was attributed to the augmentation of the ruminal resistant

starch (RRS) for lowering concentrations of VFA in the rumen. Indeed, our results showed that the LA or HA treatment reduced the concentrations of acetate, iso-butyrate and iso-valerate, although propionate, butyrate and valerate were not significantly different in this study. In addition, LA and HA treatments may increase the activity of lactate-utilizing bacteria for reducing the concentration of lactate in rumen. Compared with VFA normally associated with ruminal fermentation, the pK_a of lactate is lower (3.9 for lactate versus 4.9 for VFA) [28], and reduction of ruminal lactate production was more conducive to increasing rumen pH values. Mao et al. [28] indicated that disodium fumarate did not affect the production of total VFA or its individual acids, and ruminal pH increased linearly as the amount of disodium fumarate added increased, while lactate production decreased linearly. The results of rumen metabolomics in the present study showed that the relative abundance of DL-lactate, as a differential metabolite among the three diets, was decreased in LA and HA groups compared with the CON group. Thus, the higher ruminal pH of beef cattle fed corn grain steeped in 1% LA or 1% HA may result from the low production of VFA and lactate in the rumen.

LPS is a component of the cell wall of Gram-negative bacteria, and it has been widely recognized that ruminal LPS concentrations increase after grain engorgement. Zhou et al. [29] found that the rumen pH of cows fed a

Table 5 The cytokines concentrations in plasma of beef steers fed different diets

Items	Diet[1]			P-value
	LA	HA	CON	
IL-1β[2] (ng/L)	95.99 ± 7.33^a	96.03 ± 9.5^a	109.25 ± 9.36^b	0.0287
IL-6 (ng/L)	10.19 ± 2.66	9.85 ± 3.69	10.68 ± 1.77	0.8777
IL-8 (ng/L)	110.67 ± 20.15^A	109.94 ± 30.99^A	163.87 ± 37.30^B	0.0108
TNF-α (ng/L)	146.30 ± 40.35^a	176.88 ± 23.57^{ab}	198.85 ± 24.15^b	0.0289

$^{A-B}$Means within a row differ ($P < 0.01$)
$^{a-b}$Means within a row differ ($P < 0.05$)
[1]LA is the treatment diet based on corn grain steeped for 48 h in an equal quantity of tap water containing 1% lactic acid (wt/vol), HA is the treatment diet based on corn grain steeped for 48 h in an equal quantity of tap water containing 1% hydrochloric acid (wt/vol), and CON is the control diet containing corn grain steeped for 48 h in an equal quantity of tap water
[2]IL-1β interleukin-1β, IL-6 interleukin-6, IL-8 interleukin-8, TNF-α tumour necrosis factor-alpha

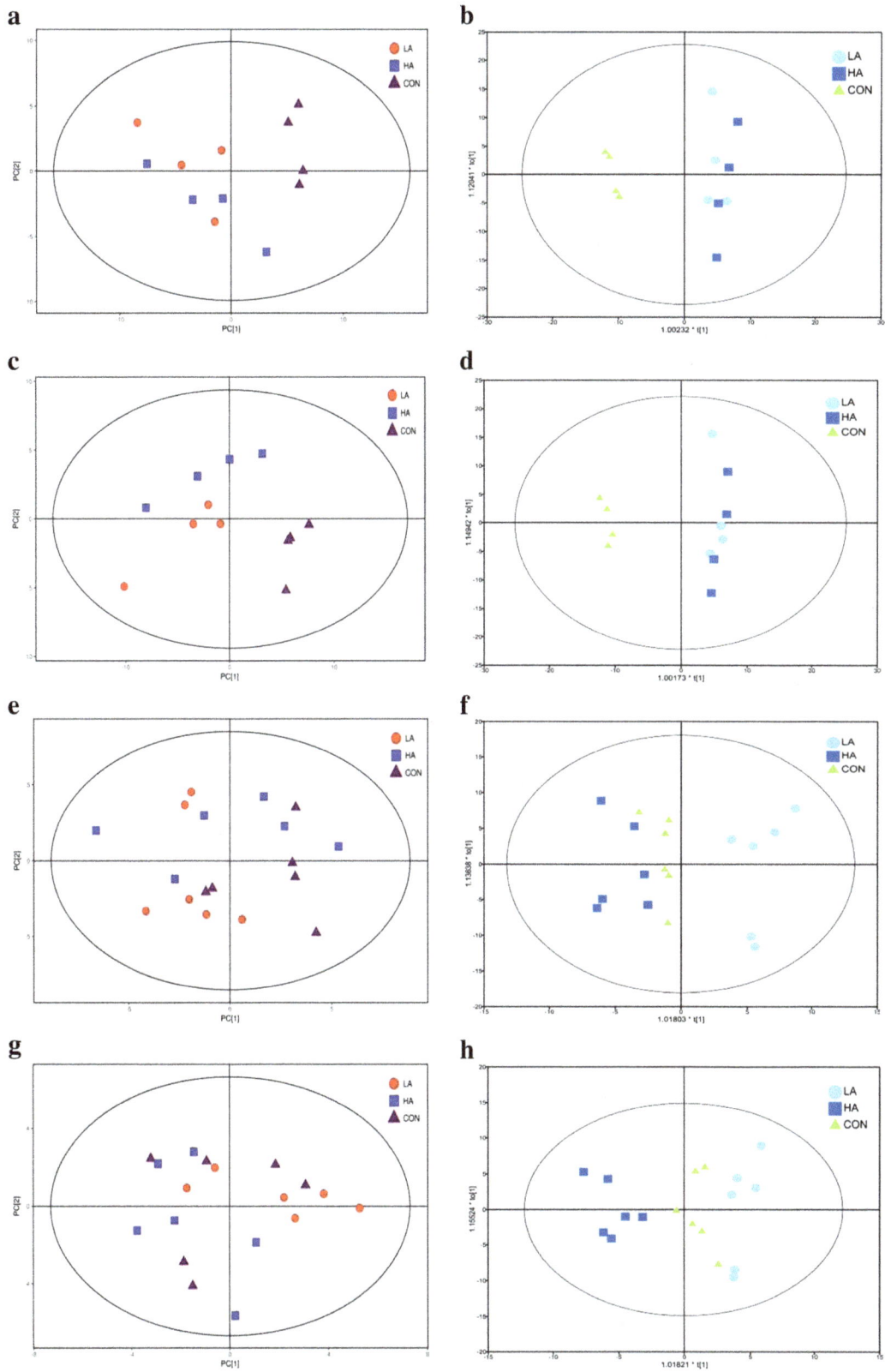

Fig. 1 (See legend on next page.)

Fig. 1 PCA score plots (**a**, **c**, **e**, and **g**) and OPLS-DA score plots (**b**, **d**, **f** and **h**) of rumen and plasma samples obtained from beef steers fed the 3 different diets. On the score plot, each point represents an individual sample. **a** and **b** are derived from the POS of the rumen fluid, and **c** and **d** are derived from the NEG of the rumen fluid. **e** and **f** are derived from the POS of plasma, and **g** and **h** are derived from the NEG of plasma. LA (circle) is the treatment diet based on corn grain steeped for 48 h in an equal quantity of tap water containing 1% lactic acid (wt/vol), HA (square) is the treatment diet based on corn grain steeped for 48 h in an equal quantity of tap water containing 1% hydrochloric acid (wt/vol), and CON (triangle) is the control diet containing corn grain steeped for 48 h in an equal quantity of tap water

high-concentrate diet with a concentrate to roughage ratio of 65:35 and a low-concentrate diet with a concentrate to roughage ratio of 46:54 were 6.21 and 6.62, and the rumen LPS concentrations were 29,065 EU/mL and 11,664 EU/mL, respectively. A decline in ruminal pH causes death and cell lysis of gram-negative bacteria, resulting in an increase in the ruminal LPS concentration [30]. As stated above, the LA and HA treatments contributed to reducing the concentrations of organic acids in the rumen (i.e., acetate, iso-butyrate, iso-valerate and DL-lactate) and thus increased the ruminal pH. Therefore, treatment with 1% LA or 1% HA tended ($P = 0.0572$) to decrease the concentrations of LPS in the

rumen. In addition, LPS released during the growth of bacteria may account for as much as 60% of that released in the rumen [4]. During rapid growth, autolytic enzymes are required to help cells expand and grow; however, excessive autolytic activity can lead to bacterial cell apoptosis and lysis. *E. coli*, a Gram-negative bacterium in the rumen, is the major contributor to the rumen LPS pool [31, 32]. There is a specific maltose-binding protein located in the periplasmic space of the cell wall of *E. coli* [33]. The finding from our study was that the relative abundance of D-maltose was the lowest in the LA group, followed by that in the HA group, and the highest in the CON group, which was consistent with

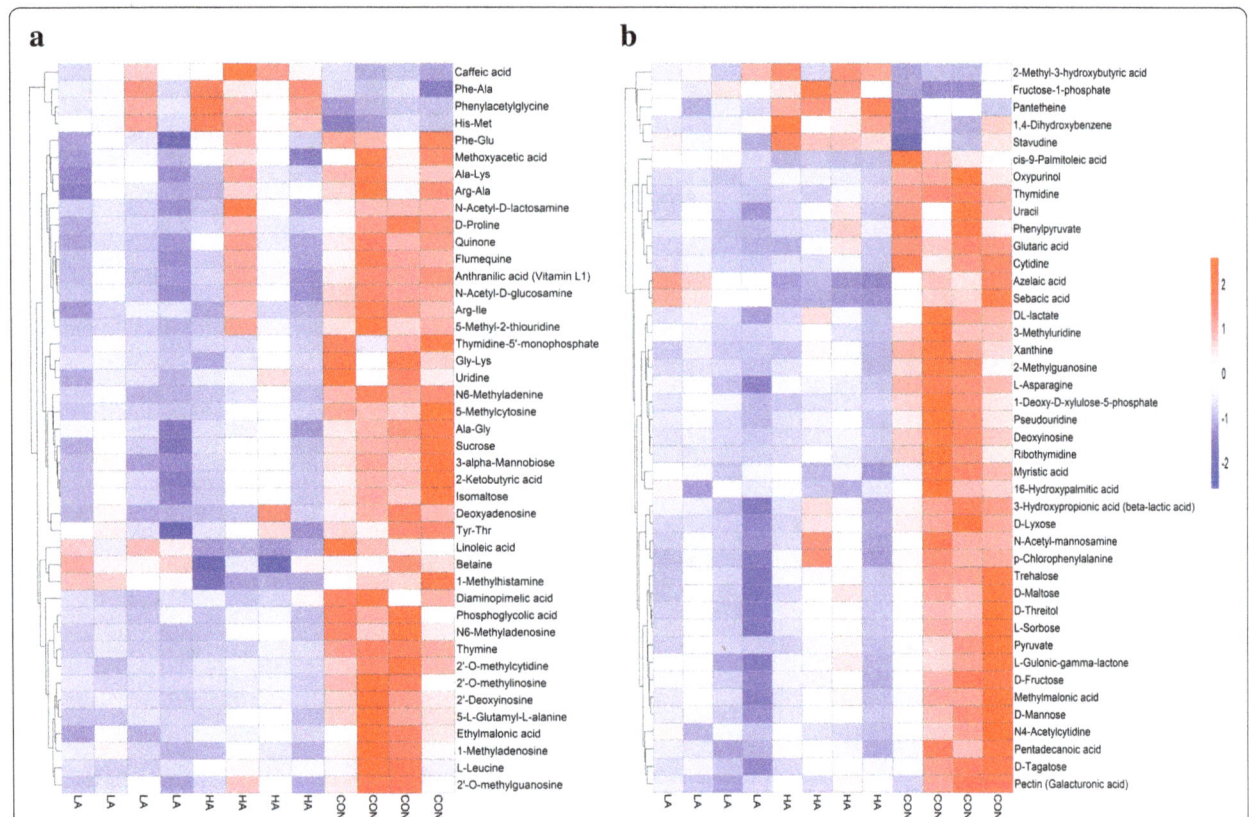

Fig. 2 Hierarchical clustering analysis of different rumen fluid metabolites in beef cattle fed with 3 different diets. **a** and **b** are derived from the POS and NEG of the rumen fluid, respectively. Each row represents one metabolite, and each column represents one sample. Cells are coloured based on the signal intensity measured in the rumen. Red represents high rumen levels, blue shows low signal intensity, and white cells show intermediate levels. LA is the treatment diet based on corn grain steeped for 48 h in an equal quantity of tap water containing 1% lactic acid (wt/vol), HA is the treatment diet based on corn grain steeped for 48 h in an equal quantity of tap water containing 1% hydrochloric acid (wt/vol), and CON is the control diet containing corn grain steeped for 48 h in an equal quantity of tap water

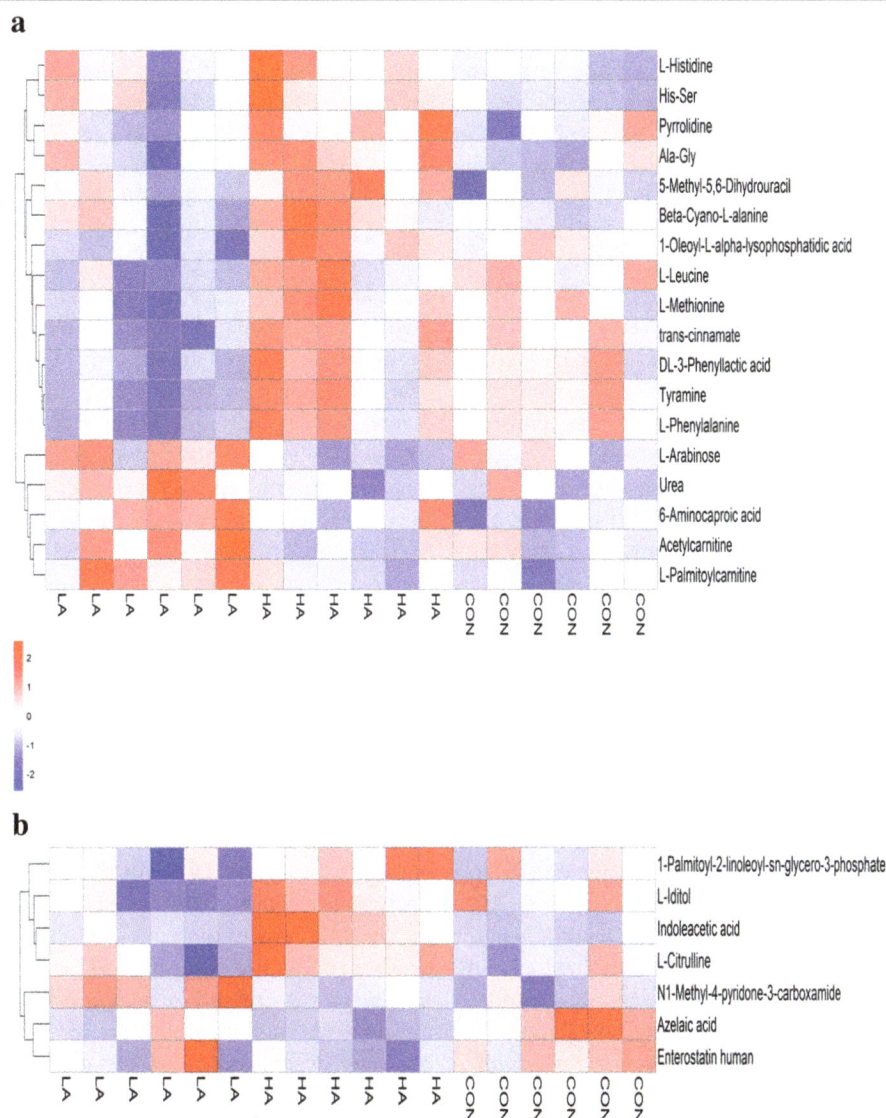

Fig. 3 Hierarchical clustering analysis of different plasma metabolites in beef cattle fed with 3 different diets. **a** and **b** are derived from the POS and NEG of the plasma, respectively. Each row represents one metabolite, and each column represents one sample. Cells are coloured based on the signal intensity measured in plasma. Red represents high plasma levels, blue shows low signal intensity, and white cells show intermediate levels. LA is the treatment diet based on corn grain steeped for 48 h in an equal quantity of tap water containing 1% lactic acid (wt/vol), HA is the treatment diet based on corn grain steeped for 48 h in an equal quantity of tap water containing 1% hydrochloric acid (wt/vol), and CON is the control diet containing corn grain steeped for 48 h in an equal quantity of tap water

the change in the ruminal LPS concentrations among the 3 groups.

LPS produced in the digestive tract can be translocated into the bloodstream; thus, the concentration of blood LPS increases [4]. As a result, immune responses are activated by circulating LPS, and the blood concentrations of acute phase proteins, such as SAA, Hp, LBP and CRP, increase [34–36]. In the current study, compared with the CON group, the concentrations of plasma LPS in the LA group decreased significantly. Meanwhile, the plasma concentrations of SAA, CRP

and Hp in the LA and HA groups were significantly lower than those in the CON group, and the plasma LBP concentrations in the LA group were also significantly lower than those in the CON group. The present findings are consistent with those of Iqbal, who found lower concentrations of plasma Hp and SAA in dairy cows fed LA-treated barley [37]. Moreover, the translocation of LPS into the systemic circulation stimulates the release of pro-inflammatory cytokines, such as IL-1, IL-6 and TNF-α, by mononuclear phagocytes [29]. Our study demonstrated that the release of cytokines (IL-1β,

IL-8 and TNF-α) in plasma decreased when beef steers were fed the LA- or HA-treated corn. IL-6 is a complex cytokine that enhances or limits the immune response [29]. The concentrations of IL-6 were not different among the three groups, which may be attributed to the fact that IL-6 is a duplicitous cytokine that plays both pro- and anti-inflammatory roles. These findings indicated that steeping corn in LA or HA had beneficial effects in beef steers fed high-concentrate diets.

In addition to evaluate the effect of feeding corn steeped in LA or HA on inflammatory responses in beef steers, we also explored the evidence of LA or HA treatment on changing the metabolomic profiles in the rumen and plasma by UHPLC-QTOF-MS metabolomics analysis. Based on each PCA scot plot and OPLS-DA scot plot, the experimental results clearly showed significant differences in the ruminal metabolites among the basal diet and the two experimental diets. This study indicated that the metabolic profile changed in association with the LA- and HA-treated corn.

The main component of most cereal grains is starch, and in this present study the content of starch in corn grain steeped with 1% LA, 1% HA and water was 63.4, 66.0 and 71.8% of the dry matter (DM), respectively. Non-structural carbohydrates have fast rumen degradation as opposed to the slow degradation rate of structural carbohydrates. Non-structural carbohydrates are easily degraded in the rumen by protozoa that engulf starch granules or amylolytic bacteria that secrete α-amylases [11]. Oligosaccharides, dextrines and small amounts of free glucose are the end products of amylose and amylopectin debranching before treatment with other oligosaccharides [38]. In the current study, all qualitative metabolites of carbohydrates were found at the highest relative abundance in the CON-fed beef steers compared with the LA- or HA-fed beef steers; these metabolites included 3-Alpha-mannobiose, D-fructose, D-lyxose, D-maltose, D-mannose, D-tagatose, D-threitol, isomaltose, L-sorbose, sucrose, and trehalose. In addition, pyruvate, an intermediate product of starch degradation in the rumen [39], was approximately 2-fold higher in the CON group than in the LA group or the HA group (Additional file 1: Table S1). These results indicated that corn starch can escape microbial degradation in the rumen after being steeped in LA or HA. Östman et al. [13] observed that the hydrolysis indices (HI) of a starch and gluten mixture containing lactic acid were 22% lower than the corresponding mixture without the acid. Khol-Parisini et al. [40] also found that treatment of barley grain in 1% LA modulated the in situ rumen degradation kinetics of starch in cows. In general, the results obtained by our study further provided powerful evidence of the modulation of starch characteristics by chemical processing.

Reduced starch degradation in the rumen would allow more starch to be digested in the intestine. The change in the site of starch digestion from the rumen to the intestine could also be responsible for some of the changes seen in protein metabolism [41]. Interestingly, most amino acid metabolites of beef cattle fed the CON diet were also found at the highest relative abundance in this study; these metabolites included Ala-Gly, Ala-Lys, Arg-Ala, Arg-Ile, D-proline, Gly-Lys, L-asparagine, L-leucine, Phe-Glu and Tyr-Thr. The structural elements of grains are the pericarp, the inner germ (embryo) and endosperm. The endosperm contains the major part of starch and consists of four layers, starting with the outer aleurone layer, followed by the subaleurone layer (peripheral endosperm), corneous endosperm, and the inner floury endosperm. The peripheral endosperm and corneous endosperm embody starch granules and are enclosed by an impenetrable protein matrix [11]. Because of the structural cross-linking changes in the protein matrix, proteolysis of the protein may be reduced and in turn reduce the degradation of the starch surrounded by the protein matrix. Iqbal [42] suggested that the composition and the ability of the protein matrix to be accessed by microorganisms are major control points for the rate of starch degradation. Our results also indicated that the concentrations of iso-butyrate and iso-valerate in rumen fluid of beef cattle in the LA group and the HA group were significantly lower than those in the CON group. This was in agreement with the study by Deckardt et al. [43] that reported that iso-butyrate and iso-valerate decreased in response to LA treatment. All amino acids that are not linked or guarded from attack are extensively degraded in the rumen to ammonia, carbon dioxide, VFA, and branch-chain fatty acids [44]. Therefore, the decrease in branch-chain fatty acids indicated reduced degradation of amino acids from dietary protein.

A stable and healthy population of ruminal microbiota is critical for maintaining the health of ruminants. Saleem and colleagues conducted several metabolic experiments to reveal alterations in rumen metabolism with an increased proportion of cereal grain in the diet of dairy cows [33, 45, 46]. All their results showed that the rumen concentrations of urea, hypoxanthine, xanthine, uracil and LPS were elevated with increasing proportions of grain. The above metabolites are degradation products of rumen bacteria. These findings indicated that a high-grain diet may disturb the microbial community. Mao et al. [30] also confirmed that high maize feeding has a significantly negative impact on the ruminal biodiverse ecosystem. An interesting observation arising from our study was that the relative abundance of uracil and xanthine in the LA or HA groups was lower than that of the CON group, which may alleviate the negative effects of high-cereal diets on rumen microorganisms.

Regarding the results of the plasma differential metabolites, the most interesting finding was related to the phenylalanine metabolism pathway. Our previous experiments showed that phenylalanine metabolism was the common key metabolic pathway involved in a high-grain diet (unpublished). L-phenylalanine and many metabolites involved in this pathway were significantly increased with the high-grain diet; for example, there was higher plasma content of phenyllactic acid and tyramine (unpublished). Our current results further confirmed the role of this metabolic pathway in the metabolism of high-grain diets. Compared with the CON group, the relative abundance of L-phenylalanine, DL-3-phenyllactic acid and tyramine significantly decreased in the LA group. The main metabolic pathway of phenylalanine is to produce tyrosine by the catalysis of phenylalanine hydroxylase. Tyramine is produced in the metabolic process of tyrosine. Tyramine is a kind of biogenic amine; low concentrations of biogenetic amines are essential for the normal growth and differentiation of cells, but higher concentrations can delay epithelial regeneration and induce epithelial damage [30]. Our findings indicated that LA treatment can improve the negative effects of a high-grain diet, which is beneficial to the health of beef cattle. In contrast to the LA group, these relative abundances in the HA group increased. The specific mechanism explaining the incremental effect of HA treatment on L-phenylalanine, DL-3-phenyllactic acid and tyramine in this research is not very well understood at present. In addition, the relative abundance of His-Ser and L-Histidine in plasma was the lowest in the CON group compared with that in the LA and HA groups. Histidine metabolism is related to another biogenic amine (i.e., histamine). Histamine in the rumen is derived from the amino carboxylic acid of histidine via a decarboxylation reaction [3]. The relative abundance (%) of ruminal histamine in the three groups was 23.04 (LA), 31.15 (HA) and 45.81 (CON) ($P = 0.18$, VIP = 1.39). The pH value of the gastrointestinal tract is a crucial factor affecting the activity of microbial amino acid decarboxylase. In fact, it was reported that the activity of the amino acid decarboxylase was greater when pH decreased [47]. The results of the present study indicated that the LA or HA treatment increased the rumen pH and inhibited the decarboxylation of histidine in the rumen, which helped more histidine to be absorbed.

When nitrogen availability in the rumen is relatively high when compared to carbohydrate availability, a great amount of ammonia is produced inside rumen, and the main nitrogen flow goes from the rumen to the bloodstream. In this case, there will be great concentrations of blood urea [44]. In this study, the relative abundance of plasma urea in the LA treatment was higher, which indicated that the degradation of grain after LA treatment in the rumen decreased. However, the increase in the concentration of urea in the blood causes losses of nitrogen. The disadvantage effect of LA treatment needs further experimental confirmation. However, the relative abundance of urea in plasma was the lowest in the HA group, which indicated that most of urea is excreted to the rumen so that it can be utilized in the synthesis of proteins that will contribute to the host's amino acid needs, and in this study the relative abundance of all amino acids in the plasma was the highest in the HA group.

Conclusions

Overall, based on UHPLC-QTOF/MS metabolomics and multivariate analyses, this study showed that steeping corn in 1% LA or 1% HA modulated the metabolic profiles of the rumen. Feeding beef steers corn steeped in 1% LA or 1% HA was associated with lower relative abundance of carbohydrate metabolites, amino acid metabolites, xanthine, uracil, DL-lactate, acetate, iso-butyrate and iso-valerate in the rumen as well as with higher ruminal pH and a tendency for lower ruminal LPS concentrations. Moreover, the data showed lower concentrations of plasma CRP, SAA, Hp, IL-1β, and IL-8 in beef steers fed diets containing corn treated with 1% LA or 1% HA. The 1% LA treatment decreased the concentrations of plasma LPS, LBP and TNF-α and the relative abundance of L-phenylalanine, DL-3-phenyllactic acid and tyramine in plasma. The 1% HA treatment decreased the relative abundance of urea in plasma and increased the relative abundance of all amino acids in plasma. These findings indicated that LA or HA treatment of corn modulated the degradation characteristics of starch, which contributed to improving the rumen and plasma metabolic profiles and to decreasing inflammatory responses in beef steers fed a high-concentrate diet.

Additional files

Additional file 1: Table S1. Relative distributions (%) of the different ruminal metabolites among the three groups.

Additional file 2: Table S2. Relative distributions (%) of the different plasma metabolites among the three groups.

Additional file 3: Table S3. pH values of corn and total mixed ration in three treatments.

Abbreviations
ADMI: Average daily dry matter intake; CE: Collision energy; CON: Control diet; CRP: C-reactive protein; DM: Dry matter; DMI: Dry matter intake; ESI: Electron spray ionization; HA: Hydrochloric acid; HI: Hydrolysis indices; Hp: Haptoglobin; IDA: Information-dependent basis; IL-1β: Interleukin-1β; IL-6: Interleukin-6; IL-8: Interleukin-8; ISVF: Ion spray voltage floating; LA: Lactic acid; LBP: Lipopolysaccharide-binding protein; LPS: Lipopolysaccharide; NEG: Negative ion mode; OPLS-DA: Orthogonal

partial least-squares discriminant analysis; PCA: Principal component analysis; POS: Positive ion mode; RRS: Ruminal resistant starch; RT: Retention time; SAA: Serum amyloid A; SARA: Subacute ruminal acidosis; TMR: Total mixed ration; TNF-α: Tumour necrosis factor-alpha; UHPLC-QTOF/MS: Ultra-high-performance liquid tandem chromatography-quadrupole time-of-flight mass spectrometry; VFA: Volatile fatty acid; VIP: Variable importance for the projection

Acknowledgements
The authors thank Mr. Zhongrong Liu and Mr. Zhongchen Qiu for their assistance in animal management and blood and rumen fluid sampling.

Funding
This work was supported by funds from Chongqing Municipal Committee of Science and Technology (Grant 1: cstc2017jcyjBX0015; Grant 2: cstc2016shmszx80078).

Authors' contributions
YY and GD conceived the study and designed the experiment. YY conducted the animal experiment, performed the laboratory analyses and drafted the manuscript. ZW, JC, ZZ and JL participated in the animal experiment and the laboratory assays. All authors read and approved the final manuscript.

Consent for publication
Not applicable.

Competing interests
The authors declare that they have no competing interests.

References
1. Reynolds CK. Production and metabolic effects of site of starch digestion in dairy cattle. Anim Feed Sci Technol. 2006;130:78–94.
2. Kleen JL, Hooijer GA, Rehage J, Noordhuizen JPTM. Subacute ruminal acidosis (SARA): a review. J Vet Med A. 2003;50:406–14.
3. Plaizier JC, Krause DO, Gozho GN, McBride BW. Subacute ruminal acidosis in dairy cows: the physiological causes, incidence and consequences. Vet J. 2009;176:21–31.
4. Dong GZ, Liu SM, Wu YX, Lei CL, Zhou J, Zhang S. Diet-induced bacterial immunogens in the gastrointestinal tract of dairy cows: impacts on immunity and metabolism. Acta Vet Scand. 2011;53:48.
5. Gressley TF, Hall MB, Armentano LE. Ruminant nutrition symposium: productivity, digestion, and health responses to hindgut acidosis in ruminants. J Anim Sci. 2011;89:1120–30.
6. Plaizier JC, Khafipour E, Li S, Gozho GN, Krause DO. Subacute ruminal acidosis (SARA), endotoxins and health consequences. Anim Feed Sci Technol. 2012;172:9–21.
7. Ametaj BN, Emmanuel DGV, Zebeli Q, Dunn SM. Feeding high proportions of barley grain in a total mixed ration perturbs diurnal patterns of plasma metabolites in lactating dairy cows. J Dairy Sci. 2009;92:1084–91.
8. Martínez TF, Moyano FJ, Díza M, Barroso FG, Alarcón FJ. Use of tannic acid to protect barley meal against ruminal degradation. J Sci Food Agric. 2005; 85:1371–8.
9. Schmidt J, Tóth T, Fábián J. Rumen fermentation and starch degradation by Holstein steers fed sodium hydroxide or formaldehyde-treated wheat. Acta Vet Hung. 2006;54:201–12.
10. Montano MF, Chai W, Zinn-Ware TE, Zinn RA. Influence of malic acid supplementation on ruminal pH, lactic acid utilization, and digestive function in steers fed high-concentrate finishing diets. J Anim Sci. 1999;77:780–4.
11. Deckardt K, Khol-Parisini A, Zebeli Q. Peculiarities of enhancing resistant starch in ruminants using chemical methods: opportunities and challenges. Nutrients. 2013;5:1970–88.
12. Iqbal S, Zebeli Q, Mazzolari A, Bertoni G, Dunn SM, Yang WZ, Ametaj BN. Feeding barley grain steeped in lactic acid modulates rumen fermentation patterns and increases milk fat content in dairy cows. J Dairy Sci. 2009;92: 6023–32.
13. Östman EM, Nilsson M, Liljeberg Elmståhl HGM, Molin G, Björck IME. On the effect of lactic acid on blood glucose and insulin responses to cereal products: mechanistic studies in healthy subjects and in vitro. J Cereal Sci. 2002;36:339–46.
14. Humer E, Lucke A, Harder H, Metzler-Zebeli BU, Böhm J, Zebeli Q. Effects of citric and lactic acid on the reduction of deoxynivalenol and its derivatives in feeds. Toxins. 2016;8:285.
15. Dawson RMC, Elliott DC, Elliott WH, Jones KM. Data for biochemical research. Oxford: Clarendon Press; 1989.
16. Sun LW, Zhang HY, Wu L, Shu S, Xia C, Xu C, et al. ¹H-nuclear magnetic resonance-based plasma metabolic profiling of dairy cows with clinical and subclinical ketosis. J Dairy Sci. 2014;97:1552–62.
17. Goldansaz SA, Guo AC, Sajed T, Steele MA, Plastow GS, Wishart DS. Livestock metabolomics and the livestock metabolome: a systematic review. PLoS One. 2017;12:e0177675.
18. Sun HZ, Wang DM, Wang B, Wang JK, Liu HY, Guan LL, et al. Metabolomics of four biofluids from dairy cows: potential biomarkers for milk production and quality. J Proteome Res. 2015;14:1287–98.
19. Shen JS, Chai Z, Song LJ, Liu JX, Wu YM. Insertion depth of oral stomach tubes may affect the fermentation parameters of ruminal fluid collected in dairy cows. J Dairy Sci. 2012;95:5978–84.
20. Dunn WB, Broadhurst D, Begley P, Zelena E, Francis-McIntyre S, Anderson N, et al. Procedures for large-scale metabolic profiling of serum and plasma using gas chromatography and liquid chromatography coupled to mass spectrometry. Nat Protoc. 2011;6:1060–83.
21. Wang JL, Zhang T, Shen XT, Liu J, Zhao DL, Sun YW, et al. Serum metabolomics for early diagnosis of esophageal squamous cell carcinoma by UHPLC-QTOF/MS. Metabolomics. 2016;12:116.
22. Smith CA, Want EJ, O'Maille G, Abagyan R, Siuzdak G. XCMS: processing mass spectrometry data for metabolite profiling using nonlinear peak alignment, matching, and identification. Anal Chem. 2006;78:779–87.
23. Kuhl C, Tautenhahn R, Böttcher C, Larson TR, Neumann S. CAMERA: an integrated strategy for compound spectra extraction and annotation of liquid chromatography/mass spectrometry data sets. Anal Chem. 2012;84:283–9.
24. Tao HX, Xiong W, Zhao GD, Peng Y, Zhong ZF, XU L, et al. Discrimination of three siegesbeckiae herba species using UPLC-QTOF/MS based metabolomics approach. Food Chem Toxicol. 2018;119:400-6.
25. Mickdam E, Khiaosa-ard R, Metzler-Zebeli BU, Humer E, Harder H, Khol-Parisini A, et al. Modulation of ruminal fermentation profile and microbial abundance in cows fed diets treated with lactic acid, without or with inorganic phosphorus supplementation. Anim Feed Sci Technol. 2017;230:1–12.
26. Dijkstra J, Ellis JL, Kebreab E, Strathe AB, López S, France J, et al. Ruminal pH regulation and nutritional consequences of low pH. Anim Feed Sci Technol. 2012;172:22–33.
27. González LA, Manteca X, Calsamiglia S, Schwartzkopf-Genswein KS, Ferret A. Ruminal acidosis in feedlot cattle: interplay between feed ingredients, rumen function and feeding behavior (a review). Anim Feed Sci Technol. 2012;172:66–79.
28. Mao SY, Zhang G, Zhu WY. Effect of disodium fumarate on ruminal metabolism and rumen bacterial communities as revealed by denaturing gradient gel electrophoresis analysis of 16S ribosomal DNA. Anim Feed Sci Technol. 2008;140:293–306.
29. Zhou J, Dong GZ, Ao CJ, Zhang S, Qiu M, Wang X, et al. Feeding a high-concentrate corn straw diet increased the release of endotoxin in the rumen and pro-inflammatory cytokines in the mammary gland of dairy cows. BMC Vet Res. 2014;10:172.
30. Mao SY, Huo WJ, Zhu WY. Microbiome-metabolome analysis reveals unhealthy alterations in the composition and metabolism of ruminal

microbiota with increasing dietary grain in a goat model. Environ Microbiol. 2016;18:525–41.

31. Khafipour E, Li S, Plaizier JC, Krause DO. Rumen microbiome composition determined using two nutritional models of subacute ruminal acidosis. Appl Environ Microbiol. 2009;75:7115–24.

32. Diez-Gonzalez F, Callaway TR, Kizoulis MG, Russell JB. Grain feeding and the dissemination of acid-resistant *Escherichia coli* from cattle. Science. 1998; 281:1666–8.

33. Saleem F, Ametaj BN, Bouatra S, Mandal R, Zebeli Q, Dunn SM, et al. A metabolomics approach to uncover the effects of grain diets on rumen health in dairy cows. J Dairy Sci. 2012;95:6606–23.

34. Emmanuel DGV, Dunn SM, Ametaj BN. Feeding high proportions of barley grain stimulates as an inflammatory response in dairy cows. J Dairy Sci. 2008;91:606–14.

35. Gozho GN, Krause DO, Plaizier JC. Rumen lipopolysaccharide and inflammation during grain adaptation and subacute ruminal acidosis in steers. J Dairy Sci. 2006;89:4404–13.

36. Zebeli Q, Ametaj BN. Relationships between rumen lipopolysaccharide and mediators of inflammatory response with milk fat production and efficiency in dairy cows. J Dairy Sci. 2009;92:3800–9.

37. Iqbal S, Zebeli Q, Mazzolari A, Duun SM, Ametaj BN. Feeding rolled barley grain steeped in lactic acid modulated energy status and innate immunity in dairy cows. J Dairy Sci. 2010;93:5147–56.

38. Cerrilla MEO, Martínez GM. Starch digestion and glucose metabolism in the ruminant: a review. Interciencia. 2003;28:380–6.

39. Nagaraja TG, Titgemeyer EC. Ruminal acidosis in beef cattle: the current microbiological and nutritional outlook. J Dairy Sci. 2007;90:E17–38.

40. Khol-Parisini A, Humer E, Sizmaz Ö, Abdel-Raheem SM, Gruber L, Gasteiner J, et al. Ruminal disappearance of phosphorus and starch, reticuloruminal pH and total tract nutrient digestibility in dairy cows fed diets differing in grain processing. Anim Feed Sci Technol. 2015;210:74–85.

41. Humer E, Zebeli Q. Grains in ruminant feeding and potentials to enhance their nutritive and health value by chemical processing. Anim Feed Sci Technol. 2017;226:133–51.

42. Iqbal S, Terrill SJ, Zebeli Q, Mazzolari A, Dunn SM, Yang WZ, et al. Treating barley grain with lactic acid and heat prevented sub-acute ruminal acidosis and increased milk fat content in dairy cows. Anim Feed Sci Technol. 2012; 172:141–9.

43. Deckardt K, Metzler-Zabeli BU, Zebeli Q. Processing barley grain with lactic acid and tannic acid ameliorates rumen microbial fermentation and degradation of dietary fibre in vitro. J Sci Food Agric. 2016;96:223–31.

44. Millen DD, Arrigoni MDB, Pacheco RDL. Rumenology. 1st ed. Switzerland: Springer International Publishing AG; 2016. p. 32–95.

45. Saleem F, Bouatra S, Guo AC, Psychogios N, Mandal R, Dunn SM, et al. The bovine ruminal fluid metabolome. Metabolomics. 2013;9:360–78.

46. Ametaj BN, Zebeli Q, Saleem F, Psychogios N, Lewis MJ, Dunn SM, et al. Metabolomics reveals unhealthy alterations in rumen metabolism with increased proportion of cereal grain in the diet of dairy cows. Metabolomics. 2010;6:583–94.

47. Aschenbach JR, Gabel G. Effect and absorption of histamine in sheep rumen: significance of acidotic epithelial damage. J Anim Sci. 2000;78:464–70.

19

Improved detection of biomarkers in cervico-vaginal mucus (CVM) from postpartum cattle

Mounir Adnane[1,2], Paul Kelly[1], Aspinas Chapwanya[3], Kieran G. Meade[4,5*] and Cliona O'Farrelly[1]

Abstract

Background: In the postpartum cow, early diagnosis of uterine disease is currently problematic due to the lack of reliable, non-invasive diagnostic methods. Cervico-vaginal mucus (CVM) is an easy to collect potentially informative source of biomarkers for the diagnosis and prognosis of uterine disease in cows. Here, we report an improved method for processing CVM from postpartum dairy cows for the measurement of immune biomarkers. CVM samples were collected from the vagina using gloved hand during the first two weeks postpartum and processed with buffer alone or buffer containing different concentrations of the reducing agents recommended in standard protocols: Dithiothriotol (DTT) or N-Acetyl-L-Cysteine (NAC). Total protein was measured using the bicinchoninic acid (BCA) assay; interleukin 6 (IL-6), IL-8 and α1-acid glycoprotein (AGP) were measured by ELISA.

Results: We found that use of reducing agents to liquefy CVM affects protein yield and the accuracy of biomarker detection. Our improved protocol results in lower protein yields but improved detection of cytokines and chemokines. Using our modified method to measure AGP in CVM we found raised levels of AGP at seven days postpartum in CVM from cows that went on to develop endometritis.

Conclusion: We conclude that processing CVM without reducing agents improves detection of biomarkers that reflect uterine health in cattle. We propose that measurement of AGP in CVM during the first week postpartum may identify cows at risk of developing clinical endometritis.

Keywords: Cervico-vaginal mucus, Postpartum, Reducing agent, Endometritis, Biomarker

Background

The postpartum bovine uterus is susceptible to diverse pathologies including viral and bacterial infection as well as endometritis, all of which impact negatively on the health, productivity and fertility of cows [1–3]. Current diagnostic methods for predicting uterine inflammation such as uterine cytology and biopsy require specialist expertise and invasive tools. In contrast, CVM could provide a useful resource for the analysis of uterine health. CVM is composed of a mixture of oviductal, uterine, cervical and vaginal secretions and their production is influenced by health status, the microbiome and pregnancy [2, 4, 5].

Biomolecules in CVM reflect the health and secretions of the uterus and CVM could therefore substitute for more invasive analyses.

Cytokines, e.g. interleukin 6 (IL-6), chemokines e.g. IL-8 and acute phase proteins e.g. α1-Acid Glycoprotein (AGP), produced by endometrial epithelial cells and local immune populations are increased in inflamed uterine tissue and in vitro models [6–8]. These inflammatory biomarkers have been detected in uterine mucus and CVM [8–10]. Due to the physical properties of mucus, processing of CVM with reducing agents is routinely recommended before the analysis of soluble-phase biomarkers [4, 9, 11]. N-acetyl-L-cysteine (NAC) and Dithiothreitol (DTT) are commonly used to homogenize mucus by reducing the disulfide bonds of mucins [12–14]. However, many immune biomarkers also have disulfide bonds [15, 16] and their detection is likely to be compromised by use of reducing agents. The overall objective of the

* Correspondence: kieran.meade@teagasc.ie
[4]Animal & Bioscience Research Department, Animal & Grassland Research and Innovation Centre, Teagasc, Grange, Co. Meath, Ireland
[5]Immunogenetics & Animal Health, Animal & Grassland Research and Innovation Centre, Teagasc, Grange, Co. Meath, Ireland
Full list of author information is available at the end of the article

current study is to improve processing of postpartum CVM for measuring candidate biomarkers that may predict uterine inflammation and disease.

Results

Total protein and SDS-PAGE electrophoresis
The standard protocol (with DTT) resulted in higher levels of total protein than the modified protocol (Fig. 1a).

To determine which protocol best preserves the integrity of proteins with molecular weights of the candidate cytokines, protocols were compared based on the protein pattern within the range of 10–35 KDa (Fig. 2b) and results are shown in Fig. 2a (red box). Furthermore, processing mucus without reducing agent gave denser band patterns for proteins with large molecular weight (> 130 KDa).

Measurement IL6, IL8 and AGP
Our modified protocol resulted in the detection of significantly higher levels of both IL-6 and IL-8 in postpartum CVM per ml of mucus ($P < 0.001$) (Fig. 1b and c). Accounting for total protein concentration in each sample, the concentration of IL-6 and IL-8 per mg of total protein was also significantly higher in the modified protocol compared to the standard protocol ($P < 0.001$) (Fig. 1d and e).

AGP levels in CVM processed without reducing agent were higher in cows which went on to develop clinical endometritis compared to healthy cows ($P < 0.05$) (Fig. 3).

Discussion
Here we demonstrated that use of reducing agents during the processing of CVM impacts on the accurate detection of proteins and biomarkers. The standard protocol (with DTT) resulted in higher levels of total protein than the modified protocol, since DTT is a powerful reducing agent that homogenizes mucus through the reduction of disulfide bounds. Due to the low level of circulating estrogen, postpartum mucus is characterized by high viscosity [5, 17]. For this reason, we believe that DTT used in standard protocols resulted in higher levels of total protein as it helped to breakdown disulfide bonds of the viscous mucins and made protein accessible for measurement by BCA assay. Thus, if for any reason the use of reducing agent is required, it is important to determine the lowest concentration that does not interfere with the marker being assessed.

SDS-PAGE confirmed that use of reducing agent affected the stability of proteins in CVM. CVM samples were run using non-reducing loading buffer to limit interaction with the effect of the reducing agent used for processing mucus. Within the molecular weight range of the main inflammatory cytokines, the protein bands were denser in the modified protocol indicating that they contain higher levels of proteins at this molecular weight. SDS-PAGE shown that for high molecular weight proteins, such as acute phase proteins and mucins, both reducing agents should be avoided as these

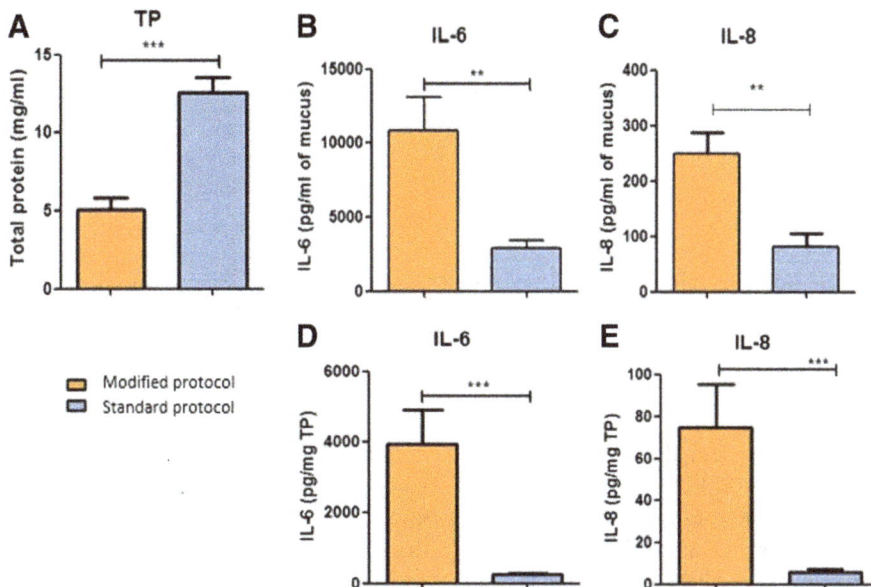

Fig. 1 a Total protein levels in CVM from 16 cows at first two weeks postpartum were measured using BCA assay and compared between our modified protocol and the standard protocol (with DTT). **b, c** IL-6 and IL-8 levels were measured by ELISA and results were presented according to the volume of mucus analyzed. **d, e** taking in consideration the amount of total protein (TP) in CVM, IL-6 and IL-8 levels were presented per mg of total protein in the mucus. Results are presented as mean ± SEM and analyzed by t test. Significant differences between groups are calculated. ***$P < 0.001$, **$P < 0.01$ and *$P < 0.05$

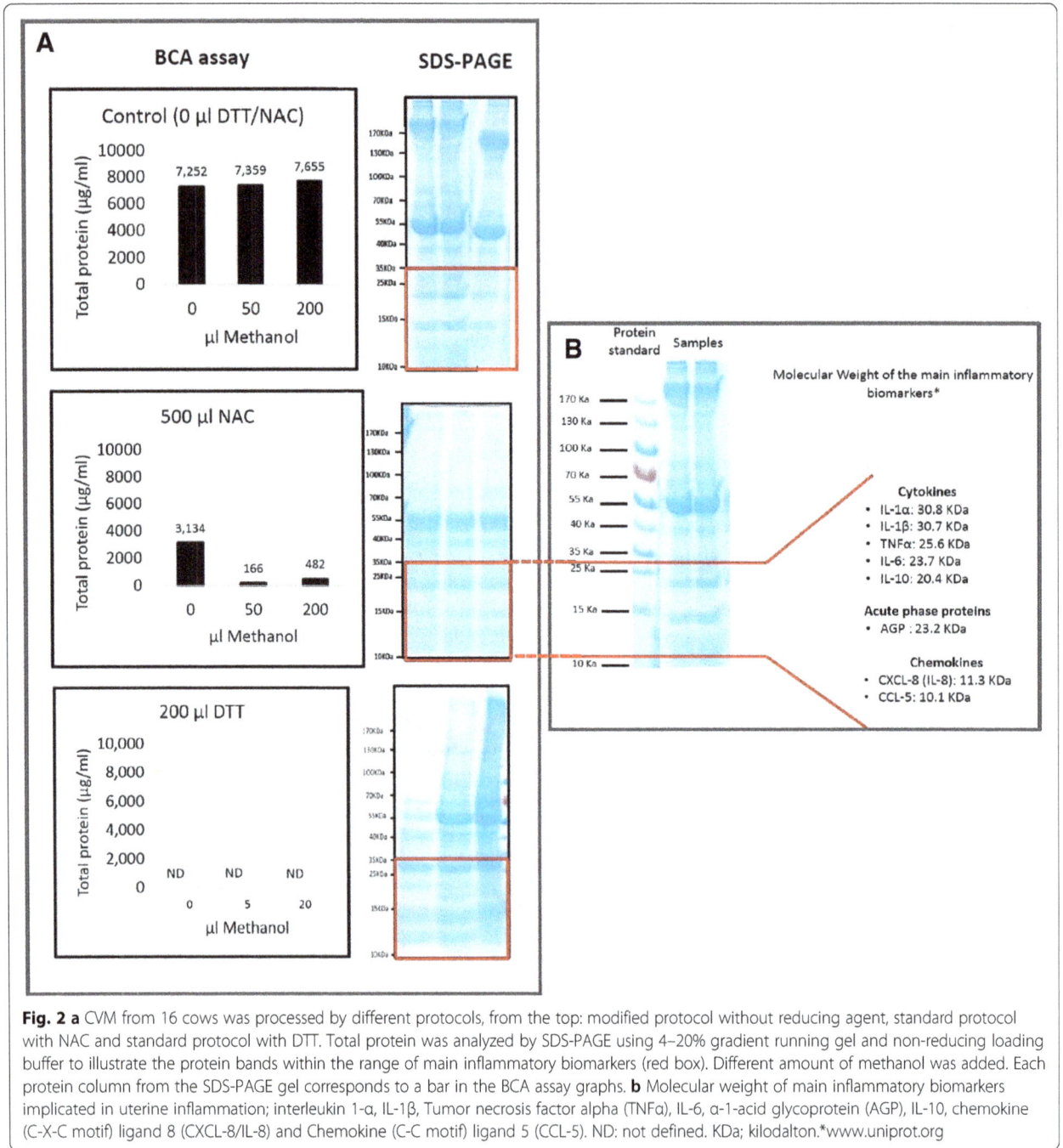

Fig. 2 a CVM from 16 cows was processed by different protocols, from the top: modified protocol without reducing agent, standard protocol with NAC and standard protocol with DTT. Total protein was analyzed by SDS-PAGE using 4–20% gradient running gel and non-reducing loading buffer to illustrate the protein bands within the range of main inflammatory biomarkers (red box). Different amount of methanol was added. Each protein column from the SDS-PAGE gel corresponds to a bar in the BCA assay graphs. **b** Molecular weight of main inflammatory biomarkers implicated in uterine inflammation; interleukin 1-α, IL-1β, Tumor necrosis factor alpha (TNFα), IL-6, α-1-acid glycoprotein (AGP), IL-10, chemokine (C-X-C motif) ligand 8 (CXCL-8/IL-8) and Chemokine (C-C motif) ligand 5 (CCL-5). ND: not defined. KDa; kilodalton.*www.uniprot.org

biomarkers would be reduced and degraded and could not be accurately measured by ELISA.

Our modified protocol resulted in the detection of significantly higher levels of both IL-6 and IL-8 in postpartum CVM. As proteins, cytokines, chemokines and acute phase proteins contain disulfide bonds between the cysteine residues, and the use of reducing agent may affect the stability of these proteins and decrease their detection using ELISA technique. For example, the IL-6 protein contains 4 cysteine residues, which are conserved between different species (i.e. human and cow) and are connected by 2 disulfide bonds (Cys 44-Cys 50 and Cys 73-Cys 83) [18, 19]. The two disulfide bridges can be reduced and alkylated under chemical reduction [20] or non-denaturing conditions [21]. Therefore, reducing agents decrease the stability of IL-6 and may decrease its detection by antibodies, since the disulfide bonds seem to be responsible for maintaining structural integrity of receptor binding sites rather than conformational stability [20, 21]. In a previous study at

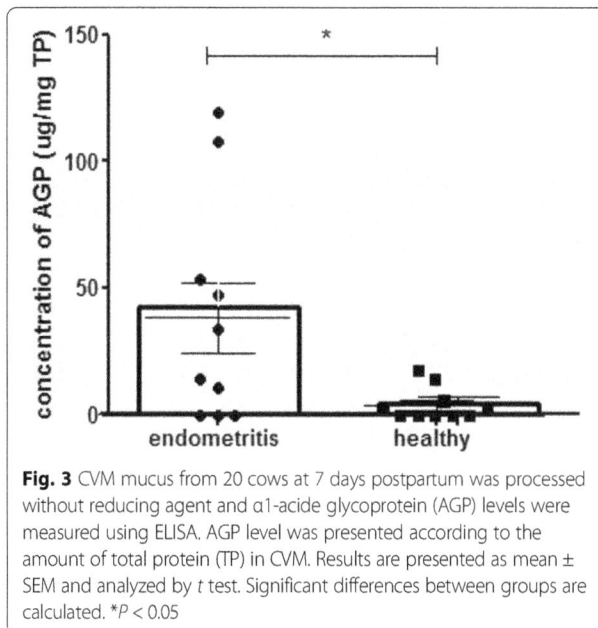

Fig. 3 CVM mucus from 20 cows at 7 days postpartum was processed without reducing agent and α1-acide glycoprotein (AGP) levels were measured using ELISA. AGP level was presented according to the amount of total protein (TP) in CVM. Results are presented as mean ± SEM and analyzed by t test. Significant differences between groups are calculated. *$P < 0.05$

high concentrations, DTT decreased by 43% the detectable concentration of IL-6 standard [15]. Likewise, IL-8 contains 4 cysteines that form 2 disulfide bonds and it is rapidly inactivated when their disulfide bonds are reduced [22] which would decrease its detection by ELISA.

Increased AGP levels are associated with uterine infection and have been proposed to be prognostic of endometritis [23, 24]. To validate our improved protocol to identify pathological inflammation, AGP was measured in CVM collected at seven days postpartum from 20 cows, 10 of which went on to develop endometritis. CVM processed with the modified protocol resulted in higher levels of AGP in cows which went on to develop clinical endometritis, compared to healthy cows. AGP has two disulfide bounds between cysteine (Cys) Cys5-Cys165 and Cys72-Cys147 [25] and use of reducing agent would breakdown these bounds and decrease the detection of AGP. Likewise, DTT was confirmed to reduce the molecular weight of pig AGP when it used for 2D electrophoresis [26]. Using our improved protocol to measure AGP in CVM to predict cows at risk of developing clinical endometritis, we found that the test could have a sensitivity of 70% and specificity of 100%. Thus, all cows positively identified at day 7 developed clinical endometritis at 21 DPP. These findings could help to reduce antibiotic use by reducing the numbers of cattle treated using CVM diagnosis alone, as well as reducing the costs associated with reproductive diseases.

Conclusion

Here we show that processing CVM without any reducing agent allows for more accurate measurement of inflammatory biomarkers in early postpartum mucus. We also show that raised AGP levels predict cows at risk of developing clinical endometritis. Thus, the improved detection of biomarkers in CVM from the postpartum cow represents a technique with significant practical utility for early disease diagnosis in a non-invasive and welfare friendly manner.

Methods
Herd identification
Material was obtained from 36 mixed-parity Holstein-Friesian dairy cows during their first two weeks postpartum. Animals used in this study belong to three commercial dairy farms. Among these cows, 16 belong to one farm were used to measure IL-6 and IL-8, and 20 cows from two different farms were diagnosed for clinical endometritis by scoring mucus aspect and odor at day 21 postpartum [24] and used to measure AGP levels. All animals were examined in their normal farming conditions and they remain on farm after the samples were collected.

Vaginal mucus collection
Vaginal examination and mucus collection were performed according to a previously described protocol [27]. Briefly, using examination sleeves, the perineum was wiped and washed with 70% ethanol to remove fecal material. An examination sleeve was covered with clean surgical glove and the gloved hand was inserted through the vulva into the vagina and CVM was collected and scored as described by Williams et al. (2005). CVM was collected in an empty sterile 20 ml tube and immediately placed on ice and transported to the laboratory within 4-6 h.

CVM processing
CVM collected from 16 cows from one farm was used to optimize the technique and processed using two different protocols: the standard protocol using reducing agent (DTT) [9, 15] and our modified protocol without reducing agent. In the modified protocol, CVM was centrifuged at 3000 x g for 15 min at 4 °C and 500 µl of the upper part was collected and mixed with 1000 µl of sterile PBS in 2 ml eppendorf tubes and vortexed. In the standard protocol (with DTT), CVM was processed according to Cronin et al. (2010). Briefly, 2 g of frozen CVM was added to 10 ml of cytolyt solution (40% methanol: 60% distilled water) and mixed with 1 mM DTT to disrupt the mucus. Tubes were then centrifuged at 3000 x g for 15 min at 4 °C. The supernatant was collected and multiple aliquots of 1 ml were prepared and used to measure IL-6, IL-8 and AGP by ELISA.

CVM from one cow was selected for further analysis using a second reducing agent NAC [15]. In the modified protocol, 500 mg from frozen CVM were mixed with 1000 µl of PBS. In the standard protocol, 500 mg of frozen CVM were mixed with 1000 µl PBS in 5 ml tubes and different volumes of 1 mM NAC (50, 100, 200, 300, 500 µl) or DTT (20, 50, 100, 200 µl) were added to each tube. Tubes were centrifuged at 3000 x g at 4 °C for 15 min and the supernatant was collected and aliquoted into Eppendorf tubes and stored at − 80 °C for further analysis by BCA assay for total protein and SDS-PAGE.

Measurement of total protein

Total protein levels in CVM were measured using a bicinchoninic acid (BCA) assay using a commercially available kit (Pierce™ BCA Protein Assay Kit (#23227, ThermoScientific®, 3747 N. Meridian Rd. Rockford, IL 61101, United States) according to manufacturer's guidelines. Briefly, the contents of one albumin standard (BSA) ampule was diluted with PBS into eight standards and one blank. The working reagent was prepared by mixing 50 parts of BCA reagent A with 1 part of BCA reagent B (50:1, Reagent A: B). Then, 25 µL of each standard or sample were added in duplicate into a microplate well. 200 µL of the working reagent was added to each well and the 96 well plate was mixed thoroughly on a plate shaker for 30 s. The plate was then covered and incubated at 37 °C for 30 min. After cooling the plate, the absorbance was measured at 562 nm on a plate reader (GloMax®-Multi Detection System, Promega Corporation, 2800 Woods Hollow Road Madison, WI 53711 USA).

SDS-PAGE electrophoresis

To visualize the range of proteins in mucus, a gradient running gel 4–20% was chosen in combination with a 5% stacking gel. To not interact with the effect of the reducing agent added to the mucus, samples were loaded in the gel cassette using non-reducing loading buffer. Gels were run at 110 V for 90 min and then stained with Coomassie blue G250 for one hour with gentle agitation. After overnight distaining in 10% glacial acetic acid, gels were scanned and interpreted.

Measurement IL6, IL8 and AGP

After thawing, samples were centrifuged at 3000 x g for 15 min at 4 °C and the supernatant was used to measure biomarkers. Levels of IL-6, IL-8 and AGP in CVM were measured using commercial ELISA kits (human IL-8 ELISA kit: R&D Systems Inc., Minneapolis, Minnesota, USA; bovine IL-6 ELISA kit: #ESS0029, ThermoScientific®, 3747 N. Meridian Rd. Rockford, IL 61101; Cow AGP: #AGP-11, Life Diagnostics Inc.®, P.O. Box 5205, West Chester, Pa. 19380) according to the guidelines provided by the manufacturers and modified according to previous studies in bovine [6, 9, 10]. Human IL-8 ELISA was used because no bovine specific IL-8 ready-to-use ELISA is commercially available. Furthermore, the antibodies used in the kit have been confirmed to cross-react

with bovine IL-8 [6, 28]. To validate the usefulness of the improved protocol to detect uterine health problems, AGP was measured in CVM from 20 animals of which 10 developed clinical endometritis.

Statistical analysis

Statistical analysis was performed using GraphPad® Prism 5 software (GraphPad Software, Inc. 7825 Fay Avenue, Suite 230 La Jolla, CA 92037 USA). A Students t test was used to compare results between two groups, while one-way ANOVA with Bonferroni post-comparison test was used to compare between three or more groups. Results were presented as mean ± SEM and considered statistically significant at P-value < 0.05.

Abbreviations
AGP: α1-acid glycoprotein; BCA assay: Bicinchoninic acid assay; CVM: Cervico-vaginal mucus; Cys: Cysteine; DTT: Dithiothriotol; ELISA: Enzyme-linked immunosorbent assay; IL: Interleukin; NAC: N-Acetyl-L-Cysteine; SDS-PAGE: Sodium dodecyl sulfate polyacrylamide gel electrophoresis; SEM: Standard error of the mean

Acknowledgements
The authors acknowledge helpful collaboration from the O'Farrelly and Meade laboratory members.

Funding
This work was supported by scholarship from The Algerian Ministry for High Education and Scientific Research and University of Tiaret, Algeria by funding the one-year research visit of MA, and by a grant ofarrecl-HRB-HRA_POR/2012/37 for funding the reagents and material used during the experiment. Funding to KGM and CO'F was also provided by the Irish Department of Agriculture, Food and the Marine (Stimulus 13/S/472).

Authors' contributions
Conceived and designed the experiments: COF, KGM, AC and MA. Sample collection and performed the experiments: MA. Analyzed the data: MA. Contributed to the writing: COF, KGM, AC, MA and PK. All authors have read and approved the final manuscript.

Consent for publication
Not applicable.

Competing interests
The authors declare that they have no competing interests.

Author details
[1]Comparative Immunology Group, School of Biochemistry and Immunology, Trinity College, Dublin, Ireland. [2]Institute of Veterinary Sciences, Tiaret, Algeria. [3]Department of Clinical Sciences, Ross University School of Veterinary Medicine, Basseterre, West Indies, St. Kitts and Nevis. [4]Animal & Bioscience Research Department, Animal & Grassland Research and Innovation Centre,

Teagasc, Grange, Co. Meath, Ireland. [5]Immunogenetics & Animal Health, Animal & Grassland Research and Innovation Centre, Teagasc, Grange, Co. Meath, Ireland.

References

1. Sheldon IM, Lewis GS, LeBlanc S, Gilbert RO. Defining postpartum uterine disease in cattle. Theriogenology. 2006;65:1516–30.

2. Sheldon IM, Price SB, Cronin J, Gilbert RO, Gadsby JE. Mechanisms of infertility associated with clinical and subclinical endometritis in high producing dairy cattle. Reprod Domest Anim. 2009;44(Suppl 3):1–9.

3. LeBlanc SJ, Duffield TF, Leslie KE, Bateman KG, Keefe GP, Walton JS, Johnson WH. Defining and diagnosing postpartum clinical endometritis and its impact on reproductive performance in dairy cows. J Dairy Sci. 2002;85:2223–36.

4. Zegels G, Van Raemdonck GA, Tjalma WA, Van Ostade XW. Use of cervicovaginal fluid for the identification of biomarkers for pathologies of the female genital tract. Proteome Sci. 2010;8:63.

5. Lopez-Gatius F, Miro J, Sebastian I, Ibarz A, Labernia J. Rheological properties of the anterior vaginal fluid from superovulated dairy heifers at estrus. Theriogenology. 1993;40:167–80.

6. Foley C, Chapwanya A, Callanan JJ, Whiston R, Miranda-CasoLuengo R, Lu J, Meijer WG, Lynn DJ, O'Farrelly C, Meade KG. Integrated analysis of the local and systemic changes preceding the development of post-partum cytological endometritis. BMC Genomics. 2015;16:811.

7. Chapwanya A, Meade KG, Foley C, Narciandi F, Evans AC, Doherty ML, Callanan JJ, O'Farrelly C. The postpartum endometrial inflammatory response: a normal physiological event with potential implications for bovine fertility. Reprod Fertil Dev. 2012;24:1028–39.

8. Brodzki P, Kostro K, Brodzki A, Wawron W, Marczuk J, Kurek L. Inflammatory cytokines and acute-phase proteins concentrations in the peripheral blood and uterus of cows that developed endometritis during early postpartum. Theriogenology. 2015;84:11–8.

9. Cronin JG, Hodges R, Pedersen S, Sheldon IM. Enzyme linked immunosorbent assay for quantification of bovine interleukin-8 to study infection and immunity in the female genital tract. Am J Reprod Immunol. 2015;73:372–82.

10. Kim IH, Kang HG, Jeong JK, Hur TY, Jung YH. Inflammatory cytokine concentrations in uterine flush and serum samples from dairy cows with clinical or subclinical endometritis. Theriogenology. 2014;82:427–32.

11. Van Raemdonck GA, Tjalma WA, Coen EP, Depuydt CE, Van Ostade XW. Identification of protein biomarkers for cervical cancer using human cervicovaginal fluid. PLoS One. 2014;9:e106488.

12. Kelly MM, Leigh R, Horsewood P, Gleich GJ, Cox G, Hargreave FE. Induced sputum: validity of fluid-phase IL-5 measurement. J Allergy Clin Immunol. 2000;105:1162–8.

13. Saraswathy Veena V, Sara George P, Jayasree K, Sujathan K. Comparative analysis of cell morphology in sputum samples homogenized with dithiothreitol, N-acetyl-L cysteine, Cytorich((R)) red preservative and in cellblock preparations to enhance the sensitivity of sputum cytology for the diagnosis of lung cancer. Diagn Cytopathol. 2015;43:551–8.

14. Louis R, Shute J, Goldring K, Perks B, Lau L, Radermecker M, Djukanovic R. The effect of processing on inflammatory markers in induced sputum. Eur Respir J. 1999;13:660–7.

15. Woolhouse IS, Bayley DL, Stockley RA. Effect of sputum processing with dithiothreitol on the detection of inflammatory mediators in chronic bronchitis and bronchiectasis. Thorax. 2002;57:667–71.

16. Carr TF, Spangenberg A, Hill JL, Halonen MJ, Martinez FD. Effect of Dithiothreitol on sputum Interleukin-13 protein measurement. J Allergy Clin Immunol. 2015;135:AB179.

17. Tsiligianni T, Amiridis GS, Dovolou E, Menegatos I, Chadio S, Rizos D, Gutierrez–Adan A. Association between physical properties of cervical mucus and ovulation rate in superovulated cows. Can J Vet Res. 2011;75:248–53.

18. Simpson RJ, Hammacher A, Smith DK, Matthews JM, Ward LD. Interleukin-6: structure-function relationships. Protein Sci. 1997;6:929–55.

19. Clogston CL, Boone TC, Crandall BC, Mendiaz EA, Lu HS. Disulfide structures of human interleukin-6 are similar to those of human granulocyte colony stimulating factor. Arch Biochem Biophys. 1989;272:144–51.

20. Zhang JG, Moritz RL, Reid GE, Ward LD, Simpson RJ. Purification and characterization of a recombinant murine interleukin-6. Isolation of N- and C-terminally truncated forms. Eur J Biochem. 1992;207:903–13.

21. Rock FL, Li X, Chong P, Ida N, Klein M. Roles of disulfide bonds in recombinant human interleukin 6 conformation. Biochemistry. 1994;33:5146–54.

22. Baggiolini M, Walz A, Kunkel SL. Neutrophil-activating peptide-1/interleukin 8, a novel cytokine that activates neutrophils. J Clin Invest. 1989;84:1045–9.

23. Sheldon IM, Noakes DE, Rycroft A, Dobson H. Acute phase protein responses to uterine bacterial contamination in cattle after calving. Vet Rec. 2001;148:172–5.

24. Williams EJ, Fischer DP, Pfeiffer DU, England GC, Noakes DE, Dobson H, Sheldon IM. Clinical evaluation of postpartum vaginal mucus reflects uterine bacterial infection and the immune response in cattle. Theriogenology. 2005;63:102–17.

25. Schmid K, Burgi W, Collins JH, Nanno S. The disulfide bonds of alpha1-acid glycoprotein. Biochemistry. 1974;13:2694–7.

26. Heegaard PMH, Miller I, Sorensen NS, Soerensen KE, Skovgaard K. Pig α(1)-acid glycoprotein: characterization and first description in any species as a negative acute phase protein. PLoS One. 2013;8:e68110.

27. Adnane M, Chapwanya A, Kaidi R, Meade KG, O'Farrelly C. Profiling inflammatory biomarkers in cervico-vaginal mucus (CVM) postpartum: potential early indicators of bovine clinical endometritis? Theriogenology. 2017;103:117–22.

28. Shuster DE, Kehrli ME Jr, Rainard P, Paape M. Complement fragment C5a and inflammatory cytokines in neutrophil recruitment during intramammary infection with Escherichia coli. Infect Immun. 1997;65:3286–92.

20

Detection of a streptomycin-resistant *Mycobacterium bovis* strain through antitubercular drug susceptibility testing of Tunisian *Mycobacterium tuberculosis* complex isolates from cattle

Saif Eddine Djemal[1], Cristina Camperio[2], Federica Armas[2], Mariam Siala[1,3], Salma Smaoui[4,5,6], Feriele Messadi-Akrout[3,4], Radhouane Gdoura[1] and Cinzia Marianelli[2]*

Abstract

Background: A rising isolation trend of drug-resistant *M. bovis* from human clinical cases is documented in the literature. Here we assessed *Mycobacterium tuberculosis* complex isolates from cattle for drug susceptibility by the gold standard agar proportion method and a simplified resazurin microtitre assay (d-REMA). A total of 38 *M. tuberculosis* complex strains, including *M. bovis* (*n* = 36) and *M. caprae* (*n* = 2) isolates, from cattle in Tunisia were tested against isoniazid, rifampin, streptomycin, ethambutol, kanamycin and pyrazinamide.

Results: *M. caprae* isolates were found to be susceptible to all test drugs. All *M. bovis* strains were resistant to pyrazinamide, as expected. In addition, one *M. bovis* isolate showed high-level resistance to streptomycin (MIC > 500.0 μg/ml). Concordant results with the two methods were found. The most common target genes associated with streptomycin resistance, namely the *rrs*, *rpsL* and *gidB* genes, were DNA sequenced. A non-synonymous mutation at codon 43 (K43R) was found in the *rpsL* gene. To the best of our knowledge, this is the first report describing the isolation of a streptomycin-resistant *M. bovis* isolate from animal origin.

Conclusions: Antitubercular drug susceptibility testing of *M. bovis* isolates from animals should be performed in settings where bTB is endemic in order to estimate the magnitude of the risk of drug-resistant tuberculosis transmission to humans.

Keywords: Bovine tuberculosis, *Mycobacterium tuberculosis* complex, Streptomycin resistance, Resazurin microtitre assay, d-REMA

Background

The World Organization for Animal Health (OIE) has recognized bovine tuberculosis (bTB) as an important animal disease and zoonosis [1]. bTB causes significant economic losses to farmers due to livestock deaths, reduced productivity and restrictions for trading animals.

The main causal agents of bTB are *Mycobacterium bovis* and, to a lesser extent, *Mycobacterium caprae*, both members of the *Mycobacterium tuberculosis* complex. These

pathogens may also cause tuberculosis in humans (hereafter referred to as zoonotic tuberculosis, zTB) although the true incidence of this disease in human beings is unknown [2–4]. An extensive meta-analysis found the proportion of zTB to be ≤1.4% in countries outside Africa and 2.8% on average in African countries [3]. The vast majority of cases were due to *M. bovis* and the contribution of cases due to *M. caprae* was not quantified. Substantial evidence suggests that zTB might be underestimated because of two major issues hindering understanding of the true burden of this disease: first, the absence of systematic surveillance for *M. bovis* and *M. caprae* as cause of tuberculosis in humans in all low-income and high tuberculosis

* Correspondence: cinzia.marianelli@iss.it
[2]Department of Food Safety, Nutrition and Veterinary Public Health, Istituto Superiore di Sanità, Viale Regina Elena 299, 00161 Rome, Italy

burden countries where bTB is endemic; and second, the inability of laboratory procedures most commonly used to diagnose human tuberculosis to identify and differentiate these pathogens from *M. tuberculosis*, with the result that all cases may be assumed to be caused by *M. tuberculosis* [5].

Animal test-and-slaughter schemes have successfully reduced the prevalence of bTB in most industrialized countries. The situation is profoundly different in unindustrialized countries where the WHO, in conjunction with FAO and OIE, has classified bTB as a neglected zoonosis. In South Africa, as in other regions in Africa, the lack of bTB control programmes [6] makes communities with high HIV/AIDS infection rates and those living in close contact with infected animals or animal products more vulnerable to zTB [7, 8].

M. bovis has one of the broadest host ranges of any known zoonotic pathogen and is globally distributed [9]. The phylogenomic analysis of *M. bovis* genomes has recently revealed large-scale polymorphisms, which may contribute to the differential adaptability of the pathogen [10]. *M. bovis* is naturally resistant to pyrazinamide (PZA) [11], which is one of the first-line antibiotics used to treat tuberculosis in humans. The lack of prompt identification of *M. bovis* human cases may result in improper treatments and have lethal consequences [12]. Several studies have documented additional drug resistances in human *M. bovis* isolates over the last two decades, such as toward isoniazid (INH) [13, 14], streptomycin (STR) [15] or multiple drugs [16–19].

M. caprae, on the other hand, is evolutionarily older than *M. bovis* and accounts for a smaller burden of zTB. Moreover, it is not globally distributed but primarily restricted to European countries [3, 20, 21]. Its resistance against first-line drugs is rarely documented [22]. Nevertheless, *M. caprae* has received more attention in recent years, particularly due to the increasing number of *M. caprae* outbreaks in wild or domestic animals which pose a threat to human health [21].

While mycobacteria isolated from human cases are generally assessed for drug susceptibility, studies on antitubercular drug susceptibility testing of *M. bovis* and *M. caprae* isolated from animals are limited [23–29]. So far, *M. bovis* isolates from cattle resistant to INH and rifampin (RIF) have been documented in Italy [23] and Brazil [29], to the best of our knowledge. Monitoring of antitubercular drug resistance of *M. bovis* isolated from animals may thus contribute to reducing the risk of drug-resistant *M. bovis* transmission from animals to humans and among human beings.

In Tunisia, bTB is enzootic and the consumption of raw milk and unpasteurized dairy products is common. A previous study demonstrated that raw milk consumers are at high risk of being infected with *M. bovis* [30]. Numerous

clinical cases of human extrapulmonary tuberculosis due to *M. bovis* have been recently documented in Tunisia [31, 32] and the consumption of unpasteurized dairy products has been indicated as the most likely source of transmission [33]. Implementation of effective and comprehensive strategies to control bTB and to prevent zTB are therefore of primary importance in the country.

In this study we assessed drug susceptibility of *M. bovis* and *M. caprae* isolates from cattle in Tunisia towards six antitubercular drugs – PZA, INH, RIF, ethambutol (EMB), STR and kanamycin (KAN) – by both the gold standard agar proportion method and the simplified resazurin microtitre assay (REMA), the dichotomous REMA (d-REMA) recently proposed by Marianelli and colleagues [27].

Methods

All experimental procedures here described were carried out at the Department of Food Safety, Nutrition and Public Animal Health of the Istituto Superiore di Sanità (ISS), (Italy).

M. tuberculosis complex strains

A total of 38 *M. tuberculosis* complex strains, including *M. bovis* (*n* = 36) and *M. caprae* (*n* = 2) previously isolated in Tunisia and molecular typed by spoligotyping and MIRU-VNTR analysis [34], were provided by the University of Sfax (Tunisia). The isolates were analysed for susceptibility to INH, RIF, STR, EMB, KAN and PZA at ISS.

Isolates were subcultured in Middlebrook 7H9 medium (Biolife, Italy) with 10% oleic acid-albumin-dextrose-catalase (OADC) enrichment (Becton Dickinson and Company) before being tested. To aid the dispersion of bacterial clumps, 3-mm glass beads were added to the tubes and bacterial suspensions were vigorous vortexed for 15 s. Any remaining large bacterial clumps were allowed to settle. Bacterial suspensions were then adjusted to match a 1.0 McFarland turbidity standard.

All *M. bovis* and *M. caprae* strains were isolated from lymph node and tissue samples showing tuberculosis-compatible lesions. Samples were collected at the abattoir during the *postmortem* inspection, in accordance with national laws.

d-REMA testing

The d-REMA test, based on Palomino and colleagues [35] and slightly modified by Marianelli and colleagues [27] by testing only two concentrations per drug – the cut-off value for drug resistance, R, and the cut-off value of REMA drug susceptibility, S – was used. The R values were determined in studies where REMA was validated against the gold standard method [35–40]. The test was carried out in triplicate in 96-well plates, as previously described [27]. Briefly, drug solutions were prepared at concentrations of 20 mg/ml (PZA; Sigma-Aldrich, UK) and 2 mg/ml in distilled water (INH, STR, EMB, and KAN; Sigma-Aldrich, UK) or methanol (RIF; Sigma-Aldrich, UK), filter sterilised, and frozen until used.

One hundred microliters of Middlebrook 7H9 medium supplemented with OADC was inoculated into each well. One hundred microliters of each bacterial suspension – previously adjusted to a 1.0 McFarland standard and then diluted 1:20 in the same medium – was then inoculated. The antitubercular drugs were subsequently added at two final concentrations, R and S: 0.1 (S) and 0.25 (R) µg/ml for INH, 0.25 (S) and 0.5 (R) µg/ml for RIF, 0.5 (S) and 1.0 (R) µg/ml for STR, 2.5 (S) and 3.125 (R) µg/ml for EMB, 2.5 (S) and 3.125 (R) µg/ml for KAN and 100.0 (S) and 800.0 (R) µg/ml for PZA.

Plates were covered with lids, placed in a plastic bag and incubated at 37 °C for 7 days. Finally, 30 µl of freshly prepared 0.01% resazurin solution (Acros Organics, USA) was added to each well. The plates were incubated overnight at 37 °C and assessed for colour development. The d-REMA testing was repeated twice for those isolates showing either resistance or ambiguous chromatic change.

The drug-sensitive *M. bovis* ATCC 19210 (used as negative control) and two resistant *M. avium* strains (used as positive controls) – one resistant to EMB and one resistant to both INH and EMB – from the Italian bacteria collection were included in the study.

Agar proportion method in Middlebrook 7H11 medium

The agar proportion method was performed in Middlebrook 7H11 agar. The test was carried out according to the approved standard (M24A) for Susceptibility Testing of Mycobacteria, Nocardiae, and Other Aerobic Actinomycetes published by the Clinical and Laboratory Standards Institute (CLSI) [41]. Briefly, the turbidity of the inoculum was adjusted to match a 1.0 McFarland standard, and diluted 1:100 and 1:10,000. One hundred microliters of these solutions was then inoculated into 35 mm plates with and without the test drug. The test was carried out in duplicate. The following final critical drug concentrations were used: 0.2 µg/ml for INH; 1.0 µg/ml for RIF; 7.5 µg/ml for EMB; 2.0 µg/ml for STR; 6.0 µg/ml for KAN; and 100 µg/ml for PZA [40, 41]. After 3 weeks of incubation at 37 °C, the number of colony forming units (CFU) growing on the drug-containing medium was compared with those growing on the drug-free medium and expressed as a percentage of the latter. The isolate was considered resistant if the number of colonies on a medium containing an antimicrobial agent, relative to the number observed on a drug-free medium was ≥1%. Negative (*M. bovis* ATCC 19210) and positive (*M. avium* strains) controls were also included in the test. The test was repeated twice for isolates showing resistance.

Streptomycin MIC determination by REMA

MIC testing was to be performed only in case of drug resistance. Since *M. bovis* growth was observed only in the presence of STR and the control drug PZA (as expected) in both the d-REMA assay and agar proportion method (see Results), MIC testing was carried out only for STR. The test was performed in triplicate according to Palomino and colleagues [35]. Ten dilutions were tested (2.5, 5.0, 10.0, 20.0, 50.0, 100.0, 200.0, 300.0, 400.0 and 500.0 µg/ml). The MIC was defined as the lowest drug concentration that prevented resazurin colour change from blue to pink. The *M. bovis* ATCC control was also tested.

DNA sequencing

Sequencing, too, was to be performed only in case of drug resistance. Since resistance was observed only in *M. bovis* and towards STR and the control drug PZA (see Results), the most common target genes associated with resistance to STR encoding 16S rRNA (*rrs*), ribosomal protein S12 (*rpsL*) [42] and a 7-methylguanosine methyltransferase (*gidB*) [43], were investigated in the resistant *M. bovis* isolate and in three randomly selected STR-susceptible *M. bovis* isolates..

DNA was extracted from cultures using a commercial kit (InstaGene Matrix; Bio-Rad Laboratories, Italy). The whole *rrs*, *rpsL* and *gidB* genes were PCR amplified and sequenced. The reference sequence accession number (AC) NC_002945 of *M. bovis* AF2122/97 available at NCBI was used to design primers for the PCR amplification and sequencing of the *rrs* gene. The *rpsL* and *gidB* genes were PCR amplified and sequenced according to Feuerriegel and colleagues [44].

PCR products were analysed by 2% agarose gel electrophoresis, stained with GelRed Nucleic Acid Stain (Biotium Inc., Hayward, CA), purified by ExoSAP-IT PCR Product Cleanup (Affymetrix, CA) and sequenced by using PCR and, if required, sequencing primers. Sequences were analysed using the ABI Prism SeqScape Software, version 2.0 (Applied Biosystems, Foster City, CA). All consensus sequences generated were then compared to the published, drug-sensitive *M. bovis* AF2122/97 reference strain, to detect genetic variation. Mutations were confirmed through resequencing. To distinguish silent from missense mutations, amino acid sequences were theoretically deduced. PCR and DNA sequencing primers used, PCR conditions followed and size of the amplicons obtained are listed in Table 1.

Results

Drug susceptibility

M. bovis and *M. caprae* isolates, as well as control strains, have been tested here against PZA, INH, RIF, EMB, STR and KAN by both the agar proportion method and d-REMA.

Table 1 PCR and DNA sequencing primers

Primer	Sequence (5' → 3')	PCR conditions[a]			Size (bp)	Reference
		D (s)	A (°C, s)	E (s)		
rrs-F	CGT GGC CGT TTG TTT TGT CA	· 30	55, 30	90	1744	This study
rrs-R	AAG TCC GAG TGT TGC CTC AG					
rrs-seq1	GAA GAA GCA CCG GCC AAC TA					
rrs-seq2	TTG TAC CGG CCA TTG TAG CA					
rpsL-F	ATG AGA CGA ATC GAG TTT GAG	30	55, 30	60	632	[44]
rpsL-R	GCT CAA GCG CAC CAT AAA CAA					
gidB-F	CGC CGA GTC GTT GTG CT	30	55, 30	60	892	[44]
gidB-R	AGC CTG GCC CGA CCT TA					

[a]Each PCR assay includes an initial denaturation step at 95 °C for 15 min to fully denature the template, followed by 35 cycles of denaturation, annealing and extension steps. D, denaturation time in s at 95 °C; A, annealing conditions (temperature and time in s); extension time in s

d-REMA results were obtained after 8 days of incubation. All *M. bovis* strains, including the *M. bovis* ATCC control, were resistant to PZA as expected. One out of 36 *M. bovis* isolates showed an additional drug resistance, the STR resistance as shown in Fig. 1, lines G–I. On the other hand, *M. caprae* isolates showed sensitivity against all drugs. The susceptibility of the *M. bovis* ATCC control to all test drugs (Fig. 1, lines J–L) was confirmed, as was the resistance of the two *M. avium* controls towards either INH or INH and ETB (Fig. 1, lines M–O). Results were confirmed through retesting the resistant isolate and controls by both drug susceptibility methods.

After 3 weeks of incubation, d-REMA results were confirmed by the agar proportion method.

The STR-resistant *M. bovis* isolate was subjected to further investigation, including STR MIC determination and DNA sequencing. The isolate showed resistance to all ten test dilutions: the blue-to-mauve colour change occurred at MIC values ranging from 2.5 to 500.0 µg/ml. We were thus unable to determine the STR MIC, as shown in Fig. 2, lines A–C. The *M. bovis* ATCC control, on the other hand, did not grow at any drug concentration tested (Fig. 2, lines D–F).

DNA sequencing of drug target genes

We then PCR amplified and sequenced the most common target genes associated with resistance to STR, namely the *rrs*, *rpsL* and *gidB* genes, in the resistant *M. bovis* isolate and in three randomly selected STR-susceptible *M. bovis* isolates.

A consensus sequence for each gene was generated and compared to the reference strain *M. bovis* AF2122/ 97 available in GenBank to detect genetic variation. A non-synonymous mutation – nucleotide substitution AAG → AGG at codon 43 (mutation K43R) – was found only in the *rpsL* gene of the STR-resistant isolate. No mutations were detected in the other two target genes in either of the isolates.

Discussion

Tunisia is an endemic country for bTB. Despite the implementation of a national bTB control programme, many intensive farms belonging to the state or parastatal sector still do not meet sanitary standards for effective prophylaxis; others, belonging to private owners, also have scarce, if any, veterinary activity [45]. bTB therefore continues to be widespread in this country, mainly in the private sector, which owns more than 70% of the cattle livestock [46]. In Tunisia, the consumption of raw milk and unpasteurized dairy products is common. A previous study has shown that raw milk may spread *M. bovis*, putting consumers of raw milk or derivatives at high risk of being infected [30]. Although the exact contribution of *M. bovis* to the burden of zTB in Tunisia remains unknown, recent studies have indicated *M. bovis* as the main etiological agent of human extrapulmonary tuberculosis [31–33]. In consideration of the above, bTB represents a serious public health problem in Tunisia and effective disease control programmes have to be implemented urgently.

So far, no study has been conducted to estimate the risk of drug or multi-drug resistant bTB transmission to humans in Tunisia where *M. bovis* infection is endemic in livestock. On that account, we assessed at ISS the susceptibility of 36 *M. bovis* and two *M. caprae* isolates from Tunisia to six antibiotic drugs commonly administered to patients with tuberculosis. We detected STR resistance in one *M. bovis* isolate and characterized the nucleotide mutation associated with that resistance.

Drug susceptibility testing is rarely carried out on *M. bovis* and other *Mycobacterium* isolates of animal origin although bTB poses a serious threat to human health particularly in low-income and bTB endemic countries [3, 5]. Multidrug-resistant *M. bovis* outbreaks in humans have occurred, some with serious consequences [16, 18, 47]. *M. bovis* isolates from cattle with drug resistance towards either RIF or INH or towards both drugs have already been documented [23, 29]. Drug susceptibility surveillance of

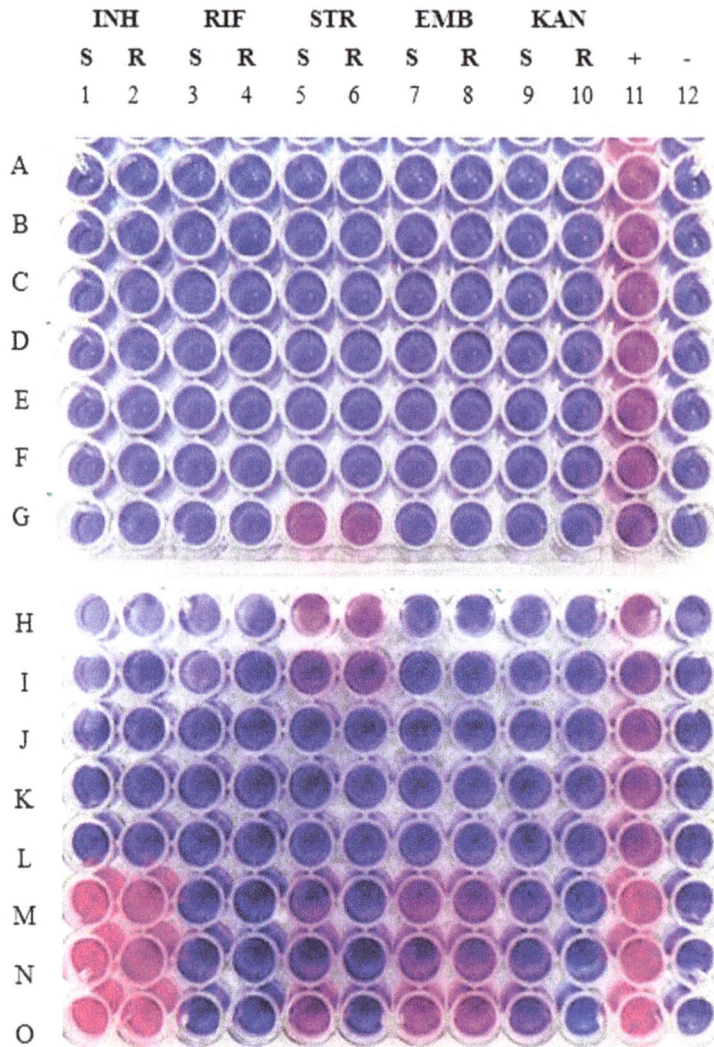

Fig. 1 Drug susceptibility profiles. Drug susceptibility to INH, RIF, STR, EMB and KAN by d-REMA. A–C, D–F, G–I: sample triplicates. J–L: *M. bovis* ATCC control in triplicates; M–O: *M. avium* control resistant to both INH and EMB in triplicates. 1–2: 0.1 (S) and 0.25 (R) µg/ml for INH; 3–4: 0.25 (S) and 0.5 (R) µg/ml for RIF; 5–6: 0.5 (S) and 1.0 (R) µg/ml for STR; 7–8: 2.5 (S) and 3.125 (R) µg/ml for EMB; 9–10: 2.5 (S) and 3.125 (R) µg/ml for KAN; (+) positive control containing no drug; (–) negative control containing uninoculated media

M. bovis from animals may therefore contribute to preventing the transmission of multidrug-resistant strains to humans and to controlling possible outbreaks.

Several techniques for testing mycobacterial drug susceptibility are available for *M. tuberculosis* and include the conventional assays [48] and the more rapid colorimetric [49] and molecular methods [50]. The WHO has recently recommended the use of colorimetric assays, which are highly sensitive, specific, rapid and inexpensive methods employing specific reagents to produce a change in colour [51].

Among these, REMA, an indirect method based on the reduction of the coloured dye resazurin added to liquid culture medium on a microtitre plate after exposure of mycobacterial strains to antituberculosis drugs in vitro,

was successfully tested on human isolates of *M. tuberculosis* [35, 36, 39, 40]. The REMA plate method proposed by Palomino and colleagues [35] has recently been simplified by Marianelli and colleagues by testing only two concentrations per drug per isolate, the R and S cut-off values, the d-REMA assay [27]. Here, we used both d-REMA and the agar proportion method to assess the bovine isolates for drug susceptibility. We found agreement between the results from the two methods. *M. caprae* isolates were sensitive to all test drugs and *M. bovis* isolates were all PZA resistant, as expected. In addition, one *M. bovis* isolate showed high-level resistance to STR (MIC > 500.0 µg/ml). To our knowledge, this is the first study describing an STR-resistant *M. bovis* isolate of animal origin. We further investigated the STR resistance

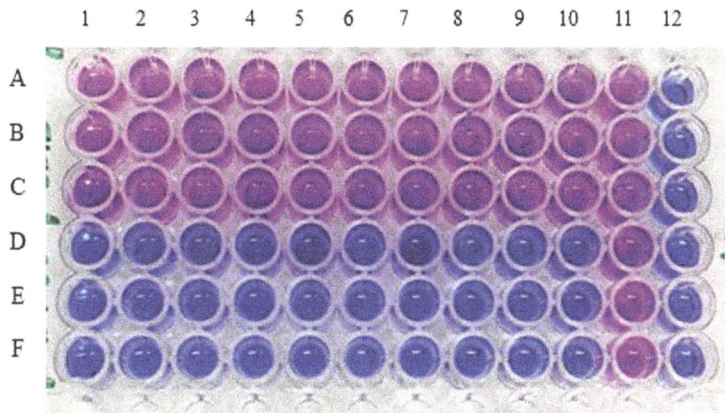

Fig. 2 STR MIC results. A–C STR-resistant *M. bovis* isolate in triplicates. D–F: *M. bovis* ATCC control in triplicates. 1–10: 2.5, 5.0, 10.0, 20.0, 50.0, 100.0, 200.0, 300.0, 400.0 and 500.0 µg/ml for STR; (+) positive control containing no drug; (–) negative control containing uninoculated media

here observed by sequencing the most common STR target genes, namely the *rrs*, *rpsL* and *gidB* genes. We found a non-synonymous mutation AAG → AGG at codon 43 (K43R) in the *rpsL* gene.

The K43R substitution in the *rpsL* gene is the single most frequent mutation associated with high-level STR resistance in *M. tuberculosis* [42, 52]. It may explain why our *M. bovis* isolate still grew at the highest STR concentration tested (500.0 µg/ml), preventing us from determining the STR MIC. Although the most common target genes associated with resistance to STR have been analysed, we cannot exclude, however, that other mechanisms may be involved. The isolation from human clinical cases of highly STR-resistant *M. tuberculosis* strains carrying the K43R substitution is largely documented in the literature [53–55]. The isolation of STR monoresistant [15] and multidrug-resistant *M. bovis* isolates from humans [16–19] is also described. Drug- and multidrug-resistant tuberculosis is one of the major threats to human medicine. It leads to treatment failures and, in the worst cases, to untreatable infections that cause death.

A recent 15-year laboratory-based surveillance programme conducted in Mexico City on mycobacterial isolates from human clinical samples showed a rising trend of *M. bovis* isolates which caused a higher proportion of pulmonary tuberculosis than previously observed in that area [15]. Additionally, the authors described an increasing rate of primary STR monoresistance in *M. bovis* isolates from humans over time, perhaps as a result of STR usage in cattle [15].

Our results, coupled with the literature, suggest that overuse and misuse of antibiotics in cattle from bTB endemic areas may lead to the development of drug-resistant *M. bovis* strains which, consequently, put humans at risk for primary drug resistant zTB. More effective strategies to reduce antibiotic use in farm animals should therefore be implemented urgently.

Conclusions

We describe, for the first time, the detection of high-level STR resistance in *M. bovis* from animal origin. Our results suggest that antitubercular drug susceptibility testing of *M. bovis* isolates from animals should be performed in settings where bTB is endemic in order to estimate the magnitude of the risk of drug-resistant tuberculosis transmission to humans.

Abbreviations

bTB: bovine tuberculosis; CFU: Colony forming units; CLSI: Clinical and Laboratory Standards Institute; d-REMA: dichotomous resazurin microtitre assay; EMB: Ethambutol; FAO: Food and Agriculture Organization; INH: Isoniazid; KAN: Kanamycin; MIC: Minimal inhibitory concentration; NCBI: National Center for Biotechnology Information; OADC: Oleic acid-albumin-dextrose-catalase; OIE: World Organization for Animal Health; PZA: Pyrazinamide; REMA: Resazurin microtitre assay; RIF: Rifampin; STR: Streptomycin; WHO: World Health Organization; zTB: zoonotic tuberculosis

Acknowledgements

We are grateful to Umberto Agrimi (Istituto Superiore di Sanità, Rome, Italy) for covering the publication fees.

Funding

This work was entirely financed by ISS, Italy. Travel and accommodation expenses for the training of the PhD student Saif Eddine Djemal at the ISS laboratories were covered by the University of Sfax, Tunisia. The authors received no specific funds for this work.

Authors' contributions

SED was involved in performing d-REMA tests and supplied *M. bovis* animal isolates. CC and FA were both involved in the molecular gene characterization and supplied *M. bovis* control strains. MS, SS, FM-A, RG were involved in the conceptualization of the study and supplied *M. bovis* animal isolates. CM was involved in the conceptualization of the study; she also designed the study, performed drug susceptibility tests (d-REMA tests and agar proportion method) and molecular gene characterization, supplied *M. bovis* control strains, analysed data, wrote the manuscript, funded and supervised the entire research study. All authors read and approved the final manuscript.

Consent for publication

Not applicable.

Competing interests

The authors declare that they have no competing interests.

Author details

[1]Department of Life Sciences, Research Laboratory of Environmental Toxicology-Microbiology and Health (LR17ES06), Faculty of Sciences, University of Sfax, Sfax, Tunisia. [2]Department of Food Safety, Nutrition and Veterinary Public Health, Istituto Superiore di Sanità, Viale Regina Elena 299, 00161 Rome, Italy. [3]Department of Biology, Preparatory Institute for Engineering Studies, University of Sfax, Sfax, Tunisia. [4]Department of Microbiology, Regional Hygiene Care Mycobacteriology Laboratory, Hedi-Chaker University Hospital, Sfax, Tunisia. [5]Department of Biology B, Faculty of Pharmacy, University of Monastir, Monastir, Tunisia. [6]Department of Microbiology, National Reference Laboratory of Mycobacteria, Research Unit (UR12SP18), A. Mami University Hospital of Pneumology, Ariana, Tunisia.

References

1. World Organization for Animal Health (OIE). Bovine Tuberculosis. http://www.oie.int/fileadmin/Home/eng/Media_Center/docs/pdf/Disease_cards/BOVINE-TB-EN.pdf. Accessed 10 Jan 2018.

2. Thoen CO, LoBue PA, de Kantor I. Why has zoonotic tuberculosis not received much attention? Int J Tuberc Lung Dis. 2010;14:1073–4.

3. Müller B, Dürr S, Alonso S, Hattendorf J, Laisse CJ, Parsons SD, et al. Zoonotic Mycobacterium bovis-induced tuberculosis in humans. Emerg Infect Dis. 2013;19:899–908.

4. Pérez-Lago L, Navarro Y, García-de-Viedma D. Current knowledge and pending challenges in zoonosis caused by Mycobacterium bovis: a review. Res Vet Sci. 2014;97 Suppl:S94–S100.

5. Olea-Popelka F, Muwonge A, Perera A, Dean AS, Mumford E, Erlacher-Vindel E, et al. Zoonotic tuberculosis in human beings caused by Mycobacterium bovis - a call for action. Lancet Infect Dis. 2017;17:e21–5.

6. Cosivi O, Grange JM, Daborn CJ, Raviglione MC, Fujikura T, Cousins D, et al. Zoonotic tuberculosis due to Mycobacterium bovis in developing countries. Emerg Infect Dis. 1998;4:59–70.

7. Ayele WY, Neill SD, Zinsstag J, Weiss MG, Pavlik I. Bovine tuberculosis: an old disease but a new threat to Africa. Int J Tuberc Lung Dis. 2004;8:924–37.

8. Michel AL, Geoghegan C, Hlokwe T, Raseleka K, Getz WM, Marcotty T. Longevity of Mycobacterium bovis in raw and traditional souring milk as a function of storage temperature and dose. PLoS One. 2015. https://doi.org/10.1371/journal.pone.0129926.

9. O'Reilly LM, Daborn CJ. The epidemiology of Mycobacterium bovis infections in animals and man: a review. Tuber Lung Dis. 1995;76 Suppl 1:1–46.

10. Patané JS, Martins J, Beatriz Castelão A, Nishibe C, Montera L, Bigi F, et al. Patterns and processes of Mycobacterium bovis evolution revealed by phylogenomic analyses. Genome Biol Evol. 2017. https://doi.org/10.1093/gbe/evx022.

11. Scorpio A, Zhang Y. Mutations in pncA, a gene encoding pyrazinamidase/nicotinamidase, cause resistance to the antituberculous drug pyrazinamide in tubercle bacillus. Nat Med. 1996;2:662–7.

12. Allix-Béguec C, Fauville-Dufaux M, Stoffels K, Ommeslag D, Walravens K, Saegerman C, et al. Importance of identifying Mycobacterium bovis as a causative agent of human tuberculosis. Eur Respir J. 2010;35:692–4.

13. Bilal S, Iqbal M, Murphy P, Power J. Human bovine tuberculosis - remains in the differential. J Med Microbiol. 2010;59:1379–82.

14. McLaughlin AM, Gibbons N, Fitzgibbon M, Power JT, Foley SC, Hayes JP, et al. Primary isoniazid resistance in Mycobacterium bovis disease: a prospect of concern. Am J Respir Crit Care Med. 2012;186:110–1.

15. Bobadilla-del Valle M, Torres-González P, Cervera-Hernández ME, Martínez-Gamboa A, Crabtree-Ramírez B, Chávez-Mazari B, et al. Trends of Mycobacterium bovis isolation and first-line anti-tuberculosis drug susceptibility profile: a fifteen-year laboratory-based surveillance. PLoS Negl Trop Dis. 2015. https://doi.org/10.1371/journal.pntd.0004124.

16. Guerrero A, Cobo J, Fortún J, Navas E, Quereda C, Asensio A, et al. Nosocomial transmission of Mycobacterium bovis resistant to 11 drugs in people with advanced HIV-1 infection. Lancet. 1997;350:1738–42.

17. Long R, Nobert E, Chomyc S, van Embden J, McNamee C, Duran RR, et al. Transcontinental spread of multidrug-resistant Mycobacterium bovis. Am J Respir Crit Care Med. 1999;159:2014–7.

18. Rivero A, Márquez M, Santos J, Pinedo A, Sánchez MA, Esteve A, et al. High rate of tuberculosis reinfection during a nosocomial outbreak of multidrug-resistant tuberculosis caused by Mycobacterium bovis strain B. Clin Infect Dis. 2001;32:159–61.

19. Etchechoury I, Valencia GE, Morcillo N, Sequeira MD, Imperiale B, López M, et al. Molecular typing of Mycobacterium bovis isolates in Argentina: first description of a person-to-person transmission case. Zoonoses Public Health. 2010;57:375–81.

20. Kubica T, Rüsch-Gerdes S, Niemann S. Mycobacterium bovis subsp. caprae caused one-third of human M. bovis-associated tuberculosis cases reported in Germany between 1999 and 2001. J Clin Microbiol. 2003;41:3070–7.

21. Prodinger WM, Indra A, Koksalan OK, Kilicaslan Z, Richter E. Mycobacterium caprae infection in humans. Expert Rev Anti-Infect Ther. 2014;12:1501–13.

22. Prodinger WM, Eigentler A, Allerberger F, Schönbauer M, Glawischnig W. Infection of red deer, cattle, and humans with Mycobacterium bovis subsp. caprae in western Austria. J Clin Microbiol. 2002;40:2270–2.

23. Sechi LA, Zanetti S, Sanguinetti M, Molicotti P, Romano L, Leori G, et al. Molecular basis of rifampin and isoniazid resistance in Mycobacterium bovis strains isolated in Sardinia, Italy. Antimicrob Agents Chemother. 2001;45:1645–8.

24. Parreiras PM, Lobato FC, Alencar AP, Figueiredo T, Gomes HM, Boéchat N, et al. Drug susceptibility of Brazilian strains of Mycobacterium bovis using traditional and molecular techniques. Mem Inst Oswaldo Cruz. 2004;99:749–52.

25. Daly M, Diegel KL, Fitzgerald SD, Schooley A, Berry DE, Kaneene JB. Patterns of antimicrobial susceptibility in Michigan wildlife and bovine isolates of Mycobacterium bovis. J Vet Diagn Investig. 2006;18:401–4.

26. Romero B, Aranaz A, Bezos J, Alvarez J, de Juan L, Tariq Javed M, et al. Drug susceptibility of Spanish Mycobacterium tuberculosis complex isolates from animals. Tuberculosis. 2007;87:565–71.

27. Marianelli C, Armas F, Boniotti MB, Mazzone P, Pacciarini ML, Di Marco Lo Presti V. Multiple drug-susceptibility screening in Mycobacterium bovis: new nucleotide polymorphisms in the embB gene among ethambutol susceptible strains. Int J Infect Dis. 2015;33:39–44.

28. Krajewska-Wedzina M, Zabost A, Augustynowicz-Kopec E, Weiner M, Szulowski K. Evaluation of susceptibility to antimycobacterial drugs in Mycobacterium tuberculosis complex strains isolated from cattle in Poland. J Vet Res. 2017;61:23–6.

29. Franco MMJ, Ribeiro MG, Pavan FR, Miyata M, Heinemann MB, de Souza Filho AF, et al. Genotyping and rifampicin and isoniazid resistance in Mycobacterium bovis strains isolated from the lymph nodes of slaughtered cattle. Tuberculosis. 2017;104:30–7.

30. Ben Kahla I, Boschiroli ML, Souissi F, Cherif N, Benzarti M, Boukadida J, et al. Isolation and molecular characterisation of Mycobacterium bovis from raw milk in Tunisia. Afr Health Sci. 2011;11:S1–5.

31. Ghariani A, Jaouadi T, Smaoui S, Mehiri E, Marouane C, Kammoun S, et al. Diagnosis of lymph node tuberculosis using the GeneXpert MTB/RIF in Tunisia. Int J Mycobacteriol. 2015;4:270–5.

32. Siala M, Smaoui S, Taktak W, Hachicha S, Ghorbel A, Marouane C, et al. First-time detection and identification of the Mycobacterium tuberculosis complex members in extrapulmonary tuberculosis clinical samples in South Tunisia by a single tube tetraplex real-time PCR assay. PLoS Negl Trop Dis. 2017. https://doi.org/10.1371/journal.pntd.0005572.

33. Fliss M, Meftahi N, Dekhil N, Mhenni B, Ferjaoui M, Rammeh S, et al. Epidemiological, clinical, and bacteriological findings among Tunisian patients with tuberculous cervical lymphadenitis. Int J Clin Exp Pathol. 2016;9:9602–11.

34. Djemal SE, Siala M, Smaoui S, Kammoun S, Marouane C, Bezos J, et al. Genetic diversity assessment of Tunisian Mycobacterium bovis population isolated from cattle. BMC Vet Res. 2017;13:393.

35. Palomino JC, Martin A, Camacho M, Guerra H, Swings J, Portaels F. Resazurin microtiter assay plate: simple and inexpensive method for detection of drug resistance in Mycobacterium tuberculosis. Antimicrob Agents Chemother. 2002;46:2720–2.

36. Martin A, Camacho M, Portaels F, Palomino JC. Resazurin microtiter assay plate testing of Mycobacterium tuberculosis susceptibilities to second-line

drugs: rapid, simple, and inexpensive method. Antimicrob Agents Chemother. 2003;47:3616–9.

37. Montoro E, Lemus D, Echemendia M, Martin A, Portaels F, Palomino JC. Comparative evaluation of the nitrate reduction assay, the MTT test, and the resazurin microtitre assay for drug susceptibility testing of clinical isolates of *Mycobacterium tuberculosis*. J Antimicrob Chemother. 2005;55:500–5.

38. Jadaun GP, Agarwal C, Sharma H, Ahmed Z, Upadhyay P, Faujdar J, et al. Determination of ethambutol MICs for *Mycobacterium tuberculosis* and *Mycobacterium avium* isolates by resazurin microtitre assay. J Antimicrob Chemother. 2007;60:152–5.

39. Affolabi D, Sanoussi N, Odoun M, Martin A, Koukpemedji L, Palomino JC, et al. Rapid detection of multidrug-resistant *Mycobacterium tuberculosis* in Cotonou (Benin) using two low-cost colorimetric methods: resazurin and nitrate reductase assays. J Med Microbiol. 2008;57:1024–7.

40. Campanerut PA, Ghiraldi LD, Spositto FL, Sato DN, Leite CQ, Hirata MH, et al. Rapid detection of resistance to pyrazinamide in *Mycobacterium tuberculosis* using the resazurin microtitre assay. J Antimicrob Chemother. 2011;66:1044–6.

41. Clinical and Laboratory Standards Institute. Susceptibility testing of mycobacteria, nocardia, and other aerobic actinomycetes; approved standard, 2nd ed document M24-A. Wayne: CLSI; 2011.

42. Sreevatsan S, Pan X, Stockbauer KE, Williams DL, Kreiswirth BN, Musser JM. Characterization of rpsL and rrs mutations in streptomycin-resistant *Mycobacterium tuberculosis* isolates from diverse geographic localities. Antimicrob Agents Chemother. 1996;40:1024–6.

43. Okamoto S, Tamaru A, Nakajima C, Nishimura K, Tanaka Y, Tokuyama S, et al. Loss of a conserved 7-methylguanosine modification in 16S rRNA confers low-level streptomycin resistance in bacteria. Mol Microbiol. 2007;63:1096–106.

44. Feuerriegel S, Köser CU, Niemann S. Phylogenetic polymorphisms in antibiotic resistance genes of the *Mycobacterium tuberculosis* complex. J Antimicrob Chemother. 2014;69:1205–10.

45. Lamine-Khemiri H, Martínez R, García-Jiménez WL, Benítez-Medina JM, Cortés M, Hurtado I, et al. Genotypic characterization by poligotyping and VNTR typing of *Mycobacterium bovis* and *Mycobacterium caprae* isolates from cattle of Tunisia. Trop Anim Health Prod. 2014;46:305–11.

46. Bahri S, Kallel A, Gouia A. Lutte contre la tuberculose bovine: programme et réalization. El Baytari. 1991;3:1–3.

47. Hughes VM, Skuce R, Doig C, Stevenson K, Sharp JM, Watt B. Analysis of multidrug-resistant *Mycobacterium bovis* from three clinical samples from Scotland. Int J Tuberc Lung Dis. 2003;7:1191–8.

48. Canetti G, Fox W, Khomenko A, Mahler HT, Menon NK, Mitchison DA, et al. Advances in techniques of testing mycobacterial drug sensitivity and the use of sensitivity tests in tuberculosis control programmes. Bull World Health Organ. 1969;41:21–43.

49. Palomino JC, Martin A, Portaels F. Rapid drug resistance detection in *Mycobacterium tuberculosis*: a review of colourimetric methods. Clin Microbiol Infect. 2007;13:754–62.

50. Laurenzo D, Mousa SA. Mechanisms of drug resistance in *Mycobacterium tuberculosis* and current status of rapid molecular diagnostic testing. Acta Trop. 2011;119:5–10.

51. World Health Organization. Noncommercial culture and drug-susceptibility testing methods for screening patients at risk for multidrug-resistant tuberculosis: policy statement. 2011. http://apps.who.int/iris/bitstream/10665/44601/1/9789241501620_eng.pdf?ua=1&ua=1. Accessed 10 Jan 2018.

52. Finken M, Kirschner P, Meier A, Wrede A, Böttger EC. Molecular basis of streptomycin resistance in *Mycobacterium tuberculosis*: alterations of the ribosomal protein S12 gene and point mutations within a functional 16S ribosomal RNA pseudoknot. Mol Microbiol. 1993;9:1239–46.

53. Tudó G, Rey E, Borrell S, Alcaide F, Codina G, Coll P, et al. Characterization of mutations in streptomycin-resistant *Mycobacterium tuberculosis* clinical isolates in the area of Barcelona. J Antimicrob Chemother. 2010;65:2341–6.

54. Cuevas-Córdoba B, Cuellar-Sánchez A, Pasissi-Crivelli A, Santana-Álvarez CA, Hernández-Illezcas J, Zenteno-Cuevas R. rrs and rpsL mutations in streptomycin-resistant isolates of *Mycobacterium tuberculosis* from Mexico. J Microbiol Immunol Infect. 2013;46:30–4.

55. Sun H, Zhang C, Xiang L, Pi R, Guo Z, Zheng C, et al. Characterization of mutations in streptomycin-resistant *Mycobacterium tuberculosis* isolates in Sichuan, China and the association between Beijing-lineage and dual-mutation in gidB. Tuberculosis (Edinb). 2016;96:102–6.

21

Transcriptomic analysis reveals that enterovirus F strain SWUN-AB001 infection activates JNK/SAPK and p38 MAPK signaling pathways in MDBK cells

Bin Zhang[1,2,3]* [ID], Xinnuo Chen[1], Hua Yue[1,2,3], Wenqiang Ruan[1], Sinan Qin[1] and Cheng Tang[1,2,3]*

Abstract

Background: Enteroviruses (*Picornaviridae* family) have been widely detected in the feces from cattle with diarrhea. However, the mechanisms responsible for the pathogenicity of enteroviruses in cattle remain unclear. Recently, we isolated a novel EV-F7 strain called SWUN-AB001 from diarrheal yak (*Bos grunniens*) feces. To explore the pathogenic mechanisms of this novel virus, we used a transcriptomics approach to find genes with differential expression patterns in Madin-Darby bovine kidney (MDBK) cells during infection with SWUN-AB001 over time.

Results: MDBK cells were sampled at 12 and 24 h post-infection (hpi) to represent the early and late stages of a SWUN-AB001 infection. Compared with the non-infected cells, 19 and 1050 differentially expressed genes (DEGs) were identified at 12 and 24 hpi, respectively. These DEGs were associated with disease, signal transduction, cellular process and cytokine signaling categories. At 24 hpi, the pathway enrichment analysis revealed that signal pathways such as c-Jun NH2-terminal kinase/ stress-activated protein kinase (JNK/SAPK) and mitogen-activated protein kinase (MAPK) pathways and cytokine-cytokine receptor interactions were associated with the interactions occurring between EV-F7 and MDBK cells. Our additional western blot analysis showed that the phosphorylation levels of JNK/SAPK and p38 MAPK proteins increased significantly in the MDBK cells at 24 hpi. The result indicated that infection with EV-F7 could activate JNK/SAPK and p38 MAPK pathways in MDBK cells, and possibly trigger large-scale cytokine production.

Conclusion: Our transcriptome analysis provides useful initial data towards better understanding of the infection mechanisms used by EV-F7, while highlighting the potential molecular relationships occurring between the virus and the host's cellular components.

Keywords: EV-F7, Transcriptomic analysis, DEG, JNK/SAPK, p38 MAPK

Background

Enterovirus (EV) genus (*Picornaviridae* family) members are small, non-enveloped icosahedral viruses with positive, single-stranded RNA genomes [1]. These viruses are widely distributed in humans and some other animal species as A-L groups, with *Rhinovirus* species belonging to groups A-C [1]. Approximately 50–80% of EV infections are asymptomatic or cause only mild, self-limiting illnesses [2–4]. However, some species can cause severe and potentially fatal infections. For instance, EV 71 (*Enterovirus A* species, serotype 71), a neurotropic virus, causes hand, foot, and mouth disease, and herpangina in children [5], while Coxsackievirus B3 (CV-B3; *Enterovirus B* species, serotype 3) the primary cause of viral myocarditis in humans, leads to cardiomyocyte death and life-endangering disease [6].

To aid their replication and survival, many EV species have evolved diverse strategies to evade the host's innate immune responses [5, 7–11]. Infection with EV 71 can activate the c-Jun NH2-terminal kinase (JNK) and p38

---

* Correspondence: binovy@sina.com; tangcheng@163.com
[1]College of Life Science and Technology, Southwest Minzu University, No.16, South 4th Section 1st Ring Road, Chengdu 610041, China
Full list of author information is available at the end of the article

mitogen-activated protein kinase (MAPK) signaling pathways, thereby contributing to increased viral replication and secretion of cytokines such as interleukin (IL)-2, IL-6, IL-10, and tumor necrosis factor (TNF)-α [8, 10]. Similarly, activation of both JNK and p38 MAPK pathways requires active replication of CV-B3, which likely contributes to viral progeny release [9]. Together, these studies indicate that the MAPK pathway plays important roles in the pathology of EV 71 and CV-B3 infections.

EVs were first identified in cattle in the late 1950s [12]. Originally two serotypes, BEV-1 and BEV-2 were described, with serotypes BEV-A and BEV-B being identified later on. BEV-A has since been renamed EV-E, while another serotype, BEV-F, has been renamed EV-F [13]. Currently, EV-E and EV-F species contain four (E1-E4) and seven (F1-F7) genotypes, respectively [13, 14]. Bovine EV infections are very common and viruses from this genus have been detected in cattle with severe enteric and respiratory diseases, as well as in the feces of presumably healthy animals [15–17]. Experimental infection trials using EV-E1 have failed to induce the clinical signs of disease, although the virus was detected in the intestinal tracts, brains, and hearts from the infected cattle [15]. Therefore, it remains unclear as to whether EV infection is a clinically relevant disease.

In our previous study, we isolated a novel EV-F7 strain (SWUN-AB001) from the feces of a yak (*Bos grunniens*) with severe diarrhea in the Qinghai-Tibetan Plateau [14]. The diarrheal fecal sample contained a higher prevalence of EV than the samples from healthy yaks in some regions, indicating that EV infections are potentially correlated with diarrhea in these animals [14]. Therefore, to explore the pathogenic mechanism of the novel SWUN-AB001 EV-F7 strain, we analyzed the transcriptomic profiles of infected Madin-Darby bovine kidney (MDBK) cells during the early and late infection periods with this strain. Our findings build on current knowledge about virus-host interactions and the molecular mechanisms of the cell signaling pathways that are activated during EV infections in yaks.

Methods

Cell culture and the SWUN-AB001 viral strain

MDBK cells (CCL-22, ATCC) were maintained in Dulbecco's Modified Eagle's Medium supplemented with 10% fetal bovine serum (Gibco) at 37 °C in a 5% CO_2 enriched atmosphere. The novel EV-F7 SWUN-AB001 strain ($TCID_{50} = 10^{7.02}$/mL) was originally isolated from the feces of a diarrheal yak in the Qinghai-Tibetan Plateau [14].

Immunofluorescence assays

For the infection assays, the MDBK cells seeded in 24-well plates were incubated to 80% confluence. The cells were then infected with EV-F7 SWUN-AB001 at a multiplicity of infection (MOI) of 0.01, and re-incubated for 6, 12, 24 and 36 h. Mock infected cells were included as the controls.

As a measure of viral replication, an immunofluorescence assay was used to confirm the status of the viral infection at 6, 12, 24 and 36 h post-infection using a previously described method [11]. Briefly, the MDBK cells grown on Lab Tek chamber slides (Nunc) for 24 h were infected with SWUN-AB001 (MOI, 0.01) for 6, 12, 24 or 36 h, after which they were fixed with HistoChoice Clearing Agent (Sigma-Aldrich), permeabilized using 0.1% Triton X-100, and blocked with 1% bovine serum albumin. The cells were then incubated with an anti-viral antiserum prepared from EV-F7 strain SWUN-AB001 (1:100), followed by incubation with Alexa Fluor 488 anti-rabbit secondary antibody (1:1000) (Invitrogen). Nuclei acids were counterstained with 4′,6-diamidino-2-phenylindole (DAPI, Invitrogen). The chamber slides were then mounted onto glass slides using Fluorescence Mounting Medium (Dako), and the cells were observed and imaged using a fluorescence microscope (OLYMPUS IX73). All experiments were performed at least three times.

cDNA library construction and sequencing

Cellular RNA was extracted from both infected (EV-12 h and EV-24 h) and non-infected (Con-12 h and Con-24 h) MDBK cells. Samples were collected in triplicate. Total RNA was isolated using TRIzol reagent (Life Technologies). For mRNA purification, the RNA samples were treated with DNase I, and poly (A) mRNA was purified using oligo-d (T) magnetic beads (Dynabeads). The purified mRNA was fragmented by divalent cation treatment at elevated temperature. The first-strand cDNA was transcribed from the cleaved RNA fragments using reverse transcriptase and random hexamer primers, followed by second-strand cDNA synthesis using DNA polymerase I and RNase H. The double-stranded cDNA was end-repaired using the Klenow fragment of T4 DNA polymerase and T4 polynucleotide kinase. A single adenine base was added via Klenow 3′–5′ exo-polymerase activity, followed by ligation to an adaptor or index adaptor using T4 DNA ligase. Adaptor-ligated fragments were separated and cDNA fragments of the correct size (200 ± 25 bp) were excised from an agarose gel. PCR was performed to selectively enrich and amplify the cDNA fragments.

Following validation on the Agilent 2100 Bioanalyzer and ABI StepOnePlus Real-Time PCR System, the resultant 12 libraries were sequenced using a single 50 bp end-read protocol using the BGISEQ-500 sequencing platform, as per the manufacturer's instructions.

Pre-processing of sequencing reads, and gene expression and differential gene analyses

Raw reads were subjected to a BGISEQ-500 quality control test using Soapnuke software (https://github.com/

BGI-flexlab/SOAPnuke). Using the default parameters, we removed the "dirty" reads containing adaptor sequences, sequences where > 10% of the base calls were identified as unknown ("N"), and low quality reads. Only the remaining "clean" reads were used for the downstream analyses. The clean reads were aligned with the *Bos taurus* genome (*B. taurus* UMD 3.1.1, ftp://ftp.ncbi.nlm.nih.gov/genomes/Bos_taurus/) using Hisat2 (version 2.0.4) [18]. The *B. taurus* genome gene annotation (Bos UMD3.1 NCBI release) was also downloaded from the National Center for Biotechnology Information's website. Following alignment, RNA-Seq by expectation maximization (or RSEM) [19] was used to normalize gene expression using the expected fragments per kilobase of transcript per million fragments sequenced (FPKM) method [20]. For the biological duplicate samples, NOIseq was used to calculate the $\log_2$ fold change ($\log_2$FC) and probability for each gene in every comparison using strict criteria ($\log_2$FC > 1 or $\log_2$FC < − 1, probability > 0.7) [21]. Principal component analysis (PCA) can be used to reveal gene expression differences in biological samples based on the R language ggplot2 package, and this approach was used to analyze the sample data from the MDBK cells under the four different culture conditions (EV-12 h, Con-12 h, EV-24 h, and Con-24 h).

Gene ontology (GO) and pathway enrichment analysis

GO analysis was used to determine the main biological functions of the DEGs. The hypergeometric test and false discovery rate (FDR) correction methods were used in the GO enrichment analysis to gain insights into the DEG functions. All of the GO annotation information was obtained from the Nr database, and we used GO::TermFinder (http://smd.stanford.edu/help/GO-TermFinder/GO_Term Finder_help. shtml) to obtain information about the gene classes. In all cases, $p < 0.05$ was considered to be statistically significant. Statistical analyses relating to the hypergeometric test and the FDR method were conducted using the R package, and all the GO analyses used a custom-made perl script. Pathway enrichment analysis was performed using KAAS (KEGG Automatic Annotation Server, http://www.genome.jp/tools/kaas/) to functionally annotate the genes using BLAST comparisons against the manually curated KEGG database. The threshold of significance was defined as $p < 0.05$.

Western blot analysis

MDBK cells were seeded in 24-well plates and incubated to 80% confluence. The cells were infected with the EV-F7 SWUN-AB001 strain (MOI, 0.01) and incubated for a further 12 h or 24 h. Western blot analysis was then performed as previously described [11], using polyclonal antibodies against JNK/SAPK, phospho-JNK/SAPK, p38 MAPK, phospho-p38 MAPK, or GADPH (Cell Signaling

Technology). Horse radish peroxidase-conjugated goat anti-mouse and goat anti-rabbit IgG antibodies were obtained from Abbkine. Densitometry values for the immunoblot signals were obtained from three separate experiments using FusionCapt Advance software (Vilber Lourmat).

Quantitative real-time polymerase chain reaction (qRT-PCR) verification of the BGI-500 sequencing data

DEGs were examined by qRT-PCR to confirm the accuracy of the sequencing data. The primer sequences for the eight selected DEGs are listed in Additional file 1: Table S1. The total RNA extracted from the infected and non-infected MDBK cells at 12 h and 24 h using TRIzol reagent (Life Technologies) was reverse transcribed using the SuperScript III First-Strand Synthesis System (Life Technology). For the qRT-PCR analysis, the SYBR Premix Ex Taq II (Tli RNaseH Plus) Kit (TakaRa) and the ABI 7500 FAST real-time PCR system (ABI) were used according to each manufacturer's instructions. The relative expression level of each gene was calculated using the $2^{-\Delta\Delta Ct}$ method.

Results

Viral infection

To identify the early and late viral infection periods, MDBK cells were infected with EV-F7 SWUN-AB001 (MOI, 0.01) for 6, 12, 24 or 36 h, and then examined using an immunofluorescence assay. Viral particles were observed at 12 h post-infection (hpi), at which point about 20% of the cells were positive for EV-F7 but lacked any obvious cytopathic effects (CPEs) (Fig. 1). The result was consistent with how the early part of an infection with this virus proceeds. A significant increase in the number of viral particles was observed at 24 hpi, and this was associated with an obvious decrease in the number of viable cells, indicating the onset of CPEs. At 36 hpi, the numbers of both viable cells and viral particles had significantly decreased. Therefore, the 12 and 24 hpi time points were selected to represent the early and late periods of infection with the SWUN-AB001 strain in MDBK cells, respectively.

Transcriptome sequencing and gene expression

To investigate the changes occurring in the cellular components during the early and late expression periods of the EV-F7 infection, we used RNA-Seq to analyze the transcription levels in the MDBK-infected cells. The sequencing libraries were prepared in triplicate from the MDBK cells under four different culture conditions (EV-12 h, Con-12 h, EV-24 h, and Con-24 h), which were sequenced using BGI-500.

After removing the low-quality reads, we obtained an average of 23,950,659 and 23,953,122 clean reads from the EV-12 h and EV-24 h libraries, respectively, and an

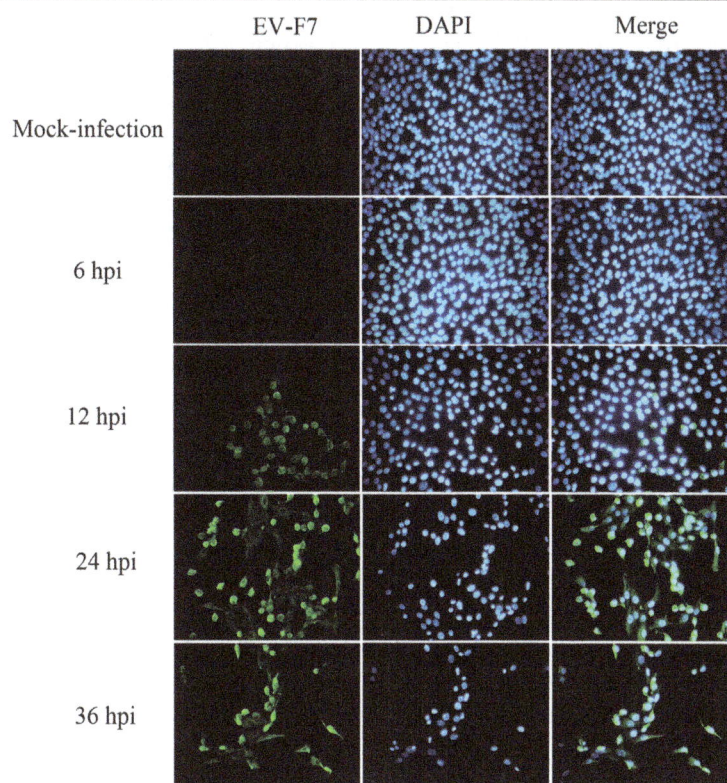

Fig. 1 Immunofluorescence detection of EV-F7 infection in MDBK cells at 6, 12, 24 and 36 h post-infection

average of 23,947,049 and 23,953,282 clean reads from the Con-12 h and Con-24 h libraries, respectively. The proportion of clean reads in all the samples was greater than 99%, thus demonstrating the reliability of the sequencing data quality (Table 1). Alignment analysis of the sequences from the infected and control samples collected at 12 hpi showed that 78.88 and 83.32%, respectively, mapped to the *B. taurus* genome. Of the reads from the 24 hpi samples, 55.76% of those from the infected samples and 82.67% of those from the control samples mapped to the *B. taurus* genome. Thus, the serious CPE observed at 24 hpi likely affected the amount of data obtained from the MDBK cells at this time point.

Gene expression levels were measured by short-read mapping and are presented in reads per kilobase per million mapped reads (RPKM), adjusted by a normalization factor. In total, we detected 21,317 expressed genes or transcripts in all 12 samples, and within each sample

Table 1 Summary of the sequencing reads from the MDBK cells with and without EV-F7 infection

Sample name	MDBK cells (NO.)	Clean Data (bp)	Clean Reads Number	Clean Rate (%)	Genome Mapping (%)	Detected Gene NO.
EV-12 h	1	1,197,656,100	23,953,122	99.98	80.08	18,658
	2	1,197,653,000	23,953,060	99.98	79.71	18,659
	3	1,197,532,950	23,950,659	99.97	75.88	18,681
Con-12 h	1	1,197,568,750	23,951,375	99.97	82.73	18,794
	2	1,197,511,250	23,950,225	99.97	83.32	18,802
	3	1,197,352,450	23,947,049	99.95	82.67	18,832
EV-24 h	1	1,197,609,800	23,952,196	99.97	55.76	17,918
	2	1,197,651,050	23,953,021	99.98	57.68	17,995
	3	1,197,636,250	23,952,725	99.98	60.44	18,109
Con-24 h	1	1,197,637,200	23,952,744	99.98	82.38	18,848
	2	1,197,664,100	23,953,282	99.98	82.67	18,818
	3	1,197,565,550	23,951,311	99.97	82.65	18,820

17,918 to 18,848 expressed genes or transcripts were detected, respectively (Table 1, Additional file 2: Table S2). According to the transcriptome features, the standard uniquely mapped reads approach we used was adjusted by constructing the sequence clusters prior to read mapping. The PCA results revealed that the transcriptome profiles of the samples that were subjected to the same culture conditions were clustered together (Additional file 3: Figure S1), which confirms the reproducibility of the transcriptomic sequencing at the different time points we analyzed.

DEG identification in the MDBK cells

We next aimed to identify the DEGs in the MDBK cells under different culture conditions. A gene was deemed to be differentially expressed if the fold change of the FPKM expression values was at least two, and the divergence probability was at least 0.7. Using both comparisons for each library pair (Con-12 h vs EV-12 h, and Con-24 h vs EV-24 h), 19 and 1050 genes were identified as being differentially expressed at the 12 h and 24 h time points, respectively. Of the 19 DEGs identified at 12 hpi, 12 were up-regulated and seven were down-regulated (Additional file 4: Table S3). A smaller number of DEGs were identified at the 12 h time point probably relates to the fact that 99% of the MDBK cells did not become infected with SWUN-AB001 at a MOI of 0.01, and the effects of the viral infection on the host cells were, therefore, quite limited. Of the 1050 DEGs identified at 24 hpi, 103 were up-regulated and 947 were down-regulated (Additional file 4: Table S3). Fifty-three genes with unknown functions were identified, including 46 significantly down-regulated and seven up-regulated genes. In addition, 216 non-coding (nc) RNAs were identified among the down-regulated DEGs at 24 hpi. ncRNAs are important functional RNA molecules that play critical roles during almost every viral infection process, including regulating viral growth, replication and cell death [22, 23]. In the case of an EV 71 infection, virus-induced cellular ncRNAs are known to modulate the cellular and infection processes and contribute to pathogenesis by targeting either host mRNAs or virus RNAs [24]. Accumulating evidence supports the importance of ncRNAs in EV 71 infection-related pathogenesis, and these molecules might be a viable target of anti-virus strategies [25, 26]. Therefore, we speculated that the numerous down-regulated ncRNAs might play roles in the EV-F7 infection-related pathogenesis or induce a range of anti-virus responses in the host cells. Hence, it is likely that the roles played by ncRNAs in EV-F7 infections will attract further study.

As illustrated by the Venn-diagram in Additional file 5: Figure S2, 12 DEGs overlapped the 12 h and 24 h time points. Of these, 10 genes were up-regulated and two

were down-regulated (Additional file 5: Figure S2). The expression level of *TNFAIP3* (A20), which codes for TNF-α-induced protein 3, was found to have increased at both time points, an interesting finding considering the critical role of TNFAIP3 in terminating NF-κB pathway activation [27]. Early growth response 1 (*EGR1)* mRNA levels were also found to have significantly increased at both 12 and 24 hpi. EGR1, a multifunctional transcription factor, regulates diverse biological functions, including inflammation, apoptosis, differentiation, tumorigenesis, and even viral infections [28]. It is possible that infection with EV71 activates *EGR1* expression to enhance viral replication [29]; if so, this suggests that *EGR1* possibly plays a role in EV-F7 infections.

Functional analysis and biological enrichment of DEGs

To gain insight into the biological roles of the DEGs in our study, we performed a gene ontology (GO) enrichment analysis on each of the genes. The GO analysis revealed that the DEGs were enriched in many GO categories, including biological processes, cellular components, and molecular function (Additional file 6: Table S4). At 12 hpi, four up-regulated genes (*RRM2, EGR1, SPRY4,* and *PIM1)* had GO annotations ($p < 0.05$) (Fig. 2a), such as 'regulation of cell cycle G1/S phase transition' (GO:1902806), 'regulation of protein kinase activity' (GO:0043549) and 'regulation of transferase activity' (GO:0051338), while the only down-regulated gene (*STARD3*) is listed as having 'sterol transporter activity' (GO:0015248) and 'lipid transporter activity' (GO:0005319) (Fig. 2b). At 24 hpi, there was an increase in the number of DEGs in the host cells and they were found to be involved in multiple biological processes. Among the up-regulated genes, 57, 58 and 61 were mapped to 'cellular component', 'molecular function' and 'biological process', respectively ($p < 0.05$) (Fig. 2c and d). The GO terms relating to the biological processes of genes up-regulated at 24 hpi included 'single-organism cellular process' (GO: 0044763), 'biological regulation' (GO: 0065007) and 'response to stimulus' (GO: 0050896). However, 374 down-regulated genes at the 24 hpi time point were identified as being enriched in GO terms like 'insulin-like growth factor binding' (GO: 0005520) and 'intrinsic component of membrane' (GO: 0031224). In addition, we found 4, 6, 12 and 32 of the genes down-regulated 24 hpi were associated with 'adaptive immune response' (GO: 0002250), 'regulation of B cell proliferation' (GO: 0030888), 'positive regulation of signaling' (GO: 0009967) and 'immune system process' (GO: 0002376), respectively, ($p < 0.05$).

KEGG analysis of the DEGs assigned only four and two KEGG pathways to the up- and down-regulated DEGs at 12 hpi, respectively. At 24 hpi, 103 of the up-regulated genes were significantly enriched in 31 pathways that were mainly related to disease, signal

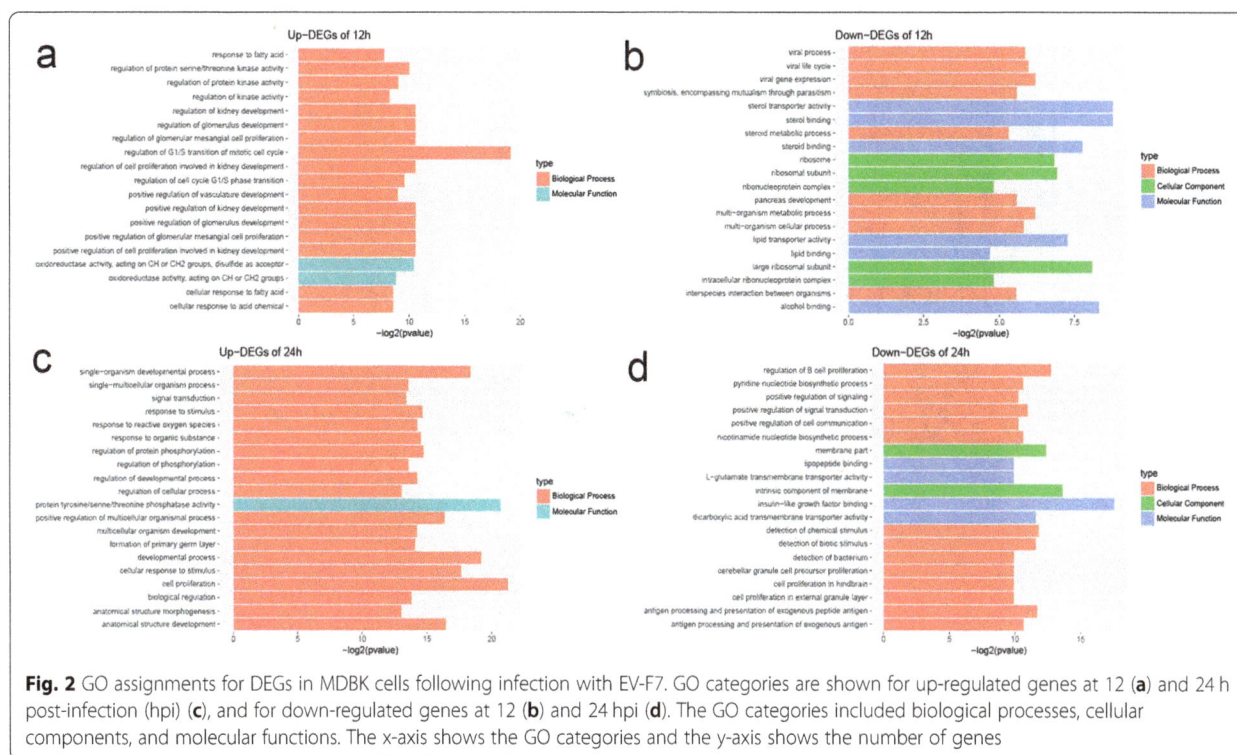

Fig. 2 GO assignments for DEGs in MDBK cells following infection with EV-F7. GO categories are shown for up-regulated genes at 12 (**a**) and 24 h post-infection (hpi) (**c**), and for down-regulated genes at 12 (**b**) and 24 hpi (**d**). The GO categories included biological processes, cellular components, and molecular functions. The x-axis shows the GO categories and the y-axis shows the number of genes

transduction, and cytokine signaling ($p < 0.05$) (Fig. 3, Additional file 7: Table S5). Nine DEGs showed MAPK signaling pathway enrichment; this pathway can activate signaling cascades to produce inflammation and mediate viral replication. These DEGs, which include *c-fos*, *c-Myc*, *c-Jun*, *DUSP1* and *HSP70*, are reportedly involved in initiating the MAPK signaling pathway [30]. Host MAPK signaling pathway activation by viruses triggers the ERK1/2, JNK, and p38 MAPK signaling pathways, thereby contributing to inflammatory cytokine secretion, induction of apoptosis in the infected cells, and enhanced viral replication [30]. The DEGs involved in cytokine signaling were also up-regulated following infection with EV-F7, and included *IL-6*, *IL-11*, leukemia inhibitory factor (*LIF*), granulocyte-macrophage colony-stimulating (*GM-CSF*), and colony stimulating factor 2 (*CSF2*). EV infections often induce high levels of inflammatory cytokines in host cells, resulting in a cytokine storm, as has been observed with EV 71 and CV-B4 infections [8, 31].

In this study, the 947 down-regulated genes were significantly enriched in 33 signaling pathways and were mainly associated with disease, cellular processes, and signal transduction ($p < 0.05$), such as the p53 signaling pathway (Fig. 3; Additional file 7: Table S5). The p53 signaling pathway has been implicated in a large number of biological processes, including cell growth arrest and apoptosis in response to DNA damage [32]. Many viruses, including CV-B3 and poliovirus [33, 34], have

evolved strategies to inhibit p53 surveillance pathways and prevent early apoptosis, allowing for effective viral replication. In this study, the KEGG analysis revealed that the p53 signaling pathway was significantly inhibited in the MDBK cells at 24 hpi, suggesting that EV-F7 interferes with this signaling pathway.

SWUN-AB001 activates JNK/SAPK and p38 MAPK signaling pathways in infected MDBK cells

The transcriptomic results suggest that the MAPK signaling pathway is involved in infection with EV-F7. To assess whether this pathway was activated in EV-F7-infected MDBK cells, the total and phosphorylated amounts of JNK/SAPK and p38 MAPK at 12 and 24 hpi were measured by western blotting, with unstimulated cells used as the mock-stimulated control. MDBK cells were infected with EV-F7 at an MOI of 0.01 and then incubated for 12 h or 24 h. The EV-F7 infection had no obvious effect on the total or phosphorylated amounts of JNK/SAPK and p38 MAPK at 12 hpi, but the amounts of phosphorylated JNK/SAPK and p38 were seen to significantly increase at 24 hpi (Fig. 4). These results indicate that phosphorylation of JNK/SAPK and p38 MAPK might play an important role in EV-F7 replication.

Transcriptome data verification by qRT-PCR

To further evaluate our DEG library, six up-regulated and two down-regulated DEGs were selected for qRT-PCR analysis. Seven housekeeping genes (*β-actin*, *GAPDH*,

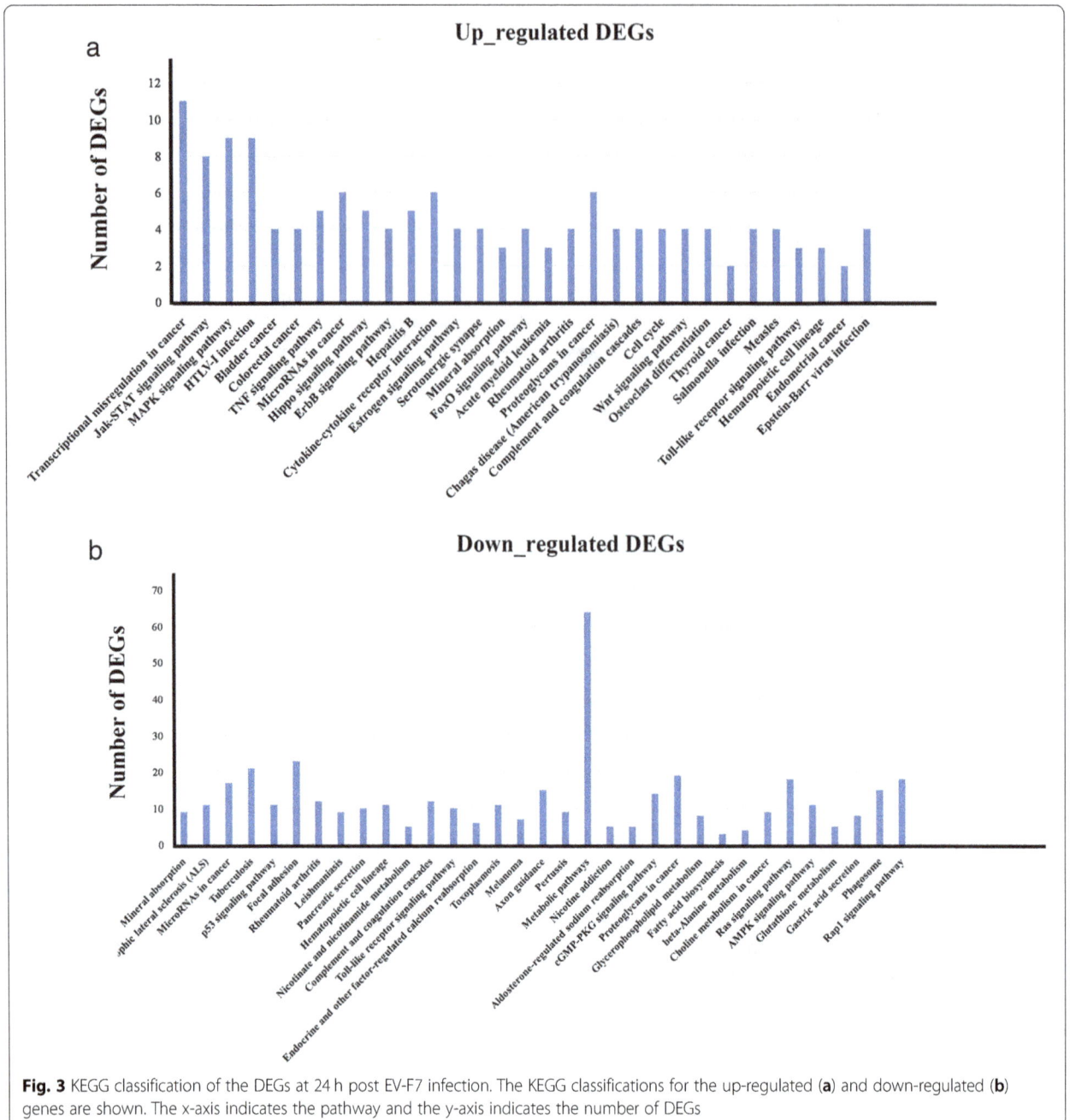

Fig. 3 KEGG classification of the DEGs at 24 h post EV-F7 infection. The KEGG classifications for the up-regulated (**a**) and down-regulated (**b**) genes are shown. The x-axis indicates the pathway and the y-axis indicates the number of DEGs

B2M, *BLM*, *TBP*, *SDHA*, and *BLM*), which were selected from the RNA-Seq data, were tested as potential internal reference genes. The results from this analysis showed that only *B2M* was stably expressed across the different time points in both the control and the infection group (Additional file 2: Table S2). *B2M* was, therefore, selected as the internal reference for the qRT-PCR experiments, the results of which showed the same patterns of expression as those observed with the DEG data, which confirms the validity of the sequencing results (Fig. 5).

Discussion

The novel EV-F7 SWUN-AB001 strain was previously isolated from the feces of a diarrheal yak [14]. In the present study, RNA-Seq analysis of EV-F7-infected MDBK cells was conducted to identify the genes involved in viral infection mechanisms, immune response evasion, and cell signaling pathway activation. We first confirmed that the initial period of infection occurred at 12 hpi using immunofluorescence assays that showed the presence of viral particles in the MDBK cells in the

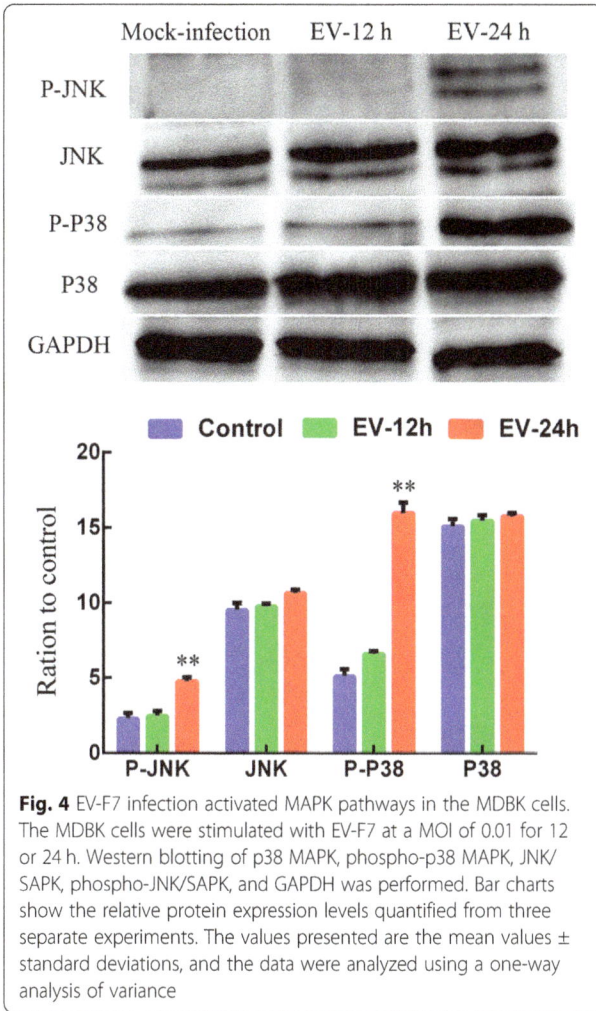

Fig. 4 EV-F7 infection activated MAPK pathways in the MDBK cells. The MDBK cells were stimulated with EV-F7 at a MOI of 0.01 for 12 or 24 h. Western blotting of p38 MAPK, phospho-p38 MAPK, JNK/SAPK, phospho-JNK/SAPK, and GAPDH was performed. Bar charts show the relative protein expression levels quantified from three separate experiments. The values presented are the mean values ± standard deviations, and the data were analyzed using a one-way analysis of variance

absence of any CPEs. At 24 hpi, CPE was evident in the MDBK cells, with an obvious decrease in the number of viable cells, suggesting that this was the late infection stage. The early and late infection periods are important stages for viral replication, and identifying genes with differential expression patterns in host cells during these periods should augment current understanding about the mechanisms involved in viral infections. Therefore, high-throughput mRNA sequencing of infected MDBK cells at 12 and 24 hpi was used to identify DEGs.

During the early period of EV-F7 infection of MDBK cells, the virus-infected host cells triggered the host's innate immune signaling pathways. The transcriptome analysis revealed that 19 DEGs, including 12 that were up-regulated and seven that were down-regulated, were identified at 12 hpi. The DEGs involved in viral replication included *TNFAIP3*, *EGR1*, and *Pim-1*. TNFAIP3, which is also called A20, is a cytoplasmic protein that plays a key role in negatively regulating the inflammatory response and the NF-κB signaling pathway [24]. Some viruses such as CV-B3, influenza virus and bovine

viral diarrhea virus can induce A20 expression during their infections [35–37]. In the present study, *TNFAIP3* expression levels increased significantly at both 12 and 24 hpi in the MDBK cells, a result supported by our independent qRT-PCR analysis. Genes in the NF-κB signaling pathway were not identified as being differentially expressed by KEGG analysis in the current study. Therefore, we speculate that TNFAIP3 might participate in inhibiting the NF-κB signaling during the early and late infection stages of an EV-F7 infection, implying that it might also regulate innate immune signaling or be involved in antivirus responses to infection with EV-F7. *EGR1* expression is associated with many different viral infection types, including EV71, HIV, herpes simplex virus-1, and Kaposi's sarcoma-associated herpesvirus [29]. EGR1 activates microRNA-141 expression and suppresses production of eukaryotic initiation factor 4E, thereby disrupting host protein synthesis and promoting replication of EV71 [38]. Pim1 is a constitutively active serine/threonine kinase known to be involved in cell survival in that it increases the threshold for apoptosis. Inhibition of Pim1 kinase activity in human rhinovirus-16 infected primary bronchial epithelial cells was found to enhance the onset of cell death, resulting in reduced viral replication [39]. In the current study, both *EGR1* and *Pim1* were significantly up-regulated in MDBK cells at 12 and 24 hpi and might, therefore, be involved in viral replication.

At 24 h post EV-F7 infection, 1050 DEGs, including 103 that were up-regulated and 947 that were down-regulated were identified. These genes are involved in the pathways associated with disease, signal transduction, cytokine signaling, and cellular processes, all of which might be involved in viral pathogenesis. Several of the up-regulated DEGs we identified at 24 hpi were proinflammatory cytokines and chemokines (e.g., *IL-6*, *IL-11*, *LIF*, *GM-CSF* and *CSF2*), and the mRNA expression levels of *IL-6* and *LIF* were confirmed to be accurate by qRT-PCR analysis. These results suggest that the release of cytokines and chemokines contributes to EV-F7 infection in MDBK cells. Cytokines and chemokines are usually induced by oxidant stress and viral infection, and they can cause host cell damage, chronic inflammation, and other immune responses [40]. EV71 infection induces high levels of inflammatory cytokines in host cells such as IL-6, IL-10 and TNF-α, with a cytokine storm recognized as the main cause of severe cardiopulmonary manifestations during this infection [8]. In addition, CV-B4 infections, which are mainly associated with meningoencephalitis, neonatal myocarditis and type-1 diabetes, can also promote cytokine and chemokine production in host cells (e.g., IL-6, LIF, and GM-CSF) [31]. Therefore, we speculate that the pathogenicity of EV-F7 might be partly the result of excessive

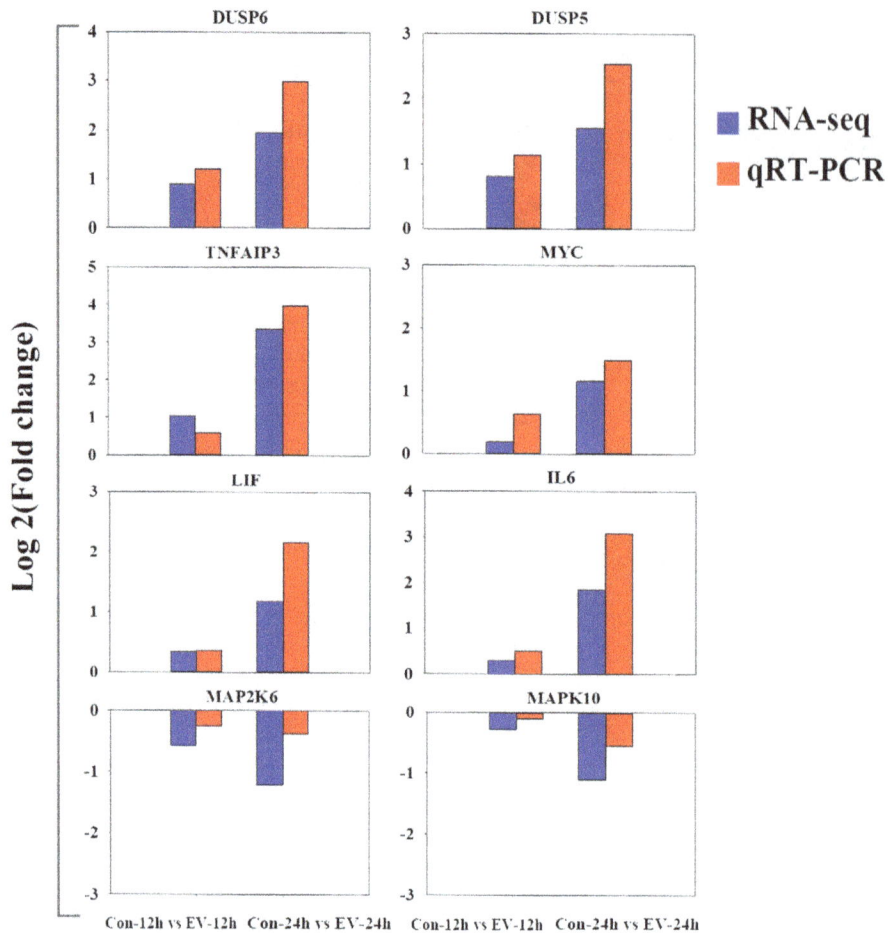

Fig. 5 Validation of the expression patterns of six differentially expressed genes by qRT-PCR analysis. The log₂ (fold change) values derived from the RNA-seq analysis of eight genes were compared with those obtained by qRT-PCR

induction of harmful proinflammatory responses and, if correct, this would represent a significant leap forward in our understanding of the viral pathogenesis of EV-F7 infections.

The family of serine/threonine protein kinases known as MAPKs are widely conserved among eukaryotes, and are involved in many cellular processes, including inflammation, proliferation, differentiation, movement, and cell death [8]. MAPK pathways are required for inflammation, infection, and replication of EV during infection, specifically the JNK/SAPK and p38 MAPK pathways [30]. The use JNK or P38 MAPK inhibitors during an EV71 infection has been found to severely inhibit the induction of cytokines and viral replication in the host cells, indicating that both signaling pathways are beneficial to EV71 infections, and can positively regulate the secretion of inflammatory cytokines in host cells [8, 41]. In the current study, the MAPK signaling pathways were enriched in the up-regulated DEGs, suggesting that they are involved in viral replication and

host inflammation during EV-F7 infection of MDBK cells. To further investigate the potential mechanism, we examined the JNK/SAPK and P38 MAPK signaling pathways during EV-F7 infection of MDBK cells. The results showed that phosphorylation of the JNK/SAPK and P38 MAPK proteins increased significantly at 24 hpi, indicating that these pathways play important roles in infection with EV-F7.

Conclusions

In this study, we extensively characterized the transcriptome of MDBK cells during the early and late stages of infection with EV-F7. A great number of the genes found to be differentially expressed upon infection with EV-F7 in MDBK cells, and were functionally annotated, were associated with disease, signal transduction, cellular process and cytokine signaling. Our findings provide a rational basis for conducting additional research aimed at unraveling the mechanisms underlying the infectious nature of this novel EV-F strain from yak.

Additional files

Additional file 1: Table S1. Sequences of the PCR primers used in this study.

Additional file 2: Table S2. List of all the genes expressed in the MDBK cells.

Additional file 3: Figure S1. Principal component analysis of the four treatment groups for MDBK cells (three biological replicates for each treatment).

Additional file 4: Table S3. List of the differentially expressed genes in the MDBK cells following infection with EV-F7 at 12 and 24 h post-infection.

Additional file 5: Figure S2. Venn diagram of the up- and down-regulated genes identified following comparisons of the Con-12 h vs. EV-12 h and Con-24 h vs. EV-24 h treatment groups.

Additional file 6: Table S4. Enrichment analysis of the biological processes GO terms assigned to DEGs in the MDBK cells 12 and 24 h post infection with EV-F7.

Additional file 7: Table S5. KEGG pathway analysis of the differentially expressed genes in MDBK cells following EV-F7 infection at 12 and 24 h post-infection.

Abbreviations

BEV: Bovine enterovirus; CPEs: Cytopathic effects; CSF2: Colony stimulating factor 2; CV-B3: Coxsackievirus B3; CV-B4: Coxsackievirus B4; DEGs: Differentially expressed genes; EGR1: Early growth response-1; EV: *Enterovirus*; EV-E: Enterovirus E; EV-F: Enterovirus F; FDR: False discovery rate; FPKM: Expected fragments per kilobase of transcript per million fragments sequenced; GM-CSF: Granulocyte-macrophage colony-stimulating factor; GO: Gene ontology; hpi: Hours post-infection; IL: Interleukin; JNK/ STAT: c-Jun NH2-terminal kinase/ stress-activated protein kinase; LIF: Leukemia inhibitory factor; Log$_2$FC: log$_2$ fold change; MAPK: Mitogen-activated protein kinase; MDBK: Madin-Darby bovine kidney; MOI: Multiplicity of infection; ncRNA: Non-coding RNA; PCA: Principal component analysis; qRT-PCR: Quantitative real-time polymerase chain reaction; TNF: Tumor necrosis factor

Acknowledgements
Not applicable

Funding
The study was supported by grants from the National 13th Five-Year Plan National Science and Technology Support Program (grant number 2016YFD0500907), the National Natural Science Foundation of China (grant no. 31772766) and the Innovative Research Team Program of Department of Education of Sichuan Province (grant number 13TD0057). The funders had no role in study design, data collection and analysis, decision to publish, or preparation of the manuscript.

Authors' contributions
Conceived and designed the experiments: BZ, and CT. Performed the experiments: BZ, XC, HY, WR and SQ. Wrote the paper: BZ, CT and XC. All authors have read and approved the final manuscript.

Consent for publication
Not applicable.

Competing interests
The authors declare that they have no competing interests.

Author details
[1]College of Life Science and Technology, Southwest Minzu University, No.16, South 4th Section 1st Ring Road, Chengdu 610041, China. [2]Key laboratory of Ministry of Education and Sichuan Province for Qinghai-Tibetan Plateau Animal Genetic Resource Reservation and Utilization, Chengdu 610041, China. [3]Animal Disease Prevention and Control Innovation Team in the Qinghai-Tibetan Plateau of State Ethnic Affairs Commission, Chengdu 610041, China.

References
1. Zell R, Delwart E, Gorbalenya AE, Hovi T, King AMQ, Knowles NJ, Lindberg AM, Pallansch MA, Palmenberg AC, Reuter G, Simmonds P, Skern T, Stanway G, Yamashita T. ICTV report consortium. ICTV virus taxonomy profile: Picornaviridae. J Gen Virol. 2017;98(10):2421–2.
2. Boros A, Pankovics P, Knowles NJ, Reuter G. Natural interspecies recombinant bovine/porcine enterovirus in sheep. J Gen Virol. 2012;93(9):1941–51.
3. Nollens HH, Rivera R, Palacios G, Wellehan JF, Saliki JT, Caseltine SL, Smith CR, Jensen ED, Hui J, Lipkin WI, Yochem K, Well S, St L, Venn-Waston S. New recognition of enterovirus infections in bottlenose dolphins (*Tursiops truncatus*). Vet Microbiol. 2009;139(1–2):170–5.
4. Palacios G, Oberste MS. Enteroviruses as agents of emerging infectious diseases. J Neuro-Oncol. 2005;11:424–33.
5. Pathinayake PS, Hsu AC, Wark PA. Innate immunity and immune evasion by enterovirus 71. Viruses. 2015;7(12):6613–30.
6. Garmaroudi FS, Marchant D, Hendry R, Luo H, Yang D, Ye X, Shi J, McManus BM. Coxsackievirus B3 replication and pathogenesis. Future Microbiol. 2015; 10(4):629–53.
7. Lei X, Xiao X, Wang J. Innate immunity evasion by enteroviruses: insights into virus-host interaction. Viruses. 2016;8:22.
8. Peng H, Shi M, Zhang L, Li Y, Sun J, Wang X, Xu X, Zhang X, Mao Y, Ji Y, Jiang J, Shi W. Activation of JNK1/2 and p38 MAPK signaling pathways promotes enterovirus 71 infection in immature dendritic cells. BMC Microbiol. 2014;14:147.
9. Si X, Luo H, Morgan A, Zhang J, Wong J, Yuan J, Esfandiarei M, Gao G, Cheung C, McManus BM. Stress-activated protein kinases are involved in coxsackievirus B3 viral progeny release. J Virol. 2005;79(22):13875–81.
10. Wang C, Gao L, Jin Y, Cardona CJ, Xing Z. Regulation of host responses and viral replication by the mitogen-activated protein kinases in intestinal epithelial cells infected with enterovirus 71. Virus Res. 2015;197(2):75–84.
11. Wang C, Zhou R, Zhang Z, Jin Y, Cardona CJ, Xing Z. Intrinsic apoptosis and proinflammatory cytokines regulated in human astrocytes infected with enterovirus 71. J Gen Virol. 2015;96(10):3010–22.
12. Moll T, Finlayson AV. Isolation of cytopathogenic viral agent from feces of cattle. Science. 1957;126(3270):401–2.
13. Zell R, Krumbholz A, Dauber M, Hoey E, Wutzler P. Molecular-based reclassification of the bovine enteroviruses. J Gen Virol. 2006;87(2):375–85.
14. He H, Tang C, Chen X, Yue H, Ren Y, Liu Y, Zhang B. Isolation and characterization of a new enterovirus F in yak feces in the Qinghai-Tibetan plateau. Arch Virol. 2017;162(2):523–7.
15. Blas-Machado U, Saliki JT, Sanchez S, Brown CC, Zhang J, Keys D, Woolums A, Harvey SB. Pathogenesis of a bovine enterovirus-1 isolate in experimentally infected calves. Vet Pathol. 2011;48(6):1075–84.
16. Li Y, Chang J, Wang Q, Yu L. Isolation of two Chinese bovine enteroviruses and sequence analysis of their complete genomes. Arch Virol. 2012;157(12): 2369–75.
17. McCarthy FM, Smith GA, Mattick JS. Molecular characterisation of Australian bovine enteroviruses. Vet Microbiol. 1999;68(1–2):71–81.
18. Kim D, Langmead B, Salzberg SL. HISAT: a fast spliced aligner with low memory requirements. Nat Methods. 2015;12(4):357–60.
19. Li B, Dewey CN. RSEM: accurate transcript quantification from RNA-Seq data with or without a reference genome. BMC Bioinformatics. 2011;12:323.
20. Mortazavi A, Williams BA, McCue K, Schaeffer L, Wold B. Mapping and quantifying mammalian transcriptomes by RNA-Seq. Nat Methods. 2008;5(7): 621–8.
21. Tarazona S, Furio-Tari P, Turra D, Pietro AD, Nueda MJ, Ferrer A, Conesa A. Data quality aware analysis of differential expression in RNA-seq with NOISeq R/bioc package. Nucleic Acids Res. 2015;43(21):e140.
22. Ha M, Kim VN. Regulation of microRNA biogenesis. Nat Rev Mol Cell Biol. 2014;15(8):509–24.

23. Kagami H, Akutsu T, Maegawa S, Hosokawa H, Nacher JC. Determining associations between human diseases and non-coding RNAs with critical roles in network control. Sci Rep. 2015;5:14577.

24. Ho BC, Yang PC, Yu SL. MicroRNA and pathogenesis of enterovirus infection. Viruses. 2016;8(1):11.

25. Ho BC, Yu IS, Lu LF, Rudensky A, Chen HY, Tsai CW, Chang YL, Wu CT, Chang LY, Shih SR, Lin SW, Lee CN, Yang PC, Yu SL. Inhibition of miR-146a prevents enterovirus-induced death by restoring the production of type I interferon. Nat Commun. 2014;5:3344.

26. Zheng Z, Ke X, Wang M, He S, Li Q, Zheng C, Zhang Z, Liu Y, Wang H. Human microRNA hsa-miR-296-5p suppresses enterovirus 71 replication by targeting the viral genome. J Virol. 2013;87(10):5645–56.

27. Pujari R, Hunte R, Khan WN, Shembade N. A20-mediated negative regulation of canonical NF-kappaB signaling pathway. Immunol Res. 2013; 57(1–3):166–71.

28. Thiel G, Cibelli G. Regulation of life and death by the zinc finger transcription factor Egr-1. J Cell Physiol. 2002;193(3):287–92.

29. Song Y, Cheng X, Yang X, Zhao R, Wang P, Han Y, Luo Z, Cao Y, Zhu C, Xiong Y, Liu Y, Wu K, Wu J. Early growth response-1 facilitates enterovirus 71 replication by direct binding to the viral genome RNA. Int J Biochem Cell Biol. 2013;62:36–46.

30. Krishna M, Narang H. The complexity of mitogen-activated protein kinases (MAPKs) made simple. Cell Mol Life Sci. 2008;65(22):3525–44.

31. Brilot F, Chehadeh W, Charlet-Renard C, Martens H, Geenen V, Hober D. Persistent infection of human thymic epithelial cells by coxsackievirus B4. J Virol. 2002;76(10):5260–5.

32. Meek DW. Tumour suppression by p53: a role for the DNA damage response? Nat Rev Cancer. 2009;9(10):714–23.

33. Gao G, Wong J, Zhang J, Mao I, Shravah J, Wu Y, Xiao A, Li X, Luo H. Proteasome activator REGgamma enhances coxsackieviral infection by facilitating p53 degradation. J Virol. 2010;84(21):11056–66.

34. Weidman MK, Yalamanchili P, Ng B, Tsai W, Dasgupta A. Poliovirus 3C protease-mediated degradation of transcriptional activator p53 requires a cellular activity. Virology. 2001;291(2):260–71.

35. Gui J, Yue Y, Chen R, Xu W, Xiong S. A20 (TNFAIP3) alleviates CVB3-induced myocarditis via inhibiting NF-kappaB signaling. PLoS One. 2012;7(9):e46515.

36. Maelfait J, Roose K, Bogaert P, Sze M, Saelens X, Pasparakis M, Carpentier I, van Loo G, Beyaert R. A20 (Tnfaip3) deficiency in myeloid cells protects against influenza a virus infection. PLoS Pathog. 2012;8(3):e1002570.

37. Villalba M, Fredericksen F, Otth C, Olavarria V. Transcriptomic analysis of responses to cytopathic bovine viral diarrhea virus-1 (BVDV-1) infection in MDBK cells. Mol Immunol. 2016;71:192–202.

38. Ho BC, Yu SL, Chen JJ, Chang SY, Yan BS, Hong QS, Singh S, Kao CL, Chen HY, Su KY, Li KC, Cheng CL, Cheng HW, Lee JY, Lee CN, Yang PC. Enterovirus-induced miR-141 contributes to shutoff of host protein translation by targeting the translation initiation factor eIF4E. Cell Host Microbe. 2011;9(1):58–69.

39. De Vries M, Smithers NP, Howarth PH, Nawijn MC, Davies DE. Inhibition of Pim1 kinase reduces viral replication in primary bronchial epithelial cells. Eur Respir J. 2015;45(6):1745–8.

40. Mastruzzo C, Crimi N, Vancheri C. Role of oxidative stress in pulmonary fibrosis. Monaldi Arch Chest Dis. 2002;57(3–4):173–6.

41. Shi W, Hou X, Li X, Peng H, Shi M, Jiang Q, Liu X, Ji Y, Yao Y, He C. Differential gene expressions of the MAPK signaling pathway in enterovirus 71-infected rhabdomyosarcoma cells. Braz J Infect Dis. 2013;17(4):410–7.

22

Neutralizing antibodies against Simbu serogroup viruses in cattle and sheep, Nigeria, 2012–2014

Daniel Oluwayelu[1], Kerstin Wernike[2*] ⓘ, Adebowale Adebiyi[1], Simeon Cadmus[3] and Martin Beer[2]

Abstract

Background: Simbu serogroup viruses of the *Orthobunyavirus* genus (Family *Peribunyaviridae*) include teratogenic pathogens that cause severe economic losses, abortions, stillbirths and congenital abnormalities in ruminants worldwide. Although they were initially isolated from ruminants and *Culicoides* biting midges about five decades ago in Nigeria, there is no current information on their prevalence and geographical distribution despite reports of abortions and congenital malformations in the country's ruminant population. Here, apparently healthy cattle and sheep obtained from eight states in the three major vegetation zones of Nigeria were screened for the presence of specific neutralizing antibodies against Schmallenberg virus (SBV), Simbu virus (SIMV) and Shamonda virus (SHAV).

Results: Using a cross-sectional design, 490 cattle and 165 sheep sera were collected between 2012 and 2014 and tested by a commercial SBV ELISA kit which enables the detection of antibodies against various Simbu serogroup viruses. The seropositivity rates for cattle and sheep were 91.2% and 65.4%, respectively. In cattle, there was no association between ELISA seropositivity and vegetation zone. However, the prevalence of anti-Simbu serogroup antibodies was significantly higher in Ebonyi State compared to other states in the rainforest vegetation zone. The seroprevalence was significantly higher in sheep obtained from live animal markets compared to farms (OR = 5.8). Testing of 20 selected ELISA-positive sera by serum neutralisation test showed that all were positive for one or more of SBV, SIMV and SHAV with the highest titres obtained for SHAV. Antibodies to SBV or a closely related virus were detected in the Sudan savannah and rainforest zones, anti-SIMV antibodies were detected only in the rainforest zone, while anti-SHAV antibodies were found in the three vegetation zones.

Conclusion: The findings of this study reveal that following the early isolation of Simbu serogroup viruses in Nigeria in the 1960s, members of this virus group are still circulating in the country. Specifically, SBV, SIMV and SHAV or closely related viruses infect cattle and sheep across the three vegetation zones of Nigeria suggesting that insect vector activity is extensive in the country. The exact vegetation zone where the animals became exposed to the viruses could, however, not be determined in this study.

Keywords: Schmallenberg virus, Simbu virus, Shamonda virus, Cattle, Sheep, ELISA, Neutralizing antibodies

Background

The order *Bunyavirales* currently consists of more than 350 viruses that are distributed among 13 genera in nine families, thus making it one of the largest orders of RNA viruses. Of these genera, the *Orthobunyavirus* genus (family *Peribunyaviridae*) is the largest and most diverse with more than 170 viruses. The majority of these viruses are assigned to one of 18 serogroups, including the Simbu serogroup, based on serologic relatedness of complement fixing, hemagglutination inhibiting and neutralizing antibodies [1–3]. Members of this genus are arthropod-borne viruses (arboviruses) that are mostly transmitted by mosquitoes, sandflies or *Culicoides* biting midges, possess a tripartite RNA genome and share common genetic features but are serologically unrelated to viruses in other genera of the *Peribunyaviridae*, and many are pathogenic to humans and animals [4, 5].

* Correspondence: kerstin.wernike@fli.de
[2]Institute of Diagnostic Virology, Friedrich-Loeffler-Institut, Südufer 10, 17493 Greifswald - Insel Riems, Germany

In particular, the Simbu serogroup comprises at least 25 viruses that are currently divided into seven species, namely: Akabane virus (AKAV), Manzanilla virus, Oropouche virus, Sathuperi virus (SATV), Shamonda virus (SHAV), Shuni virus (SHUV) and Simbu virus (SIMV) [1, 2]. Several of these Simbu serogroup viruses are known to be teratogenic in ruminants [6] causing abortions, stillbirths and congenital abnormalities. While some members such as SHAV, SHUV, Sabo, and Sango viruses are less frequently examined, AKAV, Aino virus (AINV) and Schmallenberg virus (SBV) are the most studied in this serogroup [7–10].

Virus isolation or serological methods have been used to detect Simbu serogroup viruses in domestic animals, wildlife, mosquitoes and *Culicoides* from Africa, Asia, Australia, and the Middle East [11–19]. Although different assays including serum neutralization test (SNT), immunofluorescence (IF) assay and enzyme-linked immunosorbent assay (ELISA) have been used for the serologic detection of previous infections with these viruses [16, 20–22], specific detection of antibodies against them can be achieved by SNT [20].

Simbu serogroup viruses have been reported to cause severe economic losses to the livestock industry worldwide [23, 24]. However, information on their presence in Africa is still relatively scarce. In Nigeria, where the climate favours vector activity, early arboviral studies [7, 25] led to the isolation of Simbu serogroup viruses including SHAV, Sabo, Sango, SHUV and SATV viruses from cattle, goats and *Culicoides* biting midges. However, for about five decades there has been no information on the prevalence, geographical distribution and reproductive impact of these viruses despite reports of abortions, stillbirths and congenital malformations in the country's ruminant population [26–28]. Recent studies based on commercial ELISAs to elucidate the role of Simbu serogroup viruses in the occurrence of reproductive disorders and congenital malformations among ruminants in Nigeria provided serologic evidence of AKAV, SBV or closely related viruses [19, 29]. However, because of the antigenic cross-reactivity that exists among Simbu serogroup viruses, the current study was conducted to investigate the presence of specific neutralizing antibodies against Schmallenberg, Simbu and Shamonda viruses in apparently healthy cattle and sheep obtained from eight states spread across the three major vegetation zones of Nigeria.

Methods
Study area
This study was carried out as part of recent investigations to determine the contribution of Simbu serogroup viruses to cases of reproductive disorders and congenital malformations in the Nigerian ruminant population. Cattle sera were collected from abattoirs, live animal markets,

private/backyard farms or Fulani pastoralist herds located in eight states of Nigeria. These states include Borno (Northeast) and Sokoto (Northwest) located in the Sudan savannah vegetation zone and sharing international borders respectively with Chad and Niger Republic, two countries that are major suppliers of cattle to Nigeria. The other states, which serve as transit or sales points for cattle and sheep in their respective regions, are Benue (North-central) in the Guinea savannah zone, and Ebonyi (Southeast), Ogun, Osun, Oyo and Lagos (Southwest) in the rainforest zone. Sera from sheep were collected from live animal markets, private/backyard farms or Fulani herds located in Ogun, Osun, Oyo and Lagos States (Fig. 1). The exact origin of animals sampled at abattoirs and live animal markets in the North-central, Southeast and Southwest states could not be determined, but most of them were transported in trucks from the North-eastern and North-western regions of the country. Animals in the private/backyard farms were raised under the commonly practised semi-intensive system of management where they were fed on pastures or cut grasses during the day and kept in non-insect-proof sheds at night, while the Fulani herds grazed extensively from one location to another in their traditional manner. Thus, all the animals were exposed to insect vectors.

Study design and sample collection
A cross-sectional design was used for this study. Serum samples were collected between May 2012 and April 2014 from 490 adult cattle in the eight states as follows: Borno ($n = 84$), Sokoto ($n = 89$), Benue ($n = 83$), Ebonyi ($n = 67$), Ogun ($n = 30$), Osun ($n = 34$), Oyo (n = 30) and Lagos ($n = 73$), while 165 sera from adult sheep were collected from the southwestern states of Ogun ($n = 57$), Osun ($n = 70$), Oyo ($n = 25$) and Lagos ($n = 13$). The animals were randomly selected at the individual study sites. The separated sera were stored at -20 °C until analysed.

Serologic testing
Simbu serogroup enzyme-linked immunosorbent assay (ELISA)
All the 655 sera were initially screened for the presence of anti-Simbu serogroup antibodies using the ID Screen® Schmallenberg virus competition multi-species ELISA kit (IDvet, France), which detects antibodies against various Simbu serogroup viruses [30], according to the manufacturer's instructions. For each serum sample, the competition percentage (S/N%) was calculated. Samples presenting S/N% ≤ 40%, 40% < S/N% ≤ 50%, and > 50% were considered positive, doubtful and negative, respectively.

Serum neutralization test (SNT)
A subset of 20 serum samples selected based on their very low S/N% values (i.e. high seropositivity) in the ELISA was

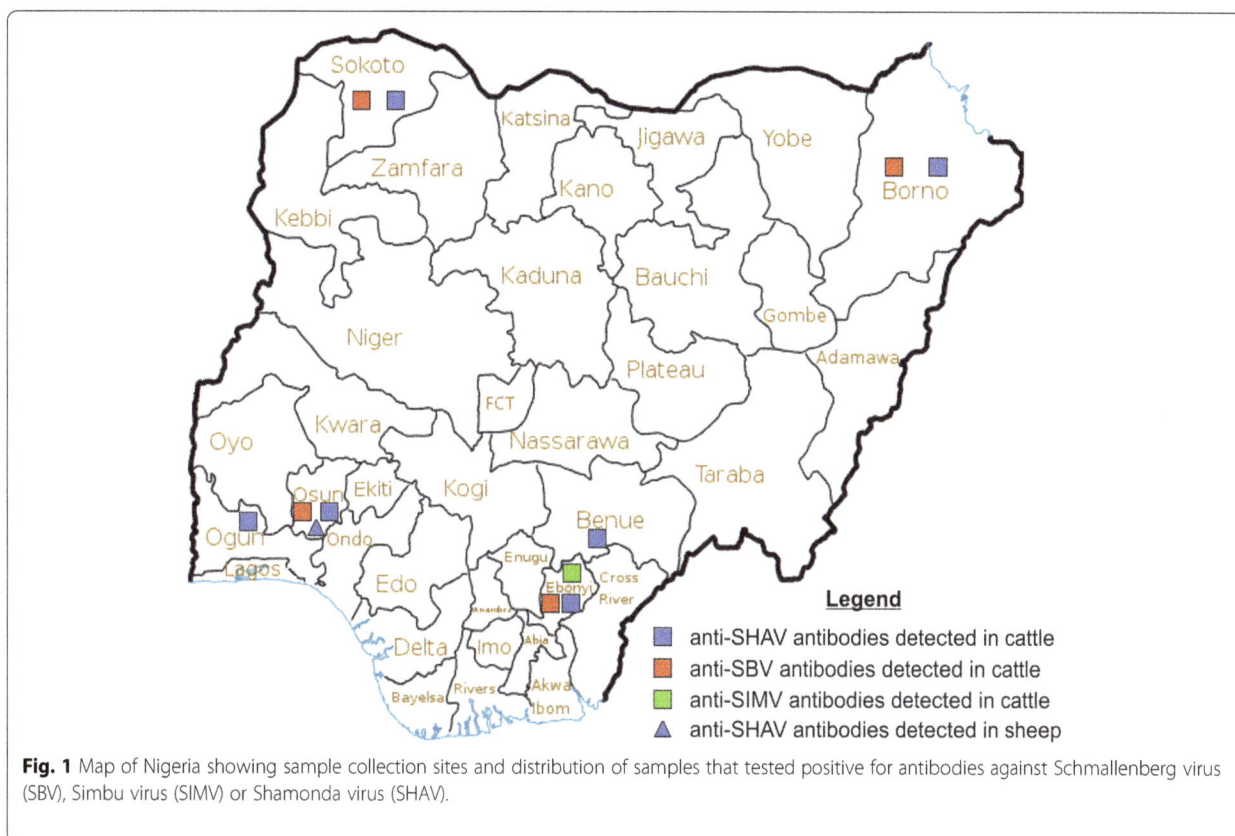

Fig. 1 Map of Nigeria showing sample collection sites and distribution of samples that tested positive for antibodies against Schmallenberg virus (SBV), Simbu virus (SIMV) or Shamonda virus (SHAV).

analysed in microneutralization tests performed as previously described [31] against SBV, SIMV and SHAV. Briefly, two-fold dilutions of sera were prepared in Minimum Essential Medium (MEM), starting with 1:20. Fifty µl of MEM containing 100 $TCID_{50}$ of SBV, SIMV or SHAV and 50 µl of the diluted sera were incubated in microtitre plates for 2 h. Thereafter, a BHK cell suspension (in 100 µl of MEM containing 10% foetal calf serum) was added and the microtitre plates were incubated for 3 days at 37 °C. Evaluation was done by assessment of the cytopathic effect. All sera were tested in three replicates and the titres were determined using the method described by Reed and Muench [32] and expressed as the 50% neutralizing dose per ml (ND_{50}/ml). The samples included 15 cattle sera (three each from the Northeast, Northwest, North-central, Southeast and Southwest regions) and five sheep sera from the Southwest.

Statistical analysis

Data were analyzed using GraphPad Prism version 5.01 (San Diego, USA). Differences in Simbu serogroup antibody seroprevalence between cattle and sheep, and female and male animals were evaluated using Chi-square (χ^2) test. Seroprevalence data based on vegetation zones, source of animals and location/state were subjected to one-way ANOVA and subsequently to Tukey's post-test for performing multiple comparisons. Statistical differences between all possible pairs of groups were assessed at $P < 0.05$.

Results

Simbu serogroup ELISA

Out of 490 cattle sera tested for anti-Simbu serogroup antibodies with the ELISA, 447 (91.2%), 20 (4.1%) and 23 (4.7%) were positive, doubtful and negative, respectively, while of the 165 sheep sera tested, 108 (65.4%), 11 (6.7%) and 46 (27.9%) were positive, doubtful and negative, respectively. In cattle, the prevalence of positive ELISA results was significantly higher in Ebonyi State compared to other states in the rainforest vegetation zone. Also, there were significant differences based on the origin of cattle tested (Table 1). Prevalence of anti-Simbu serogroup antibodies was significantly higher in sheep from live animal markets compared to farms (OR 5.8, 95% CI: 2.4–14.1), and in Ogun and Osun States relative to the other two states (Table 2).

Serum neutralization tests

Antibodies against SBV, SIMV and SHAV or closely related viruses were detected in six of the eight states (Fig. 1), with variable neutralizing antibody titres in cattle and sheep (Table 3). All the 20 sera tested were positive for antibodies against at least one of the three Simbu serogroup viruses SBV, SIMV and SHAV which were included in the present

Table 1 Results of the Simbu serogroup ELISA obtained from cattle sera

	Vegetation zone									Source			Sex		Total
	Sudan savannah		Guinea savannah	Rainforest											
	Sokoto	Borno	Benue	Ebonyi	Ogun	Osun	Oyo	Lagos	Live animal markets	Farms	Abattoirs	Female	Male	
No. sampled	89	84	83	67	30	34	30	73	326	47	117	294	196	490
Positive (%)	79 (88.8)	82 (97.6)	80 (96.4)	42 (62.7)	30 (100.0)	34 (100.0)	30 (100.0)	70 (95.9)	286 (87.7)	47 (100.0)	114 (97.4)	264 (89.8)	183 (93.4)	447 (91.2)
Doubtful (%)	–	2 (2.4)	2 (2.4)	13 (19.4)	–	–	–	3 (4.1)	17 (5.2)	–	3 (2.6)	14 (4.8)	6 (3.1)	20 (4.1)
Negative (%)	10 (11.2)	–	1 (1.2)	12 (17.9)	–	–	–	–	23 (7.1)	–	–	16 (5.4)	7 (3.6)	23 (4.7)

Table 2 Results of the Simbu serogroup ELISA obtained from sheep sera

	No. sampled	Positive (%)	Doubtful (%)	Negative (%)
State				
Ogun	57	50 (87.7)	2 (3.5)	5 (8.8)
Osun	70	51 (72.9)	6 (8.6)	13 (18.6)
Oyo	25	4 (16.0)	1 (4.0)	20 (80.0)
Lagos	13	3 (23.1)	2 (15.4)	8 (61.5)
Sex				
Female	91	56 (61.5)	7 (7.7)	28 (30.8)
Male	74	52 (70.3)	4 (5.4)	18 (24.3)
Source				
Live animal markets	64	55 (85.9)	2 (3.1)	7 (10.9)
Farms	101	53 (52.5)	9 (8.9)	39 (38.6)
Total	165	108 (65.4)	11 (6.7)	46 (27.9)

study. Of these 20 sera, 7 (35.0%), 2 (10.0%) and 20 (100.0%) were positive for SBV, SIMV and SHAV antibodies, respectively, while 6 (30.0%), 1 (5.0%) and 1 (5.0%) were positive for a combination of SBV and SHAV, SIMV and SHAV, and SBV, SIMV and SHAV antibodies, respectively (Table 3).

The geographic distribution of the 20 SNT-positive sera showed that antibodies reacting with SBV were detected in Borno and Sokoto States (Sudan savannah zone) as well as in Ebonyi and Osun States (rainforest zone), antibodies to SIMV were found only in Ebonyi State (rainforest zone), while antibodies to SHAV were found in Borno, Sokoto, Benue, Ebonyi, Osun and Ogun States, representing the three vegetation zones (Fig. 1). Information on the breed, sex and source of the SNT-positive sera are shown in Table 3.

Discussion

Simbu serogroup viruses are generally believed to be endemic in Africa. Apart from their initial isolation in Nigeria several decades ago, antibodies to several viruses in this serogroup were also detected in the 1970s in Kenya, South Africa and Nigeria [7, 12, 13, 25]. Although the ELISA-based detection of antibodies against

Table 3 Neutralizing antibody titres against the Simbu serogroup viruses Schmallenberg virus (SBV), Simbu virus (SIMV) and Shamonda virus (SHAV) in the tested Nigerian cattle and sheep sera

No.	Species	Breed	Sex	Source	State	Neutralising titre (ND$_{50}$/ml)		
						SBV	SIMV	SHAV
1	Cattle	Red Bororo	Female	Abattoir	Borno	1/57	neg.	1/180
2	Cattle	Red Bororo	Female	Abattoir	Borno	1/36	neg.	1/180
3	Cattle	Red Bororo	Female	Abattoir	Borno	1/57	neg.	1/180
4	Cattle	Sokoto Gudali	Female	Abattoir	Sokoto	1/71	neg.	1/143
5	Cattle	White Fulani	Female	Abattoir	Sokoto	neg.	neg.	1/180
6	Cattle	White Fulani	Female	Abattoir	Sokoto	1/113	neg.	1/90
7	Cattle	White Fulani	Male	Abattoir	Benue	neg.	neg.	1/904
8	Cattle	Red Bororo	Female	Abattoir	Benue	neg.	neg.	1/226
9	Cattle	Red Bororo	Female	Abattoir	Benue	neg.	neg.	1/36
10	Cattle	White Fulani	Male	Abattoir	Ebonyi	neg.	neg.	1/90
11	Cattle	Sokoto Gudali	Female	Abattoir	Ebonyi	1/57	1/113	1/71
12	Cattle	Sokoto Gudali	Male	Abattoir	Ebonyi	neg.	1/143	1/71
13	Cattle	Sokoto Gudali	Male	Farm	Ogun	neg.	neg.	1/57
14	Cattle	White Fulani	Male	Farm	Osun	neg.	neg.	1/113
15	Cattle	White Fulani	Female	Farm	Osun	1/36	neg.	1/36
16	Sheep	Ouda	Male	Live animal market	Osun	neg.	neg.	1/226
17	Sheep	Yankassa	Male	Live animal market	Osun	neg.	neg.	1/45
18	Sheep	Yankassa	Female	Live animal market	Osun	neg.	neg.	1/143
19	Sheep	Balami	Female	Live animal market	Osun	neg.	neg.	1/71
20	Sheep	Balami	Male	Live animal market	Osun	neg.	neg.	1/71

neg. = < 1/20 ND$_{50}$/ml

SBV, a newly emerged Simbu serogroup virus, was recently reported in domestic ruminants in Mozambique and Tanzania [17, 18], current knowledge about the occurrence, distribution and spread of these viruses in Nigeria is scarce. Only recently, ruminant sera were screened by commercially available SBV and AKAV antibody ELISAs in small-scale pilot studies [19, 29]. In the present follow-up study a more comprehensive sample set including bovine and ovine sera was collected. In addition, all three major vegetation zones of Nigeria were integrated in the study. Since cross-reactivity with various closely related viruses might occur when using S-segment based commercial SBV ELISAs [22], the more specific SNT was performed for further characterisation.

The results of this study show that 84.7% (555/655) of the animals tested had antibodies against Simbu serogroup viruses based on the commercial ELISA test. This high seroprevalence rate suggests that a large proportion of Nigeria's cattle and sheep population had been exposed to infection with SBV-related or other Simbu serogroup viruses. However, since it had been reported that antibodies against viruses in this serogroup frequently cross-react with more than one other member of the group [2, 18, 33], we performed SNT on the ELISA-positive serum samples for serologic confirmation. Our findings revealed that Simbu serogroup viruses continue to infect livestock in Nigeria and that at least three of them, namely SBV, SIMV and SHAV, were circulating in the country during the 2012–2014 or previous seasons. The detection of neutralizing antibodies to SBV and SHAV in cattle from Sokoto and Borno States (Sudan savannah zone) and the presence of neutralizing antibodies to SBV, SIMV and SHAV in one serum sample from Ebonyi State (rainforest zone) highlight the existence of either mixed infections with these viruses or antigenic cross-reactivity among them. Another possible explanation for this finding could be that the animals were infected with other Simbu serogroup viruses (not SBV, SIMV or SHAV) that cross-react to a certain extent with these three viruses. Such viruses, including AKAV and AINOV, are associated with epizootics of congenital malformations in ruminants and have been isolated from *Culicoides* biting midges in Africa, Australia, the Middle East and Asia [12, 34–36].

Additionally, it is noteworthy that despite being positive in the commercial ELISA test sold for anti-SBV antibody detection, only anti-SHAV antibodies were detected in cattle sera from Benue State (Guinea savannah zone) and sheep sera from Osun State (rainforest zone). This finding further corroborates reports of the existence of cross-reactivity among the Simbu serogroup viruses. Moreover, the fact that only Ebonyi State had cattle that were seropositive for SIMV suggests limited circulation of the virus although a larger sample size covering almost all the states of the country might be needed to verify this. Overall, while the seroprevalence for SIMV and SBV were low (2/20, 10.0%) to moderate (7/20, 35.0%) respectively, that of SHAV was exceptionally high (20/20, 100.0%).

The detection of neutralizing antibodies against SBV, SIMV and SHAV as well as the distribution of seropositive animals (Fig. 1) indicates that these viruses or related Simbu serogroup viruses infect cattle and sheep across the different vegetation zones of Nigeria and that the activity of insect vectors, most likely *Culicoides* biting midges, is widespread in the country. The exact vegetation zone where the animals became exposed to the viruses could, however, not be determined in this study. Further studies will be necessary to isolate and characterize Simbu serogroup viruses circulating in the Nigerian ruminant population and in the insect vectors responsible for their transmission. In addition, the possible negative impact of these viruses on reproductive performance of cattle and sheep in Nigeria needs to be investigated.

Conclusions

The findings of the present study reveal that following the early isolation of Simbu serogroup viruses in Nigeria in the 1960s, members of this virus group are still circulating in the country. Specifically, SBV, SIMV and SHAV or closely related viruses infect cattle and sheep across the three vegetation zones of Nigeria suggesting that activity of competent insect vectors is extensive in the country.

Abbreviations

AINV: Aino virus; AKAV: Akabane virus; ELISA: enzyme-linked immunosorbent assay; IF: immunofluorescence; SATV: Sathuperi virus; SBV: Schmallenberg virus; SHAV: Shamonda virus; SHUV: Shuni virus; SIMV: Simbu virus; SNT: serum neutralisation test

Funding

This work received no specific grant from any funding agency.

Authors' contributions

DO conceived the study and participated in its design, and was involved in performing the ELISA as well as drafting and making critical revisions of the manuscript. KW performed and analysed the serum neutralization tests, and participated in drafting and making critical revisions of the manuscript. AA was involved in sample collection, and in performing the ELISA and statistical analysis. SC participated in the study design and sample collection. MB conceived the study and participated in its design and coordination, and helped to draft the manuscript. All authors read and approved the final manuscript.

Consent for publication

An oral informed consent to use the animals in the study was obtained from the owners of the animals.

Competing interests
The authors declare that they have no competing interests.

Author details
[1]Department of Veterinary Microbiology, University of Ibadan, Ibadan, Oyo State, Nigeria. [2]Institute of Diagnostic Virology, Friedrich-Loeffler-Institut, Südufer 10, 17493 Greifswald - Insel Riems, Germany. [3]Department of Veterinary Public Health and Preventive Medicine, University of Ibadan, Ibadan, Oyo State, Nigeria.

References
1. Calisher CH. History, classification and taxonomy of viruses in the family Bunyaviridae. In: Elliott RM, editor. The Bunyaviridae. New York: Plenum Press; 1996. p. 1–17.
2. Plyusnin A, Beaty BJ, Elliott RM, Goldbach R, Kormelink R, Lundkvist A, et al. Bunyaviridae. In: King AMQ, Adams MJ, Carstens EB, Lefkowits EJ, editors. Virus taxonomy: ninth report of the international committee on taxonomy of viruses. London: Elsevier Academic Press; 2012. p. 725–41.
3. Elliot RM, Schmaljohn CS. Bunyaviridae. In: Knipe DM, Howley PM, editors. Field's virology. 6th ed. London: Lippincott Williams & Wilkins; 2013. p. 1244–82.
4. Elliott RM, Blakqori G. Molecular biology of Orthobunyaviruses. In: Plyusnin A, Elliott RM, editors. Bunyaviridae: molecular and cellular biology. Norfolk: Caister Academic Press; 2011. p. 1–39.
5. MacLachlan NJ, Dubovi EJ. Bunyaviridae. In: Fenner's veterinary virology. 4th ed. London: Elsevier Academic Press; 2011. p. 371–383.
6. Pawaiya RVS, Gupta VK. A review on Schmallenberg virus infection: a newly emerging disease of cattle, sheep and goats. Vet Med. 2013;58(10):516–26.
7. Causey OR, Kemp GE, Causey CE, Lee VH. Isolation of Simbu-group viruses in Ibadan, Nigeria 1964–69, including the new types Sango, Shamonda, Sabo and Shuni. Ann Trop Med Parasitol. 1972;66:357–62.
8. Inaba Y, Kurogi H, Omori T. Akabane disease: epizootic abortion, premature birth, stillbirth and congenital arthrogryposis-hydranencephaly in cattle, sheep and goats caused by Akabane virus. Aust Vet J. 1975;51:584–5.
9. Golender N, Brenner J, Motti V, Yevgeny K, Velizar B, Alexander P, et al. Malformations caused by Shuni virus in ruminants, Israel, 2014–2015. Emerg Infect Dis. 2015;21(12):2267–8.
10. Wernike K, Elbers A, Beer M. Schmallenberg virus infection. Rev Sci Tech (OIE). 2015a;34(2):363–73.
11. Hartley WJ, Wanner RA. Bovine congenital arthrogryposis in New South Wales. Aust Vet J. 1974;50(5):185–8.
12. Metselaar D, Robin Y. Akabane virus isolated in Kenya. Vet Rec. 1976;99:86.
13. Theodoridis A, Nevill EM, Els HJ, Boshoff ST. Viruses isolated from Culicoides midges in South Africa during unsuccessful attempts to isolate bovine ephemeral fever virus. Onderstepoort J Vet Res. 1979;46(4):191–8.
14. Miura Y, Inaba Y, Tsuda T, Tokuhisa S, Sato K, Akashi H, et al. A survey of antibodies to arthropod-borne viruses in Indonesian cattle. Japanese J Vet Sci. 1982;44(6):857–63.
15. Al-Busaidy S, Hamblin C, Taylor WP. Neutralising antibodies to Akabane virus in free-living wild animals in Africa. Trop Anim Health Prod. 1987;19(4):197–202.
16. Brenner J, Tsuda T, Yadin H, Chai D, Stram Y, Kato T. Serological and clinical evidence of teratogenic Simbu serogroup virus infection of cattle in Israel, 2001–2003. Vet Ital. 2004;40:119–23.
17. Blomstrom AL, Stenberg H, Scharin I, Figueiredo J, Nhambirre O, Abilio AP, et al. Serological screening suggests presence of Schmallenberg virus in cattle, sheep and goat in the Zambezia Province. Mozambique Transbound Emerg Dis. 2014;61(4):289–92.
18. Mathew C, Klevar S, Elbers ARW, van der Poel WHM, Kirkland PD, Godfroid J, et al. Detection of serum neutralizing antibodies to Simbu sero-group viruses in cattle in Tanzania. BMC Vet Res. 2015;11:208. https://doi.org/10.1186/s12917-015-0526-2.
19. Oluwayelu DO, Aiki-Raji CO, Umeh EC, Mustapha SO, Adebiyi AI. Serological investigation of Akabane virus infection in cattle and sheep in Nigeria. Adv Virol. 2016; Article ID 2936082, 4 pages. https://doi.org/10.1155/2016/2936082
20. Miura Y, Hayashi S, Ishihara T, Inaba Y, Omori T, Matumoto M. Neutralizing antibody against Akabane virus in precolostral sera from calves with congenital arthrogryposis-hydranencephaly syndrome. Archiv fur die Gesamte Virusforschung. 1974;46(3–4):377–80.
21. Howell PG, Coetzer JAW. Serological evidence of infection of horses by viruses of the Simbu-serogroup in South Africa. In: Wernery U, Wade JF, Mumford JA, Kaaden OR, editors. Equine infectious diseases. VIII proceedings of the eighth international conference: 23–16 march, 1998. Suffolk: R&W Publications (Newmarket) Ltd; 1999. p. 549.
22. Bréard E, Lara E, Comtet L, Viarouge C, Doceul V, Desprat A, et al. Validation of a commercially available indirect ELISA using a nucleocapsid recombinant protein for detection of Schmallenberg virus antibodies. PLoS One. 2013;8(1):e53446.
23. Kurogi H, Inaba Y, Goto Y, Miura Y, Takahashi H. Serologic evidence for etiologic role of Akabane virus in epizootic abortion-arthrogryposis-hydranencephaly in cattle in Japan, 1972-1974. Arch Virol. 1975;47:71–83.
24. Tarlinton R, Daly J, Dunham S, Kydd J. The challenge of Schmallenberg virus emergence in Europe. Vet J. 2012;194:10–8.
25. Lee VH. Isolation of viruses from field populations of Culicoides (Diptera: Ceratopogonidae) in Nigeria. J Med Entom. 1979;16(1):76–9.
26. Ate IU, Allam L. Multiple congenital skeletal malformations in a lamb associated with dystocia in a Yankassa ewe. Nig Vet J. 2002;23(1):61–3.
27. Bukar MM, Waziri M, Ibrahim UI. Dystocia due to arthrogryposis and associated with a mummified twin in a crossed (Yankassa/Uda) ewe: a case report. Trop Vet. 2006;24(4):85–8.
28. Ibrahim ND, Adamu S, Useh SM, Salami SO, Fatihu MY, Sambo SJ, et al. Multiple congenital defects in a Bunaji bull. Nig Vet J. 2006;27(3):80–6.
29. Oluwayelu DO, Meseko CA, Adebiyi AI. Serological screening for Schmallenberg virus in exotic and indigenous cattle in Nigeria. Sokoto J Vet Sci. 2015;13(3):14–8.
30. Wernike K, Beer M, Hoffmann B. Schmallenberg virus infection diagnosis: results of a German proficiency trial. Transbound Emerg Dis. 2017;64(5):1405–10.
31. Wernike K, Eschbaumer M, Schirrmeier H, Blohm U, Breithaupt A, Hoffmann B, et al. Oral exposure, re-infection and cellular immunity to Schmallenberg virus in cattle. Vet Microbiol. 2013;165(1–2):155–9.
32. Reed LJ, Muench H. A simple method of estimating fifty percent end points. Am J Hyg. 1938;27:493–7.
33. Goller KV, Höper D, Schirrmeier H, Mettenleiter TC, Beer M. Schmallenberg virus as possible ancestor of Shamonda virus. Emerg Infect Dis. 2012;18(10):1644–6.
34. Doherty RL, Carley JG, Standfast HA, Dyce AL, Snowdon WA. Virus strains isolated from arthropods during an epizootic of bovine ephemeral fever in Queensland. Aust Vet J. 1972;48(3):81–6.
35. Kurogi H, Akiba K, Inaba Y, Matumoto M. Isolation of Akabane virus from the biting midge Culicoides oxystoma in Japan. Vet Microbiol. 1987;15(3):243–8.
36. Zeller H, Bouloy M. Infections by viruses of the families Bunyaviridae and Filoviridae. Rev Sci Tech. 2000;19:79–91.

23

Sudden death due to gas gangrene caused by *Clostridium septicum* in goats

Abdullah Gazioglu[1], Burcu Karagülle[2], Hayati Yüksel[3], M. Nuri Açık[4*], Hakan Keçeci[5], Muhammet Bahaeddin Dörtbudak[3] and Burhan Çetinkaya[2]

Abstract

Background: Even though gas gangrene caused by *Clostridium septicum* in goats is mentioned in the classical textbooks, we have not managed to find any case description in the literature.

Case presentation: Clinical signs resembling gas gangrene such as subcutaneous bloating, edema and crepitation were detected at various body parts of nine pregnant animals at the ages of 2–3 years on a hair goat farm (*n* = 170) located in Bingol province, Eastern Turkey. Five of these suspected animals with severe clinical symptoms died within 2 days. Various samples such as internal organs, edematous skin and edema fluid collected from dead and live animals were analyzed for the presence of clostridial agents by histopathological and microbiological methods. As a result of macroscopic and microscopic examination, lesions of gas gangrene were detected. The suspected isolates were identified and confirmed as *C. septicum* by bacteriological and molecular methods.

Conclusion: The present study was the first to report identification of *C. septicum* as primary agent in the gas gangrene of goats.

Keywords: *Clostridium septicum*, Goat, Gas gangrene

Background

Gas gangrene, previously defined as malignant edema, is an important clostridial infection worldwide. It is characterized with necrosis in soft tissues, especially in muscle, toxemia, and sudden deaths in animal species, particularly ruminants and humans [1, 2]. Several clostridial agents including *Clostridium septicum*, *C. sordellii*, *C. chauvoei*, *C. perfringens* type A and *C. novyi* type A have been linked with the etiology of gas gangrene [3, 4]. Although information on the pathogenesis of the disease is limited, it is considered that vegetative or spore forms of at least one *Clostridium* species cause exogenous disease in animals by entering the body through abraded skin such as scratches, cuts and castration wounds [5]. Despite the fact that tissue traumas play an important role in the growth of *Clostridium* agents in soft tissues, especially in small ruminants, *C. septicum* can easily grow in these regions without tissue trauma as it is more aero-tolerant than other species [6].

C. septicum is the etiological agent of suppurative and necrotizing abomasitis (known as braxy) in ruminants including sheep and calves. This agent is also responsible for gas gangrene in cattle, sheep, horses and other species, and gangrenous dermatitis and cellulitis in poultry [4, 7, 8]. Extracellular toxins produced by *C. septicum* such as deoxyribonuclease, hyaluronidase, neuraminidase and alpha toxin play significant roles in the occurrence of the disease. While some of these toxins are involved in the spread of bacterium within the body, alpha toxin plays a major role in the formation of tissue necrosis [9, 10].

Even though gas gangrene caused by *C. septicum* in goats is mentioned in the classical textbooks [11], we have not managed to find any case description in the literature. In the present study, tissue and organ samples collected from a goat herd suspected of gas gangrene were examined for the presence of *C. septicum* by conventional and molecular methods.

Case presentation

Clinical signs resembling gas gangrene such as subcutaneous edema and crepitation were detected at various

* Correspondence: mehmetnuriacik@bingol.edu.tr
[4]Department of Microbiology, Faculty of Veterinary Medicine, University of Bingol, 12000 Bingol, Turkey
Full list of author information is available at the end of the article

body parts of nine pregnant animals (last month of pregnancy) at the ages of 2–3 years on a hair goat farm ($n = 170$) located in Bingol province, Eastern Turkey. The degree of clinical signs was observed to be more severe (impaired general condition, skin lesions on all legs) in five animals which died one after another within 30 h. No attempt could be made for the treatment of these animals because they were at the terminal stage of the disease when we were alert to the problem. On the other hand, the remaining four animals had a milder course of disease which suggested that they were at the onset of the disease. These animals were treated with antibiotics (Penicillin G 8.000 IU/kg + Streptomycin 10 mg/kg, s.i.d. IM, for 3 days) after collecting the required samples and the signs of gas gangrene disappeared within 3 weeks following antibiotic therapy.

Systemic necropsies were performed on dead animals within 2 h at the latest. Following necropsy, samples taken from subcutaneous fascia, lungs, mesenteric lymph nodes, cardiac muscle and kidney tissue were fixed in 10% buffered formaldehyde solution. Paraffin blocks were prepared and cut to a thickness of 5 μm from tissues taken for routine histological follow-up. The prepared paraffin sections were stained with hematoxylin-eosin and evaluated histopathologically on a light microscope (Leica, Germany).

Samples were collected in sterile conditions from edematous skin lesions and internal organs of dead animals for microbiological analyzes. In addition, edema fluid was obtained with sterile syringes from the edematous skin area while the animals were alive. The samples were transported within 2 h under cold chain conditions (+ 2/+ 8 °C in icebox) to the laboratories where culture procedures were carried out. Homogenized internal organs and edema fluid were first subjected to enrichment in Cooked Meat Medium (Oxoid) for anaerobic isolation. One ml of edema fluid was added to the bottom of a 10 ml Cooked Meat Medium with a sterile Pasteur pipette and incubated in an anaerobic jar for 24 h at 37 °C. Anaerobic media were supplied with anaerobic gas kits (Anaerocult A, Merck). In addition, samples were incubated in 5% blood

agar for 24 h at 37 °C for aerobic microorganism isolation. In the anaerobic conditions, turbidity was seen at the bottom of the Cooked Meat Medium tube. For this reason, 100 μl of Cooked Meat Medium was inoculated in 5% blood agar and incubated for 24 h at 37 °C in anaerobic conditions.

DNA samples extracted from the suspected colonies by conventional Phenol extraction method were subjected to Polymerase Chain Reaction (PCR) combined with a pair of species-specific primers derived from flagellin gene (fliC) region of C. septicum [12]. Due to the fact that C. chauvoei has similar biochemical characteristics and it has been isolated together with C. septicum in many cases, the growth colonies were also examined for the presence of C. chauvoei by a species specific PCR [12]. The presence of alpha toxin, a lethal virulence factor which plays a significant role in the pathogenesis of gas gangrene caused by C. septicum, was also investigated in the current study by employing a PCR combined with alpha toxin (Hemolysin) gene [13]. The primers and PCR conditions used in this study were presented in Table 1. In order to confirm PCR findings, one-way DNA sequence analysis was performed on two randomly selected samples from PCR products of the flagellin gene region (Sentegen Biotech Laboratory, Ankara, Turkey).

In the detailed clinical examination, poor body condition, depression, cyanosis in the eye conjunctiva, tachypnea, tachycardia and hard vesicular sounds in lung auscultation were observed in all animals with severe disease. The inner side of the affected leg was gangrenous which resulted in lameness. In addition, crepitated gas formation was observed in the palpation of the affected skin area and abundant yellow-red liquid was seen at subcutaneous tissue following puncturing.

Similar macroscopic findings were found in dead animals. No gross lesions were observed in fetal and uterus samples. Crepitation was noticed in the bloated areas of the goats while cutting the skin, in addition to the greenish appearance containing bubbles in the subcutaneous fascia (Fig. 1a). In the subcutaneous fascia, bubbles were seen with dark red-black colored fluid, extending from the inguinal region to the dorsal. Bleeding areas in dark

Table 1 Primers and PCR conditions used in this study

Target gene	Sequences (5'-3') (amplicon sizes)	PCR conditions	References
C. septicum flagellin gen (fliC)	FlaF- AGAATAAACAGAAGCTGGAGATG FlaseR-TTTATTGAATTGTGTTTGTGAAG (Amplicon:294 bp)	94 °C 1 min 55 °C 1 min 72 °C 90 s (30 cycles)	[12]
C. chauvoei flagellin gen (fliC)	FlaF- AGAATAAACAGAAGCTGGAGATG FlachR-TACTAGCAGCATCAAATGTACC (Amplicon:535 bp)		
Alpha toxin Gen	F-AATTCAGTGTGCGGCAGTAG R-CCTGCCCCAACTTCTCTTTT (Amplicon:270 bp)	94 °C 1 min 55 °C 1 min 72 °C 1 min (35 cycles)	[13]

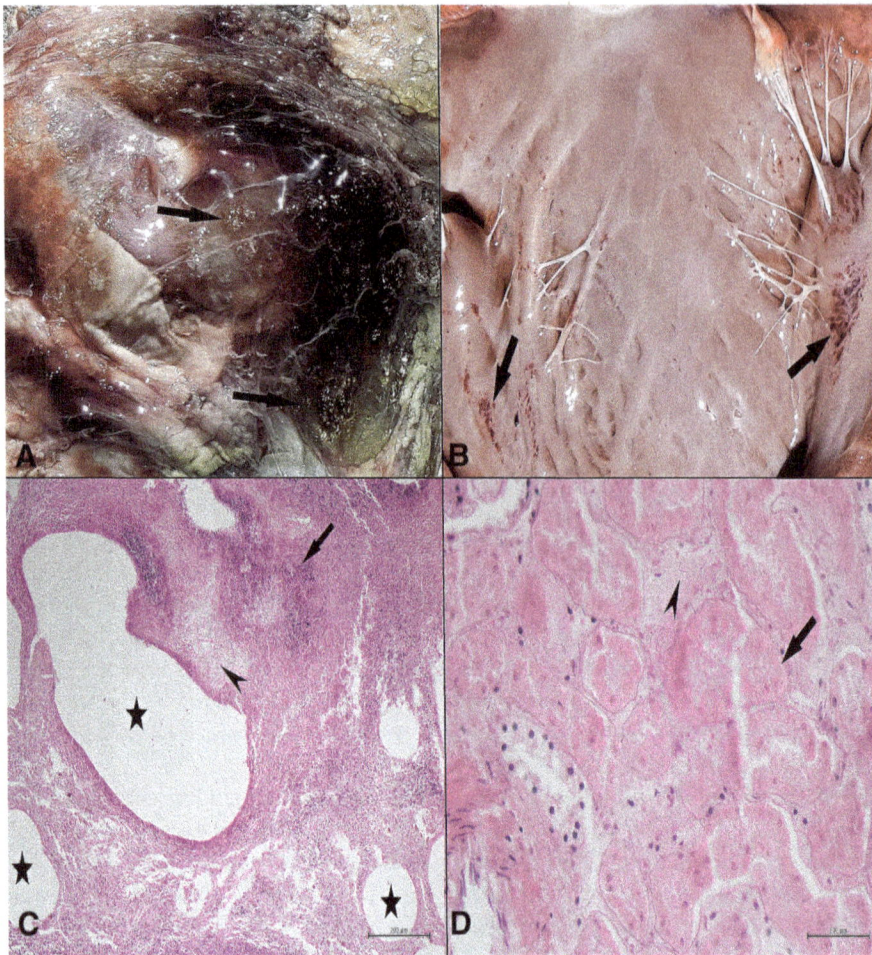

Fig. 1 Macroscopic and microscopic results obtained from various tissue and organ samples. **a** Macroscopically, greenish appearance containing bubbles in the subcutaneous fascia (arrows). **b** Petechial hemorrhages in the endocardium (arrows). **c** Microscopically, focal necrosis (arrow), gas bubbles (stars), extensive edema and rod-shaped bacterial clusters (arrow head) in the subcutaneous fascial tissue. **d** Necrosis in the proximal convoluted tubules (arrow), diffuse edema and rod-shaped bacterial clusters (arrow head) in the kidney. HEx200μ

red-black color in the subcutaneous region, and approximately 500–600 ml of bloody fluid in the abdominal cavity were observed. When the thoracic cavity was opened, it was found that there was a thick foamy fluid in the trachea and the lungs were edematous. Petechial hemorrhages were observed in the epicardium and endocardium (Fig. 1b).

In the microscopic examination; extensive edema, focal necrosis, gas bubbles in large vacuoles and rod-shaped bacterial clusters were detected in and around the subcutaneous fascial tissues (Fig. 1c). As a result of widespread edema, enlargement due to separation and a small number of neutrophil leukocyte infiltrations were observed in the fascial tissue and muscle fibers. Thrombi were detected in venous and capillary veins in these areas. In the lungs, pink homogeneous edema fluid was found in interlobular, interalveolar septum and alveolar lumens. In the kidney tissue, necrosis in the proximal convoluted

tubulus epithelium, diffuse edema and rod-shaped bacterial clusters in the intertubular areas were observed (Fig. 1d).

Following the culture of the samples taken from nine animals under anaerobic conditions, a diffuse colony appearance as well as centrally sporulated and rod-shaped Gram positive bacteria were observed on the microscope (Fig. 2a, b). However, no growth was observed in the incubation of the samples in aerobic conditions. Flagellin gene (*fliC*) specific PCR analysis of DNA's extracted from five pure colonies of all the samples produced positive results for *C. septicum* (Fig. 2c). Also, all the isolates were found to be positive for the alpha toxin (Hemolysin) gene in the PCR analysis (Fig. 2d). On the other hand, *C. chauvoei* could not be detected in any of the isolates by species-specific PCR.

The data obtained from one-way DNA sequence analysis of the PCR products of the flagellin gene region

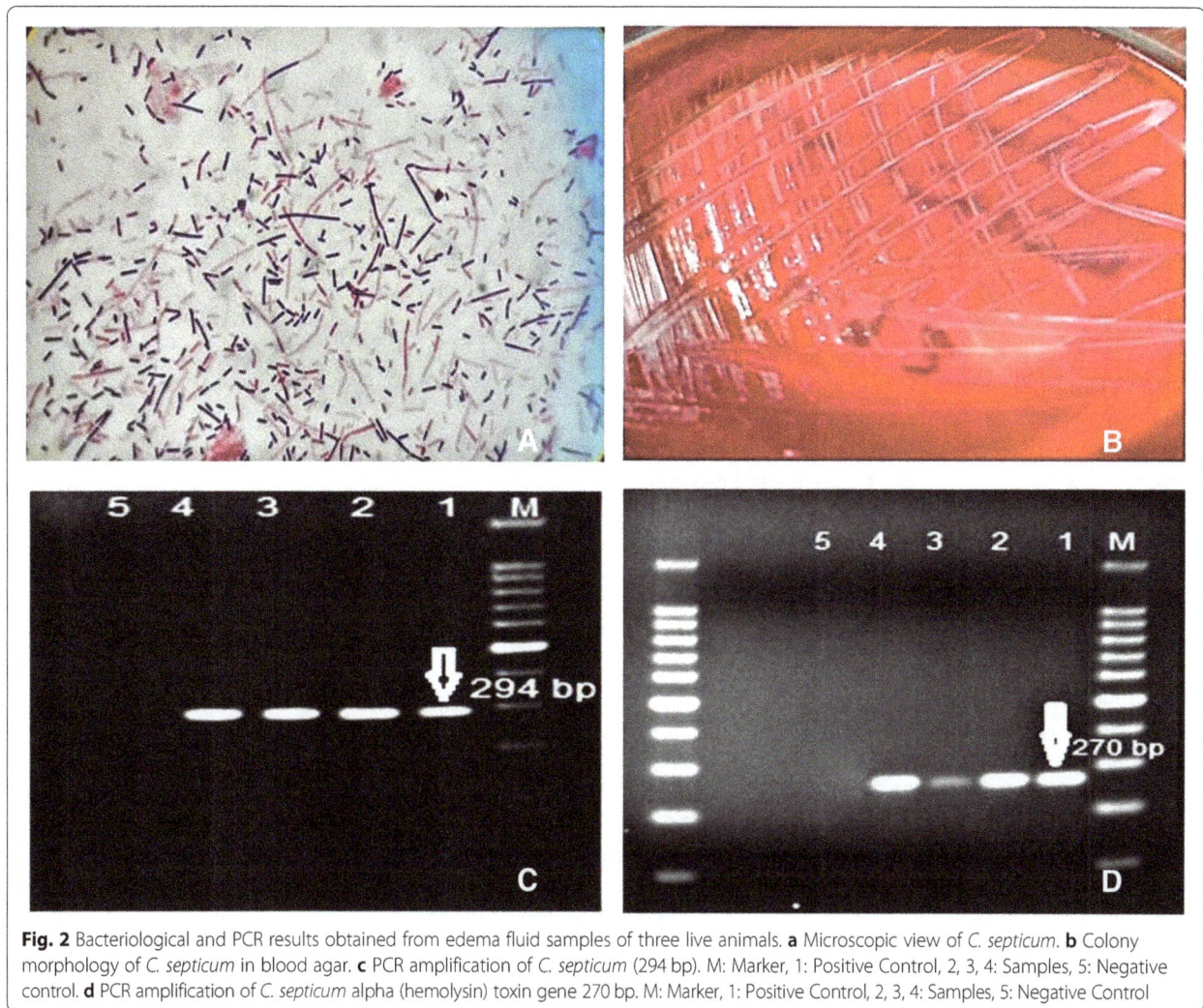

Fig. 2 Bacteriological and PCR results obtained from edema fluid samples of three live animals. **a** Microscopic view of *C. septicum*. **b** Colony morphology of *C. septicum* in blood agar. **c** PCR amplification of *C. septicum* (294 bp). M: Marker, 1: Positive Control, 2, 3, 4: Samples, 5: Negative control. **d** PCR amplification of *C. septicum* alpha (hemolysin) toxin gene 270 bp. M: Marker, 1: Positive Control, 2, 3, 4: Samples, 5: Negative Control

were aligned using the BLAST search (https://blast.ncbi.nlm.nih.gov/Blast.cgi/) with the registered *C. septicum* (HQ65058) in GenBank and the *C. botulinum* sequence (DQ844954) as an external group. Thus, a deletion at nucleotide 11 and a change in T → C at nucleotide 238 were observed in our isolates when compared to the reference sequence.

Discussion and conclusion

Gas gangrene is known as an exogenous disease in which the agent enter the body throughout skin abrasions. Because *C. septicum* is an opportunistic pathogen, it can also cause gas gangrene endogenously, especially in immunosuppressed animals [14]. No skin abrasions were noted in the clinical examination of the animals in the present study. Because most of the animals in the herd were pregnant, it was considered that the infection might be endogenous due immunosuppression caused by pregnancy. Oral administration of *C. septicum* is known to cause necrotic abomasitis in sheep, but in the necropsies of the animals examined in

this study only hemorrhagic lesions were observed in abomasum [15]. *C. septicum* can be found in the intestines of healthy ruminants. It is presumed that owing to immunosuppression, it causes septicemia by entering blood circulation and then produces myonecrosis in muscles by spreading throughout the body. In the anamnesis, information was obtained that all the animals in the herd were vaccinated with a commercial polyvalent vaccine 2 months ago. The adequate immune response was probably not developed in a limited number of animals. Also, sporadic cases of gas gangrene have been reported in cattle and sheep located in the same region. It was therefore thought that the agent was likely to be present in the environment and might have been taken by these animals.

Alpha toxin production has been shown to result in subcutaneous bloating and darkening of edematous skin areas and interstitial hemorrhage in muscle tissue [16]. Detection of similar necropsy findings in animals examined here suggests that *C. septicum* isolates produced alpha toxin which was already confirmed by toxin-specific

PCR. Indeed, studies have shown that *C. septicum* needs to produce alpha toxin in order to manifest specific clinical signs. Kennedy et al. (2005), reported a striking difference between alpha toxin positive and negative strains in terms of virulence [17]. The researchers observed fulminant gas gangrene symptoms within hours in mice infected with alpha toxin positive strains, whereas they did not find any clinical signs in mice that were infected with alpha toxin negative strains. Alpha toxin of *C. septicum* is structurally and functionally similar to epsilon toxin of *C. perfringens* type B and D and aerolysin of *Aeromonas hydrophila*. On the other hand, alpha toxin of *C. septicum* unlike *C. perfringens* alpha toxin, causes infiltration of immune system cells in the infection site. The histopathological finding that a large number of leukocytes were present in the lesioned area can be considered as another evidence for the production of alpha toxin or other toxins by *C. septicum* isolated in the current study. In infections caused by *C. septicum*, hemorrhage due to alpha toxin-induced microvascular destruction leads to decreased blood flow in the infection site. This ultimately leads to ischemia that will support the survival of *C. septicum* in the absence of external trauma [9, 10].

Clostridium septicum is generally sensitive to penicillin G, ampicillin, chloramphenicol, clindamycin, cephaloridine, oleandomycin, erythromycin, lincomycin and tetracyclines [18]. Because *C. septicum* causes sporadic disease which usually results in rapid death, there is no treatment available, in general. In this study, as the clinical signs of four affected animals had just begun which indicated the early stage of the disease, they were successfully treated with penicillin group antibiotics. This suggests that antibiotic use can be beneficial and may decrease deaths due to *C. septicum* in the early diagnosis of the disease.

C. septicum, commonly found in soil, has also been isolated from the feces of human and healthy animals [18]. The agent, as a postmortem invader, has the ability to rapidly spread throughout the body from the intestines of dead or agonized animals, especially ruminants. Due to the rapid spreading feature, it is possible to isolate *C. septicum* which can lead to misdiagnosis even if necropsy is made instantly following the death of animals. However, in this study, this agent was isolated from the skin lesions of live but diseased animals in addition to the samples taken at necropsy.

Although rapid diagnostic methods, such as fluorescent antibody test, are recommended for the diagnosis of gas gangrene, some drawbacks such as difficulty in obtaining commercially labeled antisera and potential complexity of the method restricts the use of this test. The fluorescent antibody test was not employed in this study due to the absence of labeled antisera. Instead, the isolates were examined for the presence of both *C*

septicum and *C. chauvoei* by PCR which is a rapid and sensitive molecular method and, while the former agent was identified, the latter could not be detected at all.

In conclusion, the present study reports for the first time the isolation and identification of *C. septicum* from various samples collected in a goat herd suspected of gas gangrene. It is believed that the histopathological and molecular findings of the study overruled the possible accidental presence of the agent in these goats.

Abbreviations
HE: Hematoxylin-eosin; PCR: Polymerase chain reaction

Acknowledgements
Not applicable.

Funding
No funding was received.

Authors' contributions
AG and HK performed clinical examination, sample collection and data interpretation, BK performed culture and molecular analyses, HY and MBD performed histopathological analyses and data interpretation, MNA and BC performed microbiological analyses, data interpretation and wrote the manuscript. All authors read and approved the final manuscript.

Consent for publication
Not applicable.

Competing interests
The authors declare that they have no competing interests.

Author details
[1]Department of Veterinary Science, Vocational School of Technical Sciences, University of Bingol, 12000 Bingol, Turkey. [2]Department of Microbiology, Faculty of Veterinary Medicine, University of Firat, 23119 Elazig, Turkey. [3]Department of Pathology, Faculty of Veterinary Medicine, University of Bingol, 12000 Bingol, Turkey. [4]Department of Microbiology, Faculty of Veterinary Medicine, University of Bingol, 12000 Bingol, Turkey. [5]Department of Internal Medicine, Faculty of Veterinary Medicine, University of Bingol, 12000 Bingol, Turkey.

References
1. Harwood DG. Apparent iatrogenic clostridial myositis in cattle. Vet Rec. 1984;115:412.
2. Abella BS, Kuchinic P, Hirahoka T, Howes DS. Atraumatic clostridial myonecrosis: case report and literature review. J Emerg Med. 2003;24:401–5.
3. Sterne M, Batty I. Pathogenic Clostridia. 1st Edition. Butterworth, London, Boston. 1975.
4. Silva ROS, Uzal FA, Oliveira CA, Lobato FCF. Gas Gangrene (Malignant Edema). In: Uzal FA, Songer JG, Prescott JF, Popoff MR, editors. Clostridial diseases of animals. Ames: Wiley Blackwell; 2016. p. 243–54.
5. Hatheway CL. Toxigenic clostridia. Clin Microbiol Rev. 1990;3:66–98.

6. Stevens DL, Musher DM, Watson DA, Eddy H, Hamill RJ, Gyorkey F, Rosen H, Mader J. Spontaneous, nontraumatic gangrene due to *Clostridium septicum*. Rev Infect Dis. 1990;12:286–96.

7. Schamber GJ, Berg IE, Molesworth JR. Braxy or bradsot-like abomastitis caused by *Clostridium septicum* in a calf. Can Vet J. 1986;27:194.

8. Gornatti-Churria CD, Crispo M, Shivaprasad HL, Uzal FA. Gangrenous dermatitis in chickens and turkeys. J Vet Diagn Investig. 2018;30:188–96.

9. Hang'ombe MB, Mukamoto M, Kohda T, Sugimoto N, Kozaki S. Cytotoxicity of *Clostridium septicum* alpha-toxin: its oligomerization in detergent resistant membranes of mammalian cells. Microb Pathog. 2004;37:279–86.

10. Kennedy CL, Lyras D, Cordner LM, Melton-Witt J, Emmins JJ, Tweten RK, Rood JI. Pore-forming activity of alpha-toxin is essential for *Clostridium septicum*-mediated myonecrosis. Infect Immun. 2009;77:943–51.

11. Smith MC, Sherman DM. Goat medicine. 2nd ed. New Jersey: Wiley; 2009. p. 124–6.

12. Sasaki Y, Kojima A, Aoki H, Ogikubo Y, Takikawa N, Tamura Y. Phylogenetic analysis and PCR detection of *Clostridium chauvoei*, *Clostridium haemolyticum*, *Clostridium novyi* types A and B, and *Clostridium septicum* based on the flagellin gene. Vet Microbiol. 2002;86:257–67.

13. Takeuchi S, Hashizume N, Kinoshita T, Kaidoh T, Tamura Y. Detection of *Clostridium septicum* hemolysin gene by polymerase chain reaction. J Vet Med Sci. 1997;59:853–5.

14. Srivastava I, Aldape MJ, Bryant AE, Stevens DL. Spontaneous *C. septicum* gas gangrene: a literature review. Anaerobe. 2017;48:165–71.

15. Eustis SL, Bergeland ME. Suppurative abomasitis associated with *Clostridium septicum* infection. J Am Vet Med Assoc. 1981;178:732–4.

16. Knapp O, Maier E, Mkaddem SB, Benz R, Bens M, Chenal A, Geny B, Vandewalle A, Popoff MR. *Clostridium septicum* alpha-toxin forms pores and induces rapid cell necrosis. Toxicon. 2010;55:61–72.

17. Kennedy CL, Krejany EO, Young LF, O'Connor JR, Awad MM, Boyd RL, Emmins JJ, Lyras D, Rood JI. The alpha-toxin of *Clostridium septicum* is essential for virulence. Mol Microbiol. 2005;57:1357–66.

18. Songer JG. Clostridial enteric diseases of domestic animals. Clin Microbiol Rev. 1996;9:216–34.

24

Replacement of grains with soybean hulls ameliorates SARA-induced impairment of the colonic epithelium barrier function of goats

Kai Zhang[1,2], Yuanlu Tu[1,2], Lipeng Gao[1,2], Meijuan Meng[1,2] and Yunfeng Bai[1,2]* ⓘ

Abstract

Background: The effect of soybean hull feeding on the disruption of colonic epithelium barrier function was investigated in goats fed a high-concentrate diet. Twenty-one Boer goats (live weight, 32.57 ± 2.26 kg; age, 1 year) were randomly divided into three groups: low-concentrate diet (LC), high-concentrate diet (HC), and high-concentrate diet with soybean hulls (SH).

Results: We found that the rumen fluid in the LC and SH group shown a higher pH value compared with the HC group. The mRNA and protein expression levels of extracellular regulated protein kinase (ERK), c-Jun N-terminal kinase (JNK), and p38 mitogen-activated protein kinase (MAPK) in the colonic epithelium were significantly decreased in the SH group than in the HC group. Moreover, in goats fed the HC diet, SH treatment promoted gene expression and protein abundance of claudin-1, claudin-4, occludin, and ZO-1 in the colonic epithelium. Additionally, the injury to the colonic epithelium barrier caused by the HC diet was reversed by SH treatment.

Conclusions: Our results indicated that supplemental SH feeding reverses the damage to colonic epithelium tight junctions by inhibiting the MAPK signalling pathway and has a protective effect on the colonic epithelium during SARA.

Keywords: Goat, SARA, Colonic epithelium, Soybean hulls, Tight junction proteins

Background

Ruminants are always fed a large amount of cereals in their diet to meet the nutritional requirements for energy for rapid weight gain or high milk yields. Although grains are beneficial in the short term, the risk of a metabolic disorder and systemic disease termed subacute rumen acidosis (SARA) increases after long-term feeding [1]. A ruminal pH between 5.6 and 5.8 not less than 3 h per day is the most widely accepted parameter in the diagnosis of SARA [2, 3]. This low ruminal pH perturbs the balance of rumen microbial populations, which is directly related to the release of lipopolysaccharide (LPS) endotoxin from lysed bacteria in the gastrointestinal tract and to damage to the gastrointestinal epithelium [4, 5]. Free LPS can be translocated from this dietary-induced damage to the gut mucosa, causing a systemic inflammatory response [6–8].

The colonic epithelium is a polarized monolayer of columnar epithelial cells that form a semipermeable paracellular diffusion barrier for nutrient absorption and metabolism, prohibiting the transportation of microbes and toxins under the normal physiological state [9]. However, the barrier properties of the gastrointestinal epithelium can be damaged by low pH and hyperosmolarity when SARA occurs, resulting in increased translocation of LPS into circulation [10, 11]. Epithelial cells form selective barriers with an elaborate network of

* Correspondence: blinkeye@126.com
[1]Circular Agriculture Research Center, Jiangsu Academy of Agricultural Sciences, Nanjing, China
[2]Key Laboratory of Crop and Livestock Integrated Farming, Ministry of Agriculture, Nanjing, China

intercellular protein complexes tight junctions (TJs). TJs regulate integral mechanisms of epithelial morphogenesis, which are crucial for proper function and formation of epithelial barriers. The transmembrane proteins claudins and occludin and cytoplasmic plaque composed of the proteins zonula occludins (ZO)-1, – 2, and – 3 are by far the most important for TJs [12–14].

The abnormal expression of TJ protein can lead to the impairment of the colon barrier [15]. The expression of occludin is correlated with various barrier properties [16, 17], and overexpression of occludin leads to changes in the gate and fence function of TJs in mammalian epithelial cells [18, 19]. Additionally, the occludin loop peptides can lead to the disappearance of TJs [20]. Although tight junctions exhibit normal barrier properties in occludin-knockout mice, complex abnormalities are observed, with postnatal growth retardation, the absence of fertility, the thinning of compact bone, chronic inflammation and hyperplasia [21, 22]. Claudins are identified as key molecules in the barrier function of TJs [23]. Overexpression of claudin-1 and -2 in fibroblasts can reconstitute tight junction strands [24], and multiple lines of evidence suggest that overexpression or knockout of claudins can ameliorate junctional ion permeability and barrier function [25, 26]. ZO-1, the first tight junction protein identified, interacts with multiple other junctional components, including claudins and occludin [27]. A previous study demonstrated that TJ barrier formation disintegrated because of the depletion of ZO-1 in cultured epithelial cells [28]. Moreover, claudin and occludin cannot constitute a normal TJ structure in ZO-depleted cells. MAPK signalling pathways regulate the expression of TJ proteins. We found that the activity of the MAPK signalling pathway can disrupt epithelial barrier function together with a decrease transcriptional level in claudin-1, occludin, and ZO-1 [29, 30].

Recently, replacing grain with low-starch, nonforage fibre has become a potential alternative to prevent SARA occurrence. A previous study demonstrated that partial substitution of starch with nonforage fibre sources, such as soybean hulls, positively affected milk production and composition in lactating dairy cows. Moreover, this substitution can increase feed efficiency and decrease diet feed costs [31]. Our previous studies have indicated that soybean hulls are traditionally fed as a protein supplement to improve the low rumen pH induced by SARA and do not affect the performance of goats. However, little is known about whether soybean hulls can prevent colonic epithelial barrier dysfunction when grain-induced SARA occurs. Therefore, the present study was conducted to confirm that soybean hulls increase the expression of TJs through modification and signalling pathway regulation, to provide a new treatment option for impaired barrier function.

Results

Rumen pH value analysis

The average rumen pH value in the LC and SH group was markedly higher than that in the HC group (Fig. 1, $P < 0.05$). A reduced pH value between 5.6 and 5.8 was approximately 4 h/day in the HC group of animals. The rumen pH value in the LC and SH groups was similar.

pH and LPS content in the colon and LPS and cytokines in plasma

Colonic pH value decreased from 6.90 for the LC group to 5.98 for the HC group (Table 1, $P < 0.05$). There was no significant difference in the colonic pH between the LC and SH groups.

Colonic content and plasma LPS concentration in the LC and SH group was significantly lower than that in the HC group ($P < 0.01$), while the concentration of LPS in the LC and SH groups was similar, as shown in Table 1.

IL-1β, IL-6, and TNF-α concentrations in the HC group were significantly increased ($P < 0.01$, Table 1) compared with those in the LC group, while the concentrations of IL-1β and IL-6 had no significant difference between the SH group and LC group. The level of TNF-α was significantly higher in the SH group than in the LC group ($P < 0.05$, Table 1).

Morphological analysis of the colonic epithelium

The epithelial injury scores confirmed the results of the colonic epithelium histological analysis, as shown in Fig. 2. The epithelial surface of colonic epithelium was sloughed in the HC group, while it in the LC and SH group covered by mucus and remained intact. Although the inflammatory cells in the colom serosal muscle layer of the SH group were more than the LC group, it was much less than the HC group. A

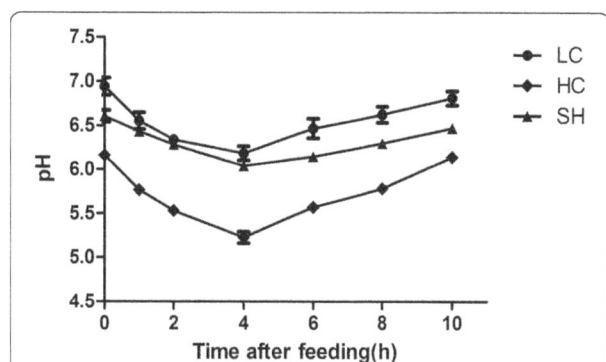

Fig. 1 Rumen pH value of different groups at different time points after feeding. The pH was markedly decreased in the HC group. With supplemental feeding with soybean hulls (SH), the pH changed significantly compared with that in the HC group. The error bars indicate the standard error of the mean. All of the data are shown as the mean ± SD. Differences between two groups were considered significant when $P < 0.05$

Table 1 LPS and primary pro-inflammatory cytokines content of different groups

Item	LC	HC	SH
pH in colon	$6.90^a\pm0.34$	$5.98^b\pm0.42$	$6.73^a\pm0.28$
LPS in colon, kEU/mL	$27.64^b\pm2.96$	$56.3^a\pm3.26$	$28.3^b\pm3.83$
LPS in plasma, EU/mL	$0.16^b\pm0.02$	$0.96^a\pm0.35$	$0.23^b\pm0.06$
IL-1β in plasma, ng/mL	$0.162^b\pm0.013$	$0.23^a\pm0.018$	$0.155^b\pm0.016$
IL-6 in plasma, pg/mL	$15.67^b\pm1.17$	$72.2^a\pm0.68$	$12.2^b\pm0.41$
TNF-α in plasma, fmol/mL	$76.73^c\pm5.32$	$570.36^a\pm24.29$	$354.15^b\pm51.42$

Different characters (a, b and c) show significant difference among diets ($p < 0.05$)

high-concentrate diet with soybean hulls feeding had a marked decrease macroscopic damage score compared with the HC group ($P < 0.05$).

Relative expression of genes related to barrier function in the colonic epithelium

Several pivotal genes were chosen to determine the colonic epithelium barrier function of the goats fed different diets, and the results are shown in Figs. 3 and 4. The mRNA relative expression of ERK1, p38 and JNK, increased in the HC group compared with that in the LC group ($P < 0.01$), while these in the LC and SH groups was similar. The expression of ERK2 was also increased markedly in the HC group ($P < 0.05$).

The mRNA expression of claudin-1, claudin-4, occlu-din, and ZO-1 was also detected by RT-qPCR. The

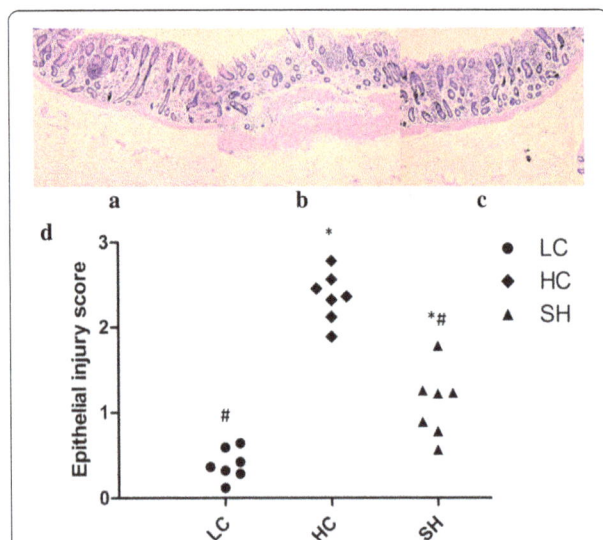

Fig. 3 The expression of genes related to the MAPK signalling pathway in the colonic epithelium of goats in the LC, HC, and SH groups. The expression of Erk1, Erk2, p38 MAPK and JNK significantly increased in the HC group compared with that in the LC group, while there was no significant difference between the SH and LC groups. The results are expressed as fold changes relative to the LC group (mean ± SEM). * indicates $P < 0.05$ and **indicates $P < 0.01$ compared with the LC; # indicates $P < 0.05$ compared with the HC. A statistically significant difference has a P value < 0.05

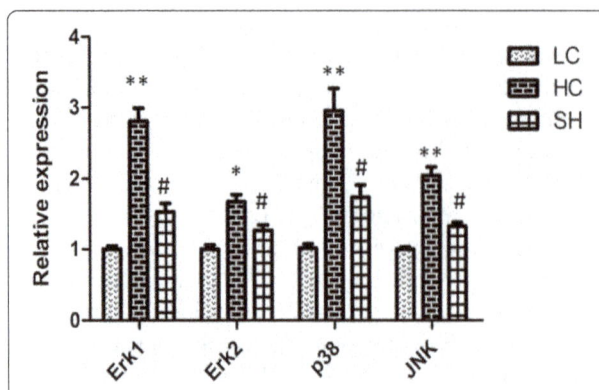

mRNA expression of these genes in the colonic epithe-lium of goats fed the HC diet declined significantly com-pared with the LC group, while the expression in the LC and SH groups was similar.

Phosphorylation levels of ERK, pERK, JNK, pJNK, p38, and pp38 in the colonic epithelium

The phosphorylation levels of ERK, p38 and JNK were determined by Western blot analysis (Fig. 5). ERK and JNK phosphorylation levels in the HC group were sig-nificantly elevated compared with those in the LC group,

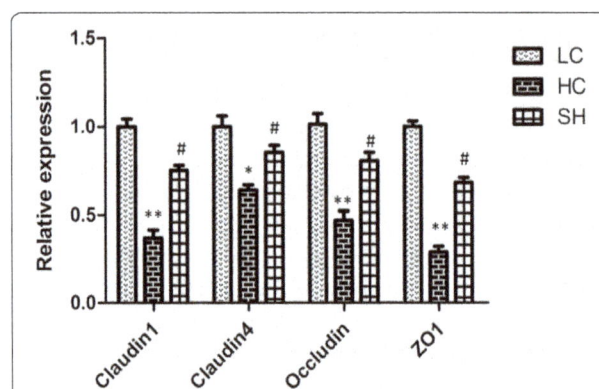

Fig. 2 Histological alteration of the colonic epithelium of goats from the different groups. Representative photomicrographs with haematoxylin and eosin staining. The colonic epithelium of the LC group was intact and showed no disruption (**a**), while the stratum corneum structure of the epithelium was severely damaged in the HC group (**b**). Slight damage of the epithelium was observed in the SH group (**c**). Epithelial injury scores are shown in **d**. * indicates $P < 0.05$ and ** indicates $P < 0.01$ compared with the LC; # indicates $P < 0.05$ compared with the HC. A statistically significant difference has a P value < 0.05

Fig. 4 The expression of genes related to tight junction proteins in the colonic epithelium of goats in the LC, HC, and SH groups. The colonic epithelium of the goat fed the HC diet had a significant decline in the mRNA expression of claudin-1, claudin-4, occludin, and ZO-1 compared with that in the LC group, while no significant change between the SH and LC groups was observed. The results are expressed as fold changes relative to those in the LC group (mean ± SEM). * indicates $P < 0.05$ and **indicates $P < 0.01$ compared with the LC; # indicates $P < 0.05$ compared with the HC. A statistically significant difference has a P value < 0.05

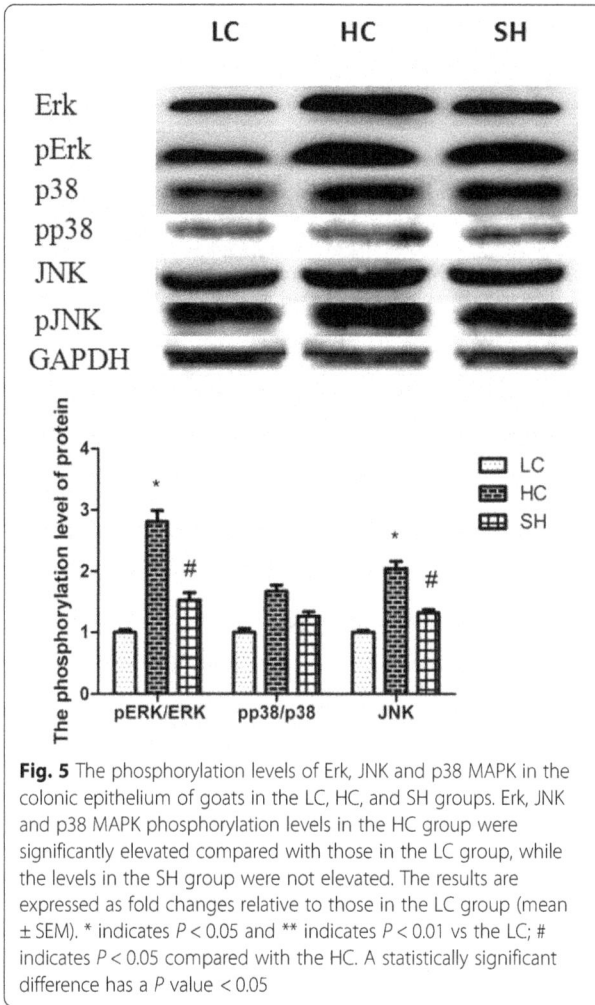

Fig. 5 The phosphorylation levels of Erk, JNK and p38 MAPK in the colonic epithelium of goats in the LC, HC, and SH groups. Erk, JNK and p38 MAPK phosphorylation levels in the HC group were significantly elevated compared with those in the LC group, while the levels in the SH group were not elevated. The results are expressed as fold changes relative to those in the LC group (mean ± SEM). * indicates $P < 0.05$ and ** indicates $P < 0.01$ vs the LC; # indicates $P < 0.05$ compared with the HC. A statistically significant difference has a P value < 0.05

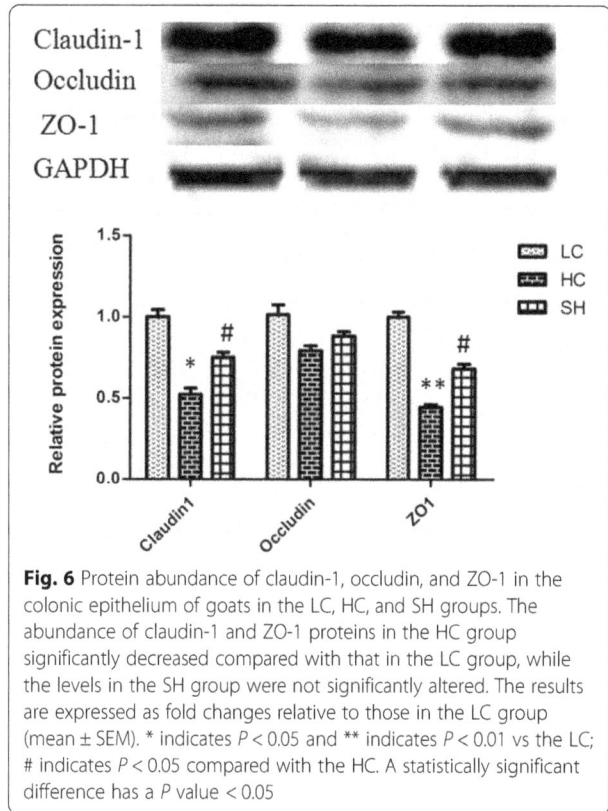

Fig. 6 Protein abundance of claudin-1, occludin, and ZO-1 in the colonic epithelium of goats in the LC, HC, and SH groups. The abundance of claudin-1 and ZO-1 proteins in the HC group significantly decreased compared with that in the LC group, while the levels in the SH group were not significantly altered. The results are expressed as fold changes relative to those in the LC group (mean ± SEM). * indicates $P < 0.05$ and ** indicates $P < 0.01$ vs the LC; # indicates $P < 0.05$ compared with the HC. A statistically significant difference has a P value < 0.05

while the levels in the SH group were not elevated. There was no significant difference in the phosphorylation levels of p38 among the treatments.

Protein levels of claudin-1, occludin, and ZO-1 in the colonic epithelium

Western blotting of claudin-1 and ZO-1 showed significantly lower protein levels in the HC group than in the LC group, whereas the levels in the SH and LC groups were similar (Fig. 6). The protein expression of occludin was not significantly different among the three groups.

Discussion

Feeding an HC diet to ruminants can not only promote microbial protein synthesis in the rumen and increase growth and milk production but also negatively affect rumen function because of the rapid accumulation of volatile fatty acids (VFA), which leads to decreased ruminal pH [32, 33]. Goats fed the HC diet for 12 weeks exhibited a low rumen pH between 5.6 and 5.8 for more than 3 h, which met

the current definition of SARA, indicating that SARA was successfully induced by the HC diet in the present study. We found that supplementing the HC diet with soybean hulls in the SH group led to a higher rumen pH than that in the HC group. Soybean hulls contain a high concentration of digestible neutral detergent fibre, which can increase chewing time and saliva secretion [31], possibly explaining how the SH diet increased the rumen pH. A low ruminal pH leads to gastrointestinal epithelial barrier dysfunction and translocation of LPS, increasing the risk of system inflammatory response [34, 35]. Based on our results, concentrations of LPS in the colon content and plasma of goats in the HC group were higher than those in the LC and SH groups.

Feeding the animals the HC diet could interrupt the barrier function of the colonic epithelium in our study. The expression of claudin-1, claudin-4, occludin, and ZO-1 declined significantly in the HC group, providing evidence that HC feeding can modulate the expression of TJ proteins and disrupt the barrier function. With the addition of SH, the expression of TJs was upregulated to a similar level to the LC group. Claudin-1, claudin-4, and ZO-1 protein abundance in the colonic epithelium of goats in the HC group was downregulated compared with that in goats treated with LC and SH. We also found that

SH diet feeding ameliorated the damage to the colonic epithelial barrier induced by HC diet feeding through histological alterations.

The MAPK signalling pathway is widely accepted to play a key role in regulating TJ protein formation [36]. JNK is essential in regulating TJ expression and epithelial barrier function. Inhibition of JNK activity in murine mammary epithelial cells can modulate of claudins expression and increase epithelial barrier function [37]. The expression of TJs can be depleted by various stimuli, such as the JNK activator anisomycin and the pro-inflammatory cytokines in human pancreatic cancer cells [38]. Proinflammatory cytokine treatment can induce the expression of TJ proteins, whereas the change was reversible by a JNK inhibitor. Some studies have indicated that HC diet feeding could enhance immune gene expression and activate the JNK signalling pathway in the liver of dairy cow [39]. These findings suggest that the concentration of pro-inflammatory cytokines and the phosphorylation level of the JNK pathway could induce disruption of the epithelial barrier. In the current study, the production of IL-1β, TNF, and IL-6 in the HC group was significantly higher than that in the LC group. The protein expression of JNK and pJNK increased, consistent with the mRNA expression of the corresponding gene by HC feeding. With SH administration, the indicators remained at levels similar to those observed in the LC group. The overexpression of the junctional membrane protein occludin can suppress ERK/MAP kinase activation [40]. In our study, the mRNA expression of ERK1 and ERK2 in goats induced by HC diet feeding declined after SH treatment, and SH treatment also decreased the phosphorylation levels of ERK. The expression of claudin-1 and -2 can be regulated by the p38 activator in vitro [41]. Decreasing expression of claudin-1 in regenerating rat liver has been reported by the p38 MAP kinase inhibitor [42], and p38 MAP kinase and Akt can inhibit the expression of claudin-4 in hepatic cell lines [43, 44]. Thus, the p38 MAP kinase pathway plays an important role in TJ formation in vivo and in vitro. SH treatment improved the downregulated expression of p38 induced by the HC diet in our study, which is consistent with the increased expression of claudin-1 and claudin-4 in our study. Hence, these lines of evidence collectively prove that the expression of TJs can be regulated by the MAPK signalling pathway in the colonic epithelium of goats.

Conclusions

According to our results, HC diet feeding for 12 weeks severely compromised the colonic epithelium and impaired epithelial barrier function. These adverse effects were attenuated after SH treatment through the phosphorylation of the MAPK signalling pathway. Thus,

a new option can be applied to remedy a series of systemic inflammatory responses induced by HC diet feeding.

Methods
Animals, diet, and experimental design
Twenty-one healthy, 1-year-old male Boer goats with an average body weight of 32.57 ± 2.26 kg, which were purchased from the Luhe Experimental Farm of Jiangsu Academy of Agricultural Sciences, were randomly divided into three groups ($n = 7$): low-concentrate diet (LC; 70% forage and 30% concentrate), high-concentrate diet (HC; 30% forage and 70% concentrate), and high-concentrate diet with soybean hulls (SH; 20% soybean hulls instead of grain). The diet nutritional composition are shown in Table 2. The goats were fed these diets at 8:00 a.m. and 6:00 p.m. and provided fresh water for 12 weeks.

Sample collection and analysis
The rumen pH values were measured on the last day of weeks 10, 11, and 12 through a cannula at 0, 1, 2, 4, 6, 8, and 10 h after feeding by a basic pH meter.

Blood samples were collected via the lacteal vein into 5-mL vacuum blood collection tubes. The blood samples were centrifuged at 1500×g for 15 min, and the plasma was collected and stored at – 20 °C for the determination of LPS, IL-1β, IL-6, and TNF-α.

The goats were slaughtered according to the law of Jiangsu Provincial People's Government, China, and the study was approved by the guidelines of Jiangsu Province Animal Regulations (Government Decree No. 45) and the Committee on the Ethics of Animal Experiments of Jiangsu Academy of Agricultural Sciences. The pH values of colonic digesta samples were immediately measured. Colons were removed and washed with PBS. Colonic tissue sample was selected from the same part of the colon aseptically and divided into two parts. One part was cut into 1 cm × 1 cm small pieces and infiltrated in 4% paraformaldehyde solution, and the other part was transferred into liquid nitrogen and then kept at – 80 °C for later detection.

Determination of LPS content in colon and plasma primary pro-inflammatory cytokines
LPS concentration in the colon content and plasma was determined with Chromogenic Endpoint Limulus Amebocyte Lysate Assay Kits (Chinese Horseshoe Crab Reagent Manufactory Co., Ltd., Xiamen, China). The detection range of LPS concentrations in colon and plasma were 0.1–1 endotoxin units (EU)/mL and 0.01–0.1 EU/mL, respectively. IL-1β, IL-6 and TNF-α concentrations in the plasma were determined by radioimmunoassay radioimmunoassay kits C09DJB, C12DJB

Table 2 Ingredients and nutrient composition of experimental diets

Ingredient,% DM	Percentage (%)of ingredients in different diets (air dry matter)		
	LC	HC	SH
Maize	12	40	21.78
Soybean meal	11.92	6.36	5.30
Wheat bran	1.00	8.98	8.57
Wheat	2.09	12.00	11.83
Soybean hull	0.00	0.00	20.00
Straw	70	30	30
Calcium hydphosphate	1.40	0.55	0.80
Limestone	0.59	1.10	0.73
Salt	0.50	0.50	0.50
Premix[a]	0.50	0.50	0.50
Total	100	100	100
Nutrient composition			
DE, MJ/kg	8.65	11.43	10.74
CP %	11.77	11.76	11.77
NDF %	44.81	26.53	36.26
ADF %	24.63	13.45	22.23
Ash %	10.26	8.13	8.52

[a]The premix consisted of the following ingredients per kg of diet: 6.60×10^4 IU of vitamin A, 8.00×10^5 IU of vitamin D3, 1.49×10^3 of vitamin E, 35.2 mg of Cu, 120 mg of Fe, 115 mg of Zn, 80 mg of Mn, 0.35 mg of Co, and 19.5 mg of Se

and C06PJB purchased from the Beijing North Institute of Biological Technology with a detection limit of 0.1–8.1 ng/mL, 50–4000 pg/mL and 9–590 fmol/mL.

Histological analysis
Fixed tissues were embedded in paraffin after immersing in a 4% paraformaldehyde solution for 72 h and dehydrating by ethanol. Five micrometer sections were cut form the paraffin on a microtome, mounted on slides, and stained with haematoxylin and eosin. The images were recorded by light microscope and high-resolution digital camera. Histological damage was assessed with a scoring epithelial injury criteria (graded 0 to 3) which has been described previously [45].

Quantitative real-time PCR
100 mg colonic tissue samples were selected for RNA extraction using RNAiso Plus (TaKaRa, Otsu, Japan). The A260/A230 and A260/A280 ratio of RNA samples were measured by spectrophotometer (Thermo Fisher Scientific Inc., Waltham, MA, United States). Then, RNA was reverse transcribed to the first-strand cDNA by PrimeScript RT kit (Cat. RR036A; TaKaRa). Primers were designed and synthesized in Shanghai Sangon Biotech Co., Ltd. and are listed in Table 3. Quantitative real-time PCR was performed using a SYBR Premix EX Taq Kit (Cat. DRR420A; TaKaRa) via an ABI 7300 system (Applied Biosystems, Foster City, CA, USA) according to the manufacturer's protocol. Glyceraldehyde-3-phosphate dehydrogenase (GAPDH) was used as the internal reference and the data was analyzed by $2^{-\Delta\Delta Ct}$ method.

Western blot analysis
Approximately100 mg of frozen grated colonic tissue was homogenized in 1 mL ice-cold RIPA protein isolation buffer (Cat. No. SN338; Sunshine Biotechnology Co., Nanjing, China) for the total protein extraction. The protein concentration was measured via the BCA Protein Assay kit (No. 23225, Thermo Fisher, USA). 50 μg of total protein was loaded onto 12% sodium dodecyl sulfate-polyacrylamide gel electrophoresis (SDS-PAGE) and transferred to a nitrocellulose membrane (Bio Trace; Pall Co., Port Washington, NY). The membranes were incubated in the appropriate primary antibodies: Jun N-terminal kinase (JNK) (No. 9252S; Cell Signaling Technology, Danvers, MA), p-JNK (No. 9255S; Cell Signaling Technology), p38 (No. 8690S; Cell Signaling Technology), p-p38 (No. 4511S; Cell Signaling Technology), Erk1/2 (No. 4695S; Cell Signaling Technology), p-Erk1/2 (No. 4370S; Cell Signaling Technology), PKC-

Table 3 The primer sequences of target and internal reference genes used in qRT-PCR

Gene	Forward primer	Reverse primer	PCR products (bp)
Erk1	CTCAGCTTACGACCATGTGC	TCAGGTCCTGCACGATGTAG	203
Erk2	CTCAGCAACGACCACATCTG	CCAGGCCAAAGTCACAGATC	151
p38	ACAACATCGTCAAGTGCCAG	CACGTAGCCAGTCATCTCCT	209
JNK	TCAGTCAGTTGAGCACCAGT	ACTTATGCCTGCTCTGCTCA	229
Claudin-1	CACCCTTGGCATGAAGTGTA	AGCCAATGAAGAGAGCCTGA	216
Claudin-4	AAGGTGTACGACTCGCTGCT	GACGTTGTTAGCCGTCCAG	238
Occludin	GTTCGACCAATGCTCTCTCAG	CAGCTCCCATTAAGGTTCCA	200
ZO-1	CGACCAGATCCTCAGGGTAA	AATCACCCACATCGGATTCT	163
GAPDH	GGGTCATCATCTCTGCACCT	GGTCATAAGTCCCTCCACGA	180

(No. 2056S; Cell Signaling Technology), occludin (No. ab167161; Abcam, Cambridge, United Kingdom), claudin-1 (No. ab15098; Abcam), and ZO-1 (No. ab214228; Abcam), then washed and incubated in corresponding horseradish peroxidase (HRP)-conjugated secondary antibodies. Finally, the membranes were washed and visualized using an enhanced chemiluminescence (ECL) Kit (Pierce, Rockford, IL). The signals were recorded using an Bio-Rad imaging system (Bio-Rad, Hercules, CA), and the results were analyzed using Quantity One software (Bio-Rad).

Statistical analysis

The results are expressed as the mean ± standard error of the mean (SEM). All data were evaluated with an unpaired Student's t-test or one-way ANOVA with the SPSS 21.0 statistical software package. A P value < 0.05 was considered to indicate a significant difference.

Abbreviations

ERK: Extracellular regulated protein kinases; HC: High-concentrate diet; JNK: c-Jun N-terminal kinase; LC: Low-concentrate diet; LPS: Lipopolysaccharide; MAPK: Mitogen-activated protein kinase; SARA: Subacute ruminal acidosis; SH: Soybean hulls; TJ: Tight junction; TNF-α: Tumour necrosis factor α

Acknowledgements

The authors would like to thank Qian Song and Jian Liu for animal feeding.

Author contributions

YFB and KZ conceived and designed the experiments. LPG, MJM, and YLT assisted in feeding animals and sampling. KZ and MJM conducted the research and analysed the data. KZ prepared the manuscript. YFB finalized the manuscript. All authors read and approved the final manuscript.

Funding

This study was supported by the Project of Jiang Su Independent Innovation (CX(15)1003) and Postdoctoral Foundation of Jiangsu Province (Grant number 1701031A). The funders had no role in the design of the study and collection, analysis, and interpretation of data and in writing the manuscript.

Consent for publication

Not applicable.

References

1. Li S, Khafipour E, Krause D, Kroeker A, Rodriguez-Lecompte J, Gozho G, et al. Effects of subacute ruminal acidosis challenges on fermentation and endotoxins in the rumen and hindgut of dairy cows. J Dairy Sci. 2012;95(1):294–303.
2. Steele MA, Croom J, Kahler M, AlZahal O, Hook SE, Plaizier K, et al. Bovine rumen epithelium undergoes rapid structural adaptations during grain-induced subacute ruminal acidosis. Am J Physiol Regul Integr Comp Physiol. 2011;300(6):R1515–23.
3. Zhou J, Dong G, Ao C, Zhang S, Qiu M, Wang X, et al. Feeding a high-concentrate corn straw diet increased the release of endotoxin in the rumen and pro-inflammatory cytokines in the mammary gland of dairy cows. BMC Vet Res. 2014;10(1):172.
4. Gott P, Hogan JS, Weiss WP. Effects of various starch feeding regimens on responses of dairy cows to intramammary lipopolysaccharide infusion. J Dairy Sci. 2015;98(3):1786–96.
5. Khafipour E, Krause DO, Plaizier JC. Alfalfa pellet-induced subacute ruminal acidosis in dairy cows increases bacterial endotoxin in the rumen without causing inflammation. J Dairy Sci. 2009;92:1712–24.
6. Plaizier J, Khafipour E, Li S, Gozho G, Krause D. Subacute ruminal acidosis (SARA), endotoxins and health consequences. Anim Feed Sci Tech. 2012;172:9–21.
7. Nagaraja TG, Titgemeyer EC. Ruminal acidosis in beef cattle: The current microbiological and nutritional outlook1,2. J Dairy Sci. 2007;90:E17–38.
8. Gozho GN, Krause DO, Plaizier JC. Ruminal lipopolysaccharide concentration and inflammatory response during grain-induced subacute ruminal acidosis in dairy cows. J Dairy Sci. 2007;90:856–66.
9. Gareau MG, Silva MA, Perdue MH. Pathophysiological mechanisms of stress-induced intestinal damage. Curr Mol Med. 2008;8:274–81.
10. Mao S, Zhang R, Wang D, Zhu W. The diversity of the fecal bacterial community and its relationship with the concentration of volatile fatty acids in the feces during subacute rumen acidosis in dairy cows. BMC Vet Res. 2012;8(237):1746–6148.
11. Liu JH, Xu TT, Liu YJ, Zhu WY, Mao SY. A high-grain diet causes massive disruption of ruminal epithelial tight junctions in goats. Am J Physiol Regul Integr Comp Physiol. 2013;305:R232–41.
12. Gassler N, Rohr C, Schneider A, Kartenbeck J, Bach A, Overmuller N, et al. Inflammatory bowel disease is associated with changes of enterocytic junctions. Am J Physiol Gastrointest Liver Physiol. 2001;281:G216–28.
13. Heller F, Florian P, Bojarski C, Richter J, Christ M, Hillenbrand B, et al. Interleukin-13 is the key effector Th2 cytokine in ulcerative colitis that affects epithelial tight junctions, apoptosis, and cell restitution. Gastroenterology. 2005;129:550–64.
14. Poritz LS, Harris LR 3rd, Kelly AA, Koltun WA. Increase in the tight junction protein claudin-1 in intestinal inflammation. Dig Dis Sci. 2011;56:2802–9.
15. John LJ, Fromm M, Schulzke JD. Epithelial barriers in intestinal inflammation. Antioxid Redox Signal. 2011;15(5):1255–70.
16. Matter K, Balda MS. Occludin and the functions of tight junctions. Int Rev Cytol. 1999;186:117–46.
17. Feldman GJ, Mullin JM, Ryan MP. Occludin: structure, function and regulation. Adv Drug Deliv Rev. 2005;57:883–917.
18. McCarthy KM, Skare IB, Stankewich MC, Furuse M, Tsukita S, Rogers RA, et al. Occludin is a functional component of the tight junction. J Cell Sci. 1996;109:2287–98.
19. Balda MS, Matter K. Tight junctions and the regulation of gene expression. Biochim Biophys Acta. 2009;1788:761–7.
20. Vietor I, Bader T, Paiha K, Huber LA. Perturbation of the tight junction permeability barrier by occludin loop peptides activates beta-catenin/TCF/LEF-mediated transcription. EMBO Rep. 2001;2:306–12.
21. Saitou M, Furuse M, Sasaki H, Schulzke JD, Fromm M, Takano H, et al. Complex phenotype of mice lacking occludin, a component of tight junction strands. Mol Biol Cell. 2000;11:4131–42.
22. Yu AL, McCarthy KM, Francis SA, McCormack JM, Lai J, Rogers RA, et al. Knock down of occludin expression leads to diverse phenotypic alterations in epithelial cells. Am J Physiol Cell Physiol. 2005;288:C1231–41.
23. Furuse M, Fujita K, Hiiragi T, Fujimoto K, Tsukita S. Claudin-1 and -2:novel integral membrane proteins localizing at tight junctions with no sequence similarity to occludin. J Cell Biol. 1998a;141:1539–50.
24. Muto S, Furuse M, Kusano E. Claudins and renal salt transport. Clin Exp Nephrol. 2012;16(1):61–7.
25. Furuse M, Furuse K, Sasaki H, Tsukita S. Conversion of zonulae occludentes from tight to leaky strand type by introducing claudin-2 into Madin-Darby canine kidney I cells. J Cell Biol. 2001;153:263–72.
26. Tsukita S, Furuse M, Itoh M. Multifunctional strands in tight junctions. Nat Rev Mol Cell Biol. 2001;2:286–93.
27. Bazzoni G, Martinez-Estrada OM, Orsenigo F, Cordenonsi M, Citi S, Dejana E. Interaction of junctional adhesion molecule with the tight junction components ZO-1, cingulin, and occludin. J Biol Chem. 2000;275:20520–6.
28. Umeda K, Ikenouchi J, Katahira-Tayama S, Furuse K, Sasaki H, Nakayama M, et al. ZO-1 and ZO-2 independently determine where claudins are

polymerized in tight junction strand formation. Cell. 2006;126:741–54.

29. Naydenov NG, Hopkins AM, Ivanov AI. C-Jun N-terminal kinase mediates disassembly of apical junctions in model intestinal epithelia. Cell Cycle. 2009;8:2110–21.

30. Lee MH, Koria P, Qu J, Andreadis ST. JNK phosphorylates beta-catenin and regulates adherens junctions. FASEB J. 2009;23:3874–83.

31. Ranathunga SD, Kalscheur KF, Hippen AR, Schingoethe DJ. Replacement of starch from corn with nonforage fiber from distillers grains and soyhulls in diets of lactating dairy cows. J Dairy Sci. 2010;93:1086–97.

32. Emmanuel DG, Dunn SM, Ametaj BN. Feeding high proportions of barley grain stimulates an inflammatory response in dairy cows. J Dairy Sci. 2008; 91:606–14.

33. Castrillo C, Mota M, Van LH, Martín-Tereso J, Gimeno A, Fondevila M, et al. Effect of compound feed pelleting and die diameter on rumen fermentation in beef cattle fed high concentrate diets. Anim Feed Sci Technol. 2013;180:34–43.

34. Penner GB, Steele MA, Aschenbach JR, McBride BW. Ruminant nutrition symposium: molecular adaptation of ruminal epithelia to highly fermentable diets. J Anim Sci. 2011;89:1108–19.

35. Klevenhusen F, Hollmann M, Podstatzky-Lichtenstein L, Krametter-Frotscher R, Aschenbach JR, Zebeli Q. Feeding barley grain-rich diets altered electrophysiological properties and permeability of the ruminal wall in a goat model. J Dairy Sci. 2013;96:2293–302.

36. Kenichi T, Takashi K, Norimasa S, Tetsuo H. Role of tight junctions in signal transduction: an update. EXCLI J. 2014;13:1145–62.

37. Carrozzino F, Pugnale P, Féraille E, Montesano R. Inhibition of basal p38 or JNK activity enhances epithelial barrier function through differential modulation of claudin expression. Am J Physiol Cell Physiol. 2009;297:C775–87.

38. Kojima T, Fuchimoto J, Yamaguchi H, Ito T, Takasawa A, Ninomiya T, et al. C-Jun N-terminal kinase is largely involved in the regulation of tricellular tight junctions via tricellulin in human pancreatic duct epithelial cells. J Cell Physiol. 2010;225:720–33.

39. Guo J, Chang G, Zhang K, Xu L, Jin D, Bilal MS, et al. Rumen-derived lipopolysaccharide provoked inflammatory injury in the liver of dairy cows fed a high-concentrate diet. Oncotarget. 2017. https://doi.org/10.18632/oncotarget.

40. Li D, Mrsny RJ. Oncogenic Raf-1 disrupts epithelial tight junctions via downregulation of occludin. J Cell Biol. 2000;148:791–800.

41. Kojima T, Yamamoto T, Murata M, Chiba H, Kokai Y, Sawada N. Regulation of the blood-biliary barrier:interaction between gap and tight junctions in hepatocytes. Med Electron Microsc. 2003;36:157–64.

42. Yamamoto T, Kojima T, Murata M, Takano K, Go M, Hatakeyama N, et al. p38 MAP-kinase regulates function of gap and tight junctions during regeneration of rat hepatocytes. J Hepatol. 2005;42:707–18.

43. Saadat I, Higashi H, Obuse C, Umeda M, Murata-Kamiya N, Saito Y, et al. Helicobacter pylori CagA targets PAR1/MARK kinase to disrupt epithelial cell polarity. Nature. 2007;447:330–3.

44. Stumpff F, Georgi MI, Mundhenk L, Rabbani I, Fromm M, Martens H, et al. Sheep rumen and omasum primary cultures and source epithelia: barrier function aligns with expression of tight junction proteins. J Exp Biol. 2011; 214:2871–82.

45. Wu X, Vallance BA, Boyer L, Bergstrom KS, Walker J, Madsen K, et al. Saccharomyces boulardii ameliorates Citrobacter rodentium-induced colitis through actions on bacterial virulence factors. Am J Physiol Gastrointest Liver Physiol. 2008;294:G295–306.

Permissions

All chapters in this book were first published in VR, by BioMed Central; hereby published with permission under the Creative Commons Attribution License or equivalent. Every chapter published in this book has been scrutinized by our experts. Their significance has been extensively debated. The topics covered herein carry significant findings which will fuel the growth of the discipline. They may even be implemented as practical applications or may be referred to as a beginning point for another development.

The contributors of this book come from diverse backgrounds, making this book a truly international effort. This book will bring forth new frontiers with its revolutionizing research information and detailed analysis of the nascent developments around the world.

We would like to thank all the contributing authors for lending their expertise to make the book truly unique. They have played a crucial role in the development of this book. Without their invaluable contributions this book wouldn't have been possible. They have made vital efforts to compile up to date information on the varied aspects of this subject to make this book a valuable addition to the collection of many professionals and students.

This book was conceptualized with the vision of imparting up-to-date information and advanced data in this field. To ensure the same, a matchless editorial board was set up. Every individual on the board went through rigorous rounds of assessment to prove their worth. After which they invested a large part of their time researching and compiling the most relevant data for our readers.

The editorial board has been involved in producing this book since its inception. They have spent rigorous hours researching and exploring the diverse topics which have resulted in the successful publishing of this book. They have passed on their knowledge of decades through this book. To expedite this challenging task, the publisher supported the team at every step. A small team of assistant editors was also appointed to further simplify the editing procedure and attain best results for the readers.

Apart from the editorial board, the designing team has also invested a significant amount of their time in understanding the subject and creating the most relevant covers. They scrutinized every image to scout for the most suitable representation of the subject and create an appropriate cover for the book.

The publishing team has been an ardent support to the editorial, designing and production team. Their endless efforts to recruit the best for this project, has resulted in the accomplishment of this book. They are a veteran in the field of academics and their pool of knowledge is as vast as their experience in printing. Their expertise and guidance has proved useful at every step. Their uncompromising quality standards have made this book an exceptional effort. Their encouragement from time to time has been an inspiration for everyone.

The publisher and the editorial board hope that this book will prove to be a valuable piece of knowledge for researchers, students, practitioners and scholars across the globe.

List of Contributors

Renata Duarte da Silva Cezar, Pollyane Raysa Fernandes de Oliveira and José Wilton Pinheiro Junior
Federal Rural University of Pernambuco (Universidade Federal Rural de Pernambuco - UFRPE), Rua Dom Manuel de Medeiros, s/n, Dois Irmãos, Recife, Pernambuco CEP 52171-900, Brazil

Norma Lucena-Silva and Maíra Arruda-Lima
Department of Immunology (Departamento de Imunologia), Research Center Aggeu Magalhães (Centro de Pesquisas Aggeu Magalhães-CPqAM) Oswaldo Cruz Foundation (Fundação Oswaldo Cruz - Fiocruz), Av. Professor Moraes Rego, s/n, Cidade Universitária, Recife, Pernambuco CEP 50.740-465, Brazil

Antônio Fernando Barbosa Batista Filho, Jonas de Melo Borges and Érica Chaves Lúcio
Academic Unit of Garanhuns (Unidade Acadêmica de Garanhuns), Federal Rural University of Pernambuco (Universidade Federal Rural de Pernambuco – UFRPE). Avenida Bom Pastor, s/n, Boa Vista, Garanhuns, Pernambuco CEP 55292-270, Brazil

Vania Lucia de Assis Santana
National Agricultural Laboratory of Pernambuco (Laboratório Nacional Agropecuário de Pernambuco- Lanagro/PE), Ministry of Agriculture, Livestock and Food Supply of Brazil (Ministério da Agricultura, Pecuária e Abastecimento – MAPA). Rua Manoel de Medeiros, s/n, Dois Irmãos, Recife, Pernambuco CEP 52171-030, Brazil

Lucilla Steinaa, Nicholas Svitek, Elias Awino, Thomas Njoroge, Rosemary Saya and Philip Toye
International Livestock Research Institute, Nairobi 00100, Kenya

Ivan Morrison
The Roslin Institute, The University of Edinburgh, Midlothian EH25 9RG, UK

Nuriqmaliza M. Kamal and Mas Jaffri Masarudin
Department of Cell and Molecular Biology, Faculty of Biotechnology and Biomolecular Sciences, Universiti Putra Malaysia, 43400 UPM Serdang, Selangor, Malaysia

M. Zamri-Saad
Research Centre for Ruminant Diseases, Faculty of Veterinary Medicine, Universiti Putra Malaysia, 43400 UPM Serdang, Selangor, Malaysia

Sarah Othman
Department of Cell and Molecular Biology, Faculty of Biotechnology and Biomolecular Sciences, Universiti Putra Malaysia, 43400 UPM Serdang, Selangor, Malaysia
Department of Cell and Molecular Biology, Faculty of Biotechnology and Biomolecular Sciences, Universiti Putra Malaysia, 43400 UPM Serdang, Selangor, Malaysia

Yolanda Corripio-Miyar, Katy Morrison and Tom N. McNeilly
Moredun Research Institute, Pentlands Science Park, Bush Loan, Midlothian, UK

Richard J. Mellanby
The Roslin Institute, Royal (Dick) School of Veterinary Studies, The University of Edinburgh, Midlothian, UK

Izabella Carolina de O. Ribeiro, Emanuelly Gomes A. Mariano, Roberta T. Careli, Franciellen Morais-Costa and Maximiliano S. Pinto
Instituto de Ciências Agrárias, Universidade Federal de Minas Gerais, Avenida Universitária, 1000, Bairro Universitário, Montes Claros, Minas Gerais CEP 39401-790, Brazil

Felipe M. de Sant'Anna and Marcelo R. de Souza
Escola de Medicina Veterinária, Universidade Federal de Minas Gerais, Av. Antonio Carlos, 6627 Pampulha, Belo Horizonte, MG CEP 31270-901, Brazil

Eduardo R. Duarte
Instituto de Ciências Agrárias, Universidade Federal de Minas Gerais, Avenida Universitária, 1000, Bairro Universitário, Montes Claros, Minas Gerais CEP 39401-790, Brazil
Instituto de Ciências Agrárias, Universidade Federal de Minas Gerais, Av Universitária 1000, Bairro Universitario, Montes Claros, MG 39400-006, Brazil

Jørgen S. Agerholm
Department of Large Animal Sciences, Faculty of Health and Medical Sciences, University of Copenhagen, Dyrlægevej 68, Frederiksberg C DK-1870, Denmark

Fiona Menzi, Vidhya Jagannathan and Cord Drögemüller
Institute of Genetics, Vetsuisse Faculty, University of Bern, Bremgartenstrasse 109a, Bern CH-3001, Switzerland

Fintan J. McEvoy
Department of Veterinary Clinical and Animal Sciences, Faculty of Health and Medical Sciences, University of Copenhagen, Dyrlægevej 16, Frederiksberg C DK-1870, Denmark

Sylvester Ochwo, Christian Ndekezi and Frank Norbert Mwiine
College of Veterinary Medicine, Animal resources and Biosecurity, Makerere University, Kampala, Uganda

Kimberly Vander Waal and Anna Munsey
College of Veterinary Medicine, University of Minnesota, 1365 Gortner Avenue, St. Paul, MN 55108, USA

Robert Mwebe, Anna Rose Ademun Okurut and Noelina Nantima
Ministry of Agriculture Animal Industry & Fisheries, Berkley Ln, Entebbe, Uganda

Mohamed E. Ahmed
EBH Research Center, Zamzam University College (ZUC), Khartoum, Sudan

Bashir Salim
Department of Parasitology, Faculty of Veterinary Medicine, University of Khartoum, Khartoum, Sudan

Martin P. Grobusch
Center of Tropical Medicine and Travel Medicine, Department of Infectious Diseases, Division of Internal Medicine, Amsterdam Medical Center, University of Amsterdam, Amsterdam, The Netherlands

Imadeldin E. Aradaib
EBH Research Center, Zamzam University College (ZUC), Khartoum, Sudan
Molecular Biology Laboratory (MBL), Department of Clinical Medicine, Faculty of Veterinary Medicine, University of Khartoum, Khartoum North, Sudan

Paolo Motta
The Roslin Institute, Royal (Dick) School of Veterinary Studies, University of Edinburgh, Edinburgh, Easter Bush, Midlothian EH25 9RG, UK
The European Commission for the Control of Foot-and-Mouth Disease (EuFMD) - Food and Agricolture Organization (FAO), Viale delle Terme di Caracalla, 00153 Rome, Italy

Thibaud Porphyre and Barend Mark Bronsvoort
The Roslin Institute, Royal (Dick) School of Veterinary Studies, University of Edinburgh, Edinburgh, Easter Bush, Midlothian EH25 9RG, UK

Saidou M. Hamman
Institute of Agricultural Research for Development, Regional Centre of Wakwa, Ngaoundere, Cameroon

Kenton L. Morgan
Institute of Ageing and Chronic Disease and School of Veterinary Science, University of Liverpool, Leahurst Campus, Neston, Liverpool, Wirral CH64 7TE, UK

Victor Ngu Ngwa
School of Veterinary Medicine and Sciences, University of Ngaoundere, Ngaoundere Cameroon

Vincent N. Tanya
Cameroon Academy of Sciences, Yaound'e Cameroon

Eran Raizman
Food and Agriculture Organization (FAO), Animal Production and Health Division, Viale delle Terme di Caracalla, 00153 Rome, Italy

Ian G. Handel
Royal (Dick) School of Veterinary Studies, University of Edinburgh, Easter Bush, Edinburgh, Midlothian EH25 9RG, UK

Kenichi Watanabe, Daisuke Kondoh, Tomoaki Kikuchi, Motoki Sasaki and Nobuo Kitamura
Division of Basic Veterinary Medicine, Obihiro University of Agriculture and Veterinary Medicine, Obihiro, Hokkaido, Japan

Tomomi Kawano, Kaoru Hatate, Norio Yamagishi and Hisashi Inokuma
Division of Clinical Veterinary Medicine, Obihiro University of Agriculture and Veterinary Medicine, Obihiro, Hokkaido, Japan

Benjamin M. C. Swift and Jonathan N. Huxley
School of Veterinary Medicine and Science, Sutton Bonington Campus, Loughborough, Leics LE12 5RD, UK

Karren M. Plain, Douglas J. Begg, Kumudika de Silva, Auriol C. Purdie and Richard J. Whittington
The University of Sydney, Farm Animal and Veterinary Public Health, Faculty of Veterinary Science, Camden, Australia

Catherine E. D. Rees
School of Biosciences, University of Nottingham, Sutton Bonington Campus, Loughborough, Leics LE12 5RD, UK

Jun Du, Xiaoyu Wang, Huixia Luo, Yujiong Wang, Xiaoming Liu and Xuezhang Zhou
Key Laboratory of the Ministry of Education for the Conservation and Utilization of Special Biological Resources of Western China, Ningxia University, Yinchuan 750021, Ningxia, China College of Life science, Ningxia University, Yinchuan 750021, Ningxia, China

C. Wolff, S. Boqvist and S. Sternberg-Lewerin
Department of Biomedical Sciences and Veterinary Public Health, Swedish University of Agricultural Sciences, Uppsala, Sweden

K. Ståhl
Department of Disease Control and Epidemiology, National Veterinary Institute, Uppsala, Sweden

C. Masembe
College of Natural Sciences, Makerere University, Kampala, Uganda

David Love, Meredyth Jones and James A. Thompson
Department of Large Animal Clinical Sciences, College of Veterinary Medicine and Biomedical Sciences, Texas A&M University, College Station, TX 77843, USA

Virginia R. Fajt
Department of Veterinary Physiology and Pharmacology, College of Veterinary Medicine and Biomedical Sciences, Texas A&M University, College Station, TX 77843, USA

Thomas Hairgrove
Texas A&M AgriLife Extension Service, Department of Animal Science, College of Agriculture and Life Sciences, Texas A&M University, College Station, TX 77843, USA

Beatriz Vidondo and Bernhard Voelkl
Veterinary Public Health Institute, University of Bern, Schwarzenburgstrasse 155, CH-3097 Liebefeld, Switzerland

W. Ray Waters, Mitchell V. Palmer, Mayara F. Maggioli and Tyler C. Thacker
National Animal Disease Center, Agricultural Research Service, United States Department of Agriculture (USDA), Ames, IA, USA

H. Martin Vordermeier, Shelley Rhodes and Bhagwati Khatri
Tuberculosis Research Group, Animal and Plant Health Agency, Addlestone, UK

Jeffrey T. Nelson, Bruce V. Thomsen, Suelee Robbe-Austerman and Doris M. Bravo Garcia
National Veterinary Services Laboratories, Animal and Plant Health Inspection Service (APHIS), USDA, Ames, IA, USA

Mark A. Schoenbaum
Veterinary Services (VS), APHIS, USDA, Fort Collins, CO, USA

Mark S. Camacho
VS, APHIS, USDA, Raleigh, NC, USA

Jean S. Ray
VS, APHIS, USDA, East Lansing, MI, USA

Javan Esfandiari, Paul Lambotte, Rena Greenwald, Adrian Grandison, Alina Sikar-Gang and Konstantin P. Lyashchenko
Chembio Diagnostic Systems, Inc., Medford, NY, USA

Sabine E. Hutter, Annemarie Käsbohrer and Katharina Brugger
Institute of Veterinary Public Health, Department for Farm Animals and Veterinary Public Health, University of Veterinary Medicine, Veterinärplatz 1, 1210 Wien, Austria

Silvia Lucia Fallas González
Laboratorio de Pruebas de Paternidad, Caja Costarricense del Seguro Social (CCSS), San José, Costa Rica

Bernal León
Servicio Nacional de Salud Animal (SENASA), Heredia, Costa Rica

Mario Baldi
Research Institute of Wildlife Ecology, University of Veterinary Medicine, Vienna, Austria

L. Mario Romero
Centro de Investigación en Enfermedades Tropicales (CIET), Universidad de Costa Rica, San Pedro de Montes de Oca, Costa Rica

Yan Gao
Centro de Investigaciones en Geografía Ambiental, Universidad Nacional Autónoma de México, 58190 Morelia, Michoacán, Mexico

Luis Fernando Chaves
Instituto Costarricense de Investigación y Enseñanza en Nutrición y Salud, Apartado Postal 4-2250, Tres Ríos, Cartago, Costa Rica
Programa de Investigación en Enfermedades Tropicales (PIET), Escuela de Medicina Veterinaria, Universidad Nacional, Heredia, Costa Rica

You Yang, Guozhong Dong, Zhi Wang, Junhui Liu, Jingbo Chen and Zhu Zhang
College of Animal Science and Technology, Southwest University, Chongqing 400716, People's Republic of China

Mounir Adnane
Comparative Immunology Group, School of Biochemistry and Immunology, Trinity College, Dublin, Ireland
Institute of Veterinary Sciences, Tiaret, Algeria

Paul Kelly and Cliona O'Farrelly
Comparative Immunology Group, School of Biochemistry and Immunology, Trinity College, Dublin, Ireland

Aspinas Chapwanya
Department of Clinical Sciences, Ross University School of Veterinary Medicine, Basseterre, West Indies, St. Kitts and Nevis

Kieran G. Meade
Animal & Bioscience Research Department, Animal & Grassland Research and Innovation Centre, Teagasc, Grange, Co. Meath, Ireland
Immunogenetics & Animal Health, Animal & Grassland Research and Innovation Centre, Teagasc, Grange, Co. Meath, Ireland

Saif Eddine Djemal and Radhouane Gdoura
Department of Life Sciences, Research Laboratory of Environmental Toxicology-Microbiology and Health (LR17ES06), Faculty of Sciences, University of Sfax, Sfax, Tunisia

Cristina Camperio, Federica Armas and Cinzia Marianelli
Department of Food Safety, Nutrition and Veterinary Public Health, Istituto Superiore di Sanità, Viale Regina Elena 299, 00161 Rome, Italy

Mariam Siala
Department of Life Sciences, Research Laboratory of Environmental Toxicology-Microbiology and Health (LR17ES06), Faculty of Sciences, University of Sfax, Sfax, Tunisia
Department of Biology, Preparatory Institute for Engineering Studies, University of Sfax, Sfax, Tunisia

Salma Smaoui
Department of Microbiology, Regional Hygiene Care Mycobacteriology Laboratory, Hedi-Chaker University Hospital, Sfax, Tunisia
Department of Biology B, Faculty of Pharmacy, University of Monastir, Monastir, Tunisia
Department of Microbiology, National Reference Laboratory of Mycobacteria, Research Unit (UR12SP18), A. Mami University Hospital of Pneumology, Ariana, Tunisia

Feriele Messadi-Akrout
Department of Biology, Preparatory Institute for Engineering Studies, University of Sfax, Sfax, Tunisia
Department of Microbiology, Regional Hygiene Care Mycobacteriology Laboratory, Hedi-Chaker University Hospital, Sfax, Tunisia

Bin Zhang, Hua Yue and Cheng Tang
College of Life Science and Technology, Southwest Minzu University, No.16, South 4th Section 1st Ring Road, Chengdu 610041, China
Key laboratory of Ministry of Education and Sichuan Province for Qinghai-Tibetan Plateau Animal Genetic Resource Reservation and Utilization, Chengdu 610041, China
Animal Disease Prevention and Control Innovation Team in the Qinghai-Tibetan Plateau of State Ethnic Affairs Commission, Chengdu610041, China

Xinnuo Chen, Wenqiang Ruan and Sinan Qin
College of Life Science and Technology, Southwest Minzu University, No.16, South 4th Section 1st Ring Road, Chengdu 610041, China

Daniel Oluwayelu and Adebowale Adebiyi
Department of Veterinary Microbiology, University of Ibadan, Ibadan, Oyo State, Nigeria

Kerstin Wernike and Martin Beer
Institute of Diagnostic Virology, Friedrich-Loeffler-Institut, Südufer 10, 17493 Greifswald - Insel Riems, Germany

Simeon Cadmus
Department of Veterinary Public Health and Preventive Medicine, University of Ibadan, Ibadan, Oyo State, Nigeria

Abdullah Gazioglu
Department of Veterinary Science, Vocational School of Technical Sciences, University of Bingol, 12000 Bingol, Turkey

Burcu Karagülle and Burhan Çetinkaya
Department of Microbiology, Faculty of Veterinary Medicine, University of Firat, 23119 Elazig, Turkey

Hayati Yüksel and Muhammet Bahaeddin Dörtbudak
Department of Pathology, Faculty of Veterinary Medicine, University of Bingol, 12000 Bingol, Turkey

M. Nuri Açık
Department of Microbiology, Faculty of Veterinary Medicine, University of Bingol, 12000 Bingol, Turkey

Hakan Keçeci
Department of Internal Medicine, Faculty of Veterinary Medicine, University of Bingol, 12000 Bingol, Turkey

Kai Zhang, Yuanlu Tu, Lipeng Gao, Meijuan Meng and Yunfeng Bai
Circular Agriculture Research Center, Jiangsu Academy of Agricultural Sciences, Nanjing, China Key Laboratory of Crop and Livestock Integrated Farming, Ministry of Agriculture, Nanjing, China

Index